7.20

PERSPECTIVES ON ENERGY

PERSPECTIVES ON ENERGY

ISSUES, IDEAS, AND ENVIRONMENTAL DILEMMAS

Edited by
LON C. RUEDISILI
UNIVERSITY OF TOLEDO

MORRIS W. FIREBAUGH
UNIVERSITY OF WISCONSIN-PARKSIDE

NEW YORK · OXFORD UNIVERSITY PRESS   LONDON   TORONTO   1975

Copyright © 1975 by Oxford University Press, Inc.
Library of Congress Catalogue Card Number: 74-22886
Printed in the United States of America

To All Those Concerned with the World's Energy Dilemmas

To my wife Susan, Stephen and Robert and future generations whose environment and welfare will depend upon the responsible energy decisions of our society   L.C.R.

To Joyce and to our children Steve and Susan, in the conviction that intelligent energy management requires more light and less heat focused on the energy issues of our day   M.W.F.

Preface

Though the immediate impact of the "energy crisis" did not strike the American public until the winter of 1973–1974, the underlying causes as well as many possible solutions have been recognized for many years. After the public clamor to "do something" to assure energy supplies and warnings of environmentalists against "rushing headlong . . ." toward various proposed solutions, the need for clear and concise information on this most confusing problem is more urgent than ever. In this book we attempt to provide such information as well as incorporate a wide spectrum of current interpretations as to its meaning and implications for public policy.

Students, and particularly students in the sciences, are accustomed to finding the "correct answer" determined by natural law. As a result, they often find the apparently contradictory statements issuing from governmental, industrial, and environmental "energy experts" both confusing and frustrating. Since the production, distribution, and utilization of energy does, in fact, involve unresolved dilemmas over which reputable authorities differ, we believe it is essential to present as wide a range of interpretation on major energy issues as feasible in one book.

The editors have attempted to present a balanced and representative analysis in areas of controversy, but we readily admit that the particular selection of articles included reflects our own interpretation of the most significant issues involved. We are attempting, in this book, to present the most current analyses in an area of rapid change and new information on both scientific and policy areas. Thus, rather than abstracting significant contributions, we have presented the material in the authors' original words in order to expedite the presentation as well as to preserve the tone and content of the more subjective interpretations. Some of the articles have been prepared specifically for this book to present a more complete picture of the dilemmas society faces in an energy age.

In the final analysis, public policy in this critical area of energy management must be determined by enlightened public opinion as it is worked out through the political process. We believe that the present generation of college students is capable of the critical "sifting and winnowing" required to reach an intelligent, informed (and, hence, effective) opinion on energy policies. We hope this book will assist them in the process.

At this point we want to acknowledge the large number of people whose fine cooperation made this book possible. First we wish to thank all the authors and publishers for permission to include their articles in this anthology. We especially appreciate the efforts of Professor Glenn T. Seaborg of the University of California, Berkeley, in preparing a very appropriate and timely Foreword. We thank Professor John S. Steinhart, University of Wisconsin-Madison, for many hours of consultation and advice on the preparation of the manuscript. We appreciate the input and advice of several of the authors included in the book as well as our Parkside colleagues and students who evaluated earlier versions of this book. The University of Wisconsin-Parkside Division of Science, the University of Wisconsin-Parkside Library-Learning Center, and the University of Wisconsin-Madison Department of Physics kindly provided valuable assistance and support in this effort for which we are grateful. Geraldine S. Covelli, Patricia E. Heckel, Nancy Lyttle, James M. Ramonowski, Gail A. Tworek, and Jacqueline Willems provided the library and clerical support crucial for the successful completion of this project. Finally, we appreciate the continued assistance and encouragement of Ellis H. Rosenberg, Jean Shapiro, and Carol Miller of Oxford University Press.

University of Toledo
University of Wisconsin-Parkside   L. C. R.
December, 1974   M. W. F.

Foreword

The year 1973 represents a turning point in our lives: we never again shall be in a position to take our supply of energy for granted. The energy problem took a long time in coming—even though general recognition came suddenly—and it will take a long time in going. We face difficult choices in the years to come, choices that will cut to the fabric of our way of life.

To make these choices wisely, and take the actions that will be necessary to implement them, we need not only a sense of determination, but also a coherent and realistic energy policy. Prerequisite to this is a public understanding of our dilemma and conflicting choices. In this book of readings, Professors Ruedisili and Firebaugh have assembled a spectrum of authoritative information and opinion on our energy problem. The balance of articles is sufficiently broad to stimulate and educate the reader, college student, or other interested citizen on this vital issue.

The readings make it clear that our only hope is to work on both elements of the energy problem—supply and demand. We must increase the supply and decrease the demand or at least reduce the rate of increase of the demand. At the same time, because it is all part of the same problem, we must not compromise the efforts that are under way to improve both our natural and our man-made environments and to protect the public health and safety. Our conflicting choices include those between the goals of abundant energy and environmental quality, and these conflicts will have to be reconciled as best they can with regard to the total social welfare.

The energy technologies that can be expected to play a major role in the American energy economy in the coming decades include the use of fossil fuels, and their improvement and interconversion, nuclear fission, nuclear fusion, geothermal energy, and solar energy. These sources, and others covered in this book's readings, should be developed. It is not a question of whether geothermal energy can produce electricity, or whether solar energy is feasible, but rather how much can we expect from these sources, when and under what conditions can we expect their utilization, and, importantly, what will it cost us to ensure their development in an expeditious, timely, and safe manner.

We are in trouble because of inadequate national planning in the past, and this includes insufficient research on and development of new energy sources. The development of new technologies, where the systems are very complex, requires much basic research before they can be utilized and institutionalized. The cost of basic research is small compared with the value of the technologies it spawns, and we should not neglect the lessons of the past in this regard, even as we search for immediate solutions to our energy problems.

We are accustomed to the idea of unlimited growth from our classical theories of economics, in which supply always rises to meet demand. But unlike capital, which is a construct of man, the natural resources of our earth are finite. In fact, we are going to be faced next with a series of resource crises—metals crises (copper, aluminum, chromium, nickel, tin, manganese, and so on), a food crisis, a water crisis—if we do not plan better for the future than we have planned in the past for our future energy and materials needs.

In trying to solve our energy problems, we must not neglect the questions of fragile values in our society. We need to understand more fully the role of our institutions in the resolution of specific value conflicts in order to understand how the patterns of these decisions, if there are patterns, themselves create new values and new institutions that perpetuate those values. Such questions as these should be addressed: In crises, do fragile values that are without institutional embodiment get lost, even though they may be deeply held by a substantial number of citizens? With what kind of analysis do we approach decisions involving competition and compromise between the satisfaction of present and future welfare? How is our concern for the future, itself a fragile value, concretely manifested?

The readings in this book should help in the understanding of and deliberations on all these issues.

University of California   Glenn T. Seaborg
Berkeley, California
July 1974

Contents

1

POSING THE DILEMMAS

Zepher I, a semi-submersible drilling rig, being towed into the North Sea to seek out sea-bottom oil. (Photo courtesy of Skyfotos, and Texaco, Inc.)

Use of energy in the grain wheat harvest along the Columbia River, Wasco County, Oregon,
(Photo courtesy of the U.S. Department of Agriculture Conservation Service: USDA-SCS
photo by G. L. Green.)

The reservoir of the Yellowtail Dam, Montana, soon to be 71 river miles long. It will generate power and supply irrigation and recreation needs. (Photo courtesty of the Bureau of Reclamation, U.S. Department of Agriculture.)

INTRODUCTION

Since the dawn of the industrial age, public policy has, in general, had only a peripheral influence on the availability and management of our energy resources. Until now our abundant supply of energy has been much more a product of geological accident, ideological and military alliances, and cultural traditions than any well-considered and farsighted governmental policy.

Such laissez faire may have been a viable attitude during that brief historical period of energy abundance, but it is rapidly becoming untenable in an age of energy scarcity. Increasingly, the availability of energy is determining public policy, influencing international relations, and altering our basic patterns of life. Proposals appear daily to tax automobiles and air conditioners according to energy consumption, eliminate energy-intensive packaging, regulate insulation and building codes, and modify existing modes and speed of transportation. The cost of energy is increasing in an already energy-intensive society, and the final price on many of our goods and services has become more and more a function of energy prices themselves. As we recognize the central and determinative role that energy plays in all aspects of our lives, we inexorably come to the conclusion that we are, in fact, entering an Energy Age.

In its pioneering study, *The Limits to Growth*,[1] the club of Rome has succinctly expressed the primary dilemma facing mankind: a finite world cannot tolerate exponential growth for very long. The conclusions of this study, independently verified by a group of British scientists in *Blueprint for Survival*,[2] emphasize the need for stabilizing population growth and economic expan-

---

1. D. H. Meadows, P. L. Meadows, J. H. Randers, and W. Behrens, *The Limits to Growth*, New York Universe Book, 1972.
2. "Blueprint for Survival," *The Ecologist*, No. 2, pp. 1–43, 1972.

sion. Though these proposals have been hotly debated, the role of energy in such a stabilized world has become even more problematic. Our "spaceship Earth" has, in actuality, "lost" very little of its initial inheritance of mineral resources (for example, a few lunar landers and space probes), and as the practice of "use and discard" necessarily gives way to the ethic of "conserve and recycle," the primary input ingredient will be energy. Seen as a closed environmental system, our industrialized society needs only a continuous and abundant energy supply to survive indefinitely. Balancing the environmental costs of such energy usage against its crucial need poses the second great dilemma we must face.

In *Realities of the Energy Crisis,* Chauncey Starr emphasizes the essential role that energy plays in an industrial society. The serious implications of our foreign oil dependence on foreign policy are suggested here on the eve of the first serious oil embargo and its eventual repercussions. Starr very appropriately points out that sociological rather than technical considerations determine existing energy policies, and that both policies have relatively long response times (10 to 25 years).

Ralph E. Lapp, in *The Chemical Century,* focuses on the dominant source of our present energy, namely, fossil fuels, and indicates that they will continue to be our number one source until the twenty-first century. The similarities of the twentieth-century fossil-fuel consumption curve and the nineteenth-century wood fuel curve are very suggestive, and this raises important questions concerning the predicted twenty-first-century nuclear-energy production curve. The roll over of the fossil-fuel curve illustrates the depletion of resource behavior resulting from our finite supply.

V. E. McKelvey, in *Mineral Resource Estimates and Public Policy,* sheds some light on the arbitrary and elastic definitions used in estimating our energy resources. The confusion often surrounding a discussion of energy supplies is dispelled by the clear distinction drawn between the terms "reserves" and "resources." In addition, by clarifying the sensitive dependence of the economic categories of mineral recovery on assumptions of price and technology, he helps elucidate many of the conflicting claims and statistics we continue to receive from various energy experts.

In *Energy Crises in Perspective,* John C. Fisher presents data indicating a reassuring, long-range view of the energy supply situation. He sees present energy shortages as well-understood and localized (in time and space) fluctuations in a historical upward trend in energy usage. The environmental movement is interpreted as one important cause of the present energy crunch, and future trends of environmental action are suggested. The close competition between fossil fuel and nuclear fission as electric energy sources is examined, as are the factors influencing the economics of each.

John and Carol Steinhart discuss the enormously complex food system with its energy implications in *Energy Use in the United States Food System.* Since failure of the highly energy-dependent food chain has been predicted to be one of the first events in an eventual social collapse,[1] a thorough understanding of this relationship is essential. The Steinharts also raise serious questions about the wisdom of attempting to export our high energy-consuming food system to other countries to ease their food shortages.

In *Energy and the Environment,* John M. Fowler very nicely summarizes the present energy situation and the many alternatives open in the future. We are introduced to the concept of energy flow through the social system and are presented with various detrimental side effects of energy production and use. These issues are investigated in more detail in subsequent sections.

In summary, we present, in Part 1, a variety of evaluations of the origin, nature, and implications of energy use in our society. Here, it is shown that the "energy content" of the goods and services is emerging as a dominant measure of value, and environmental constraints on energy consumption are indicated. A thorough understanding of these complex energy-environment interactions is essential for the development of rational energy management policies.

Realities of the Energy Crisis

CHAUNCEY STARR   1973

Between now and 2001, just thirty years away, the United States will consume more energy than it has in its entire history. By 2001 the annual United States demand for energy in all forms is expected to double, and the annual world-wide demand will probably triple. These projected increases will tax man's ability to discover, extract, and refine fuels in the huge volumes necessary, to ship them safely, to find suitable locations for several hundred new electric-power stations in the United States (thousands world wide) and to dispose of effluents and waste products with minimum harm to himself and his environment. When one considers how difficult it is at present to extract coal without jeopardizing lives or scarring the surface of the Earth, to ship oil without spillage, to find acceptable sites for power plants and to control the effluents of our present fuel-burning machines, the energy projections for 2001 indicate the need for thorough assessment of the available options and careful planning of our future course. We shall have to examine with both objectivity and humanity the necessity for the projected increase in energy demand, its relation to our quality of life, the practical options technology provides for meeting our needs, and the environmental and social consequences of these options ["Energy and Power," *Scientific American,* 225: 3 (Sept. 1971), 37–49].

The above quotation is taken from a paper I prepared more than two years ago. It describes the nature of the continuing problems we face—and which have recently reached public attention in the form of the energy crisis. The term "energy crisis" has served as a convenient layman's umbrella for encompassing a wide variety of society's concerns with the energy situation. Because these do not have a common solution, it is

Dr. Chauncey Starr is president of the Electric Power Research Institute. Previously he was dean of the School of Engineering and Applied Science at the University of California at Los Angeles (1967–1973), following a twenty-year industrial career, during which he served as vice-president of Rockwell International and as president of its Atomic International Division. Dr. Starr is past vice-president of the National Academy of Engineering, a founder and past president of the American Nuclear Society, a director of the Atomic Industrial Forum and of the American Association for the Advancement of Science, and a member of the President's Energy Research and Development Council.

From *Science and Public Affairs (Bulletin of the Atomic Scientists),* Vol. XXIX, No. 7, pp. 15–20, September 1973. Reprinted by permission of the author and *Science and Public Affairs.* Copyright 1973 by the Educational Foundation for Nuclear Science. This article is based on a lecture given at the Department of State in April 1973.

important to examine them separately and to clarify the several issues we face.

The "crisis" designation tends to be misleading, because it implies that quick-fix emergency steps should be taken to cure situations which have developed over many years. In fact, there are no quick fixes. Further, the practical realities of the situation have not yet required an immediate national "crisis" response by applying true emergency measures—such as energy rationing and cessation of energy consuming activities.

The fact that pressing localized issues have arisen should give us concern, both as indicators of widespread inadequacies and as they may portend more serious things to come. To use a medical analogy, the patient may have aches and pains, but can still do a day's work and live normally—the situation does not justify hospitalization now, but could get worse if remedial treatment is neglected.

In like manner, the most pressing energy need is for a coherent and long-range program to plan and manage our national and international energy systems. It takes ten to twenty years to significantly alter the trends of these huge systems. Waiting until the situation becomes intolerable must now be recognized as intentionally planned neglect—a societal irresponsibility difficult to condone.

Our national and international energy systems consist of a complex of interlocked activities, including fuel resources (most notably the fossil fuels: coal, oil, and gas), the distribution of these fuels either by pipeline, truck, or tanker, the distribution of electricity generated from these fuels, and finally the many end uses of energy.

Energy is consumed for residential purposes, for transportation, by the manufacturing industry, and in sundry other ways. All activities of any energy system have some environmental impact. For example, the development of fuel resources gives rise to land use and aesthetic issues. The distribution of these fuels involves transportation risks both to the public and to our ecology.

The conversion of these fuels into either electricity or into their end functions, such as automobile transportation, industrial operations and the like, creates air polluting effluents and waste heat. In addition to these more obvious environmental impacts there are secondary by-products from energy systems that are not as directly visible to the public, but which are also important societal costs, such as fires, explosions, and accidents.

The current public focus on the energy crisis arises primarily from a few immediately visible near-term events. First, because of the occasional shortages and malfunctions of the electricity delivery system, which cause dramatic blackouts and brownouts in spot areas, the public affected has a discomforting anxiety about the reliability of supply. The "crisis" nature of this issue tends to be very localized in place and time. The great majority of our population has no difficulty with getting electricity on demand.

The second near-term issue is that related to urban air pollution. However, air pollution arises from a great variety of sources, many of an industrial nature not directly related to the energy systems. The contribution to air pollution that arises from the generation of electricity is significant, but usually only a modest part of the total. Most notably the use of petroleum products for private and public motor vehicles is a major source. These two items, the continuous delivery of electricity and urban air pollution, are generally the stimuli for the public attention to energy issues.

TRADITIONAL OBJECTIVE

The continuous delivery of electricity to meet demands, without the penalties of brownouts or blackouts or other failures, has always been the traditional objective of the electric utility industry. In order to accomplish this, the industry has anticipated a decade ahead the growth in demand for electricity, so as to schedule the construction of power generation and distribution facilities to meet such foreseen needs.

Electric utilities have also tried to maintain a sufficient surplus of generation capacity to provide a reserve for unexpected breakdowns of equipment, maintenance, and other causes of disruption. In the past several years the normal anticipatory planning of the electrical industry has gone askew because of conditions not anticipated at the time when the original commitments for future plants and equipment were made. These unanticipated issues have arisen from many sources; but perhaps the two most important are, first, the recently restricted availability of suitable fuel and, second, the new environmental criteria for power plant performance.

The traditional fuel for power plants has been the fossil fuels—coal, oil, and gas. Coal, while an abundant mineral in the United States, unfortunately produces the largest over-all environmental impact. The mining operations, underground and strip mining, involve social costs associated with safety and land use which are quite substantial and require very large remedial investments. In addition, coal contains a large number of foreign elements, including sulfur compounds, which are environmental pollutants.

While the demand for coal has continuously increased for power production purposes, the result has been that its use creates problems which have yet to be solved satisfactorily. In particular, the ability to remove the sulfur contamination from the coal, either prior to its use or after its discharge as a gas in the power plant stacks, requires the commercial development of new technologies only now undergoing pilot plant trials.

The available indigenous oil in the United States has not been sufficient to meet our needs. Because of the slow development of both our onshore and offshore oil reserves, we have become one of the great importers of oil. Oil, like coal, contains sulfur, which generally requires either removal prior to combustion or in the power plant stack gas. Naturally low-sulfur oil is available in relatively small amounts.

The recent environmental restrictions on oil drilling on the continental shelf (because of the possible leakages into the marine environment), and the concern with the ecological impact of oil pipelines all have tended to slow down or inhibit the full exploration and development of oil resources.

Natural gas is the least contaminated form of fossil fuel and is, therefore, in great demand for power plant use. It is also in great demand for industrial and domestic use because of the ease with which it can be transmitted and distributed and the simplicity of combustion equipment. For complicated reasons, including pricing policies, natural gas has been mostly a by-product of oil development. At the present time there appears to be an insufficient reserve of natural gas in the United States to permit continued expansion of its use. Thus, this most environmentally acceptable of all the fossil fuels is also the most limited for the future.

NUCLEAR POWER

For all these reasons the utility industry recognized some years ago that the unique "clean air" characteristics of nuclear power would make it a very desirable addition to the available technical options for the generation of electricity. For almost two decades the utility industry has actively supported nuclear power development, and underwritten the higher costs of the first stages of commercializing nuclear power.

The rate at which nuclear power has entered into the production of electricity is, however, disappointingly less than that which was expected by the utilities. The initial delays were associated with establishing the reliability needed for commercial operations. More recently these plants have been delayed by the intervention of various public groups fearful of their potential environmental impact.

These interventions primarily serve as a means of public education and communication concerning the relative safety of nuclear power. Unfortunately, the associated delays, sometimes extending for several years, have had a serious impact on reducing the planned expansion of nuclear power availability.

Thus, as a result of the combined effects of an inadequate supply of environmentally suitable fossil fuels to meet expanding requirements, the time needed for technical development of antipollution devices to permit the use of available coal and oil, and delays in the authorization for nuclear power plants, we face a near-term situation where the generating capacity of our national electricity system does not contain everywhere an adequate reserve for meeting unique peak demands and providing protection against unexpected power plant failures.

There are many regions of the country where these issues have not been pressing. Unfortunately, there are many urban areas that have had a large expansion of electricity demand, and in these the margin of reliability has been so reduced that even minor malfunctions or unusual weather conditions can create electricity shortages with considerable public discomfort and, in some cases, public hazards.

Such problems can only be avoided by administrative removal of unproductive delays and interferences, and by the most efficient use of the available resources of fuel and power generation facilities. Because it takes a decade or more to bring new technical developments or new fuel systems into operation in our energy system, it is not likely that these near-term pressures will be rapidly removed by technologies still in the process of development.

OF BASIC CONCERN

The availability of energy has always been of basic concern because of the intimate relationship of energy to our societal development. It has become a major public issue only in the past several years, and will probably always remain with us as a primary consideration in the future. Basically our society cannot function without energy in various forms. We utilize it for elemental physical

comfort by heating and cooling, we utilize it to run our industries, and we use it for recreational purposes.

All these uses have always had some impact on the environment. As our per capita use has grown in the past several generations, and as our population has grown and also concentrated in large urban areas, these environmental impacts have become sufficiently severe that we now must begin to develop either better energy technology or some limitations on energy use, or both. It must be recognized that there is no form of energy which may be used without some environmental impact.

The issue is not one of "good or bad" but one of balancing the beneficial aspects of energy use against its undesirable environmental effects. As a nation we are presently engaged in developing a socially acceptable balance between these two issues through public debate, through technical and scientific research, and through empirical trial and error. This development of a sound social philosophy for the use and control of energy, so as to maximize the public good, may be one of the most important national issues of this decade.

FUTURE NEEDS

Long-range planning of our national and worldwide energy systems must start with some estimate of future energy demands. A conception of the future may come from a simple extension of historical trends, or may be developed from a more sophisticated analysis of changing life styles and their impact on end-use needs.

Since 1900, the average per capita energy consumption in the world and the United States has doubled every fifty years, with some short-term perturbations. There appears to be small likelihood that this long-term trend of increasing per capita use will change in the next several decades.

In spite of increased public concern with the impacts of such a growth, there is actually very little that can be done pragmatically to limit it—other than direct scarcity or rationing—because of the intimate connection between the life styles of peoples, their aspirations, and their energy supply.

The future need for energy in societal development is of two broad types, one characteristic of the highly developed sections of the world and one typical of the underdeveloped portions. During the past two centuries the industrialized nations of the world significantly increased their energy use in order to sustain their population growth and to improve the condition of their people.

It is likely that in the next century the per capita energy consumption in these advanced countries will approach an equilibrium level: first, because the quality of life for the majority of the population will be less dependent on increased energy use and, second, because environmental constraints will make energy more costly and thus encourage increased efficiency of its end use. The hoped-for population equilibrium in advanced nations will also lead to an eventual leveling off of total energy need for these countries.

While it is possible that the future creation of socially desirable high-technology energy consuming devices may maintain a continuously growing energy demand, nevertheless, the realities of resource economics will probably create a trade-off ceiling on energy demand. Only the development of new energy resources (such as fusion) that are both low cost and extensive can lift such a ceiling. Even so, the availability of investment capital—a man-made resource—may limit such growth.

For the underdeveloped part of the world, which contains most of the world's population, the situation is quite different. These peoples are still primarily engaged in maintaining a minimum level of subsistence. They have not as yet had available the power resources necessary for the transition to a literate, industrial, urban, and agriculturally advanced society. Historically such transitions have always involved both an increase in population and an increase in per capita

energy consumption. We are seeing this now in most of the underdeveloped countries. So, the inevitable population growth, combined with an increased per capita energy use, could result in an enormous world-wide energy demand.

A capsule example of what can occur is provided by Puerto Rico. It is being shifted to an industrial economy from an agrarian sugar economy by the planned investment of foreign capital. In 1940 the annual electricity consumption was about 100 kilowatt-hours per capita, comparable to India's present usage. By 1950 this had been more than doubled to 220 kilowatt-hours per capita. By 1968 this had increased to 1,800 kilowatt-hours per capita. This is an average doubling time of about 7 years. (By comparison, the United States consumption in 1968 was about 7,200 kilowatt-hours per capita with a present per capita doubling time of about 12 years. Now, in 1972, the United States level is about 8,800 kilowatt-hours per capita.) Puerto Rico is, of course, a unique case of accelerated economic development, but the twenty-fold increase in per capita electricity consumption is nevertheless startling.

At present the United States consumes about 35 per cent of the world's energy. By the year 2000 the United States share will probably drop to around 25 per cent, owing chiefly to the relative population increase of the rest of the world. The per capita increase in energy consumption in the United States is now about 1 per cent per year. Starting from a much lower base, the average per capita energy consumption throughout the world is increasing at a rate of 1.3 per cent per year.

It is evident that it may be another century before the world average even approaches the current United States level. At that time the energy gap between the United States and the underdeveloped world will still be large. With unaltered trends it would take 300 years to close the gap. By the year 2000 the world's average per capita energy consumption will have moved only from the present one-fifth of the United States average to about one-third of the present United States average. Of grave concern is the nearly static and very low per capita energy consumption of areas such as India, a country where the population growth largely negates its increased total production of energy.

If the underdeveloped parts of the world were conceivably able to reach by the year 2000 the standard of living of Americans today, the world-wide level of energy consumption would be roughly ten times the present figure. Even though this is a highly unrealistic target for thirty years hence, one must assume that world energy consumption will move in that direction as rapidly as political, economic and technical factors will allow. The problems implied by this prospect are awesome.

Increasing per capita income is essential for increasing the quality of life in underdeveloped countries, and this requires energy. It has often been suggested that energy use be arbitrarily limited everywhere because of its environmental impacts. This requires the same type of societal decision that would be associated with arbitrarily limiting water supply or food production.

Given the objective of providing the people of the world as good a life as man's ingenuity can develop, the essential role of energy availability must be recognized. With the same motivation that causes the agronomists to seek an increased yield per acre, it is the function of technology to make energy available in sufficient amount to meet all essential needs, and with sufficiently small environmental impact as to ensure that the benefits outweigh all the costs.

Because even in the industrial societies the per capita use of energy in large amounts is only a century old, and in most of the world it has not even started, we have both a growing need and an opportunity to develop long-range plans for optimally supplying this essential aid to world-wide social development.

One can better appreciate the energy problem the world faces if one simply compares the cumulative energy demand to the year 2000—when the annual rate of energy consumption will be only three times the present rate—with estimates of the economically recoverable fossil fuels.

The estimated fossil fuel reserves are greater than the estimated cumulative demand by only a factor of two. If the only energy resource were fossil fuels, the prospect would be bleak indeed.

The outlook is completely altered, however, if one includes the energy available from nuclear power. As has often been stated, nuclear fission provides another major resource, with the present light water reactors about equal to the fossil fuels, and with the breeder reactors almost 100 times as much. There is no question that nuclear power is a saving technical development for the energy prospects for mankind.

Promising, but as yet technically unsolved, is the development of a continuous supply of energy from solar sources. The enormous magnitude of the solar radiation that reaches the land surfaces of the Earth is so much greater than any of the foreseeable needs that it represents an inviting technical target.

Unfortunately, there appears to be no economically feasible concept yet available for substantially tapping that continuous supply of energy. This somewhat pessimistic estimate of today's ability to use solar radiation should not discourage a technological effort to harness it more effectively. If only a few per cent of the land area of the United States could be used to absorb solar radiation effectively (at, say, a little better than 10 per cent efficiency), we would meet most of our energy needs in the year 2000. Even a partial achievement of this goal could make a tremendous contribution.

The land area required for the commercially significant collection of solar radiation is so large, however, that a high capital investment must be anticipated. This, coupled with the cost of the necessary energy-conversion systems and storage facilities, makes solar power economically uninteresting today. Nevertheless, the direct conversion of solar energy is the only significant long-range alternative to nuclear power.

The possibility of obtaining power from

thermonuclear fusion has not been included in the listing of energy resources because of the great uncertainty about its feasibility. The term "thermonuclear fusion," the process of the hydrogen bomb, describes the interaction of very light atomic nuclei to create highly energetic new nuclei particles and radiation. Control of the fusion process involves many scientific phenomena that are not yet understood, and its engineering feasibility has not yet been seriously studied.* Depending on the process used, controlled fusion might open up not only an important added energy resource but also a virtually unlimited one. The fusion process remains a possibility with a highly uncertain outcome.

It has been proposed that tapping the heat in the rocks of the Earth's crust is feasible, and, if it is, this could be important. At present, the initial probing of this source has not yet been tried, so its pragmatic availability is yet uncertain.

It is clear from all such studies that for the next century mankind is unlikely to run out of available energy. Instead, the important issue is whether the increasing cost of energy (including environmental costs) will become a major handicap to world-wide societal improvement. Just as an increasing cost of water with increasing usage might limit the development of an area, the same could apply to the use of energy in various parts of the world.

Within nature's limitations man has tremendous scope for planning energy utilization. Some of the controlling factors that enter into energy policy depend on the voluntary decisions of the individual as well as on government actions that may restrict individual freedom. The questions of feasibility, both economic and technical, depend for their solution on the priority and magnitude of the effort applied.

---

*Editors' note: One significant engineering feasibility study (UWMAK-1) on controlled thermonuclear reactions is presently underway in the Nuclear Engineering Department at the University of Wisconsin-Madison.

BASIC POLICY QUESTIONS

The time scale and costs for implementing decisions, or resolving issues, in all areas of energy management have both short-term and long-term consequences. There are so many variables that their arrangement into a "scenario" for the future becomes a matter of individual choice and a fascinating planning game. The intellectual complexity of the possible arrangements for the future can, however, be reduced to a limited number of basic policy questions that are more sociological than technical in nature.

The only parameters under our control which can alter the nature and trends of near-term energy systems are a limited number of individual and governmental choices—life style and value oriented rather than technological in nature. The individual choice of an energy device (home heating, for example) can be made and implemented with a time constant of about a year. A choice by a societal unit (location of a power station or effluent regulation, for example) takes about a decade to make and implement. Thus, the full effect of such societal decisions often does not develop until more than a decade has passed.

In the technological domain of new economically acceptable energy devices, we are really working for the next generations rather than our own. Even nuclear power, which was certainly supported by government as enthusiastically as any technology in history, has taken 25 years to establish a commercial base—and it still has not made a real impact on our energy supply.

Of all the energy needs projected for the year 2000 nonelectric uses represent about two-thirds. These uses cover such major categories as transportation, space heating, and industrial processes. The largest energy user at that time will be the manufacturing industry, with transportation using about half as much.

Transportation is illustrative of the possibilities in societal planning. The automobile is responsible for almost half of the world's oil consumption—and a corresponding part of its air pollution. Except for the airplane, the private automobile is the most inefficient mode of using energy for travel.

For passenger travel, railroads are 2½ times as efficient as autos and 5½ times as efficient as airplanes. Buses are 4 times as efficient as autos. For freight, railroads are 3½ times as efficient as trucks and 55 times as efficient as airplanes. Clearly, to reduce energy consumption an extensive nationwide network of railroads, with local bus service, is far superior to an automobile road network. Unfortunately, the world-wide trends have been diametrically opposite, and the human preference for personal mobility has reinforced such trends.

Finally, contrary to much public comment, the development of new speculative energy resources are investments for the future, not a means of remedying the problems of today. Unfortunately, many of these as yet uncertain and undeveloped sources of energy are often misleadingly cited publicly as having a great promise for solving our present difficulties. In addition to their technical uncertainties, many of these speculative sources are likely to be limited in their contribution, even if successful.

Unfortunately, the attraction of "jam tomorrow" may persuade us to neglect the need for "bread and butter" today. Because of the very long time required for any new energy device to become part of the technological structure of our society, these speculative sources could not play a major role before the year 2000. The quality of life of the peoples of the world depends upon the availability in the near future of large amounts of low cost energy in useful form. This being the case, we must plan an orderly development and efficient use of the resources available to us now, and these are primarily fossil fuels and nuclear fission.

Given this situation, what are the possible impacts on United States relations with foreign countries. Because of our present limitations on the use of high-sulfur coal, and the present unavailability of more natural gas, a rapid shift to oil is now under way, because oil can be found with low sulfur or can be desulfurized.**

There is no emergency remedy except rationing. Because roughly half our oil goes to transportation, this is the likely area to be controlled, not electricity. United States oil production can be increased only fractionally, even if all internal controls are removed.

FOREIGN OIL

If the politically distasteful course of rationing is not taken, our 1970 foreign oil purchase of $5 billion will become $10 billion in 1975, and $15 billion by 1980, at which time half our oil consumed will be foreign. For perspective, these dollar outflows may be compared with the total United States annual capital investments of less than $100 billion. I will not dwell on the international monetary consequences.

It should also be remembered that increased fuel cost means increased cost of goods, reduced foreign sales and worsened trade balance. The foreign relations issue is, of course, aggravated by the increasing dependency of the United States on the oil-producing nations without a balancing dependency on their part. The international tensions so produced can lead to consequences of the most serious nature, a variety of scenarios can be imagined.

A parallel situation exists in Western Europe, and both France and Germany have embarked on the construction of oil storage facilities to provide at least three months reserve. (The United States now has a two to three week stored supply.) The recent North Sea discoveries will help, but not solve this problem. These countries are also developing pipeline connections to the Soviet Union and Eastern European oil fields. Western Europe and the United States may end up in conflict for limited world resources.

**Editors' note: Stated U.S. national energy policy following the 1973–1974 Arab oil embargo is to stop and reverse this shift to oil.

For the United States, the Canadian supplies are attractive, but both trade barriers and lack of incentives have made this a slowly developing course. Perhaps we should offer them deuterium oxide (heavy water) for their oil. The Canadians have no reason to be concerned with our problem and may be viewing it with some skepticism, as do many foreigners.†

After all, the environmental issues that have engendered our situation have a very dubious rationality. The public health causality relations which are the reputed origin of our pollution standards are not, in fact, based on demonstrable or credible risk-benefit analyses. They are instead judgmental levels set primarily for aesthetic or comfort purposes, with health benefit marginal at best. This is not likely to create international sympathy.

NO QUICK FIX

Although the near-term United States situation has no quick fix, the intermediate term (post-1980) has several optional aids—offshore drilling, for example. This is much less polluting than tanker imports, and given time could probably meet much of our needs. Of course, we must resolve the issue of the "law of the sea" if we wish to exploit the resources beyond the three-mile limit. Another option is   ease the environmental

†Editors' note: A recently announced (October 1974), approximately 10 billion barrels, petroleum discovery in eastern Mexico—equivalent in size to the North Slope discovery—may improve the North American supply picture.

and aesthetic constraints and reactivate coal mining, and this may occur when the public realizes the situation. Another is to speed up coal desulfurization, gasification, and liquefaction, and the recovery from oil shales and tar sands. These take both the development of commercial technology and much capital.

In the very long term (post-2000) we have the "clean air" option of nuclear power. The abundance of uranium and the fast breeder gives us potentially ample energy. Obviously, the rate at which nuclear power comes on the scene is dependent in the United States on public acceptance. It should be pointed out that this issue does not exist in most foreign countries. As a consequence, we may be buying foreign reactors eventually.

We must recognize and resolve the several very basic trade offs among environment, life styles, personal freedoms, amenities, international tensions, high energy cost and high cost of goods, public health, personal income, allocation of national resources, and perhaps others. The issue may be as basic as national security versus social costs.

For example, based on my perceptions of the alternatives, I would very much rather accept the minimal risks of large scale nuclear power than the already evident risks of international tensions from foreign oil. These issues are so important, and the energy systems so ponderous and slow to change, that our national planning must be based on comprehensively developed long-range insight rather than on fickle public emotion and short-term political expediency. Let us hope it is.

The Chemical Century

RALPH E. LAPP   1973

The American people are absolutely bewildered that this great country should be running short of energy. In part this traces back to a persistent pioneer mentality—to the unbounded days of the frontier life and to exploitation without concern for tomorrow. Americans find it difficult to believe that our natural resources are finite and subject to critical depletion.

Americans have very little appreciation of the rate at which natural resources have been exploited during this century. A convenient way to understand the energy crisis is to set down the specifics in graphic form. I have done this in Figure 1-1, beginning with the year 1860 so that there is some historical perspective on fuel use in the United States. The curve labeled total energy is the sum of heat supplied by wood burning and initially, coal. Wood was the prime source of United States energy in the mid-1800's, but this peaked a century ago and appears as a small pimple on our chart, where the bottom line is energy equal to 100 million tons of coal.

The United States shifted to a coal-burning economy prior to 1900 and averaged a production of about half a billion tons per year for many years. With the splitting of the atom it appeared that coal was headed downward on the chart, suffering the same fuel fate as wood. Beginning with World War I, oil and, to a lesser extent, gas began to provide energy inputs for the nation. The ease of getting gas and oil from the ground by means of pipes and the facility for pipeline transportation, as well as the immense automobile-generated demand for petroleum products, sent the oil/gas consumption rocketing upward so that by mid-century coal was displaced as the nation's No. 1 fuel. Despite the promise of nuclear power it did not manage to cross the x axis of our illustration, although this will happen very soon.

If we pause at this point to survey how far we have come in this century, it's clear that except for the setback of the Great Depression, during which national energy consumption followed the path of gross

Dr. Ralph E. Lapp is a consulting physicist and lecturer in radiological safety and nuclear radiation physics.

From *Science and Public Affairs (Bulletin of the Atomic Scientists)*, Vol. XXIX, No. 7, pp. 8–14, September, 1973. Reprinted by permission of the author and *Science and Public Affairs.* Copyright 1973 by the Educational Foundation for Nuclear Science. This article is based on an address delivered at the dedication of the William Rand Kenan, Jr. Laboratories of Chemistry at the University of North Carolina, Chapel Hill.

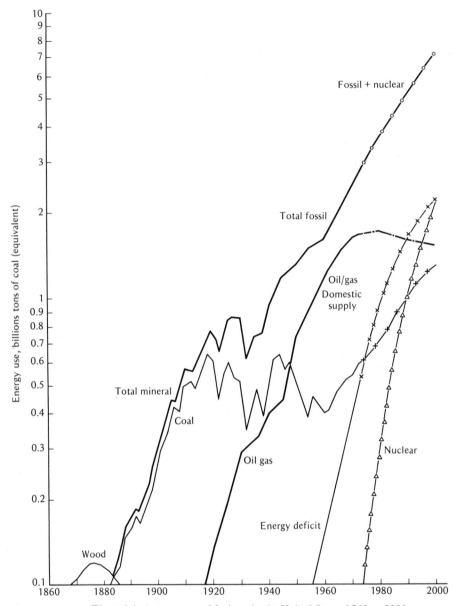

Figure 1-1. A panorama of fuel use in the United States 1860 to 2000.

national product, the pattern of energy use is one of steady growth. As given in Figure 1-1, the growth is that of chemical combustion of fossil fuels since we have omitted the contribution of hydropower. The magnitude of this 72-year-long period of combustion, beginning with 1900, is appreciated if we sum up the fossil fuel burnup. It amounts to a total of 76 billion tons of coal equivalent, of which 35 billion tons was actually coal, the rest being oil and natural gas.

This vast conflagration involved certain adverse effects which were until quite recent years tolerated by the public, presumably as a necessary evil accompanying the industrialization of America. We now have an Environmental Protection Agency charged with safeguarding the public health from ill effects of environmental hazards. The environmentalists must be given credit for focusing national attention on such hazards.

The burning of chemical fuels, whether as solids in steam boilers to generate electricity, or as liquids in internal combustion engines to provide mobility for Americans, emerges as a problem requiring technological therapy on a massive scale. It is easy to sense from the swift ascent of the oil/gas curve in Figure 1-1 that sooner or later the combustion of so much fuel would have environmental effects. Few in the scientific and technical community foresaw these adverse effects.

If we look to the future and extrapolate the national energy consumption curve to the year 2000, there are obviously many curves we may plot. The specific one I have plotted is one projected by the U.S. Department of the Interior. Table 1-1 illustrates energy consumption in the year 2000 for various annual rates of growth.

Now it can be argued that the United States does not need to sustain such a high growth rate for the next several decades. Certainly energy is wasted in our affluent society, and there are a number of ways to conserve energy. However, these are incremental in their effect and many are difficult to implement in a democratic society. Let us assume that the projection made in Figure 1-1 of 7.2 billion tons of coal equivalent is a reasonable extrapolation and, then, see what such a projection means in terms of the various fuel sector supplies.

It is now fairly well established that the growth curve for nuclear power is not an unreasonable projection. If it comes true, in the year 2000 nuclear power will be substituting for 2.1 billion tons of coal in steam boilers for the generation of electricity. About 50 per cent of all United States fuel inputs will go into the generation of electric energy, and uranium or its derivative plutonium will supply about 60 per cent of this fuel input.

The situation with respect to future production of coal is far from clear. The estimate of 1.3 billion tons of mined coal is probably not an unreasonable estimate, but there are some big if's involved.

At the present time about 50 per cent of United States coal is strip mined and this percentage is steadily increasing. The miner labor force in the United States is a fifth of its historic peak and much higher productivity can be achieved by stripping coal than by taking it from deep mines. On the other hand, strip mining has been a rather ruthless process that has left deep scars in the United States topography and has caused extensive and unnecessary environmental damage. In general, strip mining in the eastern United States states has disturbed about 0.6 acres per 1,000 tons of coal removed. However, many of these mining operations have been conducted in hilly terrain where coal beds are rather thin, running about 4 to 5 feet in some states, less in others. The high sulfur content of coal east of the Mississippi River makes much of this fuel unacceptable under strict Environmental Protection Agency source emission standards. Many utilities have been unable to obtain low sulfur coal and midwestern utilities find it difficult to buy low sulfur oil and next to impossible to

Table 1-1. Projections of United States Energy Consumption for the Year 2000 Based on Various Annual Rates of Growth (billion tons of coal equivalent).[a]

Rate	2.0	2.6	3.1	3.6	4.5 per cent per year
A.D. 2000	5.1	6.0	7.0	8.0	10.3 billion tons

[a]Source: U.S. Department of Interior. "United States Energy Through the Year 2000," Dec. 1972.

obtain more supplies of natural gas with the result that many United States firms have gone nuclear.

North Carolina, ranking twelfth in population, is the sixth leading state in rank of its nuclear power plants in operation or under construction. The availability of energy, especially electric energy, is a key to growth and is therefore a fundamental factor in how a state will prosper in the future.

The nuclear projection given in Figure 1-1 must be judged by the slope of the curve, that is, by the time-phased supply of this new energy source as compared to the oil/gas curve. It will be noted that it represents a more vigorous marketing of an energy product than any other in this century. The significance of this time-phased entry of nuclear power on the energy stage is appreciated if we make some rather optimistic assumptions about the continued oil/gas supply from domestic reserves. This has been done in Figure 1-1 on the basis that a strong exploration program results in new additions to reserves so that production does not fall much below present levels.[1]

ENERGY DEFICIT CURVE

If we add up the projected coal, oil/gas, and nuclear curves and subtract them from the fossil plus nuclear curve, we get the energy deficit curve shown in Figure 1-1. Note that the slope of this curve is roughly parallel to that of total energy growth at the turn of the century and to the initial growth of oil/gas in the 1920's. This curve represents the great energy challenge for the rest of this century. It means that new energy sources must be assessed in terms of their capability of matching the slope of this curve. We can alter somewhat the slope of the deficit curve by the expedient of using oil imports, but this has the effect of simply postponing the day of reckoning. Ultimately, we have to consider the magnitude of the imports required and the impact of acquiring such energy supplies.

1. National Petroleum Council, *U.S. Energy Outlook*, Jan. 1973.

Just what are the energy resource options available to meet the challenge of this energy deficit curve? Here I confine the options to primary sources, exempting a secondary source such as hydrogen, and to those available from domestic domains.

There are a number of geophysical options such as utilizing the thermal gradient in the ocean, tidal power, geothermal power, and, most obviously, solar power. These all deserve serious study, but it must be acknowledged that none of them appears to have a short-term potential for providing energy in a time frame capable of matching the demands of our postulated energy deficit curve.

Another qualification should be added to those already stipulated for the energy deficit curve—the form in which the energy is to be supplied. Since it is the oil/gas curve that is peaking, we must pay most attention to the energy uses served by the supply of liquid and gaseous fuels. Obviously, many of these uses involve the transportation sector and internal combustion engines. It is therefore of prime importance that substitute energy sources should be capable of pinch-hitting for motor fuels that are coming into short supply.

Given these constraints, a survey of domestically exploitable energy sources quickly narrows down to the solid fossils such as coal, subbituminous or lignite, oil shale, and tar sands. The potential energy reserves represented by these fossil solids are truly enormous and, fortunately, much western coal is low in sulfur so that it becomes environmentally acceptable. However, we should not be starry eyed over the sheer statistics of our fossil affluence. It is commonplace for many energy analysts to proclaim that we have sufficient coal resources to last us for many hundreds of years at present rates of consumption. There are two things wrong with such optimism: One, the statistics apply to potentially available reserves and do not reckon with the limiting factors that may apply to their exploitation; two, present rates, whether of coal use or of total energy use, do not take into account the demand growth for energy in the future.

Nonetheless, the United States does have tens of billions of tons of low sulfur fossil solids in the Upper Missouri Basin in the form of thick (25- to 100-foot) seams underlying flat or rolling terrain with modest overburdens of earth. For example, in eastern Montana and parts of North Dakota coal fields promise 60,000- to 100,000-ton yields per acre. These fields are being worked now at very modest rates and energy companies are moving in to acquire coal rights for immense quantities of fuel. In effect, Montana and Wyoming, as well as the Dakotas, have the potential for becoming a vast new energy center for the United States, displacing Texas and Louisiana from their ranking positions.

Because of the fundamental importance of the Upper Missouri Basin energy resources, their exploitation should not be unplanned. There is a need for a comprehensive analysis of the entire energy complex in this regional area, taking into account the environmental impact, the economic impact, the technological requirements for coal conversion into liquid and gaseous form, the water requirements, and the transportation modes for fuel or energy.[2]

The basic problem of coal conversion is one of simple chemical addition: adding hydrogen to carbon to form a hydrocarbon. This premiumization of solid fuel into liquid or gaseous form is a great challenge to chemistry and to chemical engineering.

I cannot help but think back to the wartime days on the Manhattan A-bomb project and reflect that then our problem was one of simple nuclear division—splitting uranium nuclei into two halves. It was accomplished by a massive scientific, technological, and engineering effort and there is nothing even remotely comparable to it evident today. The President's energy message of April 1973 gave little indication that this coal conversion process is being attacked on a scale

capable of meeting the demands of the energy deficit curve plotted in Figure 1-1.

I admit that the chemical conversion problem differs from that of the wartime nuclear project objective in that the latter did not have to meet the test of economic competitiveness with existing fuels. The problem in converting solid fossils to premium fuels is one of making technology pay off with a product that will be within the purchasing power of the United States consumer.

I think we should not compel the technology to meet the present price of oil or gas. The American motor fuel purchaser can afford to pay more on a per gallon basis, and cents per gallon add up to dollars per barrel. A 6 cents-per-gallon differential on motor fuel is $2.52 per barrel, and that is a hefty incremental price to shore up a domestic synthetic fuel industry. We are not talking about doubling the price of motor fuel or of getting anywhere close to European prices.

Given the great importance of motor fuel to our mobility and of gaseous fuel to home heating and industrial processes the cents-per-gallon concept suggests a way out of this energy crisis, namely, by pretaxation. For example, an immediate 2-cents-per-gallon tax on vehicular fuel would yield $2 billion per year. This could be allocated to a wide variety of means to solve the energy crisis but, most importantly, it would supply badly needed funds to research and development on energy systems. Many of the candidate schemes for coal conversion need to be subjected to the crucible test of going to scale. Such tests involve large sums of money and they are simply not available today. Moreover, the research and development effort demands a sense of urgency and concomitant technological daring that is lacking today.

However optimistic one may be about fuel synthesis, it is unrealistic to think that we can make up the short-term energy deficit by this technique. We have waited too long and have proceeded too timidly to bring this new energy source on stream in time to provide much energy in this decade.

2. Montana Department of Natural Resources and Conservation, *Coal Development in Eastern Montana,* report of Montana Coal Task Force, Jan. 1973.

There seems to be little alternative—even assuming a more vigorous exploitation of domestic oil and natural gas, which will take some years to bring to reality—to importing oil as the short-term solution to our energy ills.

Looking at the full scope of the twentieth century, it is clear that total energy consumption will be equivalent to the burn up of about 207 billion tons of coal. Since only about a tenth of this will be provided from nuclear energy, it is no mislabeling to call this the Chemical Century, at least as far as energy is concerned.

Probably about 60 billion tons of coal will be burned directly as a heat source. If the energy deficit is to be made up by exploitation of solid fossils, then still larger tonnages of carboniferous material will have to be mined. If *all* of the energy deficit were made up by mining coal, it would add another 40 billion tons to the 60 billion mined for direct burning. But if we consider the fuel requirements for premiumizing the coal or oil shale, then the mined tonnages become much larger. One is then faced with the limiting factor of mining.

However one views the twentieth century, it is significant that the United States is returning to its dependence on fuels taken from the earth as solids. We can sense the drama of the phase change of United States fuels if we look beyond the year 2000. After all the world will not end then, and it may be instructive to project ahead to the year 2100. No disclaimers are necessary at this point since all we shall attempt is a single scenario of the future in order to provide a larger picture of the situation.

THE YEAR 2100

An energy forecast to the year 2100 is presented in Figure 1-2. The right-hand scale represents United States population, and it is assumed that this grows very slowly, that is, increasing from 280 million in 2000 to only 400 million in 2100. The energy growth rate fitted to this projection corresponds to 1.4 per cent per annum for the

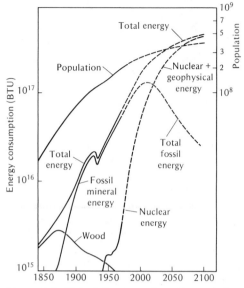

Figure 1-2. United States energy and population growth, 1840–2100.

first half of the century and to 0.5 per cent for the second half.

In effect, the energy consumption in the next century is assumed to be "bent over" rather sharply, that is, flattening out after the heady growth of the twentieth century. This is, then, a rather conservative projection. Figure 1-2 has a left-hand scale in which energy units are BTU rather than tons of coal equivalent. One quintillion is 10^{18} BTU, a convenient unit for the twenty-first century.

It is interesting to note that if the United States were to continue energy use at a 3 per cent per annum growth rate, then by the year 2050 the total energy use would be 880 quadrillion (10^{15}) BTU, or almost 1 quintillion BTU per year. This would extrapolate to 4 quintillion BTU per year by the year 2100.

If one integrates the energy under the total energy curve for the full twenty-first century, it adds up to 37 quintillion BTU or, in terms of coal equivalent, 1,500 billion tons. This amounts to almost seven times more energy than consumed in the previous century. This multiplier is not astonishingly large, but it involves such high demands on United States natural resources that we are forced to conclude that the use of fossil fuels will peak about 2010 and decline as shown in Figure 1-2. By the end of this century, we will have very probably severely depleted our premium fossil fuel reserves, and we will be limited to the sheer quantity of solids that can be mined.

Figure 1-2 reveals that the United States energy history will probably consist of a series of rise-and-fall sequences, the first being a brief wood-burning era, followed by a more protracted fossil-fuel period with the transition in the next century to nonchemical energy release of a geophysical or nuclear nature. Here we are dealing with primary energy sources, and I make no distinction between nuclear fission or fusion. It must be realized that these heat sources are limited to stationary plants and to the production of transmittable energy in the form of electricity. If in the future stored energy techniques do not make for economical and practicable mobile units, then it is likely that hydrogen will emerge as a secondary energy source, being generated from the dissociation of water in huge nuclear-electric stations.

The rise curve postulated for the growth of nuclear power assumes that fission reactors are designed, constructed and operated in such a way that the public safety is assured. The issue of nuclear power safety has been called into question in the public domain and the Atomic Energy Commission has under preparation an extensive exposition of this issue.[3] It served as a basis for hearings before the Joint Committee on Atomic Energy held in the fall of 1973. The Atomic Energy Commission also has under way a study* of the probability and conse-

quences of reactor accidents designed to update the 1957 study, which considered the consequences of what is known as a Class 9 accident.[4] It is to be regretted that the Commission has been so tardy in putting forth a detailed analysis of the consequences of a Class 9 accident. However, it does appear that the available evidence argues against a simple scale up of the 1957 report consequences.

It is to be expected that future reactors will be liquid-metal cooled, and that after the year 2000 thermal efficiencies in the range of 50 per cent will be achieved. This reduction in condenser discharge heat will abate the thermal pollution problem, but the continued electrification of the United States economy will mean high thermal burdens for the environment. The advent of high temperature gas-cooled reactors coupled to gas turbines will be a significant step forward in deflecting the thermal discharge from the water environment. Though the evolution of increasing quantities of heat may pose a problem for certain localities such as the Los Angeles basin, this should not be a limiting factor on a national scale. The total energy output of the year 2100 is almost two orders of magnitude lower than the 1 per cent of solar energy input generally regarded as climatologically significant.

The energy scenario for the next century represents about 8 quintillion BTU for fossil (chemical) energy and 28 quintillion BTU for nonchemical energy sources. One is tempted to say that the next century of energy belongs to the physicist as opposed to the chemist, but in reality chemistry will be required to play a tremendous role in the nuclear and geophysical energy phase of power generation in the next century.

WORLD ENERGY USE

The "bending over" of the United States energy curve in the twenty-first century is

3. Atomic Energy Commission, "The Safety of Nuclear Power Reactors and Related Facilities," WASH-1250, June 1973.
*Editors' note: See Part 3, article titled "Reactor Safety Study: An Assessment of Accident Risks in Commercial Nuclear Power Plants", of this book for a summary of this study.

4. Atomic Energy Commission, "Theoretical Possibilities and Consequences of Major Accidents in Large Nuclear Power Plants," WASH-740, 1957.

tied to a slow rate of population growth and to a very modest increase in per capita consumption of energy, characteristic of a nation that is approaching saturation in this respect. If we turn our attention to the world, we must come to grips with the vexing problem of population control in developing nations and with the fact that such nations are very far removed from saturation levels in per capita energy consumption. United Nations estimates for the year 2000 population place the figure between 6 and 7 billion people. A growth rate of somewhat less than 2 per cent per annum may be postulated for that time and, of course, a continuation of such a trend would see the population explode to some 45 billion people by the year 2100.

So as not to overstate the case for energy consumption in the next century, I have optimistically assumed that the world population will not exceed 11 billion by the year 2100. Such a projection involves severe population control and demands that the growth rate decrease to 1 per cent per annum in the first part of the twenty-first century, to 0.5 per cent at mid-century and to 0.25 per cent a hundred years from now.

I shall assume that the per capita consumption of energy on a world-wide basis in the year 2100 will be about that reached by United States citizens in 1950. Using a United Nations projection for the year 2000 and a "bending over" to the year 2100 per capita consumption multiplied by 11 billion people value, I arrived at the energy curve shown in Figure 1-3. Total energy for the next century amounts to 162 quintillion BTU, or 4.4 times more than for the United States for the same century. This projection

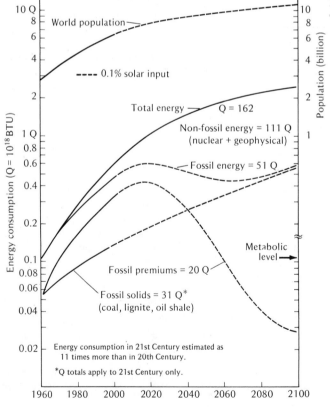

Figure 1-3. World population and energy consumption, 1960–2100.

calls for the United States, 3.6 per cent of the world's year 2100 population, to use a fifth of the world's total energy. The metabolic heat of 11 billion people is represented by an arrow on the right-hand scale.

The world's probable utilization of fossil fuels is depicted as a single curve (fossil energy) representing the use of premiums and another curve showing the exploitation of fossil solids. It is assumed that countries other than the United States will represent the market for oil and natural gas, most of which will come from the Persian Gulf. This represents a generous estimate of world oil/gas which is exploitable since it corresponds to a total of ultimately recoverable 6,000 billion barrels of oil equivalent. Fossil solids to be mined in the twenty-first century add up to 31 quintillion BTU. Considering the declining heat value of coal and the necessary resort to even less caloric fuels, this means a century-wide mining of about 1,000 billion tons of fossil solids.

The shape of the world fossil energy curve exhibits a mid-century dip due to run out of premium fuels and an assumed inability of less developed nations to switch to nonchemical fuels because of the high capital costs associated with them. The more developed nations are assumed to opt for nuclear/geophysical energy sources. Total consumption of fuels for the twenty-first century is estimated at 162 quintillion BTU, 111 quintillion BTU being of nonchemical origin.

During the twenty-first century the success of the breeder reactor or the attainment of economic fusion will serve to protect the earth's surface from mining insult. Uranium ores mined during the last quarter of this century should be quite adequate to last throughout the twenty-first century.

I am mindful that the burn up of the world's unique inheritance of hydrocarbons in the form of oil and natural gas must be viewed as an atrocity. The combustion of such molecular architecture just to produce heat must be reckoned a chemical crime when one considers the needs of the petrochemical industry.

As Figure 1-4 illustrates, the profile of United States energy use from 1900, pro-

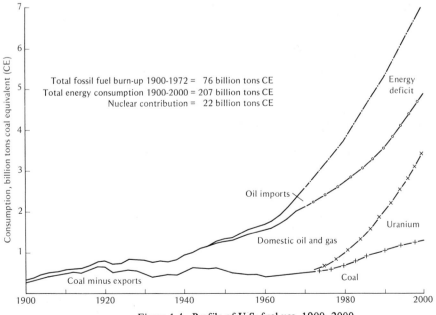

Total fossil fuel burn-up 1900-1972 = 76 billion tons CE
Total energy consumption 1900-2000 = 207 billion tons CE
Nuclear contribution = 22 billion tons CE

Energy deficit

Oil imports

Domestic oil and gas

Uranium

Coal minus exports

Coal

Consumption, billion tons coal equivalent (CE)

Figure 1-4. Profile of U.S. fuel use, 1900–2000.

jected to the year 2000, reflects a dramatic change in resource utilization. If the projection is correct, the United States started out the century utilizing predominantly fossil solid fuels and has since shifted to dependence on the fossil fluids, the run out of which forces us back to mining at the end of this century. In any event a look back at the twentieth century, taken from the vantage point of the year 2000, will show that energywise the economic propellant of this century was chemical energy.

One can, of course, imagine a wide variety of energy scenarios for the next century. The specific one I have chosen to illustrate has the virtue of being a modest example. Greater energy affluence on the planet would involve major environmental considerations and presumably greater inequities in the per capita consumption of energy. On the other hand, energy impoverishment would condemn billions of people to a lower standard of life.

A chemical challenge of great magnitude is involved in providing clean energy in the future. We hardly need to be reminded of the current dilemma of the electric utility industry, which has close at hand immense quantities of coal that can not be used because of high sulfur content. The removal of sulfur in the boiler feed, or its segregation in new types of heat sources, or its scrub out in the stack gases are still unsolved problems.

If we look at the profile of federal funding for research and development in the energy field, we note that the chemist is short changed. The fiscal 1974 budget contains a recommendation of $132 million for the Department of the Interior for research and development on coal, oil, and natural gas, while the Atomic Energy Commission has earmarked for it a total of $574 million. If we are going to get on with the job of tapping the full potential of our fossil fuels, we need to energize our research and development effort on a much larger scale.

Mineral Resource Estimates and Public Policy

V. E. McKELVEY   1972

Not many people, I have found, realize the extent of our dependence on minerals. It was both a surprise and a pleasure, therefore, to come across the observations of George Orwell in his book *The Road to Wigan Pier.* When describing the working conditions of English miners in the 1930's he evidently was led to reflect on the significance of coal:

Our civilization . . . is founded on coal, more completely than one realizes until one stops to think about it. The machines that keep us alive, and the machines that make the machines are all directly or indirectly dependent upon coal. . . . Practically everything we do, from eating an ice to crossing the Atlantic, and from baking a loaf to writing a novel, involves the use of coal, directly or indirectly. For all the arts of peace coal is needed; if war breaks out it is needed all the more. In time of revolution the miner must go on working or the revolution must stop, for revolution as much as reaction needs

coal. . . . In order that Hitler may march the goosestep, that the Pope may denounce Bolshevism, that the cricket crowds may assemble at Lords, that the Nancy poets may scratch one another's backs, coal has got to be forthcoming.

To make Orwell's statement entirely accurate—and ruin its force with complications—we should speak of mineral *fuels,* instead of coal, and of other minerals also, for it is true that minerals and mineral fuels are the resources that make the industrial society possible. The essential role of minerals and mineral fuels in human life may be illustrated by a simple equation

$$L = \frac{R \times E \times I}{P}$$

in which the society's average level of living (L), measured in its useful consumption of

Dr. V. E. McKelvey is director of the U.S. Geological Survey. Most of his research has concerned the geology and origin of marine phosphorites and associated rocks, but he has had a continuing interest in broader problems related to the distribution and magnitude of mineral and energy resources. He has been a member of the U.S. Delegation to the United Nations' Seabeds Committee since its inception in 1968, and most of his recent papers relate to the appraisal of seabed resources and problems related to their development.

Originally presented at Harvard University in February 1971 as the Seventh McKinstry Memorial Lecture, and subsequently published in the *American Scientist*, Vol. 60, No. 1, pp. 32–40, Jan.–Feb. 1972. Reprinted with light editing and by permission of the author and *American Scientist.* Copyright 1972 by The Society of the Sigma Xi.

goods and services, is seen to be a function of its useful consumption of all kinds of raw materials (R), including metals, nonmetals, water, soil minerals, biologic produce, and so on; times its useful consumption of all forms of energy (E); times its useful consumption of all forms of ingenuity (I), including political and socioeconomic as well as technologic ingenuity; divided by the number of people (P) who share in the total product.

This is a statement of the classical economists' equation in which national output is considered to be a function of its use of capital and labor, but it shows what capital and labor really are. Far from being mere money, which is what it is popularly thought to mean, capital represents accumulated usable raw materials and things made from

them, usable energy, and especially accumulated knowledge. And the muscle power expended in mere physical toil, which is what labor is often thought to mean, is a trivial contribution to national output compared to that supplied by people in the form of skills and ingenuity.

This is only a conceptual equation, of course, for numerical values cannot be assigned to some of its components, and no doubt some of them—ingenuity in particular—should receive far more weight than others. Moreover, its components are highly interrelated and interdependent. It is the development and use of a high degree of ingenuity that makes possible the high consumption of minerals and fuels, and the use of minerals and fuels are each essential to

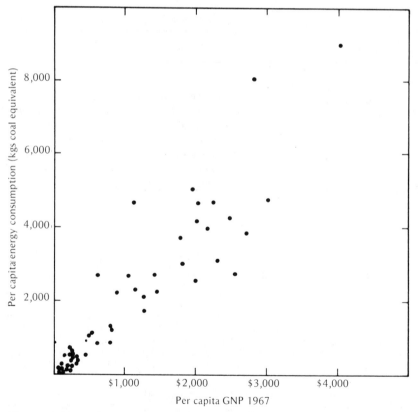

Figure 1-5. Per capita energy consumption compared to per capita gross national product in countries for which statistics are available in the United Nations *Statistical Yearbook* for 1967.

the availability and use of the other. Nevertheless, the expression serves to emphasize that level of living is a function of our intelligent use of natural resources, and it brings out the importance of the use of energy and minerals in the industrial society. As shown in Figure 1-5, per capita gross national product among the countries of the world is, in fact, closely related to their per capita consumption of energy. Steel consumption also shows a close relation to per capita gross national product (GNP) (Figure 1-6), as does the consumption of many other minerals.

Because of the key role that minerals and fuels play in economic growth and in economic and military security, the extent of

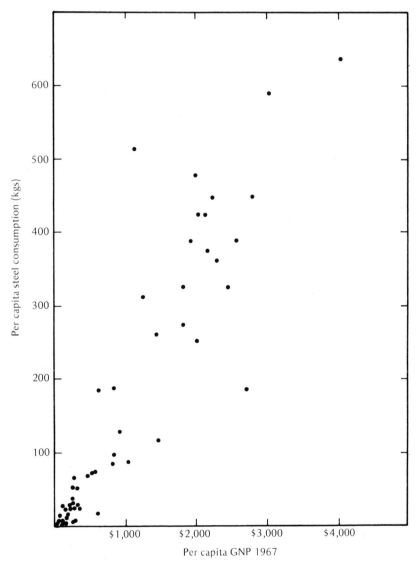

Figure 1-6. Per capita steel consumption compared to per capita gross national product in countries for which statistics are available in the United Nations *Statistical Yearbook* for 1967.

their resources is a matter of great importance to government, and questions concerning the magnitude of resources arise in conjunction with many public problems. To cite some recent examples, the magnitude of low-cost coal and uranium reserves has been at the heart of the question as to when to press the development of the breeder reactor—which requires a research and development program involving such an enormous outlay of public capital that it would be unwise to make the investment until absolutely necessary. Similarly, estimates of potential oil and gas resources are needed for policy decisions related to the development of oil shale and coal as commercial sources of hydrocarbons, and estimates are needed also as the basis for decisions concerning prices and import controls.

Faced with a developing shortage of natural gas, the Federal Power Commission is presently very interested in knowing whether or not reserves reported by industry are an accurate indication of the amount of natural gas actually on hand; it also wants to know the extent of potential resources and the effect of price on their exploration and development. At the regional or local level, decisions with respect to the designation of wilderness areas and parks, the construction of dams, and other matters related to land use involve appraisal of the distribution and amount of the resources in the area. The questions of the need for an international regime governing the development of seabed resources, the character such arrangement should have, and the definition of the area to which it should apply also involve, among other considerations, analysis of the probable character, distribution, and magnitude of subsea mineral resources.

And coming to the forefront is the most serious question of all—namely, whether or not resources are adequate to support the continued existence of the world's population and indeed our own. The possibility to consider here goes much beyond Malthus's gloomy observations concerning the propensity of a population to grow to the limit of its food supply, for both population and

level of living have grown as the result of the consumption of nonrenewable resources, and both are already far too high to maintain without industrialized high-energy, and high mineral-consuming agriculture, transportation, and manufacturing. I will say more about this question later, but to indicate something of the magnitude of the problem let me point out that, in attaining our high level of living in the United States, we have used more minerals and mineral fuels during the last thirty years than all the people of the world used previously. This enormous consumption will have to be doubled just to meet the needs of the people now living in the United States through the remainder of their lifetimes, to say nothing about the needs of succeeding generations, or the increased consumption that will have to take place in the lesser developed countries if they are to attain a similar level of living.

CONCEPTS OF RESERVES AND RESOURCES

The focus of most of industry's concern over the extent of mineral resources is on the magnitude of the supplies that exist now or that can be developed in the near term, and this is of public interest also. Many other policy decisions, however, relate to the much more difficult question of potential supplies, a question that to be answered properly must take account both of the extent of undiscovered deposits as well as deposits that cannot be produced profitably now but may become workable in the future. Unfortunately, the need to take account of such deposits is often overlooked, and there is a widespread tendency to think of potential resources as consisting merely of materials in known deposits producible under present economic and technologic conditions.

In connection with my own involvement in resource appraisal, I have been developing over the last several years a system of resource classification and terminology that brings out the classes of resources that need to be taken into account in appraising future supplies, which I believe helps to put the supply problem into a useful perspective. Before describing it, however, I want to emphasize that the problem of estimating potential resources has several built-in uncertainties that make an accurate and complete resource inventory impossible, no matter how comprehensive its scope.

One such uncertainty results from the nature of the occurrence of mineral deposits, for most of them lie hidden beneath the earth's surface and are difficult to locate and to examine in a way that yields accurate knowledge of their extent and quality. Another source of uncertainty is that the specifications of recoverable materials are constantly changing as the advance of technology permits us to mine or process minerals that were once too low in grade, too inaccessible, or too refractory to recover profitably. Still another results from advances that make it possible to utilize materials not previously visualized as usable at all.

For these reasons the quantity of usable resources is not fixed but changes with progress in science, technology, and exploration and with shifts in economic conditions. We must expect to revise our estimates periodically to take account of new developments. Even incomplete and provisional estimates are better than none at all, and if they differentiate known, undiscovered, and presently uneconomic resources they will help to define the supply problem and provide a basis for policy decisions relating to it.

The need to differentiate the known and the recoverable from the undiscovered and the uneconomic requires that a resource classification system convey two prime elements of information: the degree of certainty about the existence of the materials and the economic feasibility of recovering them. These two elements have been recognized in existing terminology, but only incompletely. Thus, as used by both the mining and the petroleum industries, the term *reserves* generally refers to economical-ly recoverable material in identified deposits, and the term *resources* includes in addition deposits not yet discovered as well as identified deposits that cannot be recovered now (e.g., Blondel and Lasky, 1956).

The degree of certainty about the existence of the materials is described by terms such as *proved, probable,* and *possible,* the terms traditionally used by industry, and *measured, indicated,* and *inferred,* the terms devised during World War II by the Geological Survey and the Bureau of Mines to serve better the broader purpose of national resource appraisal. Usage of these degree-of-certainty terms is by no means standard, but all of their definitions show that they refer only to deposits or structures known to exist.

Thus, one of the generally accepted definitions of *possible* ore states that it is to apply to deposits whose existence is known from at least one exposure, and another definition refers to an ore body sampled only on one side. The definition of *inferred* reserves agreed to by the Survey and the Bureau of Mines permits inclusion of completely concealed deposits for which there is specific geologic evidence and for which the specific location can be described, but it makes no allowance for ore in unknown structures of undiscovered districts. The previous definitions of both sets of terms also link them to deposits minable at a profit; the classification system these terms comprise has thus neglected deposits that might become minable as the result of technologic or economic developments.

To remedy these defects, I have suggested that existing terminology be expanded into the broader framework shown in Figure 1-7, in which degree of certainty decreases from left to right and feasibility of recovery decreases from top to bottom. Either of the series of terms already used to describe degree of certainty may be used with reference to identified deposits and applied not only to presently minable deposits but to others that have been identified with the same degree of certainty. Feasibility-of-recovery categories are designated by the

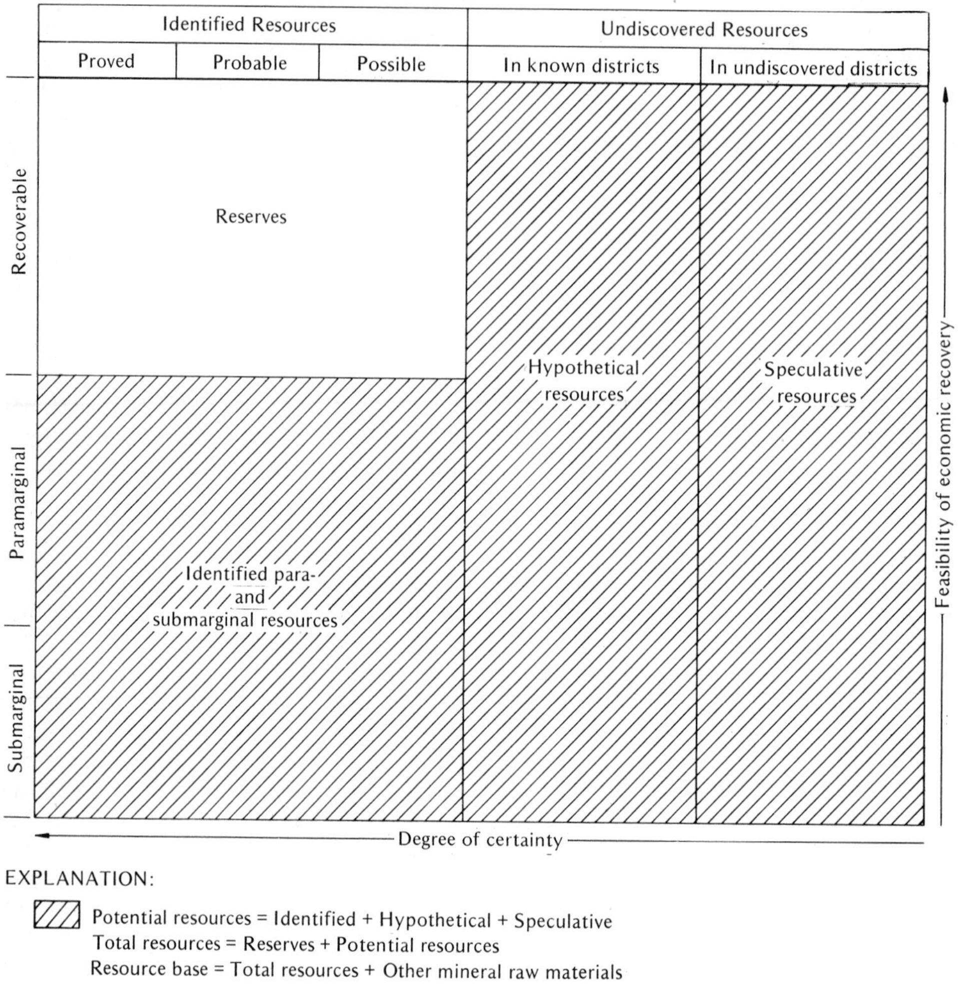

EXPLANATION:

▨ Potential resources = Identified + Hypothetical + Speculative
Total resources = Reserves + Potential resources
Resource base = Total resources + Other mineral raw materials

Figure 1-7. Classification of mineral resources being used by the U.S. Geological Survey in assessing total mineral resources in the United States. [After V. E. McKelvey, 1973, "Mineral potential of the United States," p. 69: In E. N. Cameron (ed.), *The Mineral Position of the United States, 1975–2000.* University of Wisconsin Press, Madison, Wisconsin.]

terms *recoverable, paramarginal,* and *submarginal.*

Paramarginal resources are defined here as those that are recoverable at prices as much as 1.5 times those prevailing now. (I am indebted to Stanley P. Schweinfurth for suggesting the prefix *para* to indicate that the materials described are not only those just on the margin of economic recoverability, the common economic meaning of the term *marginal.*) At first thought this price factor may seem to be unrealistic. The fact is, how-

ever, that prices of many mineral commodities vary within such a range from place to place at any given time, and a price elasticity of this order of magnitude is not uncommon for many commodities over a space of a few years or even months, as shown by recent variations in prices of copper, mercury, silver, sulphur, and coal. Deposits in this category thus become commercially available at price increases that can be borne without serious economic effects, and chances are that improvements in existing technology will make them available at prices little or no higher than those prevailing now.

Over the longer period, we can expect that technologic advances will make it profitable to mine resources that would be much too costly to produce now, and, of course, that is the reason for trying to take account of submarginal resources. Again, it might seem ridiculous to consider resources that cost two or three times more than those produced now as having any future value at all. But keep in mind, as one of many examples, that the cut off grade for copper has been reduced progressively not just by a factor of 2 or 3 but by a factor of 10 since the turn of the century and by a factor of about 250 over the history of mining. Many of the fuels and minerals being produced today would once have been classed as submarginal under this definition, and it is reasonable to believe that continued technologic progress will create recoverable reserves from this category.

EXAMPLES OF ESTIMATES OF POTENTIAL RESOURCES

For most minerals, the chief value of this classification at present is to call attention to the information needed for a comprehensive appraisal of their potential, for we have not developed the knowledge and the methods necessary to make meaningful estimates of the magnitude of undiscovered deposits, and we do not know enough about the cost of producing most presently noncommercial deposits to separate paramarginal from submarginal resources. Enough information is available for the mineral fuels, however, to see their potential in such a framework.

The fuel for which the most complete information is available is the newest one—uranium. As a result of extensive research sponsored by the Atomic Energy Commission, uranium reserves and resources are reported in several cost-of-recovery categories, from less than $8 to more than $100 per pound of uranium oxide (U_3O_8). For the lower-cost ores, the Atomic Energy Commission makes periodic estimates in two

degree-of-certainty categories, one that it calls *reasonably assured reserves* and the other it calls *additional resources,* defined as uranium surmised to occur in unexplored extensions of known deposits or in undiscovered deposits in known uranium districts. Both the Atomic Energy Commission and the Geological Survey have made estimates from time to time of resources in other degree-of-certainty and cost-of-recovery categories.

Ore in the less-than-$8-per-pound class is minable now, and the Atomic Energy Commission estimates reasonably assured reserves to be 143,000 tons and additional resources to be 167,000 tons of U_3O_8—just about enough to supply the lifetime needs of reactors in use or ordered in 1968 and only half that required for reactors expected to be in use by 1980. The Geological Survey, however, estimates that undiscovered resources of presently minable quality may amount to 750,000 tons, or about two and a half times that in identified deposits, and districts. Resources in the $8-to-30-a-pound category in identified and undiscovered deposits add only about 600,000 tons of U_3O_8 and thus do not significantly increase potential reserves.

But tens of millions of tons come into prospect in the price range of $30-to-100 per pound. Uranium at such prices would be usable in the breeder reactor. The breeder, of course, would utilize not only uranium-235 but also uranium-238, which is 140 times more abundant than uranium-235. Plainly the significance of uranium as a commercial fuel lies in its use in the breeder reactor, and one may question, as a number of critics have (e.g., Inglis, 1971), the advisability of enlarging nuclear generating capacity until the breeder is ready for commercial use.

* * *

QUANTIFYING THE UNDISCOVERED

Considering potential resources in the degree-of-certainty, cost-of-recovery framework brings out the joint role that geologists, engineers, mineral technologists, and economists must play in estimating their magnitude. Having emphasized the importance of the economic and technologic side of the problem, I want now to turn to the geological side and consider the problem of how to appraise the extent of undiscovered reserves and resources.

* * *

Two principal approaches to the problem have been taken thus far. One is to extrapolate observations related to rate of industrial activity, such as annual production of the commodity. The other is to extrapolate observations that relate to the abundance of the mineral in the geologic environment in which it is found.

The first of these methods has been utilized by M. King Hubbert, Charles L. Moore, and M. A. Elliott and H. R. Linden in estimating ultimate reserves of petroleum. The essential features of this approach are to analyze the growth in production, proved reserves, and discovery per foot of drilling over time and to project these rate phenomena to terminal values in order to predict ultimate production. Hubbert has used the logistic curve for his projections, and Moore has utilized the Gompertz curve, with results more than twice as high as those of Hubbert. As Hubbert has pointed out, these methods utilize the most reliable information collected on the petroleum industry—modern records on production, proved reserves, number of wells drilled, and similar activities are both relatively complete and accurate, at least as compared with quantitative knowledge about geologic features that affect the distribution of petroleum.

The rate methods, however, have an inherent weakness in that the phenomena they analyze reflect human activities that are strongly influenced by economic, political, and other factors that bear no relation to the amount of oil or other material that lies in the ground. Moreover, they make no allowance for major breakthroughs that might transform extensive paramarginal or submarginal resources into recoverable reserves, nor

do they provide a means of estimating the potential resources of unexplored regions. Such projections have some value in indicating what will happen over the short term if recent trends continue, but they can have only limited success in appraising potential resources.

Even the goal of such projections, namely, the prediction of ultimate production, is not a useful one. Not only is it impossible to predict the quantitative effects of man's future activities but the concept implies that the activities of the past are a part of an inexorable process with only one possible outcome. Far more useful, in my opinion, are estimates of the amounts of various kinds of materials that are in the ground in various environments; such estimates establish targets for both the explorer and the technologist, and they give us a basis for choosing among alternative ways of meeting our needs for mineral supplies.

The second principal approach taken thus far to the estimation of undiscovered resources involves the extrapolation of data on the abundance of mineral deposits from explored to unexplored ground on the basis of either the area or the volume of broadly favorable rocks. In the field of metalliferous deposits, T. B. Nolan pioneered in extrapolation on the basis of area in his study of the spatial and size distribution of mineral deposits in the Boulder Dam region and in his conclusion that a similar distribution should prevail in adjacent concealed and unexplored areas. Lewis Weeks and Wallace Pratt played similar roles with respect to the estimation of petroleum resources—Weeks extrapolating on the basis of oil per unit volume of sediment and Pratt on the basis of oil per unit area. Many of the estimates of crude oil that went into the National Petroleum Council study were made by the volumetric method, utilizing locally appropriate factors on the amount of oil expected per cubic mile of sediment. Olson and Overstreet have since used the area method to estimate the magnitude of world thorium resources as a function of the size of areas of igneous and metamorphic rocks as compared to India and the United States, and A. P. Butler used the magnitude of sandstone uranium ore reserves exposed in outcrop as a basis for estimating the area in back of the outcrop that is similarly mineralized.

Several years ago, A. D. Zapp and T. A. Hendricks introduced another approach, based on the amount of drilling required to explore adequately the ground favorable for exploration and the reserves discovered by the footage already drilled—a procedure usable in combination with either the volumetric or areal approach. Recently Zimmerman and Long (*Oil and Gas Journal*, 1969) applied this approach to the estimation of gas resources in the Delaware-Val Verde basins of western Texas and southeastern New Mexico, and Haun and others used it to estimate potential natural gas resources in the Rocky Mountain region. In the field of metals, J. David Lowell has estimated the number of undiscovered porphyry copper deposits in the southwestern United States, Chile and Peru, and British Columbia as a function of the proportion of the favorable preore surface adequately explored by drilling, and F. C. Armstrong has similarly estimated undiscovered uranium reserves in the Gas Hills area of Wyoming on the basis of the ratios between explored and unexplored favorable areas.

I have suggested another variant of the areal method for estimating reserves of non-fuel minerals which is based on the fact that the tonnage of minable reserves of the well-explored elements in the United States is roughly equal to their crustal abundance in per cent times a billion or 10 billion (Figure 1-8). Obviously this relation is influenced by the extent of exploration, for it is only reserves of the long-sought and well-explored minerals that display the relation to abundance. But it is this feature that gives the method its greatest usefulness, for it makes it possible to estimate potential resources of elements, such as uranium and thorium, that have been prospected for only a short period. Sekine tested this method for Japan and found it applicable there, which surprised me a little, for I would not have thought Japan to be a large enough sample of the continental crust to bring out this relationship.

The relation between reserves and abundance, of course, can at best be only an approximate one, useful mainly in order-of-magnitude estimates, for obviously crustal abundance of an element is only one of its properties that lead to its concentration. That it is an important factor, however, may be seen not only in its influence on the magnitude of reserves but also in other expressions of its influence on the concentrations of the elements. For example, of the 18 or so elements with crustal abundances greater than about 200 parts per million, all but fluorine and strontium are rock forming in the sense that some extensive rocks are composed chiefly of minerals of which each of these elements is a major constituent. Of the less abundant elements, only chromium, nitrogen, and boron have this distinction. Only a few other elements, such as copper, lead, and zinc, even form ore bodies composed mainly of minerals of which the valuable element is a major constituent, and in a general way the grade of minable ores decreases with decreasing crustal abundance. A similar gross correlation exists between abundance of the elements and the number of minerals in which they are a significant constituent.

Members of a committee of the Geology and the Conservation of Mineral Resources Board of the Soviet Union have described a somewhat similar method for the quantitative evaluation of what they call "predicted reserves" of oil and gas, based on estimates of the total amount of hydrocarbons in the source rock and of the fraction that has migrated into commercial reservoirs—estimates that would be much more difficult to obtain for petroleum than for the elements. Probably for this reason not much use has been made of this method, but it seems likely that quantitative studies of the effects of the natural fractionation of the elements might be of some value in estimating total resources in various size and grade categories.

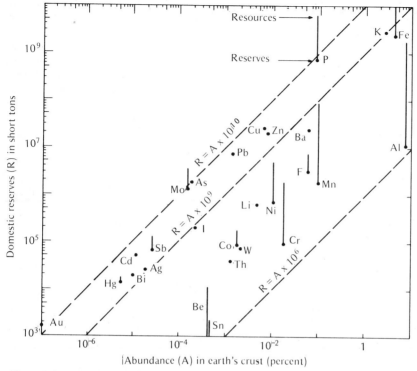

Figure 1-8. Domestic reserves of elements compared to their abundance in the earth's crust. Tonnage of ore minable now is shown by a dot; tonnage of lower-grade ores the exploitation of which depends upon future technological advances or higher prices is shown by a bar.

Some studies of the grade-frequency distribution of the elements have, in fact, been undertaken by geochemists in the last couple of decades, and, taking off from Nolan's work, several investigators have studied the areal and size-frequency distribution of mineral deposits in conjunction with attempts to apply the methods of operations research to exploration (e.g., Allais, 1957; Slichter, 1960; Griffiths, 1964; DeGeoffroy and Wu, 1969; Harris, 1969). None of these studies has been concerned with the estimation of undiscovered reserves, but they have identified two features about the distribution of mineral deposits that may be applicable to the problem.

One is that the size distribution of both metalliferous deposits, expressed in dollar value of production, and of oil and gas, expressed in volumetric units, has been found to be log normal, which means that of a large population of deposits, a few contain most of the ore (e.g., Slichter, 1960; Kaufman, 1963). In the Boulder Dam area, for example, 4 per cent of the districts produced 80 per cent of the total value of recorded production. The petroleum industry in the United States has a rule of thumb that 5 per cent of the fields account for 50 per cent of the reserves and 50 per cent of fields for 95 per cent. And in the U.S.S.R., about 5 per cent of the oil fields contain about 75 per cent of the oil, and 10 per cent of the gas fields have 85 per cent of the gas reserves.

The other feature of interest is that in many deposits the grade-tonnage distribution is also log normal, and the geochemists have found this to be the case also with the frequency distribution of minor elements.

These patterns of size- and grade-frequency distribution will not in themselves provide information on the magnitude of potential resources, for they describe only how minerals are distributed and not how much is present. But if these patterns are combined with quantitative data on the incidence of congeneric deposits in various kinds of environments, the volume or area of favorable ground, and the extent to which it has been explored, they might yield more useful estimates of potential resources than are obtainable by any of the procedures so far applied. Thus estimates of total resources described in terms of their size- and grade-frequency distributions could be further analyzed in the light of economic criteria defining the size, grade, and accessibility of deposits workable at various costs, and then partitioned into feasibility-of-recovery and degree-of-certainty categories to provide targets for exploration and technologic development as well as guidance for policy decisions.

Essential for such estimates, of course, is better knowledge than is now in hand for many minerals on the volume of ore per unit of favorable ground and on the characteristics of favorable ground itself. For petroleum the development of such knowledge is already well advanced. For example, whereas most estimates of resources have been based on an assumed average petroleum content of about 50,000 barrels per cubic mile of sediment, varied a little perhaps to reflect judgments of favorability, the range in various basins is from 10,000 to more than 2,000,000 barrels per cubic mile. As shown by the recent analysis by Halbouty and his colleagues of the factors affecting the formation of giant fields, the geologic criteria are developing that make it possible to classify sedimentary basins in terms of their petroleum potential. Knowledge of the mode of occurrence and genesis of many metalliferous minerals and of the geology of the terraines in which they occur is not sufficient to support comprehensive estimates prepared in this way. But for many kinds of deposits enough is known to utilize this kind of approach on a district or regional basis,

and I hope a start can soon be made in this direction.

NEED FOR REVIEW OF RESOURCE ADEQUACY

Let me return now to the question of whether or not resources are adequate to maintain our present level of living. This is not a new question by any means. In 1908 it was raised as a national policy issue at the famous Governors' Conference on Resources, and it has been the subject of rather extensive inquiry by several national and international bodies since then. In spite of some of the dire predictions about the future made by various people in the course of these inquiries, they did not lead to any major change in our full-speed-ahead policy of economic development. Some of these inquiries, in fact, led to immediate investigations that revealed a greater resource potential for certain minerals than had been thought to exist, and the net effect was to alleviate rather than heighten concern.

Now, however, concern about resource adequacy is mounting again. The over-all tone of the recent National Academy of Sciences' report on *Resources and Man* was cautionary if not pessimistic about continued expansion in the production and use of mineral resources, and many scientists, including some eminent geologists, have expressed grave doubts about our ability to continue on our present course. The question is also being raised internationally, particularly in developing countries where concern is being expressed that our disproportionate use of minerals to support our high level of living may be depriving them of their own future.

Personally, I am confident that for millennia to come we can continue to develop the mineral supplies needed to maintain a high level of living for those who now enjoy it and to raise it for the impoverished people of our own country and world. My reasons for thinking so are that there is a visible undeveloped potential of substantial proportions in each of the processes by which we create resources and that our experience justifies the belief that these processes have dimensions beyond our knowledge and even beyond our imagination at any given time.

Setting aside the unimaginable, I will mention some examples of the believable. I am sure all geologists would agree that minable undiscovered deposits remain in explored as well as unexplored areas and that progress in our knowledge of regional geology and exploration will lead to the discovery of many of them. With respect to unexplored areas, the mineral potential of the continental margins and ocean basins deserves particular emphasis, for the technology that will give us access to it is clearly now in sight. For many critical minerals, we already know of substantial paramarginal and submarginal resources that experience tells us should be brought within economic reach by technologic advance. The process of substituting an abundant for a scarce material has also been pursued successfully, thus far not out of need but out of economic opportunity, and plainly has much potential as a means of enlarging usable resources.

Extending our supplies by increasing the efficiency of recovery and use of raw materials has also been significant. For example, a unit weight of today's steel provides 43 per cent more structural support than it did only ten years ago, reducing proportionately the amount required for a given purpose. Similarly, we make as much electric power from one ton of coal now as we were able to make from seven tons around the turn of the century. Our rising awareness of pollution and its effects surely will force us to pay even more attention to increasing the efficiency of mineral recovery and use as a means of reducing the release of contaminants to the environment. For similar reasons, we are likely to pursue more diligently processes of recovery, reuse, and recycling of mineral materials than we have in the past.

Most important to secure our future is an abundant and cheap supply of energy, for if that is available, we can obtain materials from low-quality sources, perhaps even

country rocks, as Harrison Brown has suggested. Again, I am personally optimistic on this matter, with respect both to the fossil fuels and particularly to the nuclear fuels. Not only does the breeder reactor appear to be near enough to practical reality to justify the belief that it will permit the use of extremely low-grade sources of uranium and thorium that will carry us far into the future, but during the last couple of years there have been exciting new developments in the prospects for commercial energy from fusion. Geothermal energy has a large unexploited potential, and new concepts are also being developed to permit the commercial use of solar energy.

But many others do not share these views, and it seems likely that soon there will be a demand for a confrontation with the full-speed-ahead philosophy that will have to be answered by a deep review of resource adequacy. I myself think that such a review is necessary, simply because the stakes have become so high. Our own population, to say nothing of the world's, is already too large to exist without industrialized, high energy- and mineral-consuming agriculture, transportation, and manufacturing. If our supply of critical materials is enough to meet our needs for only a few decades, a mere tapering off in the rate of increase of their use, or even a modest cutback, would stretch out these supplies for only a trivial period. If resource adequacy cannot be assured into the far-distant future, a major reorientation of our philosophy, goals, and way of life will be necessary. And if we do need to revert to a low resource-consuming economy, we will have to begin the process as quickly as possible in order to avoid chaos and catastrophe.

Comprehensive resource estimates will be essential for this critical examination of resource adequacy, and they will have to be made by techniques of accepted reliability. The techniques I have described for making such estimates have thus far been applied to only a few minerals, and none of them has been developed to the point of general acceptance. Better methods need to be devised and applied more widely, and I hope that others can be enlisted in the effort necessary to do both.

REFERENCES

Allais, M. 1957. Methods of appraising economic prospects of mining exploration over large territories. *Management Science* 2: 285–347.

Armstrong, F. C. 1970. Geologic factors controlling uranium resources in the Gas Hills District, Wyoming. 22nd Annual Field Conference. *Wyoming Geological Association Guidebook*, pp. 31–44.

Blondel, F., and Lasky, S. F. 1965. Mineral reserves and mineral resources. *Economic Geology* 60: 686–697.

Bush, A. L., and Stager, H. K. 1956. Accuracy of ore-reserve estimates for uranium-vanadium deposits on the Colorado Plateau. *U.S. Geological Survey Bulletin* 1030-D: 137.

Buyalov, N. I., Erofeev, N. S., Kalinin, N. A., Kleschev, A. I., Kudryashova, N. M., L'vov, M. S., Simakov, S. N., and Vasil'ev, V. G. 1964. *Quantitative evaluation of predicted reserves of oil and gas.* Authorized translation Consultants Bureau.

DeGeoffroy, J., and Wu, S. M. 1970. A statistical study of ore occurrences in the greenstone belts of the Canadian Shield. *Economic Geology* 65: 496–504.

Elliott, M. A., and Linden, H. R. 1968. A new analysis of U.S. natural gas supplies. *Journal of Petroleum Technology* 20: 135–141.

Griffiths, J. C. 1962. Frequency distributions of some natural resource materials. 23rd Tech. Conf. on Pet. Prod., September 26–28. *Min. Ind. Expt. Sta. Circ.* 63: 174–198.

Griffiths, J. C. 1966. Exploration for natural resources. *Operations Research* 14: 189–209.

Halbouty, M. T., Meyerhoff, A. A., King, R. E., Dott, R. H., Sr., Klemme, H. D., and Shabad, T. 1970. World's giant oil and gas fields: world's giant oil and gas fields, geologic factors affecting their formation, and basin classification, part 1. *Geology of giant petroleum fields.* American Association of Petroleum Geologist Memoir 14: 502–528.

Halbouty, M. T., King, R. E., Klemme, H. D., Dott, R. H., Sr., and Meyerhoff, A. A. 1970. Factors affecting formation of giant oil and gas fields, and basin classification: world's giant oil and gas fields, geologic factors affecting their formation, and basin classification, Part 2. *Geology of giant petroleum fields.* American Association of Petroleum Geologists Memoir 14: 528–555.

Harris, D. P., and Euresty, D. 1969. A preliminary model for the economic appraisal of regional resources and exploration based upon geostatistical analyses and computer simulation. *Colorado School of Mines Quarterly* 64: 71–98.

Hendricks, T. A. 1965. *Resources of oil, gas, and natural gas liquids in the U.S. and the world.* U.S. Geological Survey Circular 522.

Hendricks, T. A., and Schweinfurth, S. P. 1966. Unpublished memorandum quoted in *United States petroleum through 1980* (1968). Washington, D.C.: U.S. Department of the Interior.

Hubbert, M.K. 1969. Energy resources. In *Resources and man,* pp. 157–239. National Academy of Sciences–National Research Council. San Francisco: W. H. Freeman & Co.

Inglis, D. R. 1971. Nuclear energy and the Malthusian dilemma. *Bulletin of the Atomic Scientists* 27(2): 14–18.

Kaufman, G. M. 1963. *Statistical decision and related techniques in oil and gas exploration.* Englewood Cliffs, N.J.: Prentice-Hall.

Lasky, S. G. 1950. Mineral-resource appraisal by U.S. Geological Survey. *Colorado School of Mines Quarterly* 45: 1–27. See also his (1950) How tonnage and grade relations help predict ore reserves. *Engineering Mining Journal* 151(4): 81–85.

Lowell, J. D. 1970. Copper resources in 1970. *Mining Engineering* 22(April): 67–73.

McKelvey, V. E. 1960. Relation of reserves of the elements to their crustal abundance. *American Journal of Science* (Bradley Vol.) 258-A: 234–241.

Moore, C. L. 1966. *Projections of U.S. petroleum supply to 1980.* Washington, D.C.: U.S. Department of the Interior, Office of Oil and Gas.

National Academy of Sciences–National Research Council. 1969. *Resources and man.* San Francisco: W. H. Freeman & Co.

Nolan, T. B. 1950. The search for new mineral districts. *Economic Geology* 45: 601–608.

Oil and Gas Journal. 1969. Vast Delaware-Val Verde reserve seen. 67(16): 44. (Re: Zimmerman and Long.)

Olson, J. C., and Overstreet, W. C. 1964. Geologic distribution and resources of thorium. *U.S. Geological Survey Bulletin* 1204.

Orwell, G. 1937. *The road to Wigan Pier.* American edition, 1958. New York: Harcourt, Brace & World.

Pratt, W. E. 1950. The earth's petroleum resources. In *Our oil resources,* ed. L. M. Fanning, pp. 137–153. 2nd ed. New York: McGraw-Hill.

Rodionov, D. A. 1965. *Distribution functions of the element and mineral contents of igneous rocks: a special research report,* pp. 28–29. Authorized translation Consultants Bureau.

Sekine, Y. 1963. On the concept of concentration

of ore-forming elements and the relationship of their frequency in the earth's crust. *International Geological Review* 5: 505–515.

Slichter, L. B. 1960. The need of a new philosophy of prospecting. *Mining Engineering* 12: 570–576.

Slichter, L. B., et al. 1962. Statistics as a guide to prospecting. In *Math and computer applications in mining and exploration symposium proceedings*, pp. F-1–27. Tucson: Arizona University College of Mines.

Weeks, L. G. 1950. Concerning estimates of potential oil reserves. *American Association of Petroleum Geologists Bulletin* 34: 1947–1953.

Weeks, L. G. 1956. World offshore petroleum resources. *American Association of Petroleum Geologists Bulletin* 49: 1680–1693.

Weeks, L. G. 1958. Fuel reserves of the future. *American Association of Petroleum Geologists Bulletin* 42: 431–438.

Zapp, A. D. 1962. Future petroleum producing capacity of the United States. *U.S. Geological Survey Bulletin* 1142-H.

Energy Crises in Perspective

JOHN C. FISHER   1973

How real and long lasting is the current energy crisis? There is reason to believe that domestic fossil fuels will continue to be available, at slowly declining prices (in constant dollars), in adequate amounts to support total anticipated energy consumption of the United States and the world for many centuries. There is also every reason to believe that domestic nuclear fuels will continue to be available, at slowly declining prices (in constant dollars), in adequate amounts to support total anticipated energy consumption for many millenia. If nuclear fuels or other energy sources take an increased proportion of the load, the fossil fuels will last correspondingly longer. Thus, the current energy crisis, although very real, should be viewed as a transient perturbation of the long-term trajectory of our energy economy.

CURRENT ENERGY CRISIS

The energy industries are managed by men whose decisions are based on anticipated future conditions; unexpected social or political upheavals can confound their plans and a crisis can ensue. The unanticipated closing of the Suez Canal in 1967 precipitated a crisis in Europe because a larger tanker fleet was suddenly and unexpectedly needed to carry oil around Africa to Europe. Within a few years, the tanker fleet was built and the crisis was over. The unanticipated environmental mass movement of the late 1960's and the early 1970's precipitated a crisis in the United States because more petroleum products and more refinery capacity were suddenly and unexpectedly needed as consumers shifted to oil from coal and from planned uranium. Within a few years, the refinery capacity will be built and the petroleum products will be flowing.

A large part of the United States energy crunch of the 1970's has been the result of unanticipated social factors, in particular the environmental movement. The demand for distillate oil turned out to be much higher than anticipated as coal-burning power plants shifted to oil, as electric utilities substituted oil-burning gas turbines for delayed nuclear reactors, and as automobiles began

The author is Manager of Energy Systems Planning for General Electric Company's Power Generation Business Group, New York City.

From Fisher, ENERGY CRISES IN PERSPECTIVE 1973, John Wiley & Sons, Inc. Copyright © 1973 by John Wiley & Sons, Inc. Reprinted by permission of the author and publisher.

41

burning more gasoline because of newly legislated safety and antipollution equipment.

Natural-gas price regulation was another contributing factor. In 1970 the owner of a gas well got only about a third the money from selling gas as did the owner of an oil well from selling oil with the same heating value. The motivation for developing new gas reserves eroded and the supply began to decline, while at the same time the increasing desirability of gas, relative to other fuels, caused demand to soar. With supply and demand decoupled from their customary meeting ground in the market place, they moved so far out of balance that rationing, imports, and synthetic gas have become necessary. But it appears that the underlying cause of much of the imbalance—the artificially low price set for interstate natural gas—is now more widely understood and appreciated, and the price of natural gas will be allowed gradually to seek its competitive level. With the price and market mechanism back at work, we can expect supply to increase and demand to decrease until, after several years of adjustment, the two again come into balance.

The uncertain viability of high-cost United States crude oil and refined products in the face of a potential flood of low-cost imported crude oil and refined products inhibited United States petroleum development and refinery construction, adding to the current crisis. Such uncertainty can only be dispelled, and action initiated, by a clear government policy for protecting United States energy industry investment and infrastructure.

The energy crunch of the 1970's was caused by factors that are so clearly identifiable, so clearly understood, and so readily compensatable through the ordinary response mechanisms of government, industry, and technology that there can be little doubt of its transitory nature. There are real shortages of distillate fuel now that the crunch is on, and there are sharp price rises associated with the shortages, but I expect that the crunch will be over in two or three years and

that prices will drop back to lower levels. With this comment I would like to set aside the energy crunch, and consider some of the longer-range aspects of energy resources and consumption.

PROJECTED ENERGY CONSUMPTION

In projecting energy consumption, let us first consider just the United States, which consumes much of the world's energy and is to some extent representative of the industrialized regions of the world. And before projecting into the future, let us look at the past.

The preindustrialized United States appears never to have passed through a period of low per capita energy consumption once the European settlers arrived. The first settlers found a continent completely covered with virgin forest. This standing timber was available as fuel wood for the cutting, and the small population began consuming energy at a high per capita rate, far beyond the level of energy consumption in traditional societies, as soon as they stepped ashore. In 1850, for example, per capita energy consumption was nearly half as great as it is today. Fortunately for us, coal, oil, and gas were discovered and put to use as the forests were cut down and the population grew, and the transition from fuel wood to fossil fuels was smooth.

Growth in energy consumption still continues. Five key factors affecting the growth rate can be identified and quantified; they are listed with their current annual growth rates:

(1) Total population, 1.3 per cent.

(2) Per capita residential energy, 2.8 per cent.

(3) Energy content per unit goods and services, 0.9 per cent (for example, the energy required to manufacture a modern automobile compared to the energy required to manufacture its lighter, simpler predecessor).

(4) New energy consumed by all sectors of the economy, 8.5 per cent (this includes the energy needed for computers, air condi-

tioning, clothes drying; it reflects new invention and growth of affluence).

(5) Growth of non-farm-working population, 2.9 per cent (this growth rate reflects the fact that more energy is needed to support people in the working force than people who stay at home. This extra energy provides for transportation and the maintenance of people at their places of work).

Estimates of the evolution of these growth rates allow one to project the future consumption of energy in the United States. Implicit in any such projection must be an assumption as to the future availability and price of the basic energy sources. For many decades, with occasional fluctuations one way or the other, the prices of all basic energy sources have been declining gradually relative to the prices of other goods and services. The past growth rates of energy consumption must reflect this fact to some degree. I assume that this trend of the past will continue far into the future: that energy prices will continue their slow downward trend relative to other prices, with occasional fluctuations one way or the other.

Table 1-2 gives the projected energy consumption for the United States as $300C$ for the period 1970–2000 ($C \equiv 10^{16}$ BTU). The projection for the world energy consumption, although not quite as reliable, is $1300C$ for this period. Extending the projection for the next one hundred years, I find that the United States consumption will be $1600C$ and the world consumption will be $5000C$. In the remainder of this article, I will show that current reserves and estimated resources of fuel are not only adequate for this period

but for many hundreds of years beyond the year 2000.

ENERGY SOURCES (TABLE 1-3)

A number of different sources have provided significant energy inputs at one time or another. In approximate order of their historic development, they can be classified as:

Solar energy. Conversion via fuel wood, work-animal feed, direct wind power, direct water power, and hydroelectricity.

Fossil fuels. Combustion of coal, petroleum, and natural gas.

Nuclear fuels. Fission of uranium and thorium.

Solar energy is dilute, but large in magnitude and unlimited in time. Fossil fuels are concentrated and cheap, but are exhaustible and can become exhausted after several centuries. Nuclear fuels are practically inexhaustible, particularly if breeder reactors are able to utilize the common isotopes of uranium and thorium.

In attempting to determine the adequacy or inadequacy of the world's energy resources, the first step is to make as good an estimate as possible of the available flux of solar energy, and of the fossil fuels, nuclear fuels, and geothermal heat reservoirs in the ground. The flux of solar energy is easiest to estimate, as it comes down from the sky and has been fairly well measured. Hydropower, as a special aspect of solar energy, is equally easy to estimate. The subterranean energy resources present more difficult problems, as they are usually hidden from direct view until reached by drilling, and many of the

Table 1-2. Projected Energy Consumption.

Region	30-Year Total[a] (1970–2000)	100-Year Total[a] (1970–2070)	Quality of Projection
United States	300	1600	Highest
Other industrialized	800	2800	Lower
Non-industrialized	200	800	Poor
World total	1300	5000	Intermediate

[a]Units of $C \equiv 10^{16}$ BTU.

Table 1-3. Sources of Energy for the United States 1970.[a]

Energy Source	Conventional Quantity		Energy Content (C ≡ 10^{16} BTU)	
			Units of C	Per Cent
Fossil fuel				
Coal	525	million tons	1.28	20
Petroleum	5.36	billion barrels	2.65	41
Natural gas	21.4	trillion cubic feet	2.13	33
Solar energy				
Hydroelectricity	253	billion kilowatt-hours	0.26	4
Other				
Miscellaneous			0.13	2
			6.45	100

[a]The United States consumes so much energy that the annual number of BTU's is very large. In order to bring such large numbers down to size, I define a C-unit as

$$1\ C = 10^{16}\ \text{BTU}$$

One C of heat is approximately the heat that would be generated by burning 400 million tons of coal; the approximate amount of coal consumed in the United States each year for the past half century. (Hence the use of the letter C for the unit.) It is also approximately the heat required to warm up Lake Michigan one degree Fahrenheit, for there are just about 10^{16} pints of water in the lake.

promising areas of the earth have not yet been drilled.

The quantity of solar energy reaching the earth at the upper boundary of the atmosphere is well known, amounting to 530,000C per year. Of this, on average about half reaches ground level. As rough rules of thumb, the highest solar energy flux at ground level, found in desert lands near the equator, can be taken as about a million BTU per square foot each year. The average flux over the lower 48 United States can be taken as about 0.5 million BTU per square foot each year. If the total energy supply of the United States in 1970, amounting to 6.45C, were to be obtained from solar energy with an over-all conversion efficiency of 10 per cent, an area of about 50,000 square miles would be required. This area seems large, but it is only about 3 per cent of the land now devoted to farms. It is clear that solar-energy resources are far more adequate for supplying the world with energy, al-though of course they may not be the least expensive.

Hydropower resources are limited to rivers and streams wherever they happen to be or wherever they can be diverted to. The total resource base—if all potentially developable falling water were turned to hydroelectricity—amounts to about 1C annually in North America and 9C annually in the world. This is not enough to make up the United States or world totals for energy, but is enough to allow hydroelectricity to grow for some time at its historic rate, amounting to 4 per cent of United States energy input and 6 per cent of world energy input (on a fuel-equivalent basis).

Next consider the various mineral fuels of current significance: coal, petroleum, natural gas, uranium, and thorium; and the mineral fuels of potential future significance: oil shale and tar sand. These minerals are not distributed uniformly throughout the earth, but are concentrated to varying degrees, in deposits of varying size, at different depths, in different regions of the world.

RESOURCES AND RESERVES

We must distinguish between total resources and reserves. A lack of clear thinking in this regard has led to much misunderstanding of the current energy crisis. At any given time, some of the known mineral deposits have been well mapped and have been developed for production, while others remain undeveloped. A deposit is developed when the necessary investment has been made in whatever capital equipment is required to produce the mineral from the deposit and to move it to market. For example, in the coal industry this means acquisition of mining rights, investment in mining equipment, and installation of transportation equipment to move the coal to the nearest existing railroad or waterway. Investments of this nature are made in anticipation and expectation of profitable production. The minerals that have been developed for profitable production are called "reserves." This definition of reserves is followed most closely in the oil, gas, and uranium industries, and perhaps less closely in the coal industry where well-known easily developable deposits may be counted as reserves.

Reserves amount to current inventory of minerals in the ground. As mineral production proceeds, material is withdrawn from inventory and reserves are diminished. At the same time, as investments are made in developing additional deposits, new inventory is created and reserves are increased. Current reserves at any given time reflect the interplay of these opposing tendencies. Reserves can be measured in terms of the reserve-to-annual-production ratio, which equals the number of years that the inventory would last if production continued at its present rate (and if no new inventory were developed). Economic forces keep the reserve-to-production ratio for many minerals at 10 to 20 years.

Should the reserve-to-production ratio for some mineral drop from 12 years supply to 10 years supply over some period of time,

we need not necessarily assume that ultimate mineral depletion is approaching. It may reflect only prudent trimming of inventory at a time of rising interest rates, or perhaps it may reflect growing uncertainty about a potential flood of low-cost imports.

Known undeveloped deposits are often called "submarginal," because they cannot be developed to produce minerals at a profit with today's technology, today's costs, and today's prices. Yet as the *economy of scale* reduces costs, and as *new technology* reduces costs, submarginal deposits tend progressively to be developed and are transformed into reserves. As an example, when oil is withdrawn from an oil field, it flows easily at first, then less easily, then must be pumped, and finally, with whatever state of technology exists at the time, the cost of getting it out exceeds the price that it will bring. The inventory of recoverable oil has been exhausted. The reserves are gone. Yet 70 per cent or so of the original oil in place in the field still remains there as a submarginal resource. As new recovery technology is developed, a time usually comes when it pays to redevelop the same field for additional production by water flood, fire sweep, or some other technology.

New reserves are created whenever an old oil field or coal mine is redeveloped, and in principle, the process can be continued time after time until the oil and coal are completely developed.

In addition to reserves and known submarginal deposits, there are additional deposits not yet discovered. And (particularly for oil and gas) there is substantial worldwide activity devoted to discovering them. Once discovered, some will prove to be easily developable for low-cost production, and these may be classified as unknown deposits of economically recoverable minerals. Others will prove not to be profitably developable, and these may be classified as unknown submarginal deposits. The quantities of undiscovered resources clearly are the most difficult to determine, but geological and geophysical experts have learned to make respectable estimates.

RESOURCE AND RESERVE ESTIMATES

United States Geological Survey specialists have compiled[1] estimates of United States fossil-fuel resources including coal, petroleum liquids, natural gas, and oil from shale. Their estimates are generally made on geological projections of favorable rocks and on anticipated frequency of the energy resource in the favorable rocks. The estimates of submarginal resources of oil from shale include only relatively rich deposits that might be recoverable with today's technology at less than two or three times today's oil prices, and exclude much larger quantities of lower-grade shale. I believe these Geological Survey estimates are the best objective estimates available for the United States.

Total cumulative United States consumption of fossil fuels has amounted to about $200C$ through 1970, and 1970 consumption amounted to about $6.5C$. The data in Table 1-4 show that the United States has abundant fossil fuel reserves and resources. Present fossil-fuel reserves are more than adequate to supply the $300C$ of energy required for the balance of this century. If we accept the projections that annual per capita energy consumption will level off at 450 million BTU per capita and that annual per capita fossil-fuel consumption for nonenergy purposes such as asphalt tiles, road oil and petrochemical feed stocks will level off at 50 million BTU per year; then, in an economy totally energized by fossil fuel, the annual per capita fossil-fuel consumption would level off at 500 million BTU. If we assume in addition that the United States population will stabilize at about a billion people within the next few centuries, it follows that annual fossil-fuel consumption (energy uses plus nonenergy uses) will level off at about $50C$ per year. Under these conditions, in an economy where all energy was derived from fossil fuels, United States fossil-fuel reserves and resources would be enough to last about 500 years:

$$\frac{25,000C \text{ reserves and resources}}{50C \text{ per year}} \approx 500 \text{ years.}$$

Uranium and thorium contain so much energy per pound that it makes good sense to consider very low-grade ores in estimating resources. The most comprehensive estimates have been made by an Interdepartmental Study[2] with participation by nine federal departments and agencies. The uranium reserves shown in Table 1-4 correspond to utilization of 1.5 per cent of the potential energy of the uranium, as is appropriate for today's light-water reactors. The nuclear resources correspond to utilization of 80 per cent of the potential energy of uranium and thorium as may become possible with the new technologies of breeder reactors. Although nuclear fuel reserves available with current technology are not particularly large in relationship to projected energy consumption, nuclear fuel resources are adequate for about a million years.

Thus far our consideration of fossil and nuclear fuels has focused on the United States. The Interdepartmental Study[2] considered world resources as well as United States resources, and using data available in 1962, they estimated that in terms of total resources "the United States is endowed with approximately one-fourth of the coal, one-seventh of the oil, possibly one-tenth of the natural gas, one-twelfth of shale oil, and one-seventeenth of uranium and thorium." Since the United States has about one-seventeenth of the world's land area, it appears that it may have somewhat more than its share of fossil fuels, particularly coal. I expect that as better data become available, the overall energy resources of the world will prove to be more or less uniformly distributed. Most major countries should

1. P. K. Theobald, S. P. Schweinfurth, D. C. Duncan, *Energy Resources of the United States.* Geological Survey Circular 650, U.S. Geological Survey, Washington, D.C. (1972).

2. A. B. Cambel et al., *Energy R & D and National Progress: Findings and Conclusions,* U.S. Government Printing Office, Washington D.C. (1966).

Table 1-4. United States Fossil-Fuel and Nuclear-Fuel Reserves and Resources.

Energy Source	United States Reserves[a]	United States Additional Resources[a]
Fossil fuels		
Coal	900	6,600
Petroleum	30	1,640
Natural gas	30	640
Shale oil	—	15,000
Total	960	24,000
Nuclear fuels		
Uranium	22	22,000,000
Thorium	—	34,000,000
Total	22	56,000,000

[a]Units of $C \equiv 10^{16}$ BTU.

have their proportionate share, with a few exceptions such as Japan having none, the Mideast having more than its share of oil, and the United States having more than its share of coal.

In summary, we find that fossil-fuel reserves alone are adequate for 70 years. Fossil-fuel reserves and resources alone are adequate for 500 years, and nuclear resources for a million years, provided, of course, that these reserves and resources can be utilized within the bounds of environmental acceptability. Solar energy is even more abundant: the energy reaching ground level in the United States amounts to about 10,000C annually and is assured indefinitely. World energy resources are substantially larger, in approximate ratio of the world's land area to the United States land area.

Although it is true that nuclear and solar resources exceed fossil-fuel resources by a wide margin, there is no immediate treat of fossil-fuel depletion to cause us to shift to nuclear fuel or solar energy. The United States supply of coal and shale oil is sufficient to last for many centuries if we choose to use it, and the same is true for the rest of the world. It is the over-all socioeconomics of energy supply and utilization that will determine which fuels are actually used, and on what timetable.

FUTURE OF FUELS AND TECHNOLOGIES

Noneconomic factors are of fundamental significance to the future of the energy industries. Every forecast of fuel availability, energy consumption, conversion technologies, and electrification must be based implicitly or explicitly upon an assessment of political and social forces that provide the over-all environment—the basic ground rules—within which economics and technology operate. For example, unrestricted international competition in petroleum would drop the price of crude oil fivefold or more in the United States, ruining the nation's oil and coal industry infrastructures and ruining the world's nuclear and high-performance steam-turbine infrastructures. In such a world, economics and technology would lead to the simple gas turbine as prime mover for electric power generation. Thus, it is political and social factors, in large part, that determine the over-all environment in which the technological and economic struggles for position among fuels and energy conversion technologies will be fought out in the United States.

Fuel price levels within the United States are projected to continue their historic decline (in constant dollars). For electric utility generation (but not for transportation or space heating, which already use relatively expensive clean fuel), the cost of all forms of fossil fuel is likely to double. Natural gas price at the well head—near where most gas-burning utilities get it—may approximately double when natural gas finds its competitive level. (Owing to relatively large transportation and distribution costs, doubling the well-head price in the Southwest increases the price to a residential consumer in the Northeast by only about 20 per cent.) The cost of utility coal may approximately double, because of increased transportation required to bring low-sulfur coal from long distances or because of the necessity for refining coal to purify it. The cost of utility oil may approximately double as utilities shift from high-sulfur residual oil (the dregs of refining) to the more expensive low-sulfur oil used by others.

The approximate doubling of fossil-fuel prices is a competitive boon to nuclear power, for it increases the cost of electricity from fossil fuel relative to electricity from nuclear fuel. There is a tendency among nuclear-power advocates to hope that one doubling will be followed by others, and to project much higher fossil fuel prices in the future; but I believe this view mistaken. Social factors are requiring the utility industry to join the rest of the country in burning clean fuel purchased at competitive prices. The price of clean fuel itself is likely to continue its historic downward trend.

When measured in constant dollars per kilowatt of capacity, the cost of constructing a nuclear power plant increased by perhaps 50 per cent between 1965 and 1975, during which time the cost of a fossil steam plant increased slightly and the cost of a gas turbine power plant declined. The increase in nuclear power-plant construction costs has dealt a competitive blow to nuclear power, in effect canceling the competitive advantage conferred by doubled fossil-fuel prices.

Advocates of fossil fuel may tend to hope that one 50-per cent increase in nuclear plant construction costs will be followed by

others, and to project much higher nuclear-plant costs in the future; but I believe this view mistaken. The field-construction cost lesson appears to have been learned, and there is an increasing trend toward standardization and assembly-line construction of major nuclear power-plant components. The requirements of the new licensing procedures are being better anticipated now that they are more familiar, and licensing delays may diminish. Over-all, competition between nuclear power and fossil-fuel power is likely to remain vigorous.

We note that both fossil-fuel and nuclear-fuel conversion technologies are under increased pressure to improve efficiency for generating electricity. Fossil-fuel conversion efficiency is more important than before because the refined fuel required today is more costly than the formerly acceptable unrefined fuel. Nuclear-fuel conversion efficiency is more important than before because construction costs are higher, and cost per unit electrical output can be decreased by raising the output from a plant of a given size. Both technologies are moving ahead rapidly, and from the over-all standpoint of electrification, one or the other or both in combination are expected to enable electrification to progress for decades. Interest is intense in the future market shares that will be achieved by nuclear and fossil fuels and technologies, but prudence suggests that a detailed forecast not be made. I am content to forecast continued electrification, and to watch with interest the competitive battle among the various fuels and technologies.

LONG-RANGE PERSPECTIVE

Although it is not possible from this vantage point to determine what combination of fuels and technologies will be leaders in the next century, we are as fortunate with technologies as with fuels in having many viable alternatives. We possess the technologies that will enable the cost of electricity to be reduced, the efficiency of generation to be increased, and electrification to proceed.

Although crises may be expected from

time to time, the fundamentals of the energy industries are sound. Fossil fuels, nuclear fuels and solar-energy resources are abundant. Improved technologies for generating and utilizing electricity can increase the extent of electrification. The greatest imponderables may be the future acceptability and price of the various forms of energy: acceptability as affected by ecological and environmental pressures, and price as affected by domestic and international political pressures.

Although no one can be sure what the future will bring, I would suggest a potential future for the United States in which environmental pressures require that all fuels be refined, and that all lands disturbed by mining be reclaimed; and in which political pressures require that oil imports be limited so that domestic production does not shrink, with the consequence that existing marginal producers continue to determine domestic fuel prices. Under these conditions the United States energy economy, with its moderate fuel prices and high technology content, might be expected to continue its historic evolution. Through further progress of technology and further economies of scale, particularly with the advent of synthetic oil from coal or shale, prices of refined fuels (and of electricity after adjusting for the shift to refined fuel) might be expected to continue their downward trends, when measured in constant dollars to strip away the effects of inflation.

SUPPLEMENTAL READINGS

Fisher, John C., 1974, *Energy Crisis in Perspective.* John Wiley & Sons, Inc., New York, 196 p.

Lovins, Amory B., 1974, "Word Energy Strategies," *Science in Public Affairs (Bull. of the Atomic Scientist),* May 1974, Part 1, pp. 14–32; June, 1974, Part 2, pp. 38–50.

Energy Use in the United States Food System

JOHN S. STEINHART and CAROL E. STEINHART   1974

In a modern industrial society, only a tiny fraction of the population is in frequent contact with the soil, and an even smaller fraction of the population raises food on the soil. The proportion of the population engaged in farming halved between 1920 and 1950 and then halved again by 1962. Now it has almost halved again, and more than half of these remaining farmers hold other jobs off the farm (1). At the same time the number of work animals has declined from a peak of more than 22×10^6 in 1920 to a very small number at present (2). By comparison with earlier times, fewer farmers are producing more agricultural products and the value of food in terms of the total goods and services of society now amounts to a smaller fraction of the economy than it once did.

Energy inputs to farming have increased enormously during the past fifty years (3),

and the apparent decrease in farm labor is offset in part by the growth of support industries for the farmer. With these changes on the farm have come a variety of other changes in the United States food system, many of which are now deeply embedded in the fabric of daily life. In the past fifty years, canned, frozen, and other processed foods have become the principal items of our diet. At present, the food-processing industry is the fourth largest energy consumer of the Standard Industrial Classification groupings (4). The extent of transportation engaged in the food system has grown apace, and the proliferation of appliances in both numbers and complexity still continues in homes, institutions, and stores. Hardly any food is eaten as it comes from the fields. Even farmers purchase most of their food from markets in town.

Present energy supply problems make this

Dr. J. S. Steinhart is a professor of geology and geophysics, and professor in the Institute for Environmental Studies, University of Wisconsin-Madison. He also is the associate director of the University of Wisconsin-Madison Sea Grant Program.

Dr. C. E. Steinhart, formerly a biologist with the National Institutes of Health, is now a science writer and editor.

This article is modified from *Energy: Source, Use, and Role in Human Affairs*, Duxbury Press, North Scituate, Massachusetts, 1974, and appeared in *Science*, Vol. 184, No. 4134, pp. 307–316, April 19, 1974. Reprinted with light editing and by permission of the authors, Duxbury Press, and the American Association for the Advancement of Science. Copyright 1974 by the American Association for the Advancement of Science.

growth of energy use in the food system worth investigating. It is our purpose in this article to do so. But there are larger matters at stake. Georgescu-Roegen notes that "the evidence now before us—of a world which can produce automobiles, television sets, etc., at a greater speed than the increase in population, but is simultaneously menaced by mass starvation—is disturbing" (5). In the search for a solution to the world's food problems, the common attempt to transplant a small piece of a highly industrialized food system to the hungry nations of the world is plausible enough, but so far the outcome is unclear. Perhaps an examination of the energy flow in the United States food system as it has developed can provide some insights that are not available from the usual economic measures.

MEASURES OF FOOD SYSTEMS

Agricultural systems are most often described in economic terms. A wealth of statistics is collected in the United States and in most other technically advanced countries indicating production amounts, shipments, income, labor, expenses, and dollar flow in the agricultural sector of the economy. But, when we wish to know something about the food we actually eat, the statistics of farms are only a tiny fraction of the story.

Energy flow is another measure available to gauge societies and nations. It would have made no sense to measure societies in terms of energy flow in the eighteenth century when economics began. As recently as 1940, four-fifths of the world's population were still on farms and in small villages, most of them engaged in subsistence farming.

Only after some nations shifted large portions of the population to manufacturing, specialized tasks, and mechanized food production, and shifted the prime sources of energy to move society to fuels that were transportable and usable for a wide variety of alternative activities, could energy flow be used as a measure of societies' activities. Today it is only in one-fifth of the world

where these conditions are far advanced. Yet we can now make comparisons of energy flows even with primitive societies. For even if the primitives, or the euphemistically named "underdeveloped" countries, cannot shift freely among their energy expenditures, we *can* measure them and they constitute a different and potentially useful comparison with the now traditional economic measures.

What we would like to know is this: How does our present food supply system compare, in energy measures, with those of other societies and with our own past? Perhaps then we can estimate the value of energy flow measures as an adjunct to, but different from, economic measures.

ENERGY IN THE UNITED STATES FOOD SYSTEM

A typical breakfast includes orange juice from Florida by way of the Minute Maid factory, bacon from a midwestern meat packer, cereal from Nebraska and General Mills, eggs and milk from not *too* far away, and coffee from Colombia. All of these things are available at the local supermarket (several miles each way in a 300-horsepower automobile), stored in a refrigerator-freezer, and cooked on an instant-on stove.

The present food system in the United States is complex, and the attempt to analyze it in terms of energy use will introduce complexities and questions far more perplexing than the same analysis carried out on simpler societies. Such an analysis is worthwhile, however, if only to find out where we stand. We have a food system, and most people get enough to eat from it. If, in addition, one considers the food supply problems present and future in societies where a smaller fraction of the people get enough to eat, then our experience with an industrialized food system is even more important. There is simply no gainsaying that many nations of the world are presently attempting to acquire industrialized food systems of their own.

Food in the United States is expensive by world standards. In 1970 the average annual per capita expenditure for food was about $600 (3). This amount is larger than the per capita gross domestic product of more than thirty nations of the world which contain the majority of the world's people and a vast majority of those who are underfed. Even if we consider the diet of a poor resident of India, the annual cost of his food at United States prices would be about $200—more than twice his annual income (3). It is crucial to know whether a piece of our industrialized food system can be exported to help poor nations, or whether they must become as industrialized as the United States to operate an industrialized food system.

Our analysis of energy use in the food system begins with an omission. We will neglect that crucial input of energy provided by the sun to the plants upon which the entire food supply depends. Photosynthesis has an efficiency of about 1 per cent; thus, the maximum solar radiation captured by plants is about 5×10^3 kilocalories per square meter per year (3).

Seven categories of energy use on the farm are considered here. The amounts of energy used are shown in Table 1-5. The values given for farm machinery and tractors are for the manufacture of new units only and do not include parts and maintenance for units that already exist. The amounts shown for direct fuel use and electricity consumption are a bit too high because they include some residential uses of the farmer and his family. On the other hand, some uses in these categories are not reported in the summaries used to obtain the values for direct fuel and electricity usage. These and similar problems are discussed in the references. Note the relatively high energy cost associated with irrigation. In the United States less than 5 per cent of the crop land is irrigated (1). In some countries where the "green revolution" is being attempted, the new high-yield varieties of plants require irrigation where native crops did not. If that were the case in the United States, irrigation would be the largest single use of energy on the farm.

Little food makes its way directly from field and farm to the table. The vast complex of processing, packaging, and transport has been grouped together in a second major subdivision of the food system. The seven categories of the processing industry are listed in Table 1-5. Energy use for the transport of food should be charged to the farm in part, but we have not done so here because the calculation of the energy values is easiest (and we believe most accurate) if they are taken for the whole system.

After the processing of food there is further energy expenditure. Transportation enters the picture again, and some fraction of the energy used for transportation should be assigned here. But there are also the distributors, wholesalers, and retailers, whose freezers, refrigerators, and very establishments are an integral part of the food system. There are also the restaurants, schools, universities, prisons, and a host of other institutions engaged in the procurement, preparation, storage, and supply of food. We have chosen to examine only three categories: the energy required for refrigeration and cooking, and for the manufacture of the heating and refrigeration equipment (Table 1-5). We have made no attempt to include the energy used in trips to the store or restaurant. Garbage disposal has also been omitted, although it is a persistent and growing feature of our food system; 12 per cent of the nation's trucks are engaged in the activity of waste disposal (1), of which a substantial part is related to food. If there is any lingering doubt that these activities— both the ones included and the ones left out—are an essential feature of our present food system, one need only ask what would happen if everyone should attempt to get on without a refrigerator or freezer or stove? Certainly the food system would change.

Table 1-5 and the related references summarize the numerical values for energy use in the United States food system, from 1940 to 1970. As for many activities in the past few decades, the story is one of continuing increase. The totals are displayed in Figure 1-9 along with the energy value of the food consumed by the public. The food values

Table 1-5. Energy Use in the United States Food System.[a]

Component	1940	1947	1950	1954	1958	1960	1964	1968	1970	References
On farm										
Fuel (direct use)	70.0	136.0	158.0	172.8	179.0	188.0	213.9	226.0	232.0	(13–15)
Electricity	0.7	32.0	32.9	40.0	44.0	46.1	50.0	57.3	63.8	(14, 16)
Fertilizer	12.4	19.5	24.0	30.6	32.2	41.0	60.0	87.0	94.0	(14, 17)
Agricultural steel	1.6	2.0	2.7	2.5	2.0	1.7	2.5	2.4	2.0	(14, 18)
Farm machinery	9.0	34.7	30.0	29.5	50.2	52.0	60.0	75.0	80.0	(14, 19)
Tractors	12.8	25.0	30.8	23.6	16.4	11.8	20.0	20.5	19.3	(20)
Irrigation	18.0	22.8	25.0	29.6	32.5	33.3	34.1	34.8	35.0	(21)
Subtotal	124.5	272.0	303.4	328.6	356.3	373.9	440.5	503.0	526.1	
Processing industry										
Food processing industry	147.0	177.5	192.0	211.5	212.6	224.0	249.0	295.0	308.0	(13, 14, 22)
Food processing machinery	0.7	5.7	5.0	4.9	4.9	5.0	6.0	6.0	6.0	(23)
Paper packaging	8.5	14.8	17.0	20.0	26.0	28.0	31.0	35.7	38.0	(24)
Glass containers	14.0	25.7	26.0	27.0	30.2	31.0	34.0	41.9	47.0	(25)
Steel cans and aluminum	38.0	55.8	62.0	73.7	85.4	86.0	91.0	112.2	122.0	(26)
Transport (fuel)	49.6	86.1	102.0	122.3	140.2	153.3	184.0	226.6	246.9	(27)
Trucks and trailors (manufacture)	28.0	42.0	49.5	47.0	43.0	44.2	61.0	70.2	74.0	(28)
Subtotal	285.8	407.6	453.5	506.4	542.3	571.5	656.0	787.6	841.9	
Commercial and home										
Commercial refrigeration and cooking	121.0	141.0	150.0	161.0	176.0	186.2	209.0	241.0	263.0	(13, 29)
Refrigeration machinery (home and commercial)	10.0	24.0	25.0	27.5	29.4	32.0	40.0	56.0	61.0	(14, 30)
Home refrigeration and cooking	144.2	184.0	202.3	228.0	257.0	276.6	345.0	433.9	480.0	(13, 29)
Subtotal	275.2	349.0	377.3	416.5	462.4	494.8	594.0	730.9	804.0	
Grand total	685.5	1028.6	1134.2	1251.5	1361.0	1440.2	1690.5	2021.5	2172.0	

[a]All values are multiplied by 10^{12} kilocalories.

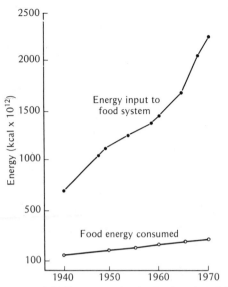

Figure 1-9. Energy use in the food system, 1940 through 1970, compared to the caloric content of food consumed.

were obtained by multiplying the daily caloric intake by the population. The differences in caloric intake per capita over this thirty-year period are small (1), and the curve is primarily an indication of the increase in population in this period.

OMISSIONS AND DUPLICATIONS FOR FOOD SYSTEM ENERGY VALUES

Several omissions, duplications, and overlaps have been mentioned. We will now examine the values in Table 1-5 for completeness and try to obtain a crude estimate of their numerical accuracy.

The direct fuel and electricity usage on the farm may be overstated by some amounts used in the farmer's household, which, by our approach, would not all be chargeable to the food system. But about 10 per cent of the total acreage farmed is held by corporate farms for which the electrical and direct fuel use is not included in our data. Other estimates of these two categories are much higher [see Table 1-5 (15,16)].

No allowance has been made for food

exported, which has the effect of overstating the energy used in our own food system. For the years prior to 1960 the United States was at times a net importer of food, at times an exporter, and at times there was a near balance in this activity. But during this period the net flow of trade was never more than a few per cent of the total farm output. Since 1960 net exports have increased to about 20 per cent of the gross farm product (1,3). The items comprising the vast majority of the exports have been rough grains, flour, and other plant products with very little processing. Imports include more processed food than exports and represent energy expenditure outside the United States. Thus, the overestimate of energy input to the food system might be 5 per cent with an upper limit of 15 per cent.

The items omitted are more numerous. Fuel losses from the well head or mine shaft to end use total 10 to 12 per cent (6). This would represent a flat addition of 10 per cent or more to the totals, but we have not included this item because it is not customarily charged to end uses.

We have computed transport energy for trucks only. Considerable food is transported by train and ship, but these items were omitted because the energy use is small relative to the consumption of truck fuel. Small amounts of food are shipped by air, and, although air shipment is energy intensive, the amount of energy consumed appears small. We have traced support materials until they could no longer be assigned to the food system. Some transportation energy consumption is not charged in the transport of these support materials. These omissions are numerous and hard to estimate, but they would not be likely to increase the totals by more than 1 or 2 per cent.

A more serious understatement of energy usage occurs with respect to vehicle usage (other than freight transport) on farm business, food-related business in industry and commercial establishments, and in the supporting industries. A special attempt to estimate this category of energy usage for 1968 suggests that it amounts to about 5 per cent of the energy totals for the food system. This estimate would be subject to an uncertainty of nearly 100 per cent. We must be satisfied to suggest that 1 to 10 per cent should be added to the totals on this account.

Waste disposal is related to the food system, at least in part. We have chosen not to charge this energy to the food system, but, if one-half of the waste disposal activity is taken as food related, about 2 per cent must be added to the food system energy totals.

We have not included energy for parts and maintenance of machinery, vehicles, buildings, and the like, or lumber for farm, industry, or packaging uses. These miscellaneous activities would not constitute a large addition in any case. We have also excluded construction. Building and replacement of farm structures, food industry structures, and commercial establishments are all directly part of the food system. Construction of roads is in some measure related to the food system, since nearly half of all trucks transport food and agricultural items [see Table 1-5 (27)]. Even home construction could be charged in part to the food system since space, appliances, and plumbing are, in part, a consequence of the food system. If 10 per cent of housing, 10 per cent of institutional construction (for institutions with food service), and 10 per cent of highway construction is included, about 10 per cent of the total construction was food related in 1970. Assuming that the total energy consumption divides in the same way that the gross national product does (which overstates energy use in construction), the addition to the total in Table 1-5 would be about 10 per cent or 200×10^{12} kcal. This is a crude and highly simplified calculation, but it does provide an estimate of the amounts of energy involved.

The energy used to generate the highly specialized seed and animal stock has been excluded because there is no easy way to estimate it. Pimentel et al. (3) estimate that 1800 kcal are required to produce 1 pound (450 grams) of hybrid corn seed. But in addition to this amount, some energy use should be included for all the schools of agriculture, agricultural experiment stations, the far-flung network of county agricultural agents [one local agent said he traveled over 50,000 automobile miles (80,000 kilometers) per year in his car], the U.S. Department of Agriculture, and the wide-ranging agricultural research program that enables man to stay ahead of the new pest and disease threats to our highly specialized food crops. These are extensive activities, but we cannot see how they could add more than a few percent to the totals in Table 1-5.

Finally, we have made no attempt to include the amount of private automobile usage involved in the delivery system from retailer to home, or other food-related uses of private autos. Rice (7) reports 4.25×10^{15} kilocalories for the energy cost of autos in 1970, and shopping constitutes 15.2 per cent of all automobile usage (8). If only half of the shopping is food-related, 320×10^{12} kilocalories of energy use is at stake here. Between 8 and 15 per cent should be added to the totals of Table 1-5, depending on just how one wishes to apportion this item.

It is hard to take an approach that might calculate smaller totals but, depending upon point of view, the totals could be much larger. If we accumulate the larger estimates from the above paragraphs as well as the reductions, the total could be enlarged by 30 to 35 per cent, especially for recent years. As it is, the values for energy use in the food system from Table 1-5 account for 12.8 percent of the total United States energy use in 1970.

PERFORMANCE OF AN INDUSTRIALIZED FOOD SYSTEM

The difficulty with history as a guide for the future or even the present lies not so much in the fact that conditions change—we are continually reminded of that fact—but that history is only one experiment of the many that might have occurred. The United States food system developed as it did for a variety of reasons, many of them not understood.

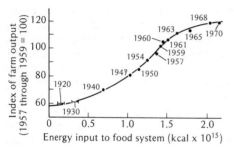

Figure 1-10. Farm output as a function of energy input to the United States food system, 1920 through 1970.

We would do well to examine some of the dimensions of this development before attempting to theorize about how it might have been different, or how parts of this food system can be transplanted elsewhere.

ENERGY AND FOOD PRODUCTION

Figure 1-10 displays features of our food system not easily seen from economic data. The curve shown has no theoretical basis but is suggested by the data as a smoothed recounting of our own history of increasing food production. It is, however, similar to most growth curves and suggests that, to the extent that the increasing energy subsidies to

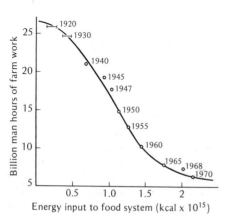

Figure 1-11. Labor use on farms as a function of energy use in the food system.

the food system have increased food production, we are near the end of an era. Like the logistic growth curve, there is an exponential phase which lasted from 1920 or earlier until 1950 or 1955. Since then, the increments in production have been smaller despite the continuing growth in energy use. It is likely that further increases in food production from increasing energy inputs will be harder and harder to come by. Of course, a major change in the food system could change things, but the argument advanced by the technological optimist is that we can always get more if we have enough energy, and that no other major changes are required. Our own history—the only one we have to examine—does not support that view.

ENERGY AND LABOR IN THE FOOD SYSTEM

One farmer now feeds 50 people, and the common expectation is that the labor input to farming will continue to decrease in the future. Behind this expectation is the assumption that the continued application of technology—and energy—to farming will substitute for labor. Figure 1-11 shows this historic decline in labor as a function of the energy supplied to the food system, again the familiar S-shaped curve. What it implies is that increasing the energy input to the food system is unlikely to bring further reduction in farm labor unless some other, major change is made.

The food system that has grown in this period has provided much employment that did not exist 20, 30, or 40 years ago. Perhaps even the idea of a reduction of labor input is a myth when the food system is viewed as a whole, instead of from the point of view of the farm worker only. When discussing inputs to the farm, Pimentel et al. (3) cite an estimate of two farm support workers for each person actually on the farm. To this must be added employment in food-processing industries, in food wholesaling and retailing, as well as in a variety of manufacturing enterprises that support the food system. Yesterday's farmer is today's canner, tractor mechanic, and fast food car-hop. The process of change has been painful to many ordinary people. The rural poor, who could not quite compete in the growing industrialization of farming, migrated to the cities. Eventually they found other employment, but one must ask if the change was worthwhile. The answer to that question cannot be provided by energy analysis anymore than by economic data, because it raises fundamental questions about how individuals would prefer to spend their lives. But if there is a stark choice between long hours as a farmer or shorter hours on the assembly line of a meat-packing plant, it seems clear that the choice would not be universally in favor of the meat-packing plant. Thomas Jefferson dreamed of a nation of independent small farmers. It was a good dream, but society did not develop in that way. Nor can we turn back the clock to recover his dream. But, in planning and preparing for our future, we had better look honestly at our collective history, and then each of us should closely examine his dreams.

THE ENERGY SUBSIDY TO THE FOOD SYSTEM

The data in Figure 1-9 can be combined to show the energy subsidy provided to the food system for the recent past. We take as a measure of the food supplied the caloric content of the food actually consumed. This is not the only measure of the food supplied, as the condition of many protein-poor peo-

ples of the world clearly shows. Nevertheless, the comparison between caloric input and output is a convenient way to compare our present situation with the past, and to compare our food system with others. Figure 1-12 shows the history of the United States food system in terms of the number of calories of energy supplied to produce 1 calorie of food for actual consumption. It is interesting and possibly threatening to note that there is no real suggestion that this curve is leveling off. We appear to be increasing the energy input even more. Fragmentary data for 1972 suggest that the increase continued unabated. A graph like Figure 1-12 could approach zero. A natural ecosystem has no fuel input at all, and those primitive people who live by hunting and gathering have only the energy of their own work to count as input.

SOME ECONOMIC FEATURES OF THE UNITED STATES FOOD SYSTEM

The markets for farm commodities in the United States come closer than most to the economist's ideal of a "free market." There are many small sellers and many buyers, and thus no individual is able to affect the price by his own actions in the market place. But government intervention can drastically alter any free market, and government intervention in the prices of agricultural products (and hence of food) has been a prominent feature of the United States food system for at least thirty years. Between 1940 and 1970, total farm income has ranged from $4.5 to $16.5 billion, and the national income originating in agriculture (which includes indirect income from agriculture) has ranged from $14.5 to $22.5 billion (1). Meanwhile, government subsidy programs, primarily farm price supports and soil bank payments, have grown from $1.5 billion in 1940 to $6.2 billion in 1970. In 1972 these subsidy programs had grown to $7.3 billion, despite foreign demand of agricultural products. Viewed in a slightly different way, direct government subsidies have accounted for 30 to 40 per cent of the farm income

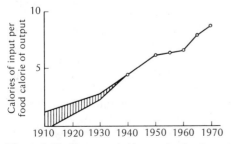

Figure 1-12. Energy subsidy to the food system needed to obtain 1 food calorie.

and 15 to 30 per cent of the National Income attributable to agriculture for the years since 1955. This point emphasizes once again the striking gap between the economic description of society and the economic models used to account for that society's behavior.

This excursion into farm price supports and economics is related to energy questions in this way: first, as far as we know, government intervention in the food system is a feature of all highly industrialized countries (and, despite the intervention, farm incomes still tend to lag behind national averages) and, second, reduction of the energy subsidy to agriculture (even if we could manage it) might decrease the farmer's income. One reason for this state of affairs is that the demand for food quantity has definite limits, and the only way to increase farm income is then to increase the unit price of agricultural products. Consumer boycotts and protests in the early 1970's suggest that there is considerable resistance to this outcome.

Government intervention in the functioning of the market in agricultural products has accompanied the rise in the use of energy in agriculture and the food supply system, and we have nothing but theoretical suppositions to suggest that any of the present system can be deleted.

SOME ENERGY IMPLICATIONS FOR THE WORLD FOOD SUPPLY

The food supply system of the United States is complex and interwoven into a highly industrialized economy. We have tried to analyze this system on account of its implications for future energy use. But the world is short of food. A few years ago it was widely predicted that the world would suffer widespread famine in the 1970's. The adoption of new high-yield varieties of rice, wheat, and other grains has caused some experts to predict that the threat of these expected famines can now be averted, perhaps indefinitely. Yet, despite increases in grain production in some areas, the world

still seems to be headed toward famine. The adoption of these new varieties of grain—dubbed hopefully the "green revolution"—is an attempt to export a part of the energy-intensive food system of the highly industrialized countries to nonindustrialized countries. It is an experiment, because, although the whole food system is not being transplanted to new areas, a small part of it is. The green revolution requires a great deal of energy. Many of the new varieties of grain require irrigation where traditional crops did not, and almost all the new crops require extensive fertilization.

Meanwhile, the agricultural surpluses of the 1950's have largely disappeared. Grain shortages in China and Russia have attracted attention because they have brought foreign trade across ideological barriers. There are other countries that would probably import considerable grain, if they could afford it. But only four countries may be expected to have any substantial excess agricultural production in the next decade. These are Canada, New Zealand, Australia, and the United States. None of these is in a position to give grain away, because each of them needs the foreign trade to avert ruinous balance of payments deficits. Can we then export energy-intensive agricultural methods instead?

ENERGY-INTENSIVE AGRICULTURE ABROAD

It is quite clear that the United States food system cannot be exported intact at present. For example, India has a population of 550×10^6 persons. To feed the people of India at the United States level of about 3,000 food calories per day (instead of their present 2,000) would require more energy than India now uses for all purposes. To feed the entire world with a United States type food system, almost 80 per cent of the world's annual energy expenditure would be required just for the food system.

The recourse most often suggested to remedy this difficulty is to export methods of increasing crop yield and hope for the best. We must repeat as plainly as possible that this is an experiment. We know that our food system works (albeit with some difficulties and warnings for the future). But we cannot know what will happen if we take a piece of that system and transplant it to a poor country, without our industrial base of supply, transport system, processing industry, appliances for home storage, and preparation, and, most important, a level of industrialization that permits higher costs for food.

Fertilizers, herbicides, pesticides, and in many cases machinery and irrigation are needed for success with the green revolution. Where is this energy to come from? Many of the nations with the most serious food problems are those nations with scant supplies of fossil fuels. In the industrialized nations, solutions to the energy supply problems are being sought in nuclear energy. This technology-intensive solution, even if successful in advanced countries, poses additional problems for underdeveloped nations. To create the bases of industry and technologically sophisticated people within their own countries will be beyond the capability of many of them. Here again, these countries face the prospect of depending upon the goodwill and policies of industrialized nations. Since the alternative could be famine, their choices are not pleasant and their irritation at their benefactors—ourselves among them—could grow to threatening proportions. It would be comfortable to rely on our own good intentions, but our good intentions have often been unresponsive to the needs of others. The matter cannot be glossed over lightly. World peace may depend upon the outcome.

CHOICES FOR THE FUTURE

The total amount of energy used on United States farms for the production of corn is now near 10^3 kilocalories per square meter per year (3), and this is more or less typical of intensive agriculture in the United States. With this application of energy we have achieved yields of 2×10^3 kilocalories per square meter per year of usable grain—bring-ing us to almost half of the photosynthetic limit of production. Further applications of energy are likely to yield little or no increase in this level of productivity. In any case, no amount of research is likely to improve the efficiency of the photosynthetic process itself. There is a further limitation on the improvement of yield. Faith in technology and research has at times blinded us to the basic limitations of the plant and animal material with which we work. We have been able to emphasize desirable features already present in the gene pool and to suppress others that we find undesirable. At times the cost of the increased yield has been the loss of desirable characteristics—hardiness, resistance to disease and adverse weather, and the like. The farther we get from characteristics of the original plant and animal strains, the more care and energy is required. Choices need to be made in the directions of plant breeding. And the limits of the plants and animals we use must be kept in mind. We have not been able to alter the photosynthetic process or to change the gestation period of animals. In order to amplify or change an existing characteristic, we will probably have to sacrifice something in the over-all performance of the plant or animal. If the change requires more energy, we could end with a solution that is too expensive for the people who need it most. These problems are intensified by the degree to which energy becomes more expensive in the world market.

WHERE TO LOOK FOR FOOD NEXT?

Our examination in the foregoing pages of the United States food system, the limitations on the manipulation of ecosystems and their components, and the risks of the green revolution as a solution to the world food supply problem suggests a bleak prospect for the future. This complex of problems should not be underestimated, but there are possible ways of avoiding disaster and of mitigating the severest difficulties. These suggestions are not very dramatic and may be difficult to accept.

Figure 1-13 shows the ratio of the energy

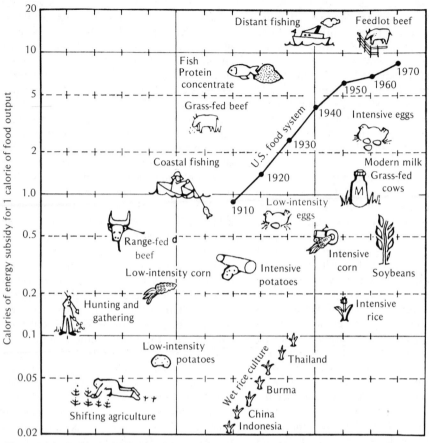

Figure 1-13. Energy subsidies for various food crops. The energy history of the United States food system is shown for comparison. (Source of data: ref. 31.)

subsidy to the energy output for a number of widely used foods in a variety of times and cultures. For comparison, the over-all pattern for the United States food system is shown, but the comparison is only approximate because, for most of the specific crops, the energy input ends at the farm. As has been pointed out, it is a long way from the farm to the table in industrialized societies. Several things are immediately apparent and coincide with expectations. High-protein foods such as milk, eggs, and especially meat have a far poorer energy return than plant foods. Because protein is essential for human diets and the amino acid balance necessary

for good nutrition is not found in most of the cereal grains, we cannot take the step of abandoning meat sources altogether. Figure 1-13 does show how unlikely it is that increased fishing or fish protein concentrate will solve the world's food problems. Even if we leave aside the question of whether the fish are available—a point on which expert opinions differ somewhat—it would be hard to imagine, with rising energy prices, that fish protein concentrate will be anything more than a by-product of the fishing industry, because it requires more than twice the energy of production of grass-fed beef or eggs (9). Distant fishing is still less likely to solve food problems. On the other hand, coastal fishing is relatively low in energy cost. Unfortunately, without the benefit of scholarly analysis, fisherman and housewives have long known this, and coastal fisheries are threatened with overfishing as well as pollution.

The position of soybeans in Figure 1-13 may be crucial. Soybeans possess the best amino acid balance and protein content of any widely grown crop. This has long been known to the Japanese, who have made soybeans a staple of their diet. Are there other plants, possibly better suited for local climates, that have adequate proportions of amino acids in their proteins? There are about 80,000 edible species of plants, of which only about 50 are actively cultivated on a large scale (and 90 per cent of the world's crops come from only 12 species). We may yet be able to find species that can contribute to the world's food supply.

The message of Figure 1-13 is simple. In "primitive" cultures, 5 to 50 food calories were obtained for each calorie of energy invested. Some highly civilized cultures have done as well and occasionally better. In sharp contrast, industrialized food systems require 5 to 10 calories of fuel to obtain 1 food calorie. We must pay attention to this difference—especially if energy costs increase. If some of the energy subsidy for food production could be supplied by onsite, renewable sources—primarily sun and wind—we might be able to continue an energy-intensive food system. Otherwise, the choices appear to be either less energy-intensive food production or famine for many areas of the world.

ENERGY REDUCTION IN AGRICULTURE

It is possible to reduce the energy required for agriculture and the food system. A series of thoughtful proposals by Pimentel and his associates (3) deserves wide attention. Many of these proposals would help ameliorate environmental problems, and any reductions in energy use would provide a direct reduc-

tion in the pollutants due to fuel consumption as well as more time to solve our energy supply problems.

First, we should make more use of natural manures. The United States has a pollution problem from runoff from animal feed lots, even with the application of large amounts of manufactured fertilizer to fields. More than 10^6 kilocalories per acre (4×10^5 kilocalories per hectare) could be saved by substituting manure for manufactured fertilizer (3) (and, as a side benefit, the soil's condition would be improved). Extensive expansion in the use of natural manure will require decentralization of feed-lot operations so that manure is generated closer to the point of application. Decentralization might increase feed-lot costs, but, as energy prices rise, feed-lot operations will rapidly become more expensive in any case. Although the use of manures can help reduce energy use, there is far too little to replace all commercial fertilizers at present (10). Crop rotation is less widely practiced than it was even twenty years ago. Increased use of crop rotation or interplanting winter cover crops of legumes (which fix nitrogen as a green manure) would save 1.5×10^6 kilocalories per acre by comparison with the use of commercial fertilizer.

Second, weed and pest control could be accomplished at a much smaller cost in energy. A 10 per cent saving in energy in weed control could be obtained by the use of the rotary hoe twice in cultivation instead of herbicide application (again with pollution abatement as a side benefit). Biologic pest control—that is, the use of sterile males, introduced predators, and the like—requires only a tiny fraction of the energy of pesticide manufacture and application. A change to a policy of "treat when and where necessary" pesticide application would bring a 35 to 50 per cent reduction in pesticide use. Hand application of pesticides requires more labor than machine or aircraft application, but the energy for application is reduced from 18,000 to 300 kilocalories per acre (3). Changed cosmetic standards, which in no way affect the taste or the edibility of foodstuffs, could also bring about a substantial reduction in pesticide use.

Third, plant breeders might pay more attention to hardiness, disease and pest resistance, reduced moisture content (to end the wasteful use of natural gas in drying crops), reduced water requirements, and increased protein content, even if it should mean some reduction in over-all yield. In the long run, plants not now widely cultivated might receive some serious attention and breeding efforts. It seems unlikely that the crops that have been most useful in temperate climates will be the most suitable ones for the tropics where a large portion of the undernourished peoples of the world now live.

A dramatic suggestion, to abandon chemical farming altogether, has been made by Chapman (11). His analysis shows that, were chemical farming to be ended, there would be much reduced yields per acre, so that most land in the soil bank would need to be put back into farming. Nevertheless, output would fall only 5 per cent, and prices for farm products would increase 16 per cent. Most dramatically, farm income would rise 25 per cent, and nearly all subsidy programs would end. A similar set of propositions treated with linear programming techniques at Iowa State University resulted in an essentially similar set of conclusions (12).

The direct use of solar energy farms, a return to wind power (modern windmills are now in use in Australia), and the production of methane from manure are all possibilities. These methods require some engineering to become economically attractive, but it should be emphasized that these technologies are now better understood than the technology of breeder reactors. If energy prices rise, these methods of energy generation would be attractive alternatives, even at their present costs of implementation.

ENERGY REDUCTION IN THE UNITED STATES FOOD SYSTEM

Beyond the farm, but still far from the table, more energy savings could be introduced. The most effective way to reduce the large energy requirements of food processing would be a change in eating habits toward less highly processed foods. The current aversion of young people to spongy, additive-laden white bread, hydrogenated peanut butter, and some other processed foods could presage such a change if it is more than just a fad. Technological changes could reduce energy consumption, but the adoption of lower energy methods would be hastened most by an increase in energy prices, which would make it more profitable to reduce fuel use.

Packaging has long since passed the stage of simply holding a convenient amount of food together and providing it with some minimal protection. Legislative controls may be needed to reduce the manufacturer's competition in the amount and expense of packaging. In any case, recycling of metal containers and wider use of returnable bottles could reduce this large item of energy use.

The trend toward the use of trucks in food transport, to the virtual exclusion of trains, should be reversed. By reducing the direct and indirect subsidies to trucks we might go a long way toward enabling trains to compete.

Finally, we may have to ask whether the ever-larger frostless refrigerators are needed, and whether the host of kitchen appliances really means less work or only the same amount of work to a different standard.

Store delivery routes, even by truck, would require only a fraction of the energy used by autos for food shopping. Rapid transit, giving some attention to the problems with shoppers with parcels, would be even more energy-efficient. If we insist on a high-energy food system, we should consider starting with coal, oil, garbage—or any other source of hydrocarbons—and producing in factories bacteria, fungi, and yeasts. These products could then be flavored and colored appropriately for cultural tastes. Such a system would be more efficient in the use of energy, would solve waste problems, and would permit much or all of the agricultural land to be returned to its natural state.

ENERGY, PRICES, AND HUNGER

If energy prices rise, as they have already begun to do, the rise in the price of food in societies with industrialized agriculture can be expected to be even larger than the energy price increases. Slesser, in examining the case for England, suggests that a quadrupling of energy prices in the next forty years would bring about a sixfold increase in food prices (9). Even small increases in energy costs may make it profitable to increase labor input to food production. Such a reversal of a fifty-year trend toward energy-intensive agriculture would present environmental benefits as a bonus.

We have tried to show how analysis of the energy flow in the food system illustrates features of the food system that are not easily deduced from the usual economic analysis. Despite some suggestions for lower-intensity food supply and some frankly speculative suggestions, it would be hard to conclude on a note of optimism. The world draw down in grain stocks which began in the mid-1960's continues, and some food shortages are likely all through the 1970's and early 1980's. Even if population control measures begin to limit world population, the rising tide of hungry people will be with us for some time.

Food is basically a net product of an ecosystem, however simplified. Food production starts with a natural material, however modified later. Injections of energy (and even brains) will carry us only so far. If the population cannot adjust its wants to the world in which it lives, there is little hope of solving the food problem for mankind. In that case the food shortage will solve our population problem.

REFERENCES AND NOTES

1. *Statistical Abstract of the United States* (Government Printing Office, Washington, D.C., various annual editions).
2. *Historical Statistics of the United States* (Government Printing Office, Washington, D.C., 1960).
3. D. Pimentel, L. E. Hurd, A. C. Bellotti, M. J. Forster, I. N. Oka, O. D. Scholes, R. J. Whitman, *Science* 182, 443 (1973).
4. A description of the system may be found in: *Patterns of Energy Consumption in the United States* (report prepared for the Office of Science and Technology, Executive Office of the President, by Stanford Research Institute, Stanford, California, Jan. 1972), Appendix C. The three groupings larger than food processing are: primary metals, chemicals, and petroleum refining.
5. N. Georgescu-Roegen, *The Entropy Law and the Economic Process* (Harvard Univ. Press, Cambridge, 1971), p. 301.
6. *Patterns of Energy Consumption in the United States* (report prepared for the Office of Science and Technology, Executive Office of the President, by Stanford Research Institute, Stanford, Calif., Jan. 1972).
7. R. A. Rice, *Technol. Rev.* 75, 32 (Jan. 1972).
8. Federal Highway Administration, Nationwide Personal Transportation Study Report No. 1 (1971) [as reported in Energy Research and Development, hearings before the Congressional Committee on Science and Astronautics, May 1972, p. 151].
9. M. Slesser, *Ecologist* 3 (No. 6), 216 (1973).
10. J. F. Gerber, personal communication (we are indebted to Dr. Gerber for pointing out that manures, even if used fully, will not provide all the needed agricultural fertilizers).
11. D. Chapman, *Environment (St. Louis)* 15 (No. 2), 12 (1973).
12. L. U. Mayer and S. H. Hargrove [*CAED Rep. No. 38* (1972)] as quoted in Slesser (9).
13. We have converted all figures for the use of electricity to fuel input values, using the average efficiency values for power plants given by C. M. Summers [*Sci. Am.* 224 (No. 3), 148 (1971)]. Self-generated electricity was converted to fuel inputs at an efficiency of 25 per cent after 1945 and 20 per cent before that year.
14. Purchased material in this analysis was converted to energy of manufacture according to the following values derived from the literature or calculated. In doubtful cases we have made what we believe to be conservative estimates: steel (including fabricated and castings), 1.7×10^7 kcal/ton (1.9×10^4 kcal/kg); aluminum (including castings and forgings), 6.0×10^7 kcal/ton; copper and brass (alloys, millings, castings, and forgings), 1.7×10^6 kcal/ton; paper, 5.5×10^6 kcal/ton; plastics, 1.25×10^6 kcal/ton; coal, 6.6×10^6 kcal/ton; oil and gasoline, 1.5×10^6 kcal/barrel (9.5×10^3 kcal/liter); natural gas, 0.26×10^3 kcal/cubic foot (9.2×10^3 kcal/m³); petroleum wax, 2.2×10^6 kcal/ton; gasoline and diesel engines, 3.4×10^6 kcal/engine; electric motors over 1 horsepower, 45×10^3 kcal/motor; ammonia,

2.7×10^7 kcal/ton; ammonia compounds, 2.2×10^6 kcal/ton; sulfuric acid and sulfur, 3×10^6 kcal/ton; sodium carbonate, 4×10^6 kcal/ton; and other inorganic chemicals, 2.2×10^6 kcal/ton.

15. Direct fuel use on farms: Expenditures for petroleum and other fuels consumed on farms were obtained from *Statistical Abstracts* (1) and the *Census of Agriculture* (Bureau of the Census, Government Printing Office, Washington, D.C., various recent editions) data. A special survey of fuel use on farms in the 1964 *Census of Agriculture* was used for that year and to determine the mix of fuel products used. By comparing expenditures for fuel in 1964 with actual fuel use, the apparent unit price for this fuel mix was calculated. Using actual retail prices and price indices from *Statistical Abstracts* and the ratio of the actual prices paid to the retail prices in 1964, we derived the fuel quantities used in other years. Changes in the fuel mix used (primarily the recent trend toward more diesel tractors) may understate the energy in this category slightly in the years since 1964 and overstate it slightly in years before 1964. S. H. Schurr and B. C. Netschert [*Energy in the American Economy, 1850–1975* (Johns Hopkins Press, Baltimore, 1960), p. 774], for example, using different methods, estimate a figure 10 per cent less for 1955 than that given here. On the other hand, some retail fuel purchases appear to be omitted from all these data for all years. M. J. Perelman [*Environment (St. Louis)* 14 (No. 8), 10 (1972)] from different data, calculates 270×10^{12} kcal of energy usage for tractors alone.

16. Electricity use on farms: Data on monthly usage on farms were obtained from the "Report of the Administrator, Rural Electrification Administration" (U.S. Department of Agriculture, Government Printing Office, Washington, D.C., various annual editions). Totals were calculated from the annual farm usage multiplied by the number of farms multiplied by the fraction electrified. Some nonagricultural uses are included which may overstate the totals slightly for the years before 1955. Nevertheless, the totals are on the conservative side. A survey of on-farm electricity usage published by the Holt Investment Corporation, New York, 18 May 1973, reports values for per farm usage 30 to 40 per cent higher than those used here, suggesting that the totals may be much too small. The discrepancy is probably the result of the fact that the largest farm users are included in the business and commercial categories (and excluded from the U.S. Department of Agriculture tabulations used).

17. Fertilizer: Direct fuel use by fertilizer manufacturers was added to the energy required for the manufacture of raw materials purchased as inputs for fertilizer manufacture. There is al-

lowance for the following: ammonia and related compounds, phosphatic compounds, phosphoric acid, muriate of potash, sulfuric acid, and sulfur. We made no allowance for other inputs (of which phosphate rock, potash, and "fillers" are the largest), packaging, or capital equipment. Source: *Census of Manufactures* (Government Printing Office, Washington, D.C., various recent editions).

18. Agricultural steel: Source, *Statistical Abstracts* for various years (1). Converted to energy values according to note (14).

19. Farm machinery (except tractors): Source, *Census of Manufactures*. Totals include direct energy use and the energy used in the manufacture of steel, aluminum, copper, brass, alloys, and engines converted according to note (14).

20. Tractors: numbers of new tractors were derived from *Statistical Abstracts* and the *Census of Agriculture* data. Direct data on energy and materials use for farm tractor manufacture was collected in the *Census of Manufactures* data for 1954 and 1947 (in later years these data were merged with other data). For 1954 and 1947 energy consumption was calculated in the same way as for farm machinery. For more recent years a figure of 2.65×10^6 kcal per tractor horsepower calculated as the energy of manufacture from 1954 data (the 1954 energy of tractor manufacture, 23.6×10^{12} kcal, divided by sales of 315,000 units divided by 28.7 average tractor horsepower in 1954). This figure was used to calculate energy use in tractor manufacture in more recent years to take some account of the continuing increase in tractor size and power. It probably slightly understates the energy in tractor manufacture in more recent years.

21. Irrigation energy: Values are derived from the acres irrigated from *Statistical Abstracts* for various years; converted to energy use at 10^6 kcal per acre irrigated. This is an intermediate value of two cited by Pimentel et al. (3).

22. Food processing industry: Source, *Census of Manufactures;* direct fuel inputs only. No account taken for raw materials other than agricultural products, except for those items (packaging and processing machinery) accounted for in separate categories.

23. Food processing machinery: Source, *Census of Manufactures* for various years. Items included are the same as for farm machinery [see note (13)].

24. Paper packaging: Source, *Census of Manufactures* for various years. In addition to direct energy use by the industry, energy values were calculated for purchased paper, plastics, and petroleum wax, according to note (14). Proportions of paper products having direct food usage were obtained from *Containers and Packaging* (U.S. Department of Commerce, Washington, D.C., various recent editions). [The

values given include only proportional values from Standard Industrial Classifications 2651 (half), 2653 (half), 2654 (all).]

25. Glass containers: Source, *Census of Manufactures* for various years. Direct energy use and sodium carbonate [converted according to note (14)] were the only inputs considered. Proportions of containers assignable to food are from *Containers and Packaging*. Understatement of totals may be more than 20 per cent in this category.

26. Steel and aluminum cans: Source, *Census of Manufactures* for various years. Direct energy use and energy used in the manufacture of steel and aluminum inputs were included. The proportion of cans used for food has been nearly constant at 82 per cent of total production (*Containers and Packaging*).

27. Transportation fuel usage: Trucks only are included in the totals given. After subtracting trucks used solely for personal transport (all of which are small trucks), 45 per cent of all remaining trucks and 38 per cent of trucks larger than pickup and panel trucks were engaged in hauling food or agricultural products, or both, in 1967. These proportions were assumed to hold for earlier years as well. Comparison with ICC analyses of class I motor carrier cargos suggests that this is a reasonable assumption. The total fuel usage for trucks was apportioned according to these values. Direct calculations from average mileage per truck and average number of miles per gallon of gasoline produces agreement to within ± 10 per cent for 1967, 1963, and 1955. There is some possible duplication with the direct fuel use on farms, but it cannot be more than 20 per cent considering on-farm truck inventories. On the other hand, inclusion of transport by rail, water, air, and energy involved in the transport of fertilizer, machinery, packaging, and other inputs of transportation energy could raise these figures by 30 to 40 per cent if ICC commodity proportions apply to all transportation. Sources: *Census of Transportation* (Government Printing Office, Washington, D.C., 1963, 1967); *Statistical Abstracts* (1); *Freight Commodity Statistics of Class I Motor Carriers* (Interstate Commerce Commission, Government Printing Office, Washington, D.C., various annual editions).

28. Trucks and trailers: Using truck sales numbers and the proportions of trucks engaged in food and agriculture obtained in note (27) above, we calculated the energy values at 75 × 10⁶ kcal per trucks for manufacturing and delivery energy [A. B. Makhijani and A. J. Lichtenberg, *Univ. Calif. Berkeley Mem. No. ERL-M310* (revised) (1971)]. The results were checked against the *Census of Manufactures* data for 1967, 1963, 1958, and 1939 by proportioning motor vehicles categories between automobiles and trucks. These checks suggest that our estimates are too small by a small amount. Trailer manufacture was estimated by the proportional dollar value to truck sales (7 per cent). Since a larger fraction of aluminum is used in trailers than in trucks, these energy amounts are also probably a little conservative. Automobiles and trucks used for personal transport in the food system are omitted. Totals here are probably significant, but we know of no way to estimate them at present. Sources: *Statistical Abstracts, Census of Manufactures,* and *Census of Transportation* for various years.

29. Commercial and home refrigeration and cooking: Data from 1960 through 1968 (1970 extrapolated) from *Patterns of Energy Consumption in the United States* (6). For earlier years sales and inventory in-use data for stoves and refrigerators were compiled by fuel and converted to energy from average annual use figures from the Edison Electric Institute [*Statistical Year Book* (Edison Electric Institute, New York, various annual editions] and American Gas Association values [*Gas Facts and Yearbook* (American Gas Association, Inc., Arlington, Virginia, various annual editions] for various years.

30. Refrigeration machinery: Source, *Census of Manufactures*. Direct energy use was included and also energy involved in the manufacture of steel, aluminum, copper, and brass. A few items produced under this SIC category for some years perhaps should be excluded for years prior to 1958, but other inputs, notably electric motors, compressors, and other purchased materials should be included.

31. There are many studies of energy budgets in primitive societies. See, for example, H. T. Odum [*Environment, Power, and Society* (Wiley, Interscience, New York, 1970)] and R. A. Rappaport [*Sci. Am.* 224 (No. 3), 104 (1971)]. The remaining values of energy subsidies in Fig. 1-13 were calculated from data presented by Slesser (9), Table 1.

32. This article is modified from C. E. Steinhart and J. S. Steinhart, *Energy: Sources, Use, and Role in Human Affairs* (Duxbury Press, North Scituate, Mass., 1974) (used with permission). Some of this research was supported by the U.S. Geological Survey, Department of the Interior, under grant No. 14-08-0001-G-63. Contribution 18 of the Marine Studies Center, University of Wisconsin–Madison. Since this article was completed, the analysis of energy use in the food system of E. Hirst has come to our attention ["Energy Use for Food in the United States," *ONRL–NSF-EP-57* (Oct. 1973)]. Using different methods, he assigns 12 per cent of total energy use to the food system for 1963. This compares with our result of about 13 per cent in 1964.

Energy and the Environment

JOHN M. FOWLER   1972

ENERGY: WHERE IT COMES FROM AND WHERE IT GOES

The world runs on energy, both literally and figuratively. It spins on its axis and travels in its orbit about the sun; the winds blow, waves crash on the beaches, volcanoes and earthquakes rock their surroundings. Without energy it would be a dead world. Energy was needed to catalyze the beginning of life; energy is needed to sustain it.

For most of life, animal and plant, energy means food, and most of life turns to the sun as ultimate source. The linked-life patterns—the ecosystems—which have been established between plants and animals are very complex; the paths of energy wind and twist and double back, but ultimately they all begin at that star that holds us in our endless circle.

When man crossed that threshold of consciousness which separated him from animals, his uses of energy began to diversify. He, too, needed food, but poorly furred as he was, he also needed warmth. With the discovery of fire he was able to warm himself. He also found that fire could make his food digestible and thus increase its efficiency as an energy source. After a while he began also to use fire to make the implements through which he slowly started to dominate nature.

Man's use of energy grew very slowly. In the beginning he required only the 2000 or so Calories* per day for food; the con-

*We will consistently deal with Calories (kilocalories), the amount of heat energy needed to raise the temperature of one kilogram of water one degree Celsius ($1°C$).

Dr. John M. Fowler is presently a visiting professor in the Department of Physics and Astronomy at the University of Maryland, College Park, Maryland. He was director, Commission on College Physics, 1965–1972, and was awarded the Millikan Medal by the American Association of Physics Teachers for contributions to the teaching of physics in 1969. During an earlier tenure at Washington University, St. Louis he helped start the Committee on Environmental Information, and the magazine *Environment*. He is currently on the Advisory Board to *Environment* and a Director of the Scientist's Institute for Public Information. His publications include *Fallout: A Study of Superbombs, Strontium 90, and Survival*, Basic Books, Inc. (1960). His latest book *Energy and the Environment* is scheduled for publication by McGraw-Hill, fall 1974.

From *The Science Teacher*, Vol. 39, No. 9, pp. 10–22, December 1972. Reprinted by permission of the author and *The Science Teacher*. Copyright 1972 by the National Science Teachers Association. This article is adapted from Dr. Fowler's Sunoco Science Seminar presentation at the 1972 convention of the National Science Teachers Association in New York City, April 8–10, 1972.

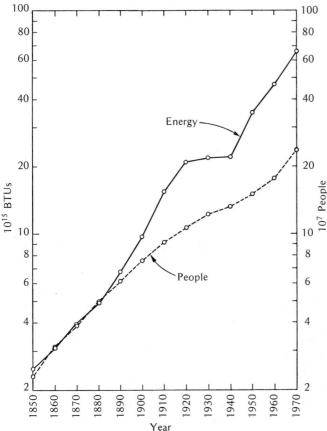

Figure 1-14. Comparison of energy and population growth, United States, 1850 to 1960.

venience of warmth added a few thousand more Calories from easily obtainable wood. The first big jump in energy use came about four or five thousand years B.C. when man domesticated several animals and—at the cost of a little food, much of which he gathered himself—was able to triple the amount of energy at his service.

The water wheel appeared in the first century B.C. and again multiplied the amount of energy available to man. Its introduction was perhaps even more significant because it was a source of inanimate energy. For a long time the water wheel was the most important source of energy for the nascent industry. It was not until the twelfth century, more or less, that the analogy between flowing air and flowing water led to the use of windmills.

The beginning of the modern era of industry coincided with the development of the steam engine. Since that time the world's use of energy, which until then had been very nearly proportional to the number of people, began to grow in the industrial countries more rapidly than the population increased. This growth, shown in Figure 1-14 for the United States, continues.

The second historical trend which, together with the increasing per capita use, has brought us to the present state, is the constant change in the energy mix. Wood was the dominant fuel in the 1850's but had lost all but 20 per cent of the market to coal by 1900. Coal in turn lost out after 75 years to petroleum products, which now account for 75 per cent of the energy, but they in turn will be (and *must* be as we shall see) replaced

Table 1-6. Energy Values for Some Fuels and Foods (Retail Prices).

Source	Energy (Calories per kilogram)	Energy (kilowatt-hours per pound)	Energy (cost per 1000 kilowatt-hours)
Coal	5,600	3060	$ 0.065
Gasoline	11,000	6000	0.008
Fuel oil	10,600	5800	0.004
Alcohol	5,500	3000	0.3
Bread	2,600	1420	0.11
Butter	9,000	4900	0.16
Sugar	4,000	2180	0.06
Beef steak	1,840	1000	1.00
Electricity	–	–	20.00

by other sources. Nuclear energy is the best present candidate.[1]

Energy as a Commodity

In the early stages of man's history, energy was food, something to be found and consumed. But as life became more complex, and early barter systems were followed by a money-based economy, energy had to be bought. At first it was purchased indirectly as food or fuel. With the introduction of electricity, energy could be piped directly into the house or factory.

Energy is a commodity; it can be measured, bought, and sold. But its price depends on the form in which it is purchased—as food or fuel or electricity. Table 1-6 is an "energy shopping list." It is clear that we pay for the good taste of energy in the form of steak but even more for the convenience of electricity.

Where It Goes

The energy crisis is not a crisis caused by the "using up" or the disappearance of energy. The First Law of Thermodynamics assures us of that. Energy is conserved, at least in the closed system of the universe. The crisis must then be found in the pathways of energy conversion.

1. See Cook, E. "The Flow of Energy in an Industrial Society," *Scientific American* 224: 134–144; September 1971.

We use energy in its kinetic form, as mechanical energy, heat, or radiant energy. The form in which it is stored is potential energy. We know from physics that the potential energy of a system is increased by ΔE when we operate against a force over a distance ΔX, that is,

$$\Delta E = \vec{F} \cdot \vec{\Delta X}$$

In the infinite variety of the universe we have, so far, discovered only three types of forces: gravitational, electrical, and nuclear (there seem to be two nuclear forces corresponding to the weak and the strong nuclear interaction). It follows, therefore, that there are three primary sources of energy: gravitational, electrical (chemical), and nuclear. On the scale of the universe these are the most important, and the weak force, gravitational, and the strong one, nuclear, give us the most visible effects.

At earth's scale we choose other primary sources of energy. Solar energy, radiated from the thermonuclear processes in the sun, is the most important of these. It gives us the kinetic energy of water power and wind power, warms us, is stored as chemical energy in growing things, and was preserved in the fossil fuels.

We store the gravitational energy of lifted water in reservoirs, but the only true primary source of gravitational energy of which we make commercial use (in a small way, admittedly) is that of the tides. Here we draw on the gravitational energy stored in the earth-moon system.

We show these and the other important primary sources of energy in Figure 1-15. The chemical energy of fossil fuels is at present far and away the most important of these, but there are two nonsolar sources, geothermal energy from the earth's heated interior (originally heated by gravitational contraction and kept warm by radioactivity) and the new entrant onto the scene, nuclear energy.

Except solar energy, the other primary sources are of little direct use to us; they must be converted to the intermediate forms and often converted again to the end uses which are also shown schematically in Figure 1-15.

The major sources of energy in this country are the chemical energy of the fossil fuels. From them we get 95.9 per cent of the inanimate energy we use. They are fuels; their chemical energy is released by burning. Thus the major conversion pathway is from primary chemical energy to intermediate thermal energy. In fact, most of the conversion pathways go through the thermal intermediate form.

Figure 1-15. Paths of energy conversion.

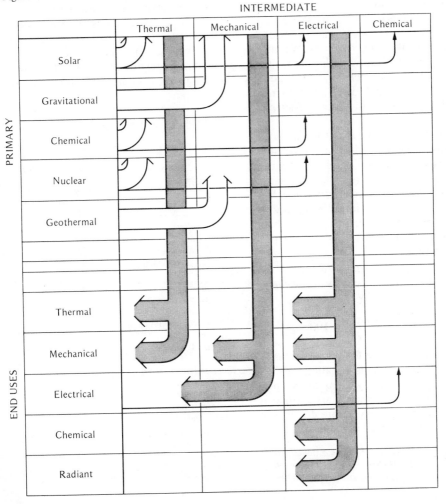

We will look later in detail at the distribution of energy among the various end uses. We know in advance, however, that the major end uses are thermal (space heating, for example) and mechanical. Mechanical energy is also a major intermediate form and is also converted to that most important intermediate form, electrical energy. The convenience of electrical energy shows up in its ready conversion to all the important end uses.

Conversion Efficiency

The most important conversion pathway is thus chemical → thermal → mechanical; and here we enter into the domain of the Second Law of Thermodynamics. It is this "thermal bottleneck" through which most of our energy flows that contributes mightily to the various facets of the energy crisis. We burn to convert, and this causes pollution. We are doomed to low efficiencies by the Second Law, and the wasted heat causes "thermal pollution." Let us consider the efficiency problem first.

Efficiency, the ratio of output work to input energy, varies greatly from conversion to conversion. Generally speaking, we can convert back and forth from electrical energy to other forms with high efficiency, but when we convert other forms of energy to heat and then try to convert heat to mechanical energy, we enter the one-way street of the Second Law.[2]

The efficiency of a "heat engine" (a device for converting heat energy to mechanical energy) is governed by the equation

$$\text{Eff.} = \left(1 - \frac{T_{out}}{T_{in}}\right) \times 100$$

where T_{in} and T_{out} are the temperatures of intake and exhaust. This equation sets an upper limit of efficiency (it is for a "perfect" Carnot cycle). Since we are forbidden $T_{out} = 0°K$ or $T_{in} = \infty$, we are doomed to the intermediate range of modest efficien-

cies. For example, most modern power plants use steam at 1,000°F (811°K) and exhaust at about 212°F (373°K) with a resulting upper limit of efficiency of 63 per cent. The actual efficiency is closer to 40 per cent. Nuclear reactors presently operate at a T_{in} of about 600°F (623°K) and T_{out} of 212°F (373°K) for an upper limit of 40 per cent. They actually operate at about 30 per cent. In an automobile the input temperature of 5,400°F (3,255°K) and output of 2,100°F (1,433°K) would allow an efficiency of 56 per cent. The actual efficiency is about 25 per cent.

So far we have talked about the efficiency of the major conversion process, heat to mechanical work. What is more important to an understanding of the entire energy picture, however, is the system efficiency; for example, the over-all efficiency with which we use the energy stored in the petroleum underground to move us down a road in an automobile. Table 1-7 shows the

Table 1-7. Energy System Efficiencies.

	Efficiency of Each Step (Per Cent)	Efficiency Including All Preceding Steps (Per Cent)
Automobile		
Production of crude petroleum	96	96
Refining of petroleum	87	83.5
Transportation of gasoline	97	81
Thermal efficiency of engine	29	23.5
Mechanical efficiency of engine	71	16.7
Rolling efficiency	30	5
Electric Power Generation		
Production of coal	96	96
Chemical energy of fuel → boiler heat	88	84.5
Boiler heat → mechanical energy	50	42.3
Mechanical energy → electrical energy	99	42
Transmission efficiency	80	33.5

2. See Summers, Claude M. "The Conversion of Energy," *Scientific American* 224: 149–160; September 1971.

system efficiency for transportation by automobile and the production of electric power.
One can see that over-all there are large leaks
in the system and that most of the available
energy is lost along the way.

"Lost" does not, of course, describe precisely what happens to energy. We know
what happens; it is converted to heat. The
inexorable Second Law describes the one-
way street of entropy. All energy conversion
processes are irreversible; even in the highly
efficient electrical generator some of the
mechanical work goes into unwanted heat.
The conversion of other forms of energy to
heat is a highly efficient process—ultimately
100 per cent. It is a downhill run. But the
reverse is all uphill; heat energy can never be
completely converted to mechanical work.
The potential energy available to us, whether
it be chemical, nuclear, or gravitational in
form, is slowly being converted to the random motion of molecules. We cannot reverse
this process, we can only slow it down.

Patterns of Consumption

Ever since President Johnson turned off the
lights in the White House there has been a
small (too small) but growing effort to save

energy. It seems reasonable that this country
and perhaps all countries will, at least for a
while, have to make a real effort in this
direction. To produce measurable effects,
however, these efforts will have to be aimed
at important sections of consumption.

A gross flow chart of energy in our economy is shown in Figure 1-16. One sees the
thermal bottleneck. Heat is the desired end
product from about half of our energy. We
do use that amount of energy efficiently. Of
the half that goes to provide mechanical
work, however, large amounts are lost in the
production of electrical energy and transportation. The net result is that over-all (and
one must remember here that we are dealing
with refined fuels delivered to the converters) our system is about 50 per cent
efficient.

ENVIRONMENTAL EFFECTS OF ENERGY USE

Patterns of Consumption

The intimate connections between energy,
our way of life, and the natural environment
occur at many places. The most important

Figure 1-16. The flow of energy in the United States.

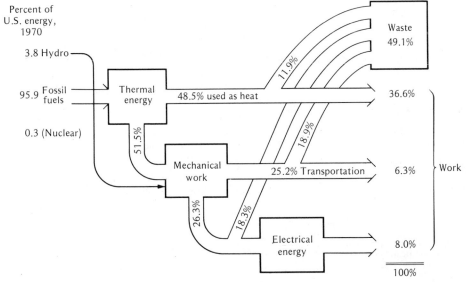

points are, of course, in the production of energy—in the mines and wells, refineries and generating plants—and at the points of consumption. Figure 1-16 gives a crude picture of consumption; we need to look at it in more detail.

Figure 1-17 gives both a crude breakdown and details of the 60,526 trillion BTU's of energy in each category. One sees that industry and transportation use the lion's share. The importance of space and water heating also shows up strongly. Predictions are that transportation and commercial use will be the fastest-growing sectors.[3]

Electrical Energy—The People's Choice

What doesn't show up in this presentation is the special case of electrical energy. It is there, contributing heavily to all categories except transportation, and it shares with transportation most of the blame for energy's role in environmental degradation.

The growth rate of electrical energy consumption, shown in Figure 1-18, is the highest of all the various forms of energy. In

3. Landsberg, H. H., and S. H. Schurr, *Energy in the United States: Sources, Uses and Policy Issues.* Random House, New York. 1960. p. 76.

discussing growth a most useful concept is "doubling time." The energy curve of Figure 1-14 shows several different periods of growth and, therefore, several different doubling times. In the late 1800's the doubling time was about 30 years; by the early 1900's this had been cut in half to about 16 years. The doubling time during the growth period from 1950 to 1960 was 25 years, and for the period 1960 to 1970 dropped to 18 years.

Electrical energy can be said to have arrived commercially with the start up of the Pearl Street Station by Thomas Edison in 1882. (The energy curve in Figure 1-14 breaks away from the "people" curve by about 1890.) The doubling time for per capita electrical energy consumption of Figure 1-18 was only 7.5 years during the start up period of 1910 to 1920, was about 14 years in the 1950's and 1960's and has decreased to about 10 years now. This means that in the period 1970 to 1980 the per capita use of electrical energy will be expected to double.

The impact of electrical energy can be better understood from the plot of total electrical energy consumption also shown in Figure 1-18. This curve has been doubling every 10 years since 1950. This means that in each of those 10-year periods the United

Figure 1-17. United States energy consumption 1971 (total 60.5 × 10¹⁵ BTU).

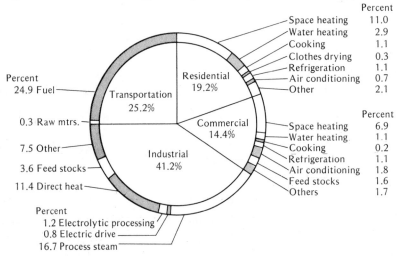

	Percent
Space heating	11.0
Water heating	2.9
Cooking	1.1
Clothes drying	0.3
Refrigeration	1.1
Air conditioning	0.7
Other	2.1

	Percent
Space heating	6.9
Water heating	1.1
Cooking	0.2
Refrigeration	1.1
Air conditioning	1.8
Feed stocks	1.6
Others	1.7

Residential 19.2%

Transportation 25.2%

Commercial 14.4%

Industrial 41.2%

Percent
24.9 Fuel
0.3 Raw mtrs.
7.5 Other
3.6 Feed stocks
11.4 Direct heat

Percent
1.2 Electrolytic processing
0.8 Electric drive
16.7 Process steam

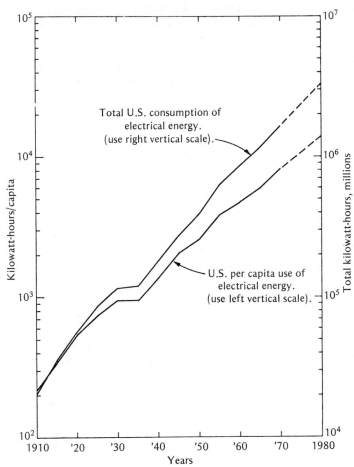

Figure 1-18. Growth of electrical energy consumption in the United States since 1910.

States used as much electrical energy as in its entire previous history.

The reasons for the rapid increase in demand for electrical energy are several. It is in many ways the most convenient of the forms of energy. It can be transported by wire to the point of consumption and then turned into mechanical work, heat, radiant energy, or other forms.

It cannot be very effectively stored, and this has also contributed to its increasing use. Generating facilities have to be designed for peak use. In the late 1950's and early 1960's this peak came in the winter, when nights were longer and more lighting and heating were needed. It was economically sound to heavily promote off-peak use, such

as summer use of air conditioners. This promotion was so effective that the summer is now the peak time, and the sales effort seems to be going into selling all-electric heating for off-peak winter use.

The rate structure has also contributed to increasing use of electrical energy. Rate reductions are offered to attract bulk consumers. Hindsight suggests that there has been an imbalance between research and promotion. The figures bear this out. Senator Lee Metcalf has reported that the utilities in 1969 spent $323.8 million on sales and advertising and $41 million for research and development.

Perhaps the most important clue to the great increase in the use of electrical energy

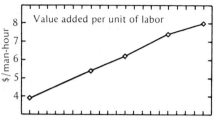

Figure 1-19. Labor productivity in the period 1947–1968.

was suggested by Barry Commoner in an address to the American Association for the Advancement of Science.[4] Commoner looked at the important economic parameter "productivity," which is defined as the ratio of value added to a product/man-hours to produce it. The strength of the economy is built on increasing productivity. The data Commoner presents, for the period 1947 to 1968 (Figure 1-19) show that labor productivity has been steadily increasing. He and his colleagues then looked at the history of electric power productivity. These two quantities, man-hours and kilowatt-hours, do play similar roles in industry—electricity amplifies the existing muscle power in many cases. This analysis of electric power productivity showed the very different results of Figure 1-20. The ratio of value added to kilowatt-hours/electric power productivity, declined sharply from 1947 to 1958 and has flattened out since then. This suggests the important

4. Commoner, Barry. "Power Consumption and Human Welfare," Paper prepared for delivery at the annual meeting of the American Association for the Advancement of Science, December 29, 1971.

conclusion (which merits much more careful study) that the increase in labor productivity has been bought, at least partially, at the expense of a decrease in power productivity. Since labor is more expensive than electric power, the effects on the economy have been beneficial. But what about effects on the environment?

Environmental Effects: Air Pollution

There are two major areas of pollution, air and thermal, which are almost completely attributable to energy consumption. Air pollution is unhappily well known to all of us through smog—that collection of irritating hydrocarbons and oxides of sulfur and nitrogen which is becoming a fixture of urban living. Table 1-8 gives a breakdown of the contributions to pollution of the various categories of polluters. One sees that the generation of electric power is the major source of sulfur oxides, while the automobile leads for three other pollutants.

It is of course not possible to deduce the importance of these pollutants from their gross weight because they have very different effects. Some, like carbon monoxide, affect health in even minute concentrations, others, like the particulates, largely add to cleaning bills. This article is not the place for a detailed discussion of the effects of air pollution.[5] We will simply summarize the costs, which come from effects on the health, damage to crops and exposed materials, and property values, by quoting the Second Annual Report of the Council on Environmental Quality, August 1971: "The annual toll of air pollution on health, vegetation, and property values has been estimated by EPA at more than 16 billion dollars annually—over $80 for each person in the United States."[6]

5. *Air Pollution,* a Scientists' Institute for Public Information Workbook. SIPI, 30 East 68th Street, New York.

6. *Environmental Quality,* the second annual report of the Council on Environmental Quality, August 1971. U.S. Government Printing Office, Washington, D.C., p. 107.

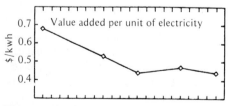

Figure 1-20. Electric power productivity in the period 1947–1968.

Table 1-8. Estimated Emissions of Air Pollutants by Weight Nationwide, 1969; Total 281.2 Million Tons[a] (in millions of tons/year).

Source	Carbon Monoxide		Particulates		Sulfur Oxides		Hydrocarbons		Nitrogen Oxides		Total
	Amount	%	Amount	%	Amount	%	Amount	%	Amount	%	Amount
Automobile	111.5	74	0.8	2	1.1	3	19.8	53	11.2	46	144.4
Power plants	1.8	1	7.2	21	24.4	73	.9	2	10.0	42	44.3
Industrial	12.0	8	14.4	41	7.5	22	5.5	15	.2	1	39.6
Refuse burning	7.9	5	1.4	4	.2	1	2.0	5	.4	2	11.9
Miscellaneous	18.2	12	11.4	32	.2	1	9.2	25	2.0	9	41.0
Total	151.4	–	35.2	–	33.4	–	37.4	–	23.8	–	281.2

[a]Council on Environmental Quality. *Environmental Quality.* p. 212. (See also footnote 6.)

The dependence of our society on the automobile for transportation presents us with a complex mix of problems; in addition to polluting the air, it uses one-quarter of our energy total in a very inefficient way, leads to the covering of our countryside with concrete, contributes to many aspects of the problems of our cities, and takes a high toll of human life. The discussion of these problems and suggestions for solutions are fascinating and important, but cannot be undertaken here.

The generation of electric power at present depends almost entirely on the burning of the fossil fuels. The sulfur oxides come from the sulfur impurities in these fuels. The burning of these fuels also converts large amounts of carbon to carbon dioxide. This familiar gas is not a pollutant in the ordinary sense, but its steady increase in the atmosphere is a cause for concern. Carbon dioxide is largely transparent to the incoming short-wave solar radiation, but reflects the longer-wave radiation by which the earth's heat is radiated outward, producing the so-called "greenhouse effect." Presently about six billion tons of carbon dioxide are being added to the earth's atmosphere per year, increasing its carbon dioxide content by 0.5 per cent per year. By the year 2000 the increase could be as much as 25 per cent. Our understanding of the atmosphere is not sufficient to predict the eventual effects on climate which might be produced by this increase and by a related increase in water vapor and dust, but small changes in the average temperature could have catastrophic effects.

Nuclear Reactors—Clean Power?

There are strong forces in this country pushing the nuclear reactor as an answer to our need for clean power sources. The reactor gains its energy from the fission of uranium-235 or plutonium-239. The energetic byproducts of this fissioning are stopped in the fuel rods, heating them, and this heat is transferred by some heat exchanger to a conventional steam-powered electric generator.

The fission products are radioactive, dangerously so. They have many different half-lives, but the whole mess averages a half-life of perhaps 100 to 150 years. The switch to nuclear energy for the generation of electricity will be accompanied by a growing problem of disposal for this radioactive waste. Snow[7] has estimated that the 16 tons of radioactive fission products from reactors accumulated in 1970 will have grown to 388 tons by 1980 and will be more than 5,000 tons by the year 2000.

7. Snow, J. "Radioactive Waste from Reactors: The Problem that Won't Go Away," *Science and Citizen (Environment Magazine)* 9: 89—95; May 1967.

Nuclear reactors are carefully designed against the release of these products which are collected and stored for safety. But the storage problem itself is a far from negligible one, with no generally agreed-on solution in sight. It has been proposed that the most dangerous wastes be dried and stored in salt mines in Kansas. There are now indications that above-ground storage will be the approved means.

So far the radioactivity associated with nuclear reactors seems to have been handled with exemplary safety. Any exposure to the general population from this source is in the range of present exposure from past nuclear testing. It is in all likelihood causing damage, but so do all the other forms of power generation.[8] What really must concern us when we consider substituting the fissioning of uranium for the burning of fuel is the possibility of accident.

When discussing accidents, we are not talking about a real nuclear explosion in which a "critical mass" of fissionable material accumulates and goes off. The low enrichment densities preclude that. But since the reactor core is a witches' caldron of radioactive waste products, any accident which opens that up and spreads it over the countryside is catastrophic. The accident that designers fear is cooling system failure. If the cooling water were somehow denied the fuel rods, in only a matter of seconds they would begin to melt, leaving the reactor core an uncontrollable blob of melting metal, heated internally so that it continues to melt. The resulting steam pressure explosions then could release the radioactivity to the environment. It is this small but troublesome chance of accident that keeps reactors away from the cities where their products, electricity and heat, are needed.

Heat as a Pollutant

As we have earlier stressed, energy conversion is largely a one-way street: All work eventually produces heat. The "heat engines," because of their inefficiency, however, are particularly troublesome. A steam power plant, which is only 30 per cent efficient, dumps two units of heat energy for every one it converts to electricity. As our appetite for electricity grows in its apparently unbounded way, so also grows the problem of heat discharged to the environment.

In the steam power plant the waste heat problem is associated with the necessity to lower the temperature of the exhaust steam (so that the piston will not have to work against an appreciable back pressure). The most inexpensive and convenient way to accomplish this is to divert water from a stream or river. Nuclear reactors, since their working parts are steam engines, have the same problems. In fact, the nuclear reactor, because of its lower efficiency, presents a more serious cooling problem. Because of the difference in power plant efficiencies (30 versus 40 per cent) and in fuel efficiency, and because the fossil-fuel plant discharges about 10 per cent to the atmosphere through its stack, the reactor dumps about twice as much heat per kilowatt hour of energy produced as does the fossil-fuel plant.

We do not need to look ahead very far to see that this heating up of the environment cannot go on. There are two different sorts of projections that make this point.

The first of these concerns the cooling-water needs. If the growth of Figure 1-18 continues, we will need one-sixth of the total fresh-water runoff of this country to cool our generating plants by 1990 and one-third by the year 2000.[9] Long before we reach that point we will have to make some hard decisions about stream and river use and plant siting if we are to preserve inland aquatic life.

The second projection is even more indicative of the problem. If we express our consumption of electricity in terms of energy released per square foot of United States

8. This is treated more fully in *The Environmental Cost of Electric Power Production.* A SIPI Workbook. SIPI, 30 East 68th St., New York.

9. Federal Power Commission Staff Study, "Selected Materials on Environmental Effects of Producing Electric Power," Joint Committee on Atomic Energy, August 1969, p. 323.

land area, we obtain for 1970, 0.017 watts per square foot. At our present doubling time of 10 years for electric power consumption, in 100 years we will have gone through 10 doubling periods, and the energy release will be 17 watts per square foot—almost the same as the 18 or 19 watts per square foot of incoming solar energy (averaged over 24 hours). Long before we reach such a level, something will have to be changed.

These two projections only serve to emphasize what should by now be obvious: energy use, particularly electric power, cannot be allowed to continue to grow as it has. There are other data that reinforce this conclusion. Electricity means power plants and transmission lines; doubling consumption means doubling these. There are now about 300,000 miles of high-voltage transmission lines occupying four million acres of countryside in the United States. By 1990 this is projected to be 500,000 miles of lines occupying seven million acres.[10]

All this serves to make the point that exponential growth cannot continue. But we could have learned that from nature. Exponential growth is unnatural; it occurs only for temporary periods when there is an uncoupling from the constraints of supply and of control. For instance, it will be demonstrated for a while by the growth of a bacterial population with plenty of food, but will eventually be turned over either by exhaustion of the food supply or by control from environmental processes which resist the population growth. We have examined some of the areas of environmental damage which may cause us to resist continued growth of energy production and consumption. What about our energy sources; are they likely to be the controlling factor?

RESOURCES AND NEW SOURCES

Before we ask for a statement of the lavish deposits nature has made to our energy ac-

count, we must shed our parochial view and briefly look at energy consumption as the world problem it is.

Energy and the Gross National Product

It can and will be argued that man can live happily and productively at rather low levels of energy consumption. The fact remains, however, that today per capita energy consumption is an indicator of national wealth and influence—of the relative state of civilization as we have defined it. That this is so is seen most clearly by plotting that talisman of success, the (per capita) gross national product (GNP) against the (per capita) energy consumption shown in Figure 1-21. There appears a rough proportionality between per capita gross national product and per capita energy consumption with the United States at the top, and countries like Portugal and India near the bottom. To the right of the "band of proportionality" lie the countries which manage a relatively large gross national product with a relatively small energy expenditure. Perhaps they are worthy of study.

There are two lines of interest which lead out from this kind of data and bear on future uses of energy. One is to look at the time dependence of gross national product/energy data. Data for the United States show two interesting effects.

We find a long period, 1920 to 1965, during which the country was progressively more efficient in its energy use or at least managed to increase its per capita gross national product more rapidly than its per capita energy expenditure. This was apparently largely due to increased efficiency of conversion and end-use techniques. This trend reversed, however, around 1965, and we now are in a period during which this ratio is rising steadily. Reasons for this seem to be in part at least due to a rise in nongross national product-connected energy uses, such as heating and air conditioning. Since these uses are on the increase and since we are near ultimate efficiency in most of our major conversion and end-use tech-

10. Energy Policy Staff, "Electric Power and the Environment," Office of Science and Technology, August 1970. U.S. Government Printing Office, Washington, D.C.

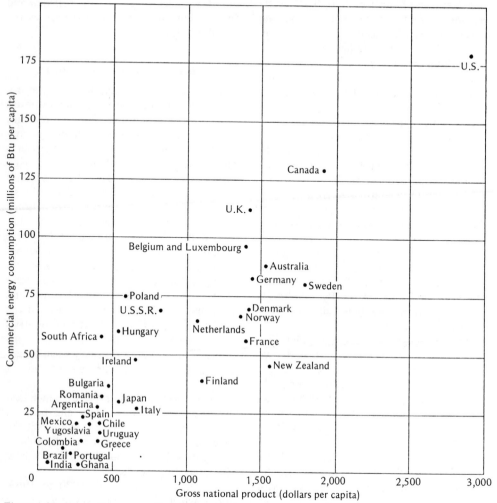

Figure 1-21. Per capita consumption of energy and gross national product for some countries of the world. (From Earl Cook, "The Flow of Energy in an Industrial Society." *Scientific American* 224: 142; September 1971.)

niques, this rise in energy consumed per dollar of gross national product is expected to continue for a while and must be built into energy use projections.

The second line of inquiry concerns ultimate world use. The United States, with 6 per cent of the world's population, uses 35 per cent of the world's energy. If we look at comparative rates of growth, we see that the United States per capita energy consumption is much larger (a factor of about 30) than that of India, for instance, and is growing more rapidly. The world figure is some five times smaller but is growing a bit more rapidly than is the United States figure.

Even if the United States were to stabilize at the present per capita figure of 250 kilowatt-hours per day, it would take about 120 years for the world per capita average to equal it and hundreds of years for India at its present rate of growth to catch up.[11] If some sort of equalization of world energy use is what we are aiming at, with the present United States figure as target, then we are talking about increasing world consumption by a factor of about 100. And this brings us to energy resources.

How Long Will They Last?

As someone said, "Prophecy is very difficult, especially when it deals with the future." Predicting the lifetime of energy resources is doubly difficult. Energy use curves must be projected and then unknown resource poten-

11. These points are discussed in more detail in Starr, C. "Energy and Power," and Cook, E. "The Flow of Energy in an Industrial Society," *Scientific American* 224: 37–49 and 134–144, respectively; September 1971.

tials guessed at. It is difficult to hope for much accuracy in either of these processes.

The estimation of resources is based on general knowledge of the kind of geological conditions associated with the resource and on detailed knowledge of the distribution and extent of a resource within a favorable geological area. Coal is the easiest to work with, for it seems almost always to appear where it is predicted. Oil and natural gas, on the other hand, are erratic in distribution within favorable areas and are found only by exploration. In addition to coal, oil, and natural gas, there are two other sources of organic carbon compounds which are potential fuel sources, the so-called *tar sands* and *oil shale*. In the tar sands, which so far have been found in appreciable amounts only in Canada, a heavy petroleum compound (tar) binds the sands together. A Canadian refin-

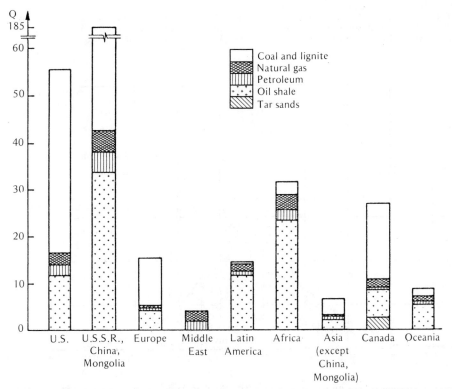

Figure 1-22. Remaining recoverable energy resources by region. [From R. H. Williams and K. Fenton, "World Energy Resources: Distribution, Utilization, and Long-Term Availability." (See ref. 12.)]

ery is currently producing oil products from this material. Oil shale is shale rock containing considerable amounts of a solid organic carbon compound (kerogen). Oil can be extracted by heating the rocks, but this process has not yet been demonstrated to have commercial viability.

In Figure 1-22 we show the estimates of world fossil fuel resources of various types and the global distribution of these resources.[12] The unit used to measure these resources is Q, 10^{18} BTU's. As a crude reference to the size of Q, it would take about that much energy to boil Lake Michigan. Perhaps more useful is the fact that United States total energy consumption in 1970 was about $0.07Q$, and world consumption about $0.2Q$.

One sees from these data that most of the remaining fossil-fuel resources, for the

United States and for the world, are in the form of coal.

Presenting data on resources does not by itself answer the question "How long will they last?" To answer that question one has to look at data on the rate at which the resources are being used. A simplified but very graphic way of displaying this has been adopted by M. King Hubbert of the U.S. Geological Survey.[13] Since supplies of fossil fuels are finite, the curve which traces their production rate will be pulse-like, that is, it will rise exponentially in the beginning, turn over when the resources come into short supply, and then decay exponentially as the resources become harder and harder to find. Such data for United States oil and United States coal are displayed in Figures 1-23 and 1-24.

The curve for United States oil is of particular interest. It shows that the actual rate

12. Williams, R. H., and K. Fenton, "World Energy Resources: Distribution, Utilization, and Long-Term Availability," Paper delivered at the annual meeting of the American Association for the Advancement of Science, December 29, 1971.

13. Hubbert, M. K. Chapter 8, "Energy Resources," in *Resources and Man*. W. H. Freeman, San Francisco, Calif., 1969.

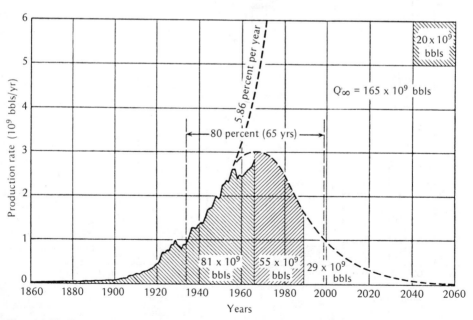

Figure 1-23. Use rate for United States oil resources. [From M. K. Hubbert, *Resources and Man*. p. 183. (See footnote 14.)]

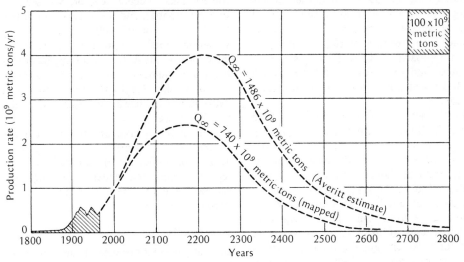

Figure 1-24. Use rate for United States coal resources. [From M. K. Hubbert, *Resources and Man.* p. 200. (See ref. 13.)]

of production from 1880 to the present does have roughly the predicted shape. If Hubbert's choice of total resources, 165×10^9 billion barrels, is correct, it also shows that the United States is probably now past the peak of oil production. The most important feature of a curve like this is its width, for that tells us how long a resource might be expected to last. We see that for oil, these data suggest that the total amount of oil to be produced by the United States is 165×10^9 billion barrels and that 80 per cent of that will be produced in the 65-year period from 1934 to 1999. Thus, within our lifetime we can expect to see a radical change in the fuel mix, with petroleum losing its dominant position. The effects of this on automobile transportation will be of major importance in our automobile-centered economy.

Figure 1-23 for coal is consistent with the previous Figure 1-22. We have a great amount of coal and are just started up the slope to the peak in production. The time scale is much larger, the time to produce 80 per cent of the United States coal is on the order of 300 to 400 years.

In combination, Figures 1-22, 1-23, and 1-24 give a representative view of the state of United States and world fossil-fuel resources. They give a qualified answer to, "How long will they last?" The answer: "Not very long," if we are talking about crude oil and natural gas, "long enough for us to find other sources," if we are talking about coal.

The answer might also have been: "Long enough for us to wreck our environment." Without much improvement in the protection we give our environment about two more doubling periods of energy consumption may be all it can take. Each doubling not only reduces the resources but, for instance, doubles the generating capacity (more plants), doubles (almost) the amount of transmission lines, doubles the coal-mining activity, doubles (unless rigid controls are implemented) the sulfur oxides and fly ash in the atmosphere, and so on.

Are There Clean Continuous Energy Sources?

While coal may provide for our energy needs for 300 to 400 years, it won't last forever; on the scale we hope is still appropriate to mankind, the period of fossil-fuel dependency will be a short blip on the sweep of time. If we are to see a world in which all people have "energy slaves" in numbers which approach the United States standard, we must find some source of energy which can last for at least a few tens of thousands of years.

In Energy: Where It Comes From and Where It Goes we cataloged the primary energy sources. Let us look at them again. The primary sources with their important components are shown in Table 1-9.

From Table 1-9 it is easy to see the necessary shape of our long-range future. Of the continuous sources only solar energy can provide energy at the level of expectation. And for solar energy, the three converters we now use—photosynthesis, hydropower, and wind power—even at maximum utilization, could just barely suffice.

If we look to the depletable sources and spend no more time discussing the fossil fuels, we see some interesting things. In the first place, we are struck by the fact that the ordinary nuclear reactor is not a long-term answer. If we successfully convert to the

Table 1-9. Primary Energy Sources with Maximum Power Available.[a]

Continuous Sources	Maximum Power— World (Units × 10^{12} watts)
Solar	28,000
Photosynthesis fuel	13
Hydro power	3
Wind power	0.1
Gravitational (tidal)	1.0
Geothermal	0.06
World cumulative demand 1960–2000	~ 500
World annual demand by year 2000	~ 15
Depletable sources	
Chemical (fossil fuels)	~ 1,000
Nuclear	
Fission ordinary reactor	~3,000
Fission breeder reactor	~ 300,000
Fusion	>5 × 10^9

[a]Adopted from Starr, C. "Energy and Power," Scientific American 224: 43; September 1971.

controversial breeder reactor, we see our first big number, and we increase our potential by a factor of 100. Finally, looking somewhat further into the future, determinedly rosy-hued, successful application of the deuterium-deuterium reaction in a fusion reactor could, due to the vast amounts of deuterium in the ocean, give us an essentially infinite source.

Let us now take off the rose-colored glasses and look somewhat more skeptically at the most important of these estimates. We refer the interested reader to the suggested general reading for more detailed reviews of these various topics.

Solar Energy

Even at the great distance of earth from sun, our share of the solar output is impressive. Solar energy, which fuels life, pours onto the earth in prodigal amount. But to harness it for the massive needs of industry is something else. Solar energy has several disadvantages. It is dilute: approximately 18 watts per square foot averaged over 24 hours. It is erratic: not only is there a familiar cycle of light and darkness but less regular cloud obscuration. This latter is a serious problem since neither heat nor electricity—the most likely forms to which solar energy will be converted—can be readily stored.

There are ambitious schemes being proposed to capture and convert significant amounts of solar energy, and the budget for solar energy research is climbing from the present "hobby" level of about $50,000 per year to several million. We are beginning to make a real effort to use this clean free energy more massively, but it will take decades to see results.[14]

Nuclear Energy—Fission

We have discussed in Environmental Effects of Energy Use some of the dangers of the ordinary nuclear reactor. This country has embarked on a program to develop the fast-breeder reactor, which uses some of the neutrons from the fissioning fuel to convert

14. For further details see Glaser, P. E. "Solar Energy—Prospects for Its Large-Scale Use," The Science Teacher 39: 36–39; March 1972.

uranium-238 to the reactor fuel plutonium-239.[15] The energy arguments for this are easily seen from Table 1-9. What is not seen are the problems this will bring. The breeder reactor poses several. It is a new and difficult technology, running at higher temperatures (more efficient but more susceptible to cooling-water accidents), it will probably use a liquid metal heat exchanger. Most serious, however, are the dangers associated with plutonium. Creating this material in large quantities means that the half-life of waste (in which there will be some plutonium) will go from about 150 years to >20,000 years. Can we guarantee safe storage for that long? It also means that the material for "atomic bombs" will now be created in ton lots and shipped across country and maybe across oceans. The political implications are obvious.

Nuclear Energy—Fusion

We have dreamed for twenty years of tapping the essentially unlimited resources of deuterium to fuel a deuterium-deuterium fusion reaction. But so far it is largely a dream. Scientifically we understand the theory, but to create a plasma of fusionable material at the $100,000,000°F$ we need and to hold it together for long enough to get appreciable power out, is presently beyond our means. Research in this area is being accelerated, and the optimists predict scientific feasibility by 1980 and pilot demonstration by the year 2000. The pessimists laugh.[16]

Fusion, if it can be tapped, offers much; not only essentially unlimited fuel, but at least a hundred-fold reduction in the radioactive waste problem. It will also be, we expect, free from threat of accident. There will still be heat to worry about, but the combination of cleanliness and safety may allow siting to take advantage of this heat.

Conclusions

Mankind is in the midst of a crisis of energy. It has several dimensions. On a short time scale, we are faced with a serious shortage of natural gas and, in certain places where demand has exceeded present capacity, with a shortage of electric power.

On an intermediate time scale, 30 to 60 years, we are faced with the necessity of finding a substitute for the petroleum products which now dominate the energy mix.

Finally, on a larger scale, 300 to 400 years, we are faced with the exhaustion of fossil fuels.

What can we do? Some of the answers are obvious. We can plan more carefully and further ahead. We can try to reduce energy expenditures and increase research and development support. As citizens we must demand this. We must also demand a role in the decision-making process and decide, for instance, whether we lean heavily on the breeder reactor with its dangers or rely on coal and coal products (gases and liquids).

The science teachers of this country have a vital role to play in the next few years. These decisions must be based on information, and it is the science teacher who must make sure that developing citizens have access to the necessary information devoid of bias and distortion. It is my hope that this summary, brief and sketchy as it is, can serve as a part of the foundation of that necessary effort.

SUGGESTED GENERAL READING

1. *Scientific American* 224; September 1971. The issue is devoted entirely to energy.
2. "The Energy Crisis." *Bulletin of the Atomic Scientists* 27; September and October 1971. Two issues on energy.
3. *The Science Teacher* 39; March 1972. Five articles on energy.
4. Romer, R. H. "Energy-Resources, Production and Environmental Effects." A Resource Letter of the American Association of Physics Teachers. *American Journal of Physics* 40: 805; June 1972.

15. Seaborg, G. T., and J. L. Bloom, "Fast Breeder Reactors," *Scientific American* 223: 13; November 1970.
16. For a review of progress toward fusion see: Auer, P. L., and R. N. Sudan, "Progress in Controlled Fusion Research," *The Science Teacher* 39: 44–50; March 1972.

2

FOSSIL-FUEL ENERGY SOURCES

Large land rig drilling the Standard Oil Company of California No. 1 Moothart Well, Cameron County, Texas. (Photo courtesy of Lon C. Ruedisili.)

TVA's Bull Run steam plant, Oak Ridge, Tennessee. This large, coal-fired unit has a capacity of 950 megawatts (electric), burning over 5 tons of coal a minute. (Photo courtesy of the Tennessee Valley Authority.)

Spoil piles from strip mining of coal in the 1930's, Montana. With no reclamation, the return of normal vegetation has been slow, even after more than 40 years. (Photo courtesy of the Montana Department of State Lands.)

INTRODUCTION

Here we look in more detail at the supply and projected production rates of the traditional fossil fuels—crude oil, natural gas, and coal—and at some of environmental dilemmas engendered by their use. And the dilemmas posed by our "warm but filthy friends," the fossil fuels, are indeed many.

Diseases such as black lung, the ravages of strip mining, and oil pollution of our beaches are immediately obvious social and environmental costs imposed by society's demand for energy. Less apparent are the long-term health hazards and economic consequences of the waste products from the combustion of fossil fuels. The long-term geologic effects of the increased production of carbon dioxide from burning fossil fuels may be another form of social Russian roulette—one cartridge labeled "the new ice age" and another labeled "the hot house effect," with the resulting submergence of most of the world's major coastal cities. Though there is uncertainty about the direction the temperature will take, it is generally agreed that such man-made climatic effects will have profound social implications.

Even our best-intentioned efforts at a technological solution for pollution from fossil fuel may backfire in totally unforeseen ways. Recent evidence indicates that the installation of smoke precipitators to remove the most obvious particulate waste products from coal-burning power plants results in a considerable increase in "acid rain" from sulfur dioxide which previously had been neutralized by the alkaline smoke particles. Precipitation falling through the sulfur dioxide gas converts it to sulfuric acid with pH values as low as 2.1. This unexpected side effect of "clean" smoke stacks may actually result in the increased destruction of coniferous forests, in streams too acid to support life, and in soils from which essential nutrients have been leached.

Though it is impossible to present a complete picture of fossil fuels and their impact on society, the selections included present an overview of the supply situation and analyses of several representative problem areas.

The present large-scale use of fossil fuels by the human species represents a unique event in the billions of years of geologic history. In perspective, most of this development has occurred during this century—a period of unprecedented exponential industrial growth. In *Survey of World Energy Resources,* M. King Hubbert presents the best current estimates of fossil fuel supplies, rates of usage, and the useful concept of "culmination time." It is apparent from the curves presented that, in the United States, the culmination time (peak in the production curve) for domestic crude-oil production (in 1970) already has been passed, that for domestic natural gas production is imminent (in 1977), and that for coal production will occur around the year 2200. Thus, from these estimates, we can easily see why a renewed interest has been shown in developing the vast coal resource of the United States. Another notable observation in this article, which was discussed extensively in the *Limits to Growth,* is the insensitivity of culmination time to the assumed total supply of a given natural resource, one of the principal consequences of which will be an inevitable stabilizing of our exponential-growth culture.

In *Scientific Aspects of the Oil Spill Problem,* Max Blumer presents the far-reaching and complex environmental effects of oil releases, particularly those effects on marine life. Again it is emphasized that even our oceans are not infinite absorbers of oil pollution and that such "cures" as detergents and disperants are often worse than the original oil spill. When the *Ocean Eagle* broke up in San Juan harbor, the detergents used to break up the resulting oil slick caused severe beach erosion by lowering the surface tension of the sand— one is reminded of the story of the Tar Baby.

Edmund A. Nephew examines our largest source of fossil fuel in *The Challenge and Promise of Coal.* Safety and economic considerations are given as the reasons for the shift from deep mining to strip mining. He discusses efficiency of recovery, reclamation costs, and expected technological improvements in mining techniques. Until such improvements occur, coal mining poses a harsh dilemma—underground mines with the resultant danger and high costs versus strip mines with the severe environmental disruption.

Alvin M. Josephy, Jr., describes the dilemmas posed by the extraction of western coal in *Agony of the Northern Plains.* Ranchers face the dilemma of selling versus staying on their land; states face the dilemma of passing strict mining laws, which may discourage production, versus increasing their industrialization with the resulting increased employment; we, as a nation, face the dilemma of continuing use of polluting high-sulfur coal versus increasing use of low-sulfur western coal with all of the problems presented here. This article very forcefully raises the question: Even though "we have more coal than the Arabs have oil," what price are we prepared to pay to exploit our vast reserves?

The difficulty of quantifying the known detrimental effects of air pollution on human health is highlighted in Robert Frank's article, *Biologic Effects of Air Pollution.* He stresses the many variables involved in investigations in this area, as well as the possibility of synergistic effects. The nature of photochemical reactions producing smog has been clarified only recently. Thus, it appears that one of the most obvious detrimental environmental costs of energy consumption remains one of the most difficult to evaluate and least understood.

The ecological balance of nature is at the same time both delicate and sturdy. Man in his ignorance or greed may abuse his natural habitat, causing enormous short-term damage, but nature is resilient and will generally recuperate in the long haul. Abuses in fossil-fuel production and consumption, the most severe environmental disruptions, will certainly disappear—the question is simply when and how. Will it be sooner, through responsible human decisions, or later, through the total depletion of available resources or zonal climatic changes? As one wit has observed, "Nature always bats last."

Survey of World Energy Resources

M. KING HUBBERT   1973

INTRODUCTION

By now, it has become generally recognized that the world's present civilization differs fundamentally from all earlier civilizations in both the magnitude of its operations and the degree of its dependence on energy and mineral resources—particularly energy from the fossil fuels. The significance of energy lies in the fact that it is involved in everything that occurs on the earth—everything that moves. In fact, in the last analysis, about as succinct a statement as can be made about terrestrial events is the following. The earth's surface is composed of the 92 naturally occurring chemical elements, all but a minute radioactive fraction of which obey the laws of conservation and of nontransmutability of classical chemistry. Into and out of this system is a continuous flux of energy, in consequence of which the material constituents undergo either continuous or intermittent circulation.

The principal energy inputs into this system are three (Figure 2-1):

(1) $174,000 \times 10^{12}$ thermal watts from the solar radiation intercepted by the earth's diametrical plane;

(2) 32×10^{12} thermal watts conducted and convected to the earth's surface from inside the earth; and

(3) 3×10^{12} thermal watts of tidal power from the combined kinetic and potential energy of the earth-moon-sun system. Of these inputs of thermal power, that from solar energy is overwhelmingly the largest, exceeding the sum of the other two by a factor of more than 5,000.

Of the solar input, about 30 per cent, the earth's albedo, is directly reflected and scattered into outer space, leaving the earth as short-wavelength radiation; about 47 per cent is directly absorbed and converted into heat; and about 23 per cent is dissipated in circulating through the atmosphere and the

Dr. M. King Hubbert is a research geophysicist with the U.S. Geological Survey, Washington, D.C. His scientific work has included: geophysical and geological exploration for oil, gas, and other minerals; structural geology and the physics of earth deformation; the physics of underground fluids; and the world's mineral and energy resources and the significance of their exploitation to human affairs. He has published 65 technical papers and is author or coauthor of several texts.

Originally presented during the Plenary Session of the 75th Annual General Meeting of the CIM, Vancouver, April 16, 1973, and subsequently published in *The Canadian Mining and Metallurgical Bulletin*, Vol. 66, No. 735, pp. 37–54, July, 1973. Reprinted by permission of the author and *The Canadian Mining and Metallurgical Bulletin* published by the Canadian Institute of Mining and Metallurgy.

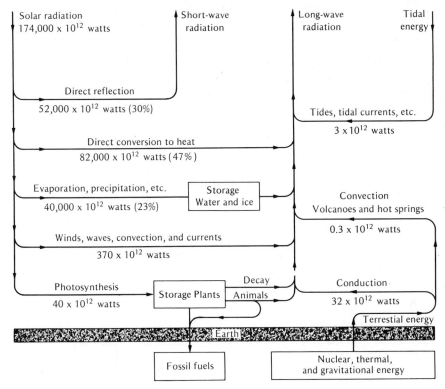

Figure 2-1. World energy flow sheet.

oceans, and in the evaporation, precipitation, and circulation of water in the hydrologic cycle. Finally, a minute fraction, about 40×10^{12} watts, is absorbed by the leaves of plants and stored chemically by the process of photosynthesis whereby the inorganic substances, water and carbon dioxide, are synthesized into organic carbohydrates according to the approximate equation

$$\text{Light energy} + CO_2 + H_2O \rightarrow [CH_2O] + O_2$$

Small though it is, this fraction is the energy source for the biological requirements of the earth's entire populations of plants and animals.

From radioactive dating of meteorites, the astronomical cataclysm that produced the solar system is estimated to have occurred about 4.5 billion years ago and microbial organisms have been found in rocks as old as 3.2 billion years. During the last 600 million years of geologic history, a minute fraction of the earth's organisms have been deposited in swamps and other oxygen-deficient environments under conditions of incomplete decay, and eventually buried under great thicknesses of sedimentary muds and sands. By subsequent transformations, these have become the earth's present supply of fossil fuels; coal, oil, and associated products.

About 2 million years ago, according to recent discoveries, the ancestors of modern man had begun to walk upright and to use primitive tools. From that time to the present, this species has distinguished itself by its inventiveness in the progressive control of an ever-larger fraction of the available energy supply. First, by means of tools and weapons, the invention of clothing, the control of fire, the domestication of plants and animals, and use of animal power, this control was principally ecological in character. Next followed the manipulation of the inorganic world, including the smelting of metals and the primitive uses of the power of wind and water.

Such a state of development was sufficient for the requirements of all premodern civilizations. A higher-level industrialized civilization did not become possible until a larger and more concentrated source of energy and power became available. This occurred when the huge supply of energy stored in the fossil fuels was first tapped by the mining of coal, which began as a continuous enterprise about nine centuries ago near Newcastle in northeast England. Exploi-

tation of the second major fossil-fuel supply, petroleum, began in 1857 in Romania and two years later in the United States. The tapping of an even larger supply of stored energy, that of the atomic nucleus, was first achieved in a controlled manner in 1942, and now the production of nuclear power from plants in the 1,000-megawatt range of capacity is in its early stages of development.

In addition to increased energy sources, energy utilization was markedly enhanced by two technological developments near the end of the last century: the development of the internal-combustion engine, utilizing petroleum products for mobile power, and the development of electrical means for the generation and distribution of power from large-scale central power plants. This also made possible for the first time the large-scale use of water power. This source of

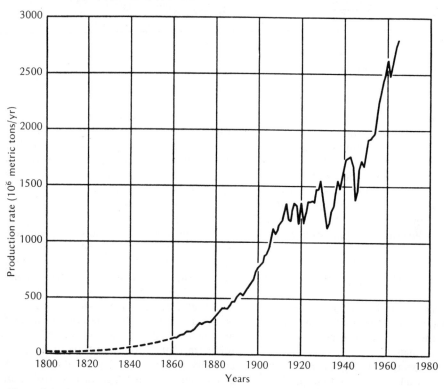

Figure 2-2. World production of coal and lignite (Hubbert, 1969, Figure 8.1).

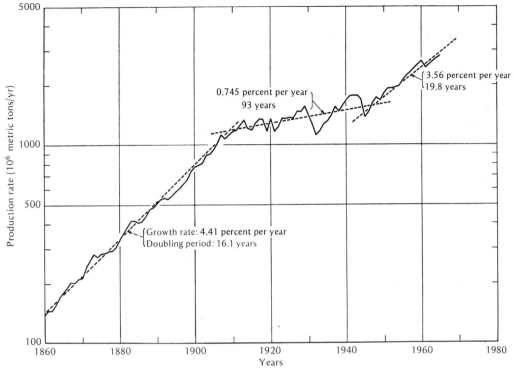

Figure 2-3. World production of coal and lignite (semilogarithmic scale) (Hubbert, 1971, Figure 4).

power derived from the contemporary flux of solar energy has been in use to some degree since Roman times, but always in small units—units rarely larger than a few hundred kilowatts. With electrical generation and distribution of hydropower, first accomplished at Niagara Falls about 1895, progressively larger hydropower stations have been installed with capacities up to several thousand megawatts.

ENERGY FROM FOSSIL FUELS

To the present the principal sources of energy for industrial uses have been the fossil fuels. Let us therefore review the basic facts concerning the exploitation and utilization of these fuels. This can best be done by means of a graphical presentation of the statistics of annual production.

World Production of Coal and Oil

Figure 2-2 shows the annual world production of coal and lignite from 1860 to 1970, and the approximate rate back to 1800, on an arithmetic scale. Figure 2-3 shows the same data on a semilogarithmic scale. The significance of the latter presentation is that straight-line segments of the growth curve indicate periods of steady exponential growth in the rate of production.

Annual statistics of coal production earlier than 1860 are difficult to assemble, but from intermittent earlier records it can be estimated that from the beginning of coal mining about the twelfth century A.D. until 1800, the average growth rate of production must have been about 2 per cent per year, with an average doubling period of about 35 years. During the 8 centuries to 1860 it is

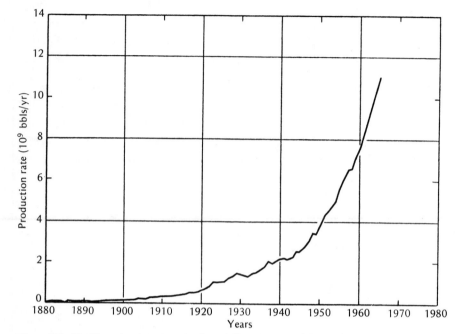

Figure 2-4. World production of crude oil (Hubbert, 1969, Figure 8.2).

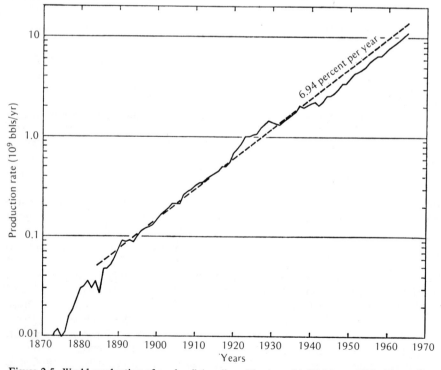

Figure 2-5. World production of crude oil (semilogarithmic scale) (Hubbert, 1971, Figure 6).

estimated that cumulative production amounted to about 7×10^9 metric tons. By 1970, cumulative production reached 140×10^9 metric tons. Hence, the coal mined during the 110-year period from 1860 to 1970 was approximately 19 times that of the preceding 8 centuries. The coal produced during the last 30-year period from 1940 to 1970 was approximately equal to that produced during all preceding history.

The rate of growth of coal production can be seen more clearly from the semilogarithmic plot of Figure 2-3. The straight-line segment of the production curve from 1860 to World War I indicates a steady exponential increase of the rate of production during this period at about 4.4 per cent per year, with a doubling period of 16 years. Between the beginning of World War I and the end of World War II, the growth rate

slowed down to about 0.75 per cent per year and a doubling period of 93 years. Finally, after World War II a more rapid growth rate of 3.56 per cent per year and a doubling period of 19.8 years was resumed.

Figure 2-4 shows, on an arithmetic scale, the annual world crude-oil production from 1880 to 1970. Figure 2-5 shows the same data plotted semilogarithmically. After a slightly higher initial growth rate, world petroleum production from 1890 to 1970 has had a steady exponential increase at an average rate of 6.94 per cent and a doubling period of 10.0 years. Cumulative world production of crude oil to 1970 amounted to 233×10^9 barrels. Of this, the first half required the 103-year period from 1857 to 1960 to produce, the second half only the 10-year period from 1960 to 1970.

When coal is measured in metric tons and

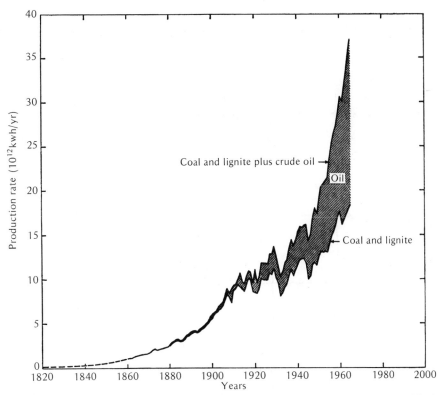

Figure 2-6. World production of thermal energy from coal and lignite plus crude oil (Hubbert, 1969, Figure 8.3).

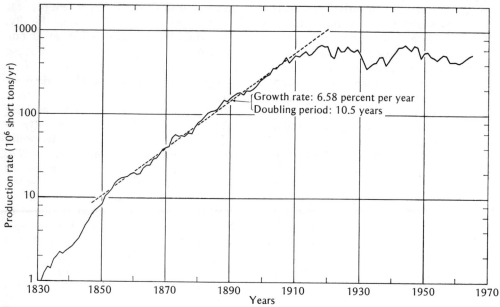

Figure 2-7. United States production of coal (semilogarithmic scale).

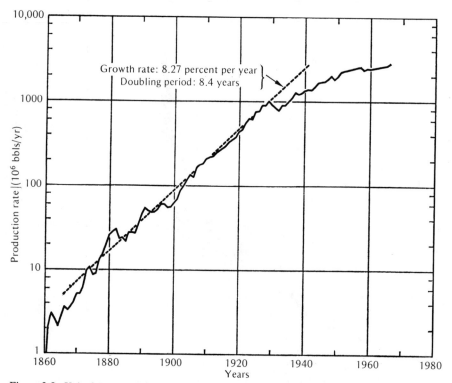

Figure 2-8. United States production of crude oil, exclusive of Alaska (semilogarithmic scale).

oil in United States 42-gallon barrels, a direct comparison between coal and oil cannot be made. Such a comparison can be made, however, by means of the energy contents of the two fuels as determined by their respective heats of combustion. This is shown in Figure 2-6, where the energy produced per year is expressed in power units of 10^{12} thermal watts. From this it is seen that until after 1900 the energy contributed by crude oil was barely significant as compared with that of coal. By 1970, however, the energy from crude oil had increased to 56 per cent of that from coal and oil combined. Were natural gas and natural-gas liquids also to be included, the energy from petroleum fluids would represent about two-thirds of the total.

United States Production of Fossil Fuels

The corresponding growths in the production of coal, crude oil, and natural gas in the United States are shown graphically in Figures 2-7 to 2-9. From before 1860 to 1907 annual United States coal production increased at a steady exponential rate of 6.58 per cent per year, with a doubling period of 10.5 years. After 1907, due largely to the increase in oil and natural-gas production, coal production fluctuated about a production rate of approximately 500×10^6 metric tons per year. After an initial higher rate, United States crude-oil production increased steadily from 1870 to 1929 at about 8.27 per cent per year, with a doubling period of 8.4 years. After 1929, the growth

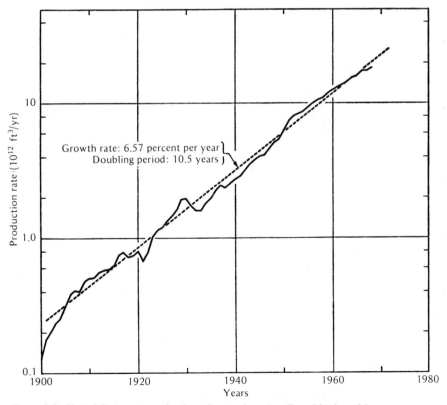

Figure 2-9. United States net production of natural gas (semilogarithmic scale).

rate steadily declined to a 1970 value of approximately zero. From 1905 to 1970 the United States production of natural gas increased at an exponential rate of 6.6 per cent per year, with a doubling period of 10.5 years.

Finally, Figure 2-10 shows the annual production of energy in the United States from coal, oil, natural gas, and hydroelectric and nuclear power from 1850 to 1970. From 1850 to 1907, this increased at a steady growth rate of 6.9 per cent per year and doubled every 10.0 years. At about 1907, the growth rate dropped abruptly to an average value from 1907 to 1960 of about 1.77 per cent per year, with a doubling period of 39 years. Since 1960, the growth rate has increased to about 4.25 per cent per year, with the doubling period reduced to 16.3 years.

DEGREE OF ADVANCEMENT OF FOSSIL-FUEL EXPLOITATION

The foregoing are the basic historical facts pertaining to the exploitation of the fossil fuels in the world and in the United States. In the light of these facts we can hardly fail to wonder: How long can this continue? Several different approaches to this problem will now be considered.

Method of Donald Foster Hewett

In 1929, geologist Donald Foster Hewett delivered before the American Institute of Metallurigical Engineers one of the more important papers ever written by a member of the U.S. Geological Survey, entitled "Cycles in Metal Production." In 1926, Hewett had made a trip to Europe during

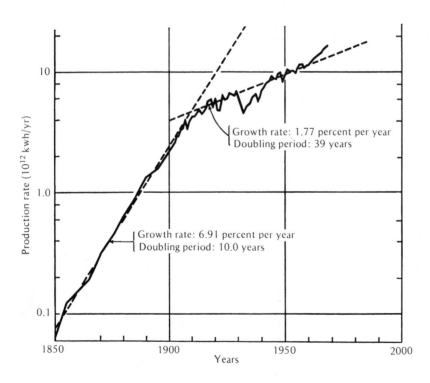

Figure 2-10. United States production of thermal energy from coal, oil, natural gas, water power, and nuclear power (semilogarithmic scale).

which he visited 28 mining districts, of which about half were then or had been outstanding sources of several metals. These districts ranged from England to Greece and from Spain to Poland. Regarding the purpose of this study, Hewett stated: "I have come to believe that many of the problems that harass Europe lie in our path not far ahead. I have therefore hoped that a review of metal production in Europe in the light of its geologic, economic and political background may serve to clear our vision with regard to our own metal production."

In this paper, extensive graphs were presented of the production of separate metals from these various districts showing the rise, and in many cases the decline, in the production rates as the districts approached exhaustion of their ores. After having made this review, Hewett generalized his findings by observing that mining districts evolve during their history through successive stages analogous to those of infancy, adolescence, maturity and old age. He sought criteria for judging how far along in such a sequence a given mining district or region had progressed, and from his study he suggested the successive culminations shown in Figure 2-11. These culminations were:

(1) the quantity of exports of crude ore;
(2) the number of mines in operation;
(3) the number of smelters or refining units in operation;
(4) the production of metal from domestic ore; and
(5) the quantity of imports of crude ore.

Although not all of Hewett's criteria are applicable to the production of the fossil fuels, especially when world production is considered, the fundamental principle is applicable—namely, that like the metals, the exploitation of the fossil fuels in any given area must begin at zero, undergo a period of more or less continuous increase, reach a culmination and then decline, eventually to a zero rate of production. This principle is illustrated in Figure 2-12, in which the complete cycle of the production rate of any exhaustible resource is plotted arithmetically as a function of time. The shape of the curve is arbitrary within wide limits, but it still must have the foregoing general characteristics.

An important mathematical property of such a curve may be seen if we consider a vertical column of base Δt extending from the time axis to the curve itself. The altitude of this column will be the production rate

$$P = \Delta Q / \Delta t$$

at the given time, where ΔQ is the quantity produced in time Δt. The area of the column will accordingly be given by the product of its base and altitude:

$$P \times \Delta t = (\Delta Q / \Delta t) \times \Delta t = \Delta Q.$$

Hence, the area of the column is a measure of the quantity produced during the time interval Δt, and the total area from the beginning of production up to any given

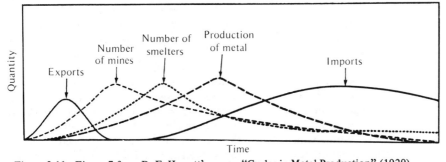

Figure 2-11. Figure 7 from D. F. Hewett's paper, "Cycles in Metal Production" (1929).

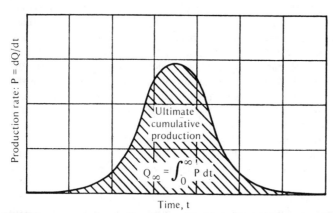

Figure 2-12. Mathematical rela tions involved in the complete cycle of production of any exhaustible resource (Hubbert, 1956, Figure 11).

time t will be a measure of the cumulative production up to that time. Clearly, as the time t increases without limit, the production rate will have gone through its complete cycle and returned to zero. The area under the curve after this has occurred will then represent the ultimate cumulative production, Q_∞. In view of this fact, if from geological or other data the producible magnitude of the resource initially present can be estimated, then any curve drawn to represent the complete cycle of production *must be consistent with that estimate*. No such curve can subtend an area greater than the estimated magnitude of the producible resource.

Utilization of this principle affords a powerful means of estimating the time scale for the complete production cycle of any exhaustible resource in any given region. As in the case of animals where the time required for the complete life cycle of, say, a mouse is different from that of an elephant, so in the case of minerals, the time required for the life cycle of petroleum may differ from that of coal. This principle also permits a reasonably accurate estimate of the most important date in the production cycle of any exhaustible resource, that of its culmination. This date is especially significant because it marks the dividing point in time between the initial period during which the production rate almost continuously increases and the subsequent period during

which it almost continuously declines. It need hardly be added that there is a significant difference between operating an industry whose output increases at a rate of 5 to 10 per cent per year and one whose output declines at such a rate.

Complete Cycle of Coal Production

Because coal deposits occur in stratified seams which are continuous over extensive areas and often crop out on the earth's surface, reasonably good estimates of the coal deposits in various sedimentary basins can be made by surface geological mapping and a limited amount of drilling. A summary of the current estimates of the world's initial coal resources has been published by Paul Averitt (1969) of the U.S. Geological Survey. These estimates comprise the total amount of coal (including lignite) in beds 14 inches (35 centimeters) or more thick and at depths as great as 3,000 feet (900 meters), and in a few cases as great as 6,000 feet. Averitt's estimates as of January 1, 1967, for the initial producible coal, allowing 50 per cent loss in mining, are shown graphically in Figure 2-13 for the world's major geographical areas. As seen in this figure, the original recoverable world coal resources amounted to an estimated 7.64×10^{12} metric tons. Of this, 4.31×10^{12}, or 56 per cent, were in the U.S.S.R., and 1.49×10^{12}, or 19 per cent, in the U.S. At the other extreme, the

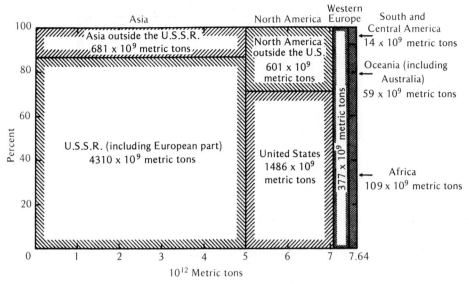

Figure 2-13. Averitt (1969) estimate of original world recoverable coal resources (Hubbert, 1969, Figure 8.24).

three continental areas, Africa, South and Central America, and Oceania, together contained only 0.182×10^{12} metric tons, or 2.4 per cent of the world's total.

Figure 2-14 shows two separate graphs for the complete cycle of world coal production. One is based on the Averitt estimate for the ultimate production of 7.6×10^{12} metric tons. These curves are also based on the assumption that not more than three more doublings, or an eightfold increase, will occur before the maximum rate of production is reached. The dashed curve extending to the top of the drawing indicates what the production rate would be were it to continue to increase at 3.56 per cent per year, the rate that has prevailed since World War II. For either of the complete-cycle curves, if we disregard the first and last ten percentiles of the cumulative production, it is evident that the middle 80 per cent of Q_∞ will probably be consumed during the three-century period from about the years 2000 to 2300.

Figure 2-15 shows the complete cycle of United States coal production for the two

values for Q_∞, 1486×10^9 and 740×10^9 metric tons. Here too the time required to consume the middle 80 per cent would be the 3 or 4 centuries following the year 2000.

A serious modification of the above coal-resource figures has been given by Averitt (cited in Theobald, Schweinfurth, and Duncan, 1972). Here, Averitt, in February 1972, has given an estimate of the amount of coal remaining in the United States that is recoverable under present economic and technological conditions. This comprises coal in seams with a minimum thickness of 28 inches and a maximum depth of 1000 feet. The amount of coal in this category is estimated to be 390×10^9 short tons or 354×10^9 metric tons. Adding the 37×10^9 metric tons of coal already produced gives 391×10^9 metric tons of original coal in this category. This amounts to only 26 per cent of the 1486×10^9 metric tons assumed previously. Of this, 9.5 per cent has already been produced. If we apply the same ratio of 26 per cent to the previous world figure of 7.6×10^{12} metric tons, that is reduced to 2.0×10^{12} metric tons. Of this, $0.145 \times$

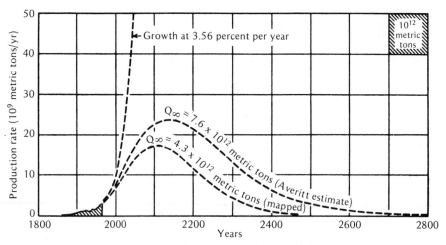

Figure 2-14. Complete cycle of world coal production for two values of Q_∞ (Hubbert, 1969, Figure 8.25).

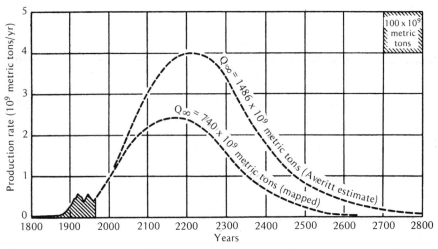

Figure 2-15. Complete cycle of United States coal production for two values of Q_∞ (Hubbert, 1969, Figure 8.26).

10^{12} metric tons, or 7.2 per cent, has already been produced.

Revisions of Figures 2-14 and 2-15 incorporating these lower estimates of recoverable coal have not yet been made, but in each instance the curve for the reduced figure will encompass an area of only about one-quarter that of the uppermost curve shown, and the probable time span for the middle 80 per cent of cumulative production will be cut approximately in half.

Estimates of Petroleum Resources

Because oil and gas occur in limited volumes of space underground in porous sedimentary rocks and at depths ranging from a few hundred feet to five or more miles, the estima-

Table 2-1. Petroleum Estimates by Geological Analogy: Louisiana and Texas Continental Shelves (crude oil, 10^9 barrels).

	U.S. Geological Survey Estimates, 1953	Cumulative Discoveries to 1971
Louisiana	4	ca. 5
Texas	9	Negligible

tion of the ultimate quantities of these fluids that will be obtained from any given area is much more difficult and hazardous than for coal. For the estimation of petroleum, essentially two methods are available: (a) estimation by geological analogy and (b) estimation based on cumulative information and evidence resulting from exploration and productive activities in the region of interest.

The method of estimating by geological analogy is essentially the following. A virgin undrilled territory, Area B, is found by surface reconnaissance and mapping to be geologically similar to Area A, which is already productive of oil and gas. It is inferred, therefore, that Area B will eventually produce comparable quantities of oil and gas per unit of area or unit of volume of sediments to those of Area A.

Although this is practically the only method available initially for estimating the oil and gas potential of an undrilled region, it is also intrinsically hazardous, with a very wide range of uncertainty. This is illustrated in Table 2-1, in which the estimates made in 1953 for the future oil discoveries on the continental shelf off the Texas and Louisiana coasts are compared with the results of subsequent drilling.

In 1953, the U.S. Geological Survey, on the basis of geological analogy between the onshore and offshore areas of the Gulf Coast and the respective areas of the continental shelf bordering Texas and Louisiana, estimated future discoveries of 9 billion barrels of oil on the Texas continental shelf and 4

billion on that of Louisiana. After approximately 20 years of petroleum exploration and drilling, discoveries of crude oil on the Louisiana continental shelf have amounted to approximately 5 billion barrels; those on the continental shelf off Texas have been negligible.

The second technique of petroleum estimation involves the use of various aspects of the Hewett criterion that the complete history of petroleum exploration and production in any given area must go through stages from infancy to maturity to old age. Maturity is plainly the stage of production culmination, and old age is that of an advanced state of discovery and production decline.

In March 1956, this technique was explicitly applied to crude-oil production in the United States by the present author (Hubbert, 1956) in an invited address, "Nuclear Energy and the Fossil Fuels," given before an audience of petroleum engineers at a meeting of the Southwest Section of the American Petroleum Institute at San Antonio, Texas. At that time the petroleum industry in the United States had been in vigorous operation for 97 years, during which 52.4 billion barrels of crude oil had been produced. A review of published literature in conjunction with inquiries among experienced petroleum geologists and engineers indicated a consensus that the ultimate amount of crude to be produced from the conterminous 48 states and adjacent continental shelves would probably be within the range of 150 to 200 billion barrels. Using these two limiting figures, the curves for the complete cycle of United States crude-oil production shown in Figure 2-16 (Hubbert, 1956) were constructed. This showed that if the ultimate cumulative production, Q_∞, should be as small as 150×10^9 barrels, the peak in the rate of production would probably occur about 1966—about 10 years hence. Should another 50×10^9 barrels be added, making $Q_\infty = 200 \times 10^9$ barrels, the date of the peak of production would be postponed by only about 5 years. It was accordingly predicted on the basis of available information that the peak in United

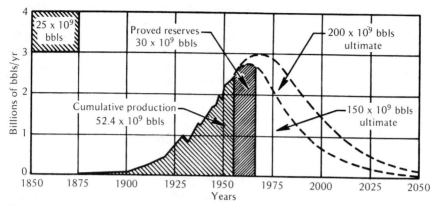

Figure 2-16. 1956 prediction of the date of peak in the rate of United States crude oil production (Hubbert, 1956, Figure 21).

States crude-oil production would occur within 10–15 years after March 1956.

This prediction proved to be both surprising and disturbing to the United States petroleum industry. The only way it could be avoided, however, was to enlarge the area under the curve of the complete cycle of production by increasing the magnitude of Q_∞. As small increases of Q_∞ have only small effects in retarding the date of peak production, if this unpleasant conclusion were to be avoided, it would be necessary to increase Q_∞ by large magnitudes. This was what happened. Within the next five years, with insignificant amounts of new data, the published values for Q_∞ were rapidly escalated to successively higher values—204, 250, 372, 400 and eventually 590 billion barrels.

In view of the fact that values for Q_∞ used in Figure 2-16 involved semisubjective judgments, no adequate rational basis existed for showing conclusively that a figure of 200×10^9 barrels was a much more reliable estimate than one twice that large. This led to the search for other criteria derivable from objective, publicly available data of the petroleum industry. The data satisfying this requirement were the statistics of annual production available since 1860, and the annual estimates of proved reserves of the Proved Reserves Committee of the American Petroleum Institute, begun in

1937. From these data cumulative production from 1860 could be computed, and also cumulative proved discoveries defined as the sum of cumulative production and proved reserves after 1937.

This type of analysis was used in the report, *Energy Resources* (Hubbert, 1962), of the National Academy of Sciences Committee on Natural Resources. The principal results of this study are shown in Figures 2-17 and 2-18, in which it was found that the rate of proved discoveries of crude oil had already passed its peak about 1957, proved reserves were estimated to be at their peak in 1962, and the peak in the rate of crude-oil production was predicted to occur at about the end of the 1960 decade. The ultimate amount of crude oil to be produced from the lower 48 states and adjacent continental shelves was estimated to be about 170 to 175 billion barrels.

The corresponding estimates for natural gas are shown in Figures 2-19 and 2-20 (Hubbert, 1962). From these figures it will be seen that the rate of proved discoveries was estimated to be at its peak at about 1961. Proved reserves of natural gas were estimated to reach their peak ($dQ_r/dt = 0$) at about 1969, and the rate of production about 1977.

At the time the study was being made, the U.S. Geological Survey, in response to a

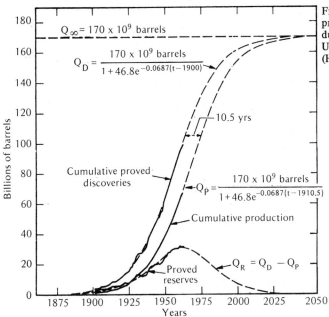

Figure 2-17. Curves of cumulative proved discoveries, cumulative production and proved reserves of United States crude oil as of 1962 (Hubbert, 1962, Figure 27).

$Q_\infty = 170 \times 10^9$ barrels

$$Q_D = \frac{170 \times 10^9 \text{ barrels}}{1 + 46.8e^{-0.0687(t-1900)}}$$

10.5 yrs

Cumulative proved discoveries

$$Q_P = \frac{170 \times 10^9 \text{ barrels}}{1 + 46.8e^{-0.0687(t-1910.5)}}$$

Cumulative production

Proved reserves

$Q_R = Q_D - Q_P$

Billions of barrels

Years

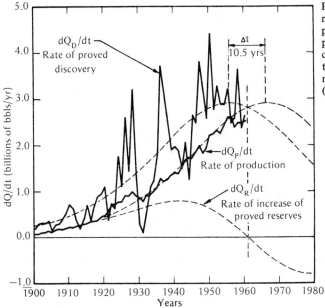

Figure 2-18. Curves showing the rates of proved discovery and of production, and rate of increase of proved reserves of United States crude oil as of 1961. Note prediction of peak of production rate near the end of 1960 decade (Hubbert, 1962, Figure 28).

dQ_D/dt
Rate of proved discovery

Δt
10.5 yrs

dQ_P/dt
Rate of production

dQ_R/dt
Rate of increase of proved reserves

dQ/dt (billions of bbls/yr)

Years

presidential directive of March 4, 1961, presented to the Academy Committee estimates of 590×10^9 barrels for crude oil and 2650 cubic feet for natural gas as its official estimates of the ultimate amounts of these fluids that would be produced from the lower 48 states and adjacent continental shelves.

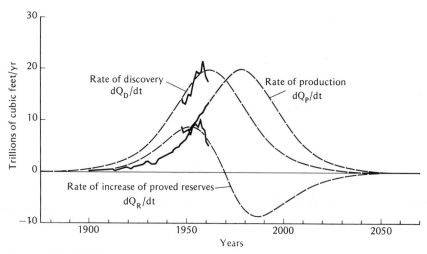

Figure 2-19. 1962 estimates of the dates of the peaks of rate of proved discovery, rate of production, and proved reserves of United States natural gas (Hubbert, 1962, Figure 46).

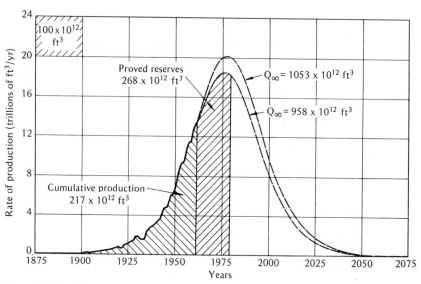

Figure 2-20. 1962 estimates of ultimate amount of natural gas to be produced in conterminous United States, and estimates of date of peak production rate (Hubbert, 1962, Figure 47).

These estimates were, by a wide margin, the highest that had ever been made up until that time. Moreover, had they been true, there would have been no grounds for the expectation of an oil or gas shortage in the United States much before the year 2000.

These estimates were cited in the Academy Committee report, but because of their wide disparity with any available evidence from the petroleum industry, they were also rejected.

As only became clear sometime later, the

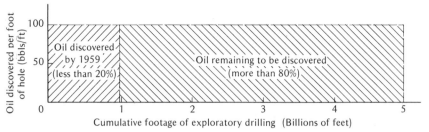

Figure 2-21. Zapp (1962) hypothesis of oil discoveries per foot versus cumulative footage of exploratory drilling for conterminous United States and adjacent continental shelves (Hubbert, 1969, Figure 8.18).

basis for those large estimates was an hypothesis introduced by the late A. D. Zapp of the U.S. Geological Survey, as illustrated in Figure 2-21 (Hubbert, 1969). Zapp postulated that the exploration for petroleum in the United States would not be completed until exploratory wells with an average density of one well per each 2 square miles had been drilled either to the crystalline basement rock or to a depth of 20,000 feet in all the potential petroleum-bearing sedimentary basins. He estimated that to drill this pattern of wells in the petroliferous areas of the conterminous United States and adjacent continental shelves would require about 5×10^9 feet of exploratory drilling. He then estimated that, as of 1959, only 0.98×10^9 feet of exploratory drilling had been done and concluded that at that time the United States was less than 20 per cent along in its ultimate petroleum exploration. He also stated that during recent decades there had been no decline in the oil found per foot of exploratory drilling, yet already more than 100×19^9 barrels of oil had been discovered in the United States. It was implied, but not expressly stated, that the ultimate amount of oil to be discovered would be more than 500×10^9 barrels.

This was confirmed in 1961 by the Zapp estimate for crude oil given to the Academy Committee. At that time, with cumulative drilling of 1.1×10^9 feet, Zapp estimated that 130×10^9 barrels of crude oil had already been discovered. This would be at an average rate of 118 barrels per foot. Then, at this same rate, the amount of oil to be discovered by 5×10^9 feet of exploratory drilling should be 590×10^9 barrels, which is the estimate given to the Academy Committee. This constitutes the "Zapp hypothesis." Not only is it the basis for Zapp's own estimates, but with only minor modifications it has been the principal basis for most of the subsequent higher estimates.

The most obvious test for the validity of this hypothesis is to apply it to past petroleum discoveries in the United States. Has the oil found per foot of exploratory drilling been nearly constant during the past? The answer to this is given in Figure 2-22 (Hubbert, 1967), which shows the quantity of oil discovered and the average amount of oil found per foot for each 10^8 feet of exploratory drilling in the United States from 1860 to 1965. This shows an initial rate of 194 barrels per foot for the first unit from 1860 to 1920, a maximum rate of 276 barrels per foot for the third unit extending from 1929 to 1935 and then a precipitate decline to about 35 barrels per foot by 1965. This is approximately an exponential decline curve, the integration of which for unlimited future drilling gives an estimate of about 165×10^9 for Q_∞, the ultimate discoveries.

The superposition of the actual discoveries per foot shown in Figure 2-22 on the discoveries per foot according to the Zapp hypothesis of Figure 2-21 is shown in Figure 2-23 (Hubbert, 1969). The difference between the areas beneath the two curves

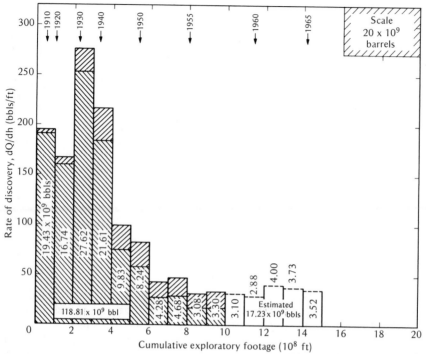

Figure 2-22. Actual United States crude-oil discoveries per foot of exploratory drilling as a function of cumulative exploratory drilling from 1860 to 1965 (Hubbert, 1967, Figure 15).

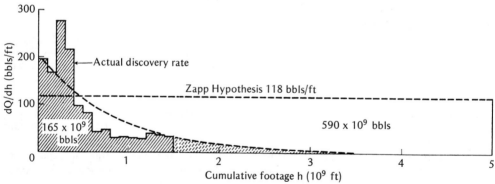

Figure 2-23. Comparison of United States crude oil discoveries according to Zapp hypothesis with actual discoveries. The difference between the areas beneath the two curves represents an overestimate of about 425 billion barrels (Hubbert, 1969, Figure 8.19).

represents the difference between the two estimates—an apparent overestimate of about 425 × 10⁹ barrels.

To recapitulate, in the Academy Committee report of 1962, the peak in United States proved crude-oil discoveries, excluding Alaska, was estimated to have occurred at about 1957, the peak in proved reserves at about 1962 and the peak in production was predicted for about 1968–1969. The peak in proved reserves did occur in 1962, and the peak in the rate of production occurred in 1970. Evidence that this is not likely to be exceeded is afforded by the fact that for the six months since March 1972, the production rates of both Texas and Louisiana, which together account for 60 per cent of the total United States crude-oil production, have been at approximately full capacity, and declining.

As for natural gas, the Academy report estimated that the peak in proved reserves would occur at about 1969 and the peak in the rate of production about 1977. As of September 1972, the peak of proved reserves for the conterminous 48 states occurred in 1967, two years ahead of the predicted date, and it now appears that the peak in the rate of natural-gas production will occur about 1974 to 1975, two to three years earlier than predicted. In the 1962 Academy report, the ultimate production of natural gas was estimated to be about 1000×10^{12} cubic feet. Present estimates by two differ-

ent methods give a low figure of 1000×10^{12} and a high figure of 1080×10^{12}, or a mean of 1040×10^{12} cubic feet.

Because of its early stage of development, the petroleum potential of Alaska must be based principally on geological analogy with other areas. The recent Prudhoe Bay discovery of a 10-billion-barrel field—the largest in the United States—has been a source of excitement for an oil-hungry United States petroleum industry, but it still represents less than a three-year supply for the United States. From present information, a figure of 30×10^9 barrels is about as large an estimate as can be justified for the ultimate crude-oil production from the land area of Alaska, although a figure greater than this is an admitted possibility. Adding this to a present figure of about 170×10^9 barrels for the conterminous 48 states gives 200×10^9 as the approximate amount of crude oil ultimately to be produced in the whole United States.

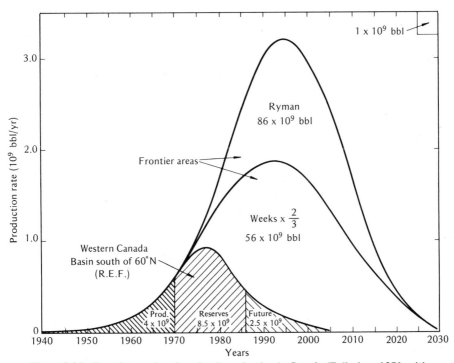

Figure 2-24. Complete cycles of crude oil production in Canada (Folinsbee, 1970, with permission).

Canada's Resources

For this article, it has not been possible to make an analysis of the oil and gas resources of Canada. However, Figure 2-24, from R. E. Folinsbee's Presidential Address before the geological section of the Royal Society of Canada (1970), provides a very good appraisal of the approximate magnitude of Canadian crude-oil resources. According to this estimate, the ultimate production of crude oil from western Canada south of latitude $60°$ will be about 15×10^9 barrels, of which 12.5×10^9 have already been discovered. The peak in the production rate for this area is estimated at about 1977. This figure also shows a maximum estimate of 86×10^9 barrels of additional oil from the frontier areas of Canada. Should this be exploited in a systematic manner from the present time, a peak production rate of about 3×10^9 barrels per year would probably be reached by about 1995.

As of 1973, however, the proved reserves for Canadian crude oil and natural-gas liquids both reached their peaks in 1969; those for natural gas in 1971. Therefore, unless development and transportation of oil and gas from the frontier provinces begins soon, there may be a temporary decline in total Canadian production of oil and gas toward the end of the present decade.

World Crude Oil Production

In this brief review, only a summary statement can be made for the petroleum resources of the world as a whole. Recent estimates by various major oil companies and petroleum geologists have been summarized by H. R. Warman (Warman, 1971) of the British Petroleum Company, who gave 226×10^9 barrels as the cumulative world crude-oil production and 527×10^9 barrels for the proved reserves at the end of 1969. This totals 753×10^9 barrels as the world's proved cumulative discoveries. For the ultimate recoverable crude oil, Warman cited the following estimates published during the period 1967–1970:

Year	Author	Quantity (10^9 barrels)
1967	Ryman (Esso)	2,090
1968	Hendricks (USGS)	2,480
1968	Shell	1,800
1969	Hubbert (NAS-NRC)	1,350–2,100
1969	Weeks	2,200
1970	Moody (Mobil)	1,800

To this, Warman added his own estimate of $1200–2000 \times 10^9$ barrels. A recent unpublished estimate by the research staff of another oil company is in the mid-range of $1900–2000 \times 10^9$ barrels.

From these estimates, there appears to be a convergence toward an estimate of 2000×10^9 barrels, or slightly less. The implication of such a figure to the complete cycle of world crude-oil production is shown in Figure 2-25 (Hubbert, 1969), using two limiting values of 1350×10^9 and 2100×10^9 barrels. For the higher figure, the world will reach the peak in its rate of crude-oil production at about the year 2000; for the lower figure, this date would be about 1990.

Another significant figure for both the United States and the world crude-oil production is the length of time required to produce the middle 80 per cent of the ultimate production. In each case, the time is about 65 years, or less than a human lifetime. For the United States, this subtends the period from about 1937 to 2003, and for the world, from about 1967 to 2032.

Another category of petroleum liquids is that of natural-gas liquids, which are produced as a by-product of natural gas. In the United States (excluding Alaska), the ultimate amount of natural-gas liquids, based on an ultimate amount of crude oil of 170×10^9 barrels, and 1040×10^{12} feet of natural gas, amounts to about 36×10^9 barrels. Corresponding world figures, based on an estimate of 2000×10^9 barrels for crude oil, would be about 400×10^9 barrels for natural-gas liquids, and 12,000 cubic feet for natural gas.

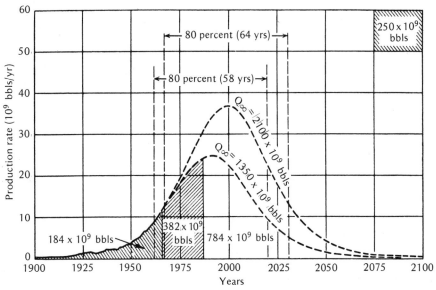

Figure 2-25. Complete cycle of world crude-oil production for two values of Q_∞ (Hubbert, 1969, Figure 8.23).

Other Fossil Fuels

In addition to coal, petroleum liquids and natural gas, the other principal classes of fossil fuels are the so-called tar, or heavy-oil, sands and oil shales. The best known and probably the largest deposits of heavy-oil sands are in the "Athabasca Tar Sands" and two smaller deposits in northern Alberta containing an estimated 300×10^9 barrels of potentially producible oil. One large-scale mining and extracting operation was begun in 1966 by a group of oil companies, and others doubtless will follow as the need for this oil develops.

Unlike tar sands, the fuel content of which is a heavy, viscous crude oil, oil shales contain hydrocarbons in a solid form known as *kerogen*, which distils off as a vapor on heating and condenses to a liquid on cooling. The extractible oil content of oil shales ranges from as high as 100 United States gallons per short ton for the richest grades to near zero as the grades diminish. When all grades are considered, the aggregate oil con-

tent of the known oil shales is very large. However, in practice, only the shales having an oil content of about 25 gallons or more per ton and occurring in beds 10 feet or more thick are considered to be economical sources at present. According to a world inventory of known oil shales by Duncan and Swanson (1965), the largest known deposits are those of the Green River Formation in Wyoming, Colorado and Utah. From these shales, in the grade range from 10 to 65 gallons per ton, the authors estimate that only 80×10^9 barrels are recoverable under 1965 economic conditions. Their corresponding figure for oil shales outside the United States is 110×10^9 barrels.

The absolute magnitude of the world's original supply of fossil fuels recoverable under present technological and economic conditions and their respective energy contents in terms of their heats of combustion are given in Table 2-2. The total initial energy represented by all of these fuels amounted to about 83×10^{21} thermal joules, or 23×10^{15} thermal kilowatt-hours.

Table 2-2. Approximate Magnitudes and Energy Contents of the World's Original Supply of Fossil Fuels Recoverable under Present Conditions.

		Energy Content		
Fuel	*Quantity*	*10^{21} Thermal Joules*	*10^{15} Thermal Kilowatt-Hours*	*Per Cent*
Coal and lignite	2.35×10^{12} metric tons	53.2	14.80	63.78
Petroleum liquids	$2{,}400 \times 10^{9}$ barrels	14.2	3.95	17.03
Natural gas	$12{,}000 \times 10^{12}$ cubic feet	13.1	3.64	15.71
Tar-sand oil	300×10^{9} barrels	1.8	0.50	2.16
Shale oil	190×10^{9} barrels	1.1	0.31	1.32
Totals		83.4	23.20	100.00

Of this, 64 per cent was represented by coal and lignite, 17 and 16 per cent, respectively, by petroleum liquids and natural gas, and 3 per cent by tar-sand and shale oil combined. Although the total amount of coal and lignite in beds 14 or more inches thick and occurring at depths less than 3,000 feet, as estimated by Averitt, are very much larger in terms of energy content, than the initial quantities of oil and gas, the coal practically recoverable under present conditions is only about twice the magnitude of the initial quantities of gas and oil in terms of energy content. Therefore, at comparable rates of production, the time required for the complete cycle of coal production will not be much longer than that for petroleum—in order of a century or two for the exhaustion of the middle 80 per cent of the ultimate cumulative production.

To appreciate the brevity of this period in terms of the longer span of human history, the historical epoch of the exploitation of the fossil fuels is shown graphically in Figure 2-26, plotted on a time scale extending from 5000 years in the past to 5000 years in the future—a period well within the prospective span of human history. On such a time scale, it is seen that the epoch of the fossil fuels can be only a transitory or ephemeral event—an event, nonetheless, which has exercised the most drastic influence on the human species during its entire biological history.

OTHER SOURCES OF INDUSTRIAL ENERGY

The remaining sources of energy suitable for large-scale industrial use are principally the following.

1. Direct use of solar radiation
2. Indirect uses of solar radiation
   (a) Water power
   (b) Wind power
   (c) Photosynthesis
   (d) Thermal energy of ocean water at different temperatures
3. Geothermal power
4. Tidal power
5. Nuclear power
   (a) Fission
   (b) Fusion

Solar Power

By a large margin, the largest flux of energy occurring on the earth is that from solar radiation. The thermal power of the solar radiation intercepted by the earth, according to recent measurements of the solar constant, amounts to about $174{,}000 \times 10^{12}$ thermal watts. This is roughly 5,000 times all other steady fluxes of energy combined. It also has the expectation of continuing at about the same rate for geological periods of time into the future.

The largest concentrations of solar radiation reaching the earth's surface occur in

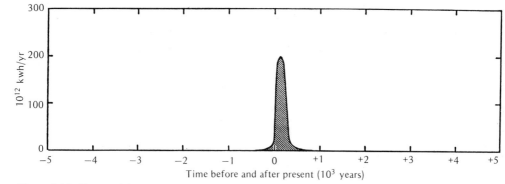

Figure 2-26. Epoch of fossil-fuel exploitation in perspective of human history from 5,000 years in the past to 5000 years in the future (modified from Hubbert, 1962, Figure 54).

desert areas within about 35 degrees of latitude north and south of the equator. Southern Arizona and neighboring areas in the southwestern part of the United States are in this belt, as well as northern Mexico, the Atacama Desert in Chile, and a zone across northern Africa, the Arabian Peninsula, and Iran. In southern Arizona, the thermal power density of the solar radiation incident upon the earth's surface ranges from about 300 to 650 calories per square centimeter per day, from winter to summer. The winter minimum of 300 calories per square centimeter per day, when averaged over 24 hours, represents a mean power density of 145 watts per square meter. If 10 per cent of this could be converted into electrical power by photovoltaic cells or other means, the electrical power obtainable from 1 square kilometer of collection area would be 14.5 megawatts. Then, for an electrical power plant of 1,000 megawatts capacity, the collection area required would be about 70 square kilometers. At such an efficiency of conversion, the collection area required to generate 350,000 megawatts of electrical power—the approximate electric-power capacity of the United States at present—would be roughly 25,000 square kilometers or 9,000 square miles. This is somewhat less than 10 per cent of the area of Arizona.

Such a calculation indicates that large-scale generation of electric power from direct solar radiation is not to be ruled out on the grounds of technical infeasibility. It is also gratifying that a great deal of interest on the part of technically competent groups in universities and research institutions has arisen during the last five years over the possibility of developing large-scale solar power.

Hydroelectric Power

Although there has been continuous use of water power since Roman times, large units were not possible until a means was developed for the generation and transmission of power electrically. The first large hydroelectric power installation was that made at Niagara Falls in 1895. There, ten 5,000-horsepower turbines were installed for the generation of alternating current power, which was transmitted a distance of 26 miles to the city of Buffalo. The subsequent growth of hydroelectric power in the United States is shown in Figure 2-27 and that for the world in Figure 2-28.

In the United States, by 1970, the installed hydroelectric power capacity amounted to 53,000 megawatts, which is 32 per cent of the ultimate potential capacity of 161,000 megawatts as estimated by the Federal Power Commission. The world installation,

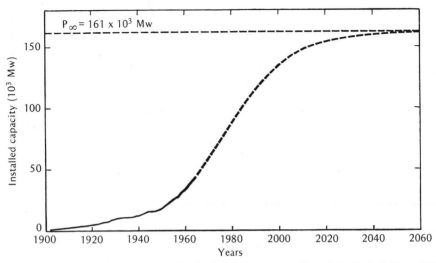

Figure 2-27. Installed and potential hydroelectric-power capacity of the United States (Hubbert, 1969, Figure 8.28).

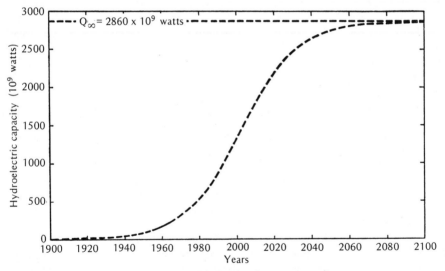

Figure 2-28. Installed and potential world hydroelectric-power capacity.

by 1967, amounted to 243,000 megawatts, which is 8.5 per cent of the world's estimated potential hydroelectric power of 2,860,000 megawatts. Most of this developed capacity is in the highly industrialized areas of North America, Western Europe, and the Far East, especially Japan.

The areas with the largest potential water-power capacities are the industrially underdeveloped regions of Africa, South America, and Southeast Asia, where combined capacities represent 63 per cent of the world total.

The total world potential water power of approximately 3×10^{12} watts, if fully developed, would be of about the same magnitude as the world's present rate of utilization of industrial power. It may also appear that this would be an inexhaustible source of power, or at least one with a time span comparable to that required to remove mountains by stream erosion. This may not be true, however. Most water-power developments require the creation of reservoirs by the damming of streams. The time required to fill these reservoirs with sediments is only two or three centuries. Hence, unless a technical solution of this problem can be found, water power may actually be comparatively short lived.

Tidal Power

Tidal power is essentially hydroelectric power obtainable by damming the entrance to a bay or estuary in a region of tides with large amplitudes, and driving turbines as the tidal basin fills and empties. An inventory of the world's most favorable tidal-power sites gives an estimate of a total potential power capacity of about 63,000 megawatts, which is about 2 per cent of the world's potential water power capacity. At present, one or more small pilot tidal-power plants of a few megawatts capacity have been built, but the only full-scale tidal plant so far built is that on the Rance estuary on the English Channel coast of France. This plant began operation in 1966 with an initial capacity of 240 megawatts and a planned enlargment to 320 megawatts.

One of the world's most favorable tidal-power localities is the Bay of Fundy region of northeastern United States and southeastern Canada. This has the world's maximum tides, with amplitudes up to 15 meters, and a combined power capacity of nine sites of about 29,000 megawatts. Extensive plans have been made by both the United States and Canada for the utilization of this power, but as yet no installations have been made.

Geothermal Power

Geothermal power is obtained by means of heat engines which extract thermal energy from heated water within a depth ranging from a few hundred meters to a few kilometers beneath the earth's surface. This is most practical where water has been heated to high temperatures at shallow depths by hot igneous or volcanic rocks that have risen to near the earth's surface. Steam can be used to drive steam turbines. At present, the major geothermal power installations are in two localities in Italy with a total capacity of about 400 megawatts, the Geysers in California with a planned capacity by 1975 of 500 megawatts, and at Wairakei in New Zealand with a capacity of 160 megawatts. The total world installed geothermal power capacity at present is approximately 1,500 megawatts.

What may be the ultimate capacity can be estimated at present to perhaps only an order of magnitude. Recently, a number of geothermal-power enthusiasts (many with financial interests in the outcome) have made very large estimates for power from this source. However, until better information becomes available, an estimation within the range of 60,000 to 600,000 megawatts, or between 2 and 20 per cent of potential water power, is all that can be justified. Also, as geothermal-power production involves "mining" quantities of stored thermal energy, it is likely that most large installations will also be comparatively short lived—perhaps a century or so.

Nuclear Power

A last major source of industrial power is that of atomic nuclei. Power may be obtained by two contrasting types of nuclear reactions: (a) the fissioning of heavy atomic isotopes, initially uranium-235; and (b) the fusing of the isotopes of hydrogen into heavier helium. In the fission process, two stages are possible. The first consists of power reactors which are dependent almost solely on the rare isotope, uranium-235, which represents only 0.7 per cent of natural uranium. The second process is that of breeding whereby either the common isotope of uranium, uranium-238, or alternatively thorium, is placed in a reactor initially fueled by uranium-235. In response to neutron bombardment, uranium-238 is converted into plutonium-239, or thorium-232 into uranium-233, both of which are fissionable. Hence by means of a breeder reactor, in principle, all of the natural uranium or thorium can be converted into fissionable reactor fuel.

Uranium-235 is sufficiently scarce that, without the breeder reactor, the time span of large-scale nuclear power production would probably be less than a century. With complete breeding, however, it becomes possible not only to consume all of the natural uranium, or thorium, but to utilize low-grade sources as well.

The energy released by the fissioning of a gram of uranium-235 or plutonium-239 or uranium-233 amounts to 8.2×10^{10} joules of heat. This is approximately equivalent to the heat of combustion of 2.7 metric tons of bituminous coal or 13.4 barrels of crude oil. For the energy obtainable from a source of low-grade uranium, consider the Chattanooga Shale, which crops out along the western edge of the Appalachian Mountains in eastern Tennessee and underlies, at minable depths, most of several midwestern states. This shale has a uranium-rich layer about 16 feet or 5 meters thick with a uranium content of 60 grams per metric ton, or 150 grams per cubic meter. This is equivalent to 750 grams per square meter of land area. Assuming only 50 per cent extraction, this would be equivalent in terms of energy content to about 1,000 metric tons of bituminous coal or to 5,000 barrels of crude oil per square meter of land area, or to one billion metric tons of coal or 5 billion barrels of oil per square kilometer. In this region, an area of only 1,600 square kilometers would be required for the energy obtainable from the uranium in the Chattanooga Shale to equal that of all the fossil fuels in the United States. Such an area would be equivalent to that of a square 40 kilometers, or 25 miles, to the side, which would represent less than 2 per cent of the area of Tennessee.

The fusion of hydrogen into helium is known to be the source of the enormous amount of energy radiating from the sun.

Fusion has also been achieved by man in an uncontrolled or explosive manner in the thermonuclear or hydrogen bomb. As yet, despite intensive efforts in several countries, controlled fusion has not been achieved. Researchers, however, are hopeful that it may be within the next few decades.

Should fusion be achieved, eventually the principal raw material will probably be the heavy isotope of hydrogen, deuterium. This occurs in sea water at an abundance of 1 deuterium atom to each 6,700 atoms of hydrogen. The deuterium-deuterium, or D-D, reaction involves several stages, the net result of which is

$$5 \, {}_1^2\mathrm{D} \rightarrow {}_2^4\mathrm{He} + {}_2^3\mathrm{He} + \mathrm{H} + 2\mathrm{n} + 24.8 \text{ Mev}$$

or, in other words, 5 atoms of deuterium, on fusion, produce 1 atom of helium-4, 1 atom of helium-3, 1 atom of hydrogen, and 2 neutrons, and in addition release 24.8 million electron volts, or 39.8×10^{-13} joules.

It can be computed that 1 liter of water contains 1.0×10^{22} deuterium atoms, which upon fusion would release 7.95×10^9 joules of thermal energy. This is equivalent to the heat of combustion of 0.26 metric tons of coal or 1.30 barrels of crude oil. Then, as 1 cubic kilometer of sea water is equivalent to 10^{12} liters, the heat released by the fusion of the deuterium contained in 1 cubic kilometer of sea water would be equivalent to that of the combustion of 1300 billion barrels of oil or 260 billion tons of coal. The deuterium in 33 cubic kilometers of sea water would be equivalent to that of the world's initial supply of fossil fuels.

ECOLOGICAL ASPECTS OF
EXPONENTIAL GROWTH

From the foregoing review, what stands out most clearly is that our present industrialized civilization has arisen principally during the last two centuries. It has been accomplished by the exponential growth of most of its major components at rates commonly in the range of 4 to 8 per cent per year, with periods of doubling from 8 to 16 years. The question now arises: What are the limits to such growth, and what does this imply concerning our future?

What we are dealing with, essentially, are the principles of ecology. It has long been known by ecologists that the population of any biologic species, if given a favorable environment, will increase exponentially with time; that is, that the population will double repeatedly at roughly equal intervals of time. From our previous observations, we have seen that this is also true of industrial components. For example (Figure 2-29), the world electric power capacity is now growing at 8 per cent per year and doubling every 8.7 years. The world automobile population and the miles flown per year by the world's civil aviation scheduled flights are each doubling every 10 years. Also, the human population is now doubling in 35 years (Figure 2-30).

The second part of this ecological principle is that such exponential growth of any biologic population can only be maintained

for a limited number of doublings before retarding influences set in. In the biological case, these may be represented by restriction of food supply, by crowding or by environmental pollution. The complete biologic growth curve is represented by the logistic curve of Figure 2-31.

That there must be limits to growth can easily be seen by the most elementary arithmetic analysis. Consider the familiar checkerboard problem of placing 1 grain of wheat on the first square, 2 on the second, 4 on the third, and doubling the number for each successive square. The number of grains on the nth square will be 2^{n-1}, and on the last or 64th square, 2^{63}. The sum of the grains on the entire board will be twice this amount less one grain, or $2^{64} - 1$. When translated into volume of wheat, it turns out that the quantity of wheat required for the last square would equal approximately 1,000 times the present world annual wheat crop, and the requirement for the whole board would be twice this amount.

It follows, therefore, that exponential

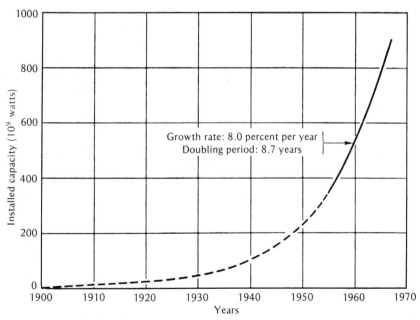

Figure 2-29. World electric generating capacity as an example of exponential growth (Hubbert, 1971, Figure 2).

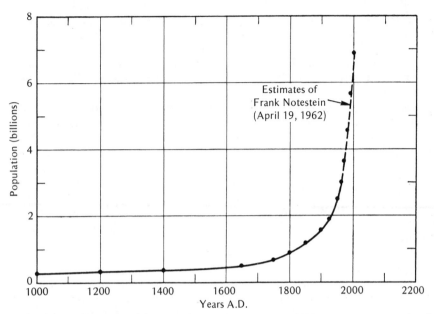

Figure 2-30. Growth of human population since the year 1000 A.D. as an example of an ecological disturbance (Hubbert, 1962, Figure 2).

growth, either biological or industrially, can be only a temporary phenomenon because the earth itself cannot tolerate more than a few tens of doublings of any biological or industrial component. Furthermore, most of the possible doublings have occurred already.

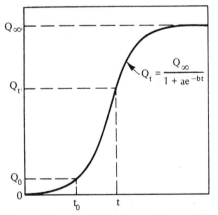

Figure 2-31. The logistic growth curve showing both the initial exponential phase and the final slowing down during a cycle of growth.

After the cessation of exponential growth, any individual component has only three possible futures:

(1) it may, as in the case of water power, level off and stabilize at a maximum;

(2) it may overshoot and, after passing a maximum, decline and stabilize at some intermediate level capable of being sustained; or

(3) it may decline to zero and become extinct.

Applied to human society, these three possibilities are illustrated graphically in Figure 2-32. What stands out most clearly is that our present phase of exponential growth based on man's ability to control ever larger quantities of energy can only be a temporary period of about three centuries' duration in the totality of human history. It represents but a brief transitional epoch between two very much longer periods, each characterized by rates of change so slow as to be regarded essentially as a period of nongrowth. Although the forthcoming period poses no insuperable physical or bio-

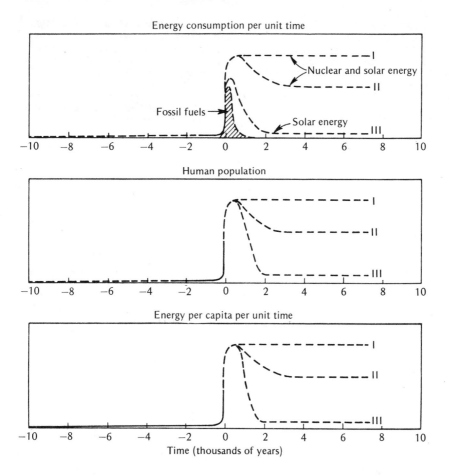

Figure 2-32. Epoch of current industrial growth in the context of a longer span of human history (Hubbert, 1962, Figure 61).

logical difficulties, it can hardly fail to force a major revision in those aspects of our current culture, the tenets of which are dependent on the assumption that the growth rates which have characterized this temporary period can somehow be sustained indefinitely.

REFERENCES

Averitt, Paul, 1969, Coal resources of the United States, January 1, 1967: *U.S. Geological Survey Bulletin 1275,* 116 p.

Duncan, D. C., and Swanson, V. E., 1965, Organic-rich shale of the United States and world land areas: *U.S. Geological Survey Circular 523,* 30 p.

Folinsbee, R. E., 1970, Nuclear energy and the fossil fuels: *Trans. Royal Society of Canada, Fourth Series,* Vol. 8, pp. 335–359.

Hewett, D. F., 1929, Cycles in metal production: *AIME Tech. Pub. 183,* 31 p.; Trans. 1929, pp. 65–93; discussion pp. 93–98.

Hubbert, M. King, 1956, Nuclear energy and the fossil fuels: *American Petroleum Institute, Drilling and Production Practice* (1956), pp. 7–25.

Hubbert, M. King, 1962, Energy resources: *National Academy of Sciences–National Research Council, Publication 1000-D,* 141 p.

Hubbert, M. King, 1967, Degree of advancement of petroleum exploration in United States: *American Assoc. of Petroleum Geologists*

Bulletin, Vol. 51, pp. 2207–2227.

Hubbert, M. King, 1969, Energy resources, *Resources and Man,* a study and recommendations by the Committee on Resources and Man of the Division of Earth Sciences, National Academy of Sciences. National Research Council: San Francisco, W. H. Freeman, pp. 157–242.

Hubbert, M. King, 1971, Energy Resources for power production, *Environmental Aspects of Nuclear Power Stations:* Vienna, International Atomic Energy Agency, pp. 13–43.

Warman, H. R., 1971, Future problems in petroleum exploration: *Petroleum Review,* Vol. 25, No. 291, pp. 96–101.

Zapp, A. D., 1962, Future petroleum producing capacity of the United States: *U.S. Geological Survey Bulletin 1142-H,* 36 p.

Scientific Aspects of the Oil Spill Problem

MAX BLUMER   1971

THE EXTENT OF MARINE OIL POLLUTION

Oil pollution is the almost inevitable consequence of our dependence on an oil-based technology. The use of a natural resource without losses is nearly impossible and environmental pollution occurs through intentional disposal or through inadvertent losses in production, transportation, refining and use. How large is the oil influx to the ocean? The washing of cargo tanks at sea, according to the director of Shell International, Marine Ltd.,[1] had the potential in 1967 of introducing 2.8 million tons into the ocean, assuming that no use was made of the load-on-top technique. With the increase in ocean oil transport from 1967 to 1970 this potential has grown to 6 million tons. The load-on-top technique is not being applied to one quarter of the oil tonnage moved by tankers; consequently, these vessels introduce about 1.5 million tons of oil into the sea. The limitations of the load-on-top technique have been described by E. S. Dillon[2]: the technique is not always used even if the equipment exists, the equipment may be inadequate, shore receiving facilities may be lacking, and principal limitations lie in the formation of emulsions in heavy seas or with heavy crude oils. Insufficient time may be available for the separation of the emulsion or the oil water interface may not be readily recognized. In addition, the most toxic components of oil are also readily soluble in water, and their disposal into the ocean could be avoided only if clean ballasting were substituted for the load-on-top technique. For these reasons it is estimated that the present practices in tanker ballasting introduce about 3 million tons of petroleum into the

1. Statement by J. H. Kirby, quoted by J. R. Wiggins, *Washington Post*, March 15, 1970.

2. Dillon, E. Scott, "Ship Construction and Operation Standards for Oil Pollution Abatement," presented to a Conference on Ocean Oil Spills, held by the NATO Committee on Challenges of Modern Society, Brussels, November 2–6, 1970.

Dr. Max Blumer is an organic geochemist and senior scientist with Woods Hole Oceanographic Institution, Massachusetts. His research interests are in organic compounds in nature, their origin, fate, and interaction with the environment.

From *Environmental Affairs*, Vol. 1, No. 1, pp. 54–73, April 1971. Reprinted with permission of the author and the Environmental Law Center of the Boston College Law School. This paper was presented to a Conference on Ocean Oil Spills, held by the NATO Committee on Challenges of Modern Society, Brussels, November 2–6, 1970. Contribution No. 2616 of the Woods Hole Oceanographic Institution.

ocean. The pumping of bilges by vessels other than tankers contributes another 500,000 tons.[3] In addition, in-port losses from collisions and during loading and unloading contribute an estimated 1 million tons.[4]

Oil enters the ocean from many other sources whose magnitude is much less readily assessed. Among these are accidents on the high seas (*Torrey Canyon*) or near shore, outside of harbors (West Falmouth, Mass.), losses during exploration (oil based drilling mud) and production (Santa Barbara, Gulf of Mexico), in storage (submarine storage tanks), and in pipeline breaks, and spent marine lubricants and incompletely burned fuels. A major contribution may come from untreated domestic and industrial wastes; it is estimated that nearly 2 million tons of used lubricating oil is unaccounted for each year in the United States alone, and a significant portion of this reaches our coastal waters.[5,6]

Thus, the total annual oil influx to the ocean lies probably between 5 and 10 million tons. A more accurate assessment of the oil pollution of the oceans and of the relative contribution of different oils to the different marine environments is urgently needed. Such an assessment might well lie within the role of the NATO Committee on Challenges of the Modern Society.

With the anticipated increase in foreign and domestic oil production, with increased oil transport and with the shift of production to more hazardous regions (Alaska, continental shelf, deep ocean), we can expect a rapid increase of the spillage rate and of the oil influx to the ocean. Floating masses of crude oil ("tar") are now commonly encoun-

tered on the oceans and crude oil is present on most beaches. Oil occurs in the stomach of surface feeding fishes[7] and finely dispersed hydrocarbons occur in marine plants (e.g., sargassum[8]) and in the fat of fish and shellfish.[6,9] Hydrocarbons from a relatively small and restricted oil spill in the coastal waters of Massachusetts, U.S.A., have spread, nine months after the accident to an area occupying 5000 acres (20 square kilometers) offshore and 500 acres (2 square kilometers) in tidal rivers and marshes. The effect on the natural populations in this area has been catastrophic. The full extent of the coverage of the ocean bottom by petroleum hydrocarbons is unknown; chemical analyses are scarce or nonexistent.

EVALUATION OF THE THREAT

Oil: Immediate Toxicity

All crude oils and all oil fractions except highly purified and pure materials are poisonous to all marine organisms. This is not a new finding. The wreck of the *Tampico* in Baja, California, Mexico (1957) "created a situation where a completely natural area was almost totally destroyed suddenly on a large scale. . . . Among the dead species were lobsters, abalone, sea urchins, starfish, mussels, clams and hosts of smaller forms."[10] Similarly, the spill of fuel oil in West Falmouth, Massachusetts, U.S.A., has virtually extinguished life in a productive coastal and intertidal area, with a complete kill extending over all phyla represented in that habitat (Hampson and Sanders[11] and

3. Statement by C. Cortelyou, Mobil Oil Company, quoted by W. D. Smith, *New York Times*, April 19, 1970.
4. Blumer, M., "Oil Pollution of the Ocean," in *Oil on the Sea*, D. P. Hoult, ed., Plenum Press, 1969.
5. Anon., "Final Report of the Task Force on Used Oil Disposal," American Petroleum Institute, New York, N.Y., 1970.
6. Murphy, T. A., "Environmental Effects of Oil Pollution," Paper presented to the Session on Oil Pollution Control, American Society of Civil Engineers, Boston, Mass., July 13, 1970.

7. Horn, M. H., Teal, J. H., and Backus, R. H., "Petroleum Lumps on the Surface of the Sea," *Science* 168, 245, 1970.
8. Youngblood, W. W., and Blumer, M., unpublished data, 1970.
9. Blumer, M., Souza, G., and Sass, J., "Hydrocarbon Pollution of Edible Shellfish by an Oil Spill," *Marine Biology* 5, 195–202, 1970; Blumer, M., Testimony before the Conservation and Natural Resources Subcommittee, Washington, D.C., July 22, 1970.
10. North, W. J., "Tampico, a Study of Destruction and Restoration," *Sea Frontiers* 13, 212–217, 1967.
11. Hampson, G. R., and Sanders, H. L., "Local Oil Spill," *Oceanus* 15, 8–10, 1969.

unpublished data). Toxicity is immediate and leads to death within minutes or hours.[12]

Principally responsible for this immediate toxicity are three complex fractions. The *low-boiling saturated hydrocarbons* have, until quite recently, been considered harmless to the marine environment. It has now been found that this fraction, which is rather readily soluble in sea water, produces at low concentration anaesthesia and narcosis and at greater concentration cell damage and death in a wide variety of lower animals; it may be especially damaging to the young forms of marine life.[13] The *low-boiling aromatic hydrocarbons* are the most immediately toxic fraction. Benzene, toluene, and xylene are acute poisons for man as well as for other organisms; naphthalene and phenanthrene are even more toxic to fishes than benzene, toluene, and xylene.[14] These hydrocarbons and substituted one-, two-, and three-ring hydrocarbons of similar toxicity are abundant in all oils and most, especially the lower-boiling oil products. Low-boiling aromatics are even more water soluble than the saturates and can kill marine organisms either by direct contact or through contact with dilute solutions. *Olefinic hydrocarbons*, intermediate in structure and properties, and probably in toxicity, between saturated and aromatic hydrocarbons are absent in crude oil but occur in refining products (e.g., gasoline and cracked products) and are in part responsible for their immediate toxicity.

Numerous other components of crude oils are toxic. Among those named by Speers and Whitehead,[15] cresols, xylenols, naph-thols, quinoline, and substituted quinolines, pyridines, and hydroxybenzoquinolines are of special concern here because of their great toxicity and their solubility in water. It is unfortunate that statements which disclaim this established toxicity are still being circulated. Simpson[16] claimed that "there is no evidence that oil spilt round the British Isles has ever killed any of these (mussels, cockles, winkles, oysters, shrimps, lobsters, crabs) shellfish." It was obvious when this statement was made that such animals were indeed killed by the accident of the *Torrey Canyon* as well as by earlier accidents; work since then has confirmed the earlier investigation. In addition, this statement, by its emphasis only on the adult forms, implies wrongly that juvenile forms were also unaffected.

Oil and Cancer

The higher boiling crude oil fractions are rich in multiring aromatic compounds. It was at one time thought that only a few of these compounds, mainly 3,4-benzopyrene, were capable of inducing cancer. As R. A. Dean[17] of British Petroleum Company stated, "no 3,4-benzopyrene has been detected in any crude oil .... [I]t therefore seems that the risk to the health of a member of the public by spillage of oil at sea is probably far less than that which he normally encounters by eating the foods he enjoys." However, at the time this statement was made, carcinogenic fractions containing 1,2-benzanthracene and alkylbenzanthracenes had already been isolated by Carruthers, Stewart, and Watkins[18] and it was known that "biological tests have shown

12. Sanders, H. L., Testimony before the Conservation and Natural Resources Subcommittee, Washington, D.C., July 22, 1970.
13. Goldacre, R. J., "The Effects of Detergents and Oils on the Cell Membrane," Suppl. to Vol. 2 of *Field Studies*, Field Studies Council London, 131–137, 1968.
14. Wilber, C. G., *The Biological Aspects of Water Pollution*, Charles C Thomas, Publisher, Springfield, Ill., 1969.
15. Speers, G. C. and Whitehead, E. V., "Crude Petroleum," in *Organic Geochemistry*, Eglinton, G. and Murphy, M. R. J., eds., Springer, Berlin, 638–675, 1969.

16. Simpson, A. C., "Oil, Emulsifiers and Commercial Shell Fish," Suppl. to Vol. 2 of *Field Studies*, Field Studies Council, London, 91–98, 1968.
17. Dean, R. A., "The Chemistry of Crude Oils in Relation to their Spillage on the Sea," Suppl. to Vol. 2 of *Field Studies*, Field Studies Council, London, 1–6, 1968.
18. Carruthers, W., Stewart, H. N. M., and Watkins, D. A. M., "1,2-Benzanthracene Derivatives in a Kuwait Mineral Oil," *Nature* 213, 691–692, 1967.

that the extracts obtained from high-boiling fractions of the Kuwait oil . . . (method) . . . are carcinogenic." Further, "Benzanthracene derivatives, however, are evidently not the only type of carcinogen in the oil. . . ." In 1968, the year when Dean claimed the absence of the powerful carcinogen 3,4-benzopyrene in crude oil, this hydrocarbon was isolated in crude oil from Libya, Venezuela, and the Persian Gulf.[19] The amounts measured were between 450 and 1,800 milligrams per ton of the crude oil.

Thus, we know that chemicals responsible for cancer in animals and man occur in petroleum. The causation of cancer in man by crude oil and oil products was observed some years ago, when a high incidence of skin cancer in some refinery personnel was observed. The cause was traced to prolonged skin contact by these persons with petroleum and with refinery products. Better plant design and education, aimed at preventing the contact, have since reduced or eliminated this hazard.[20] However, these incidents have demonstrated that oil and oil products can cause cancer in man, and have supported the conclusions based on the finding of known carcinogens in oil. These references and a general knowledge of the composition of crude oils suggest that all crude oils and all oil products containing high-boiling aromatic hydrocarbons should be viewed as potential cancer inducers.

Safeguards in plant operations protect the public from this hazard. However, when oil is spilled into the environment, we loose control over it and should again be concerned about the possible public health hazard from cancer-causing chemicals in the oil. We have shown that marine organisms ingest and retain hydrocarbons to which they are exposed. These are transferred to and retained by predators. In this way even animals that were not directly exposed to a spill can become polluted by eating contaminated chemicals. This has severe implications for commercial fisheries and for human health. It suggests that marketing and eating of oil-contaminated fish and shellfish at the very least increases the body burden of carcinogenic chemicals and may constitute a public health hazard.

Other questions suggest themselves: Floating masses of crude oil now cover all oceans and are being washed up on shores. It has been thought that such stranded lumps are of little consequence ecologically. It has been shown that such lumps, even after considerable weathering, still contain nearly the full range of hydrocarbons of the original crude oil, extending in boiling point to as low as $100°C$. Thus, such lumps still contain some of the immediately toxic lower-boiling hydrocarbons. In addition, the oil lumps contain all of the potentially carcinogenic material in the $300-500°C$ boiling fraction. The presence of oil lumps ("tar") or finely dispersed oil on recreational beaches may well constitute a severe public health hazard, through continued skin contact.

Low-Level Effects of Oil Pollution

The short-term toxicity of crude oil and of oil products and their carcinogenic properties are fairly well understood. In contrast to this we are rather ignorant about the long-term and low-level effects of oil pollution. These may well be far more serious and long lasting than the more obvious short-term effects. Let us look at low-level interference of oil pollution with the marine ecology.

Many biological processes which are important for the survival of marine organisms and which occupy key positions in their life processes are mediated by extremely low concentration of chemical messengers in the sea water. We have demonstrated that marine predators are attracted to their prey by organic compounds at concentrations below the part per billion level.[21] Such chemical

19. Graef, W., and Winter, C., "3,4 Benzopyrene in Erdoel," *Arch. Hyg.* 152/4, 289–293, 1968.
20. Eckardt, R. E., "Cancer Prevention in the Petroleum Industry," *Int. J. Cancer* 3, 656–661, 1967.

21. Whittle, K. J. and Blumer, M., "Chemotaxis in Starfish, Symposium on Organic Matter in Natural Waters," University of Alaska, Fairbanks, Alaska, 1970.

attraction—and in a similar way repulsion—plays a role in the finding of food, the escape from predators, in homing of many commercially important species of fishes, in the selection of habitats, and in sex attraction. There is good reason to believe that pollution interferes with these processes in two ways, by blocking the taste receptors and by mimicking for natural stimuli. The latter leads to false response. Those crude oil fractions likely to interfere with such processes are the high-boiling saturated and aromatic hydrocarbons and the full range of the olefinic hydrocarbons. It is obvious that a very simple—and seemingly innocuous—interference at extremely low concentration levels may have a disastrous effect on the survival of any marine species and on many other species to which it is tied by the marine food chain.

Research in this critical area is urgently needed. The experience with DDT has shown that low-level effects are unpredictable and may suddenly become an ecological threat of unanticipated magnitude.

The Persistence of Oil in the Environment

Hydrocarbons are among the most persistent organic chemicals in the marine environment. It has been demonstrated that hydrocarbons are transferred from prey to predator and that they may be retained in organisms for long time periods, if not for life. Thus, a coastal spill near Cape Cod, Massachusetts, U.S.A., has led to the pollution of shellfish by fuel oil. Transplanting of the shellfish to clean water does not remove the hydrocarbons from the tissues. Oil may contaminate organisms not only at the time of the spill; hydrocarbon-loaded sediments continue to be a source of pollution for many months after the accident.

Oil, though lighter than water, does not remain at the sea surface alone; storms, or the uptake by organisms or minerals, sink the oil. Oil at the sea bottom has been found after the accidents of the *Torrey Canyon*, at Santa Barbara, and near Cape Cod. Clay minerals with absorbed organic matter are an excellent adsorbent for hydrocarbons; they

retain oil and may transport it to areas distant from the primary spill. Thus, ten months after the accident at Cape Cod, the pollution of the bottom sediments covers an area that is much larger than that immediately after the spill. In sediments, especially if they are anaerobic, oil is stable for long time periods. Indeed, it is a key fact of organic geochemistry that hydrocarbons in anaerobic recent sediments survive for millions of years until they eventually contribute to the formation of petroleum.

COUNTERMEASURES

Compared to the number and size of accidents and disasters the present countermeasures are inadequate. Thus, in spite of considerable improvement in skimming efficiency since the Santa Barbara accident, only 10 per cent of the oil spilled from the Chevron well in the Gulf of Mexico was recovered.[22] From an ecological point of view this gain is nearly meaningless. While we may remain hopeful that the gross esthetic damage from oil spills may be avoided in the future, there is no reason to be hopeful that existing or planned countermeasures will eliminate the biological impact of oil pollution.

The most immediately toxic fractions of oil and oil products are soluble in sea water; therefore, biological damage will occur at the very moment of the accident. Water currents will immediately spread the toxic plume of dissolved oil components and, if the accident occurs in inshore waters, the whole water column will be poisoned even if the bulk of the oil floats on the surface. The speed with which the oil dissolves is increased by agitation, and in storms the oil will partly emulsify and will then present a much larger surface area to the water; consequently, the toxic fractions dissolve more rapidly and reach higher concentrations. From the point of view of avoiding the immediate biological effect of oil spills,

22. Wayland, R. G., Federal Regulations and Pollution Controls on the U.S. Offshore Oil Industry, this conference.

countermeasures are completely effective only if *all of the oil is recovered immediately* after the spill. *The technology to achieve this goal does not exist.*

Oil spills damage many coastal and marine values: water fowl, fisheries, and recreational resources; they lead to increased erosion; they diminish the water quality and may threaten human life or property through fire hazard. A judicious choice has to be made in each case: which—if any—of the existing but imperfect countermeasures to apply to minimize the overall damage or the damage to the most valuable resources. Guidelines for the use of countermeasures, especially of chemical countermeasures, exist[23] and are being improved.[24] Some comments on the ecological effects and desirability of the existing countermeasures appear appropriate.

Detergents and Dispersants

The toxic, solvent-based detergents which did so much damage in the cleanup after the *Torrey Canyon* accident are presently only in limited use. However, so-called "nontoxic dispersants" have been developed. The term "nontoxic" is misleading; these chemicals may be nontoxic to a limited number of often quite resistant test organisms but they are rarely tested in their effects upon a wide spectrum of marine organisms including their juvenile forms, preferably in their normal habitat. Further, in actual use all dispersant-oil mixtures are severely toxic, because of the inherent toxicity of the oil, and bacterial degradation of "nontoxic" detergents may lead to toxic breakdown products.

The effect of a dispersant is to lower the surface tension of the oil to a point where it will disperse in the form of small droplets. It is recommended that the breakup of the oil

slick be aided by agitation, natural or mechanical. Thus, the purpose of the detergent is essentially a cosmetic one. However, the recommendation to apply dispersants is often made in disregard of their ecological effects. Instead of removing the oil, dispersants push the oil actively into the marine environment; because of the finer degree of dispersion, the immediately toxic fraction dissolves rapidly and reaches a higher concentration in the sea water than it would if natural dispersal were allowed. The long-term poisons (for example, the carcinogens) are made available to and are ingested by marine filter feeders, and they can eventually return to man incorporated into the food he recovers from the ocean.

For these reasons I feel that the use of dispersants is unacceptable, inshore or offshore, except under special circumstances, for example, extreme fire hazard from spillage of gasoline, as outlined in the Contingency Plan for Oil Spills, Federal Water Quality Administration, 1969.[23,24]

Physical Sinking

Sinking has been recommended. "The long term effects on marine life will not be as disastrous as previously envisaged. Sinking of oil may result in the mobile bottom dwellers moving to new locations for several years; however, conditions may return to normal as the oil decays."[25] Again, these conclusions disregard our present knowledge of the effect of oil spills.

Sunken oil will kill the bottom faunas rapidly, before most mobile dwellers have time to move away. The sessile forms of commercial importance (oysters, scallops, etc.) will be killed and other mobile organisms (lobsters) may be attracted into the direction of the spill where the exposure will contaminate or kill them. The persistent fraction of the oil which is not readily attacked by bacteria contains the long term

23. Contingency Plan for Spills of Oil and Other Hazardous Materials in New England, U.S. Dept. Interior, Federal Water Quality Administration, Draft, 1969.
24. Schedule of Dispersants and Other Chemicals to Treat Oil Spills, May 15, 1970, Interim Schedule, Federal Water Quality Administration, 1970.

25. Little, A. D., Inc., "Combating Pollution Created by Oil Spills," Report to the Dept. of Transportation, U.S. Coast Guard, Vol. 1: Methods, p. 71386 (R), June 30, 1969.

poisons, for example, the carcinogens, and they will remain on the sea bottom for very long periods of time. Exposure to these compounds may damage organisms or render them unfit for human nutrition even after the area has been repopulated.

The bacterial degradation of sunken oil requires much oxygen. As a result, sediments loaded with oil become anaerobic and bacterial degradation and reworking of the sediments by aerobic benthic organisms is arrested. It is one of the key principles of organic geochemistry that hydrocarbons in anaerobic sediments persist for millions of years. Similarly, sunken oil will remain; it will slow down the resettlement of the polluted area; and it may constitute a source for the pollution of the water column and of fisheries resources for a long time after the original accident.

For these reasons I believe that sinking of oil is unacceptable in the productive coastal and offshore regions. Before we apply this technique to the deep ocean with its limited oxygen supply and its fragile faunas, we should gather more information about the interplay of the deep marine life with the commercial species of shallower waters.

Combustion

Burning the oil through the addition of wicks or oxidants appears more attractive from the point of view of avoiding biological damage than dispersion and sinking. However, it will be effective only if burning can start immediately after a spill. For complete combusion, the entire spill must be covered by the combustion promoters, since burning will not extend to the untreated areas; in practice, in stormy conditions, this may be impossible to achieve.

Mechanical Containment and Removal

Containment and removal appear ideal from the point of avoiding biological damage. However, they can be effective only if applied immediately after the accident. Under severe weather conditions floating booms and barriers are ineffective. Booms were ap-

plied during the West Falmouth oil spill; however, the biological damage in the sealed-off harbors was severe and was caused probably by the oil which bypassed the booms in solution in sea water and in the form of wind-dispersed droplets.

Bacterial Degradation

Hydrocarbons in the sea are naturally degraded by marine microorganisms. Many hope to make this the basis of an oil removal technology through bacterial seeding and fertilization of oil slicks. However, great obstacles and many unknowns stand in the way of the application of this attractive idea.

No single microbial species will degrade any whole crude oil; bacteria are highly selective, and complete degradation requires many different bacterial species. Bacterial oxidation of hydrocarbons produces many intermediates which may be more toxic than the hydrocarbons; therefore, organisms are also required that will further attack the hydrocarbon decomposition products.

Hydrocarbons and other compounds in crude oil may be bacteriostatic or bacteriocidal; this may reduce the rate of degradation, where it is most urgently needed. The fraction of crude oil that is most readily attacked by bacteria is the least toxic one, the normal paraffins; the toxic aromatic hydrocarbons, especially the carcinogenic polynuclear aromatics, are not rapidly attacked.

The oxygen requirement in bacterial oil degradation is severe; the complete oxidation of 1 gallon of crude oil requires all the dissolved oxygen in 320,000 gallons of air saturated sea water. Therefore, oxidation may be slow in areas where the oxygen content has been lowered by previous pollution and the bacterial degradation may cause additional ecological damage through oxygen depletion.

Cost Effectiveness

The high value of fisheries resources, which exceeds that of the oil recovery from the sea, and the importance of marine proteins for human nutrition demand that cost effectiveness analysis of oil spill countermeasures consider the cost of direct and indirect ecological damage. It is disappointing that existing studies completely neglect to consider these real values.[17] A similarly one-sided approach would be, for instance, a demand by fisheries concerns that all marine oil production and shipping be terminated, since it clearly interferes with fisheries interests.

We must start to realize that we are paying for the damage to the environment, especially if the damage is as tangible as that of oil pollution to fisheries resources and to recreation. Experience has shown that cleaning up a polluted aquatic environment is much more expensive than it would have been to keep the environment clean from the beginning.[26] In terms of minimizing the environmental damage, spill prevention will produce far greater returns than cleanup—and we believe that this relationship will hold in a *realistic* analysis of the over-all cost effectiveness of prevention or cleanup costs.

THE RISK OF MARINE OIL POLLUTION

The Risk to Marine Life

Our knowledge of crude-oil composition and of the effects of petroleum on marine organisms in the laboratory and in the marine environment force the conclusion that petroleum and petroleum products are toxic to most or all marine organisms. Petroleum hydrocarbons are persistent poisons. They enter the marine food chain, they are stabilized in the lipids of marine organisms, and they are transferred from prey to predator. The persistence is especially severe for the most poisonous compounds of oil; most of these do not normally occur in organisms and natural pathways for their biodegradation are missing.

Pollution with crude oil and oil fractions *damages the marine ecology* through different effects:

(1) Direct kill of organisms through coating and asphyxiation.[27]

(2) Direct kill through contact poisoning of organisms.

(3) Direct kill through exposure to the water soluble toxic components of oil at some distance in space and time from the accident.

(4) Destruction of the generally more sensitive juvenile forms of organisms.

(5) Destruction of the food sources of higher species.

(6) Incorporation of sublethal amounts of oil and oil products into organisms resulting in reduced resistance to infection and other stresses (the principal cause of death in birds surviving the immediate exposure to oil[28]).

(7) Incorporation of carcinogenic and potentially mutagenic chemicals into marine organisms.

(8) Low level effects that may interrupt any of the numerous events necessary for the propagation of marine species and for the survival of those species which stand higher in the marine food web.

The degree of toxicity of oil to marine organisms and the mode of action are fairly well understood. On the other hand, we are still far from understanding the effect of the existing and increasing oil pollution on the marine ecology on a large, especially worldwide, scale.

Few, if any, comprehensive studies of the effects of oil spills on the marine ecology have been undertaken. Petroleum and petroleum products are toxic *chemicals;* the longterm biological effect of oil and its persistence cannot be studied without chemical analyses. Unfortunately, chemical analysis

26. Ketchum, B. H., *Biological Effects of Pollution of Estuaries and Coastal Waters*, Boston Univ. Press, 1970.

27. Arthur, D. R., "The Biological Problems of Littoral Pollution by Oil and Emulsifiers—a Summing up," Suppl. to Vol. 2 of *Field Studies*, Field Studies Council, London, 159—164, 1968.
28. Beer, J. V., "Post-Mortem Findings in Oiled Auks during Attempted Rehabilitation," Suppl. to Vol. 2 of *Field Studies*, Field Studies Council, London, 123—129, 1968.

has not been used to support such studies in the past and conclusions on the persistence of oil in the environment have been arrived at solely by visual inspection. This is not sufficient; a sediment can be uninhabitable to marine bottom organisms because of the presence of finely divided oil, but the oil may not be visually evident. Marine foods may be polluted by petroleum and may be hazardous to man but neither taste nor visual observation may disclose the presence of the toxic hydrocarbons.

A coordinated biological and chemical study of the long-term effect and fate of a coastal oil spill in West Falmouth, Massachusetts, U.S.A., has shown that even a relatively low-boiling, soluble and volatile oil persists and damages the ecology for many months after the spill. In this instance about 650 tons of No. 2 fuel oil were accidentally discharged into the coastal waters off the Massachusetts coast. I wish to summarize our present findings of the effect of this accident.

Persistence and Spread of the Pollution[9,29]

Oil from the accident has been incorporated into the sediments of the tidal rivers and marshes and into the offshore sediments, down to 42 feet, the greatest water depth in the sea. The fuel oil is still present in inshore and offshore sediments, eight months after the accident. The pollution has been spreading on the sea bottom and now covers at least 5,000 acres offshore and 500 acres of marshes and tidal rivers. This is a much larger area than that affected immediately

29. This and the next two sections of the paper were written nine months after the West Falmouth oil spill. The following reports, giving the status after two years, are now available:
  (a) The Persistence and Degradation of Spilled Fuel Oil. *Science* (1972) 176, 1120–1122.
  (b) The West Falmouth Oil Spill. I. Biology. Howard L. Sanders, J. Frederick Grassle, and George R. Hampson. *WHOI 72-20.*
  (c) The West Falmouth Oil Spill. II. Chemistry. Data Available in November, 1971. M. Blumer and J. Sass. *WHOI 72-19.*
These reports are available from the National Technical Information Service, Springfield, Va. 22151.

after the accident. Bacterial degradation of the oil is slow; degradation is still negligible in the most heavily polluted areas, and the more rapid degradation in outlying, less affected, areas has been reversed by the influx of less degraded oil from the more polluted regions. The kill of bottom plants and animals has reduced the stability of marshland and sea bottom; increased erosion results and may be responsible for the spread of the pollution along the sea bottom.

Bacterial degradation first attacks the least toxic hydrocarbons. The hydrocarbons remaining in the sediments are now more toxic on an equal weight basis than immediately after the spill. Oil has penetrated the marshes to a depth of at least 1 to 2 feet; bacterial degradation within the marsh sediment is still negligible eight months after the accident.

Biological Effects of the Pollution[11,12]

Where oil can be detected in the sediments there has been a kill of animals; in the most polluted areas the kill has been almost total. Control stations outside the area contain normal, healthy bottom faunas. The kill associated with the presence of oil is detected down to the maximum water depth in the area. A massive, immediate kill occurred offshore during the first few days after the accident. Affected were a wide range of fish, shellfish, worms, crabs, and other crustaceans and invertebrates. Bottom-living fishes and lobsters were killed and washed up on the beaches. Trawls in 10 feet of water showed 95 per cent of the animals dead and many still dying. The bottom sediments contained many dead clams, crustaceans, and snails. Fish, crabs, shellfish, and invertebrates were killed in the tidal Wild Harbor River; and in the most heavily polluted locations of the river almost no animals have survived.

The affected areas have not been repopulated, nine months after the accident. Mussels that survived last year's spill as juveniles have developed almost no eggs and sperm.

Effect on Commercial Shellfish Values[9]

Oil from the spill was incorporated into oysters, scallops, soft-shell clams, and quahogs. As a result, the area had to be closed to the taking of shellfish.

The 1970 crop of shellfish is as heavily contaminated as was the 1969 crop. Closure will have to be maintained at least through this second year and will have to be extended to areas more distant from the spill than last year. Oysters that were removed from the polluted area and that were maintained in clean water for as long as 6 months retained the oil without change in composition or quantity. Thus, once contaminated, shellfish cannot cleanse themselves of oil pollution.

The tidal Wild Harbor River, a productive shellfish area of about 22 acres, contains an estimated 4 tons of the fuel oil. This amount has destroyed the shellfish harvest for two years. The severe biological damage to the area and the slow rate of biodegradation of the oil suggest that the productivity will be ruined for a longer time.

Some have commented to us that the effects measured in the West Falmouth oil spill are not representative of those from a crude oil spill and that No. 2 fuel oil is more toxic than petroleum. However, the fuel oil is a typical refinery product that is involved in marine shipping and in many marine spillages; also, the fuel oil is a part of petroleum and as such it is contained within petroleum. Therefore, its effect is typical, both for unrefined oil and for refinery products. In terms of chemical composition crude oils span a wide range; many lighter crude oils have a composition very similar to those of the fuel oils and their toxicity and environmental danger correspond respectively. However, many crude oils contain more of the persistent, long-term poisons, including the carcinogens, than the fuel oils. Therefore, crude oils can be expected to have even more serious long-term effects than the lower-boiling fuel oils.

The pollution of fisheries resources in the West Falmouth oil spill is independent of the molecular size of the hydrocarbons; the oil taken up reflects exactly the boiling point distribution of the spilled oil. Thus, spills by other oils of different boiling point distributions can be expected to destroy fisheries resources in the same manner.

We believe that the environmental hazard of oil and oil products has been widely underestimated, because of the lack of thorough and extended investigations. The toxicity and persistence of the oil and the destruction of the fisheries resources observed in West Falmouth are typical for the effects of marine oil pollution.

The Risk to Human Use of Marine Resources

The destruction of marine organisms, of their habitats and food sources, directly affects man and his intent to utilize marine proteins for the nutrition of an expanding population. However, the presence in oil of toxic and carcinogenic compounds combined with the persistence of hydrocarbons in the marine food chain poses an even more direct threat to human health. The magnitude of this problem is difficult to assess at this time. Our knowledge of the occurrence of carcinogens in oil is recent and their relative concentrations have been measured in very few oils. Also, our understanding of the fate of hydrocarbons, especially of carcinogens, in the marine food chain needs to be expanded.

Methods for the analysis of fisheries products for the presence of hazardous hydrocarbons exist and are relatively simple and the analyses are inexpensive. In spite of this, no public laboratory in the United States—and probably in the world—can routinely perform such analysis for public health authorities. There is increasing evidence that fish and shellfish have been and are now being marketed which are hazardous from a public health point of view. Taste tests, which are commonly used to test for the presence of oil pollutants in fish or shellfish, are inconclusive. Only a small fraction of petroleum has a pronounced odor; this may be lost while the more harmful long-term poisons are retained. Boiling or frying may remove the odor but will not eliminate the toxicity.

The Risk to the Recreational Use of Marine Resources

The presence of petroleum, petroleum products, and petroleum residue ("tar," "beach tar") is now common on most recreational beaches. Toxic hydrocarbons contained in crude oil can pass through the barrier of the human skin, and the prolonged skin contact with carcinogenic hydrocarbons constitutes a public health hazard. Intense solar radiation is known to be one of the contributing factors for skin cancer. The presence of carcinogens in beach tar may increase the risk to the public in a situation where a severe stress from solar radiation already exists.

The Risk to Water Utilization

Many of the toxic petroleum hydrocarbons are also water soluble. Water treatment plants, especially those using distillation, may transfer or concentrate the steam-volatile toxic hydrocarbons into the refined water streams, especially if dissolved hydrocarbons are present in the feed streams or if particulate oil finds its way into the plant intake.

CONCLUSIONS

1. Oil and oil products must be recognized as poisons that damage the marine ecology and that are dangerous to man. Fisheries resources are destroyed through direct kill of commercially valuable species, through sublethal damage and through the destruction of food sources. Fisheries products that are contaminated by oil must be considered as a public health hazard.

2. Only crude estimates exist of the extent of marine oil pollution. We need surveys that can assess the influx of petroleum and petroleum products into the ocean. They should be worldwide, and special attention should be paid to the productive regions of the ocean; data are needed on the oil influx from tankers and nontanker vessels, on losses in ports, on offshore and inshore accidents from shipping, exploration, and production, and on the influx of oil from domestic and industrial wastes.

3. The marine ecology is changing rapidly in many areas as a result of man's activities. We need to establish base-line information on composition and densities of marine faunas and floras and on the hydrocarbon levels and concentrations encountered in marine organisms, sediments and in the water masses.

4. All precautions must be taken to prevent oil spills. Prevention measures must be aimed at eliminating human error, which at the present time is the principal cause of oil spills.

5. Spill prevention must be backed by effective surveillance and law enforcement. *In terms of cost effectiveness spill prevention is far superior to cleanup.*

6. Perfection and further extension of the use of the load-on-top methods is promising as a first step in reduction of the oil pollution from tankers. The effectiveness of the technique should be more closely assessed, and improvements are necessary in interface detection, separation, and measurement of hydrocarbon content in the effluent, both in the dispersed and dissolved state. On a longer time scale, clean ballast techniques should supersede the load-on-top technique.

7. The impact of oil pollution on marine organisms and on sources of human food from the ocean has been underestimated because of the lack of coordinated chemical and biological investigations. Studies of the effect of oil spills on organisms in different geographic and climatic regions are needed. The persistence of hydrocarbon pollution in sea water, sediments, and organisms should be studied.

8. Research is urgently needed on the low-level and long-term effects of oil pollution. Does oil pollution interfere with feeding and life processes at concentrations below those where effects are immediately

measured? Are hydrocarbons concentrated in the marine food chain?

9. Carcinogens have been isolated for crude oil, but additional efforts are needed to define further the concentrations and types of carcinogens in different crude oils and oil products.

10. The public health hazard from oil-derived carcinogens must be studied. What are the levels of oil-derived carcinogens ingested by man and how wide is the exposure of the population? How much does this increase the present body burden of carcinogens? Is there direct evidence for the causation of cancer in man by petroleum and petroleum products outside of oil refinery operations?

11. Public laboratories must be established for the analysis of fisheries products for toxic and carcinogenic chemicals derived from oil and oil products, and tolerance levels will have to be set.

12. The ocean has a limited tolerance for hydrocarbon pollution. The tolerance varies with the composition of the hydrocarbons and is different in different regions and in different ecological subsystems. The tolerance of the water column may be greater than that of the sediments and of organisms. An assessment of this inherent tolerance is necessary to determine the maximum pollution load that can be imposed on the environment.

13. Countermeasures which remove the oil from the environment reduce the ecological impact and danger to fisheries resources. All efforts should be aimed at the most rapid and complete removal since the extent of the biological damage increases with extended exposure of the oil to sea water.

14. Countermeasures that introduce the entire, undegraded oil into the environment should be used only as a last resort in situations such as those outlined in the Contin-

gency Plan of the Federal Water Quality Administration, involving extreme hazard to a major segment of a vulnerable species of waterfowl or to prevent hazard to life and limb or substantial hazard of fire to property. Even in those cases assessment of the long-term ecological hazard must enter into the decision whether to use these countermeasures (detergents, dispersants, sinking agents).

15. As other countermeasures become more effective, the use of detergents, dispersants, and sinking agents should be further curtailed or abolished.

16. Efforts to intensify the natural bacterial degradation of oil in the environment appear promising and should be supported by basic research and development.

17. Ecological damage and damage to fisheries resources are direct consequences of oil spills. In the future, the cost of oil leases should include a fee for environmental protection.

18. Environmental protection funds derived from oil leases should be used to accomplish the necessary research and education in the oil pollution field.

ACKNOWLEDGMENTS

The author expresses his gratitude for continued support to the National Science Foundation, to the Office of Naval Research and to the Federal Water Quality Administration.

SUGGESTED READINGS

Federal Task Force on Alaskan Oil Development, 1972, "Final Environmental Impact Statement, Proposed Trans-Alaska Pipeline," U.S. Dept. of the Interior, Washington, D.C. (vols. 1–6). Livingston, Dennis, 1974, "Oil on the Seas," *Environment* 16: 38–43, September.

The Challenge and Promise of Coal

EDMUND A. NEPHEW   1973

Coal is by far the most abundant of the fossil fuels, both in the United States and in the world. The original, in-ground coal resource of the United States, as determined from mapping and exploration, is estimated to have been some 1,624 billion (short) tons. This amount includes all bituminous and anthracite coal in seams greater than 14 inches, and all subbituminous coal and lignite in seams greater than 30 inches found at depths up to 3,000 feet. An additional coal resource of 1,650 billion tons, bringing the total to more than 3.2 trillion tons, is considered probable from presently unexplored and unmapped areas and at greater depths—up to 6,000 feet. Even allowing for past depletion (40 billion tons of cumulative coal production) and for losses incurred during mining (a similar amount), the enormous magnitude of our remaining coal resource is evident when it is compared with the annual production in 1969 of 560.5 million tons.

Our actual coal reserves are much smaller than this total resource estimate, however, because only a fraction of the total resource can be considered minable under present economic conditions and with present technology. Most of the over-all coal resource lies either at depths too great, in seams too thin, or at locations too remote from present-day markets to permit economical recovery. Studies by Paul Averitt indicate that the recoverable coal resource in the United States under less than 1,000 feet of overburden amounts to 374 billion tons in seams of intermediate thickness and 406 billion tons in thick seams. Thus, assuming a recovery factor of 50 per cent, only 390 billion tons of the total coal resource can be considered as a proven, identified reserve. This amount represents slightly less than one-eighth of the total probable United States coal resource.

In 1968, the total United States production of coal was 556 million short tons. At this production rate the remaining proven coal reserves in the United States would suffice for nearly 700 years.

Edmund A. Nephew is a member of the research staff at the Oak Ridge National Laboratory. He is currently engaged in a National Science Foundation Program, "The Environment and Technology Assessment," which includes an investigation of the environmental inpacts of electricity production and use.

From *Technology Review*, Vol. 76, No. 2, pp. 21–29, December 1973. Reprinted by permission of the author and the *Technology Review*. Copyright 1973 by the Alumni Association of the Massachusetts Institute of Technology. This article is adapted from a paper prepared for a Special Summer Program at Massachusetts Institute of Technology, July, 1973.

This apparent abundance of coal is illusory for at least two reasons. The annual consumption rate will rise sharply as shortages in the other fuels become more pronounced, as new uses for coal are developed, and as normal industrial growth occurs. And intensive exploitation of the United States coal resource may be limited by severe stresses upon the environment which are associated with coal production and consumption.

TECHNOLOGY TURNS MINING TO THE SURFACE

An intelligent forecaster in the 1940's pondering the distribution of the coal resource as shown in Table 2-3 might have concluded logically that future advances in coal mining technology would occur in three principal areas. In order of priority, these advances would have been: the development of techniques yielding a higher percentage coal recovery, the adoption of new methods for economically mining thin seams, and mining at ever greater depths. Progress in each of these areas would result in a higher utilization of the total coal resource, thus expanding the nation's coal reserve.

Coal mining today is far removed from the static, old-fashioned art still envisaged by some people. Both underground and surface

mining have been subject to substantial technological change. But the most striking improvements in mining technology over the last three decades have not been in the field of underground coal mining. Instead we have witnessed a continuing increase in coal surface mining, which has more than compensated for a declining production from underground mines. In 1971, for the first time in history, the total coal production from strip and auger (surface) mines surpassed that from deep mines. (See Figure 2-33.)

The large production of coal by surface mining has resulted primarily from improved, heavy excavating equipment introduced in the 1960's. Mammoth drag lines with bucket capacities up to 220 cubic yards, power shovels, and bucket-wheel excavators weighing nearly 8,000 tons are now in use. These machines, exploiting economies of scale, have made possible the surface mining of coal at ever greater depths and, at the same time, have nearly doubled the output per man-day from coal stripping operations.

The coal recovery factor for surface mining, except in the case of augering, is much higher than for underground mining, approaching 80 to 90 per cent. In addition, during the process of stripping away the overburden to reach thick coal seams, thinner seams are often encountered and profit-

Table 2-3. Amount of United States Coal Resources (billions of tons).[a]

Overburden thickness	Coal seam thickness			
(feet)	14 to 28 inches	28 to 42 inches	Over 42 inches	Total
0 to 100 } 100 to 150 }	Total of surface-minable coal resources			{ 115 { 50
Less than 1000	666	374	406	1,446
1000 to 2000	33	73	49	154
2000 to 3000		5	20	25

[a]Surface mining must have a small role in the ultimate history of America's recovery of its bountiful coal resources. For less than 10 per cent of the total United States coal resource lies within 150 feet of the surface of the ground—the maximum depth at which surface mining is likely to be practicable with presently foreseeable technology.

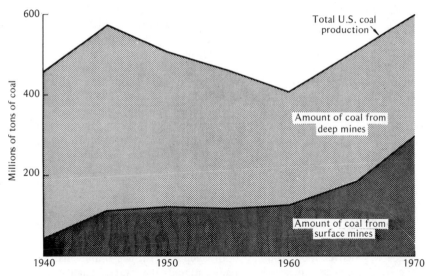

Figure 2-33. Coal represents the most abundant fossil fuel resource in the United States, but its production and use are limited by economic and environmental constraints. Though only a small fraction of all United States coal can be reached by surface mining, that method has gained ground rapidly in the past decade as a result of new technological developments. Now similarly major technological changes may be on the horizon for deep mining. (Data: Bituminous Coal Facts, 1972.)

ably extracted. Indeed, at first glance it appears that coal surface mining fulfills two of the important technological objectives for coal mining described above, and it is on this basis that strip mining has advanced so rapidly in the past three decades.

Yet during the same period, two and possibly three major technological revolutions have swept the underground mining industry, bringing worker productivities there to new highs as well. These changes began in the 1940's when hand cutting and loading were replaced by a fully mechanized procedure for cutting, drilling, and blasting the coal from the seam face. Loading machines were built to gather up the loosened coal and load it into shuttle cars for transportation to the surface. In the 1950's, several of these separate steps were unified into a single operation with the development of the continuous mining machine, which rips coal directly from the seam using a rotating cutter head equipped with replaceable teeth. The same machine transfers the coal to conveyor belts or shuttle cars, eliminating the

need for a separate loading operation. Today, more than half of the coal produced in underground mines is extracted and loaded by this method.

A third major advance in underground mining, today still in its infancy, began in the 1960's with introduction of the longwall mining technique. As will be described later in this article, the longwall system, with powered roof supports and partial or full automation of the face equipment, affords the promise of truly exciting increases in productivity—prodigious coal-cutting capacity, for example, of as much as 5,000 tons per shift with a 10-man face crew.

The end result of these innovations, both above and beneath the surface, has been to increase the average productivity from all mines from five tons per man-day in the 1940's to 20 tons per man-day in the 1970's. The success of these astonishing strides is best placed in perspective by comparing the results with current productivity levels of western European nations—two to four tons of coal per man-day.

SURFACE MINING AND THE ENVIRONMENT

The coal found in the Appalachian Mountains was originally deposited in horizontal layers. Over millions of years, erosion and weathering have cut deep valleys into the highland plateau and shaped the landscape we observe today. The coal outcrops thus exposed roughly follow the contour lines as on a topographical map. Stripping in its conventional form proceeds along the coal outcrop, with the overburden material removed from above the coal bed during each cut typically being cast down the hillside or stacked along the outer edge of the bench. The exposed coal is then removed and a second cut is made to uncover more coal. Finally, when the overburden becomes too thick for economical stripping, augers—up to seven feet in diameter—are used to drill horizontally several hundred feet into the coal seam to bring out additional coal.

The mining proceeds along the mountainside using this combination of stripping and augering. Behind is left a steep, nearly vertical highwall on the upslope side, and piles of overburden or spoil material stacked along the outer bench or cast downslope. (See Figure 2-34.)

In other parts of the United States, where coal seams are located near the surface under level terrain, strip mining is simpler. An area is cleared of overburden and then of coal, and the overburden from the next adjacent area is moved onto the cleared area. There remains eventually a surface composed of heterogeneous overburden—mostly gravel and small rock—in whatever form was dictated by the machinery from which it was dumped.

The environmental impacts of strip mining are not confined to the directly disturbed areas. Siltation of streams from erosion of spoil and acid water runoff extends the harmful effects of strip mining far beyond the actual mine site. (Acid water is water containing sulfuric acid produced by weathering of sulfur-bearing minerals.) According to a recent Appalachian Regional Commission study, contamination caused by both deep and surface mining has substantially altered the water quality of some 10,500 miles of streams in Appalachia, and acid drainage seriously pollutes some 5,700 miles of streams. Annual erosion from freshly strip-mined areas in Appalachia is as high as 27,000 tons per square mile, or up to 1,000 times greater than for undisturbed lands. There are many examples of marked decreases in the variety and abundance of aquatic life downstream from strip-mined areas, caused by erosion and acid water drainage which effectively destroyed stream habitats. Landslides, damage to timber, and as yet unknown effects such as the spread of heavy metals or carcinogenic materials by leaching of the spoil during rains add to the

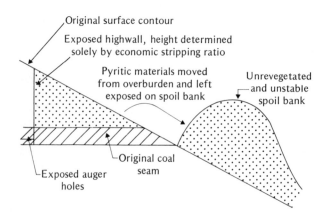

Original surface contour

Exposed highwall, height determined solely by economic stripping ratio

Pyritic materials moved from overburden and left exposed on spoil bank

Unrevegetated and unstable spoil bank

Original coal seam

Exposed auger holes

Figure 2-34. In its crudest form, strip mining in Appalachia consists of excavating surface material to the depth of the desired coal by pushing the spoil down the mountain—and leaving it there. There remains an ugly gouge in the mountain and a spoil bank contributing landslides, acid drainage, and sediment for many years to come. Fortunately, writes the author, this type of mining "has virtually disappeared" from Appalachia today.

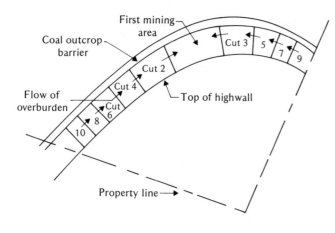

Figure 2-35. This plan view shows the modified block-and-cut method of strip mining which can provide high-quality land reclamation after the coal is removed. The overburden from the first mining area of a property is removed and then the coal is cut away. Then mining begins in cuts 2 and 3, with the overburden placed in roughly the original sequence and contour in the first mined area. Mining then proceeds through additional cuts, with the overburden in each case moved in the direction of the arrows. Finally, the spoil from the first mining area is returned to cover the scars of the last areas mined—cuts 9 and 10 in the diagram at the left.

list of possible off-site damages.

The nature of environmental damage from coal surface mining depends significantly upon the climate and terrain of the mining region. Large-scale surface mining of coal has now begun in the water-poor regions of the West and Southwest, and these operations may pose environmental hazards which we do not yet understand. For example, in the semidesert Southwest, surface mining may bring highly saline material to the surface. The difficulties of establishing new vegetation on this overburden material will be far greater than in the Appalachians. Before the long-range feasibility of surface mining in arid regions can be assessed, revegetation methods effective in regions with low average rainfall must be developed, soil conditions must be investigated, and the ground-water movements must be determined. Although additional research is needed, it is clear that strip mining—whether conducted in Appalachia, in the Southwest, or in harsh northern climates—can seriously affect ecological balance.

Erosion, acid runoff, and ground slides are inevitable consequences of conventional mountain strip mining, with no reclamation. An exhausted, abandoned mine may contribute to stream acidity, sedimentation, and landslides for many years to come. In contrast, a new method of contour strip mining,

known as the modified block-and-cut method of mining, avoids many of these adverse effects.

Briefly outlined, the block-and-cut method involves completely mining to their full depth a series of sections of a coal seam. The spoil from the first section of the seam is stored on the downslope side of its bench or off site if possible. Then this first mined area is available to hold the spoil from sections immediately adjacent on both right and left, and in turn these mined areas are used for storing the spoil from the next sections. Finally, the spoil from the first section is distributed to the last mined sections as mining of the property is completed.

This block-and-cut system reduces the area of disrupted land and improves the potential for economic land rehabilitation by eliminating the need for double handling of most of the spoil. (See Figure 2-35.)

HOW TO RESTORE STRIP-MINED LAND

Much of the environmental damage from strip mining can be reversed, though the cost of land restoration varies greatly, depending on how the mining was conducted and on the degree of restoration required.

The method used for removal and replacement of the overburden is of crucial importance both to the economics of coal production and to successful land reclamation. Economic coal recovery is possible only if the spoil is handled minimally, with the overburden placed when first removed at or near its final resting place. At the same time, successful reclamation requires that the overburden be placed in a stable condition and selectively layer-sequenced to provide the best soil materials on top, with heavy rocks and toxic substances deeply buried.

In order of increasing costs and quality of reclamation, the alternatives may be described as follows:

Basic reclamation consists simply of spoil stabilization and the prevention of off-site damage. To achieve this minimum level of reclamation, spoil material must be placed during mining operations in stable configurations which reduce the probability of landslides, with the best soil materials on top and toxic substances, including acid-forming materials, buried deeply. Access roads must be designed and maintained to prevent erosion. Following mining, water catch basins must be built to prevent stream siltation and grading, ditching, and revegetation completed to reduce erosion.

Partial reclamation—involving greater effort—is required if the reclaimed land is to become productive in a reasonable period after mining. The land must be returned to a topography suitable for the intended future uses. The spoil bank need not be returned to the approximate original contour; it can instead, for example, be graded to form "terraces" of various shapes suitable for grazing cattle and other uses; high walls need be only partially toppled to improve safety and accessibility. The spoil must be handled with care during mining so that the original sequence of strata—including upper and lower leach layers—in the overburden is preserved. The original topsoil or perhaps a superior alternative material must be put in place. Following mining the steps listed above must be taken, and then the land must be managed for a period of several years to restore the nutrient and humus levels of the soil.

Full reclamation includes all of these, plus detailed grading of the land approximately to its original contours and revegetation with native plant species—or with foreign species which will better respond to the new conditions but which are ecologically compatible with local flora.

No sensible discussion of the costs of land restoration is possible unless the reclamation objectives are clearly specified.

The amount of spoil that has to be moved during reclamation is the principal factor in costs. Return to contour typically requires movement of 75 per cent of the total spoil material, while the various terrace backfills require movement of only 25 to 40 per cent of the total spoil. Other cost variables include the average haul distance involved and the type of overburden. In all cases the cost of backfilling and revegetation is much lower if the work is performed at the time of stripping.

Recent estimates (1972) of the Appalachian Regional Commission indicate backfilling and grading costs of $1,000/acre for return-to-contour backfilling and $600 per acre for terraces. Covering the graded spoil with topsoil may cost as much as $0.50 per cubic yard, or up to $2,500 per acre. Revegetation costs, including soil treatment measures such as mulching, liming, fertilizing, and seeding, range from $100 to $500 per acre. Thus, the total cost of full reclamation—for all of these activities—appears to be of the order of $1,500 to $4,000 per acre.

This estimate agrees well with reported foreign experience and is much higher than the costs of $200 to $300 per acre sometimes quoted in the coal mining industry. Clearly, industry spokesmen are speaking of a lower quality of land reclamation.

Prior to the early 1960's, even basic reclamation, as defined above, was uncommon in the mountainous coal fields of Appalachia. The resulting widespread environmental damage has since prompted most states to require at least basic—and in some cases more complete—reclamation. Full land reclamation is not commonly practiced in the

United States, although it has been applied extensively in some countries of Europe, notably in the United Kingdom and the Federal Republic of Germany. In the United States, Pennsylvania is especially noted for stringent reclamation requirements; its law specifies some of the features of full reclamation described above.

Although judgments vary, most people would agree that basic reclamation is essential. Large segments of the general public and the mining industry would probably consider partial reclamation to be sufficient. Full reclamation may be needed in some cases, to prevent the loss of valuable land-use options, such as the development of scenic, mountainous regions for recreational purposes.

STRIPPING: LARGE AND GROWING LARGER

The lands already affected by contour and area strip mining in the United States are enormous in extent. The production of surface-mined coal in the United States has already resulted in some 3,000 square miles of disturbed land. At present coal stripping rates, more than 100 square miles of new land are being stripped annually—and surface mining is increasing because of its high profitability.

As a result, extensive tracts of unrestored land have accumulated, and whole watersheds have suffered extensive damage. Thus, the nation is presently faced with a backlog of environmental damage from coal mining in the form of unreclaimed land and degraded river systems. This past damage must be repaired and regulations must be implemented to curb further destructive mining practices. There is a growing agreement that federal legislation is needed to set surface mining and reclamation standards and to require land restoration as an integral part of the mining cycle.

Two independent estimates of the remaining strippable coal resource in the United States have recently been made. The first, by Averitt, rests primarily upon relatively comprehensive data on the coal resource located under less than 100 feet of overburden in ten selected states. From these data, obtained from mapping, exploration, and other sources, extrapolations are made for the remaining states, taking into account their geological characteristics, known total coal resource, and other factors. The conclusion is that before mining began strippable coal resources of the United States were 165 billion tons; of this some 115 billion tons were under overburden with depths up to 100 feet, and an additional 50 billion tons were under overburden of 100 to 150 feet. Production and production losses in surface mining have to date been some 5.5 billion tons; assuming an 80 per cent recovery factor, a total of 128 billion tons of recoverable strip coal remains.

A more recent study of strippable coal resources by the U.S. Bureau of Mines yields a slightly lower figure. This study uses two criteria for maximum overburden thickness and minimum seam thickness. For coal fields reasonably close to high-demand markets, such as those of Appalachia or the Midwest and Gulf states, coal in seams of 29 inches or more covered by less than 120 feet of overburden is considered to be a strippable resource. In the western states, coal in thicker seams of from 4 to 5 feet and at overburden depths generally less than 150 feet is considered strippable. (Slightly different criteria were used to define surface-minable lignite and subbituminous coal.) After allowing for past depletion, the remaining strippable resource in the United States on this basis was found to be approximately 118 billion tons.

DEEP MINES: HOSTILE AND HAZARDOUS

Mining coal in deep mines avoids much of the landscape disruption accompanying surface mining. Primarily for this reason, many people have advocated gradually eliminating all coal surface mining and expanding the capacity of underground mines. Although the distribution of the total United States coal resource will some day necessitate greater reliance on deep mining, we should not forget that this form of mining also presents serious problems.

Under the best of circumstances, the underground coal mine represents an extremely hostile and hazardous environment for the miner; this fact is reflected in the much higher injury and death rates experienced in underground mining. Present-day underground mining also suffers from low productivity and low resource utilization, as compared to surface mining. Furthermore, although they are not as visually dramatic as in strip mining, the environmental impacts from deep mining—acid mine drainage, land surface subsidence, gob disposal, and mine and waste heap fires—are far from trivial. Indeed, studies have shown that the drainage of acidic waters from underground mines presently poses a greater problem than acid mine drainage from surface mines.

Technological advances are needed in each of these areas if a shift toward deep mining is to occur. Otherwise, the social, economic and environmental trade-offs involved in such a shift may be unacceptable. We now discuss briefly the nature and magnitude of some of these trade-off problems and the prospects of a new form of mining—the longwall system—for ameliorating them.

SAFETY: ROOM FOR IMPROVEMENT

The human costs of moving from surface to underground mining take the form of higher injury and death rates and of greater occupational hazards in general. In 1971, 86 per cent of all coal mining fatalities occurred in underground mines. In the same year, only half of the total coal production came from deep mines. Over the seven-year period 1965 to 1972, 1,412 lives were lost in the underground coal mining industry in the production of 2,336 billion tons of coal. This amounts to an average of 0.606 deaths per million tons of coal—a fatality rate more than five times greater than that of the coal surface mining industry. In recent years, falls

of roof have accounted for 40 per cent of the deaths in underground mines and coal haulage accidents for about 20 per cent. Annual fatalities from dust and gas explosions fluctuate greatly, but over the years there has been a declining trend. A similar safety disparity between surface and deep mines holds for nonfatal injuries as well. (See Figure 2-36.)

During the time period from 1968 to 1971, the safety performance in deep mining of the nation's top ten coal producers ranged from 0.28 to 1.52 deaths per million man-hours. The differences are even more marked for the category of nonfatal injuries—2.72 to 72.13 injuries per million man-hours. Since the passage of the Coal Mine Health and Safety Act of 1969, most companies have greatly strengthened their safety programs, and this increased emphasis on safety may bring significant improvements in fatality and injury rates. Indeed, the wide range of safety performances cited above makes it clear that—even without a technological breakthrough—much can be done to narrow the safety gap existing between deep and surface mining.

RESOURCE UTILIZATION AND PRODUCTIVITY

Underground mining is generally more wasteful of the coal resource than surface mining. Room-and-pillar mining by conventional methods may leave as much as 50 per cent of the coal behind. Although the recovery factor for continuous mining, employing pillar retreat sections, is much higher, the average national recovery factor for underground coal mines is still only 57 per cent. The remaining 43 per cent of the original coal resource is wasted, left in the mine and rendered unfit for future recovery. This loss of reserves represents a significant disadvantage of underground coal mining. In contrast, the recovery factor for surface mining ranges from nearly 100 per cent in open-pit mines to 80 per cent for area and contour stripping and to about 50 per cent for augering.

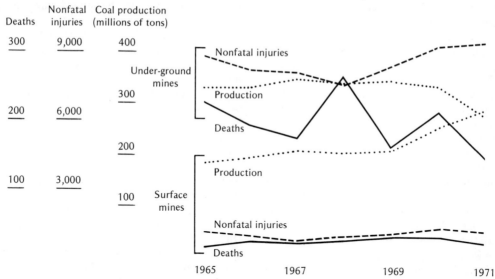

Figure 2-36. The safety record of surface mining—in terms of both deaths and nonfatal injuries per ton of coal mined—is spectacularly better than deep mining, which provides at best a hostile, hazardous environment for the miner. The record of underground mines will improve as the impact of the Coal Mine Health and Safety Act of 1969 continues to increase, but the requirements of this Act may also increase the cost of underground mining enough so that surface mining—at least in the next decade—becomes the dominant method of obtaining coal. (Data: Bituminous Coal Facts, 1972.)

The productivity of old-fashioned, room-and-pillar underground mines using hand tools ranged from about 2 to 5 tons per man-day. With the advent of mechanized, conventional mining, productivity levels of 10 tons per man-day were obtained. Finally, the introduction of the continuous miner in the 1950's has provided another substantial increase in productivity—to 15 tons per man-day. But the current productivity of surface mining is much higher: 35 tons per man-day in area and contour stripping, 35 to 45 tons per man-day in augering, and up to 80 tons per man-day in open-pit mines. New technological advances are needed to overcome the present higher production costs of deep mines.

MODERN LONG-WALL MINING

One such technological advance—long-wall mining—may be a significant new development. Although long-wall mining, involving the use of stone packs and manually moved roof supports, was once widely employed in the United States, productivity was extremely low, and it could not compete successfully with other mining systems. By necessity, however, the development of long-wall mining to meet the more adverse conditions of mining at great depths continued in Europe, where coal reserves under shallow cover have long been exhausted. Gradually, research in rock mechanics and other areas led to a better understanding of the behavior of supports, the structural properties of strata, and roof movement. These theoretical advances were of vital importance, since the success of modern long-wall mining depends greatly upon the development of an effective roof control system.

In parallel to these developments after World War II came some major technological achievements which made possible the mechanization of long-wall mining: a flexible armored face conveyor capable of being moved forward without disassembly, efficient plows and shearers to ride on rails mounted on the armored flexible conveyor, and self-advancing hydraulic roof supports.

In the long-wall system, coal is mined from panels which are established by driving access entries along both sides of the panel. The panels, from 300 to 800 feet wide and 1,500 to 5,000 feet long, are connected laterally by the long-wall face which contains the shearer, the armored flexible conveyor, and the row of side-by-side hydraulic roof support units. The shearer travels back and forth along the face, biting several feet into the coal; the coal thus cut falls to the conveyor and is carried away. When the cutting machine reaches an access entry, the cutting edge is advanced and the process continues, the whole machine returning toward the opposite access.

Following the traverse of the cutting machine, roof support units are moved forward automatically. Each of these consists of four to six hydraulic jacks mounted upon a single base plate; the jacks act to support the roof, each providing a force of from 30 to 175 tons. When the roof support units are to advance with the shearer and conveyor, each unit in turn is decompressed, moved forward, and reactivated. The roof behind is left to collapse. The main roof should fall regularly and predictably, some optimal distance behind the retreating long-wall face.

The long-wall system of mining offers important advantages which make it highly attractive. These include high productivity, high resource recovery, and improved mine safety conditions. The face equipment can be operated by a small crew of from eight to ten men, and production of up to several thousand tons of coal per day can be achieved. No coal support pillars are left behind, and the operating crew at the face is protected by the mechanical roof support units. Effective ventilation is simpler and easier to provide, reducing the likelihood of gas and dust explosions. All of these features make long-wall mining very attractive indeed—if only it could be applied universally.

Unfortunately, at its present stage of development long-wall mining is not universally applicable, and many experts in the mining industry believe that it can be employed successfully only in a limited range of stratigraphic conditions. If this is true, long-wall mining, though continuing to gain ground, will probably not become the predominant form of underground mining in this country. On the other hand, additional research and development is required before the system can reach its optimum potential. Automation of the face equipment, development of improved haulage techniques, studies of inducing roof collapse, and the development of rapid driving machines are but a few of the challenges ahead.

The distribution of United States coal resources appears to be such that the vast bulk of the nation's coal will remain forever inaccessible to recovery by surface mining methods. This means that we must improve existing underground mining techniques and develop new ones if we are to utilize our rich endowment of coal. Long-run considerations point to the inevitable resurgence of underground coal mining, and the prospects for significant progress in the technology of underground mining appear bright. A vigorous research effort is needed to allow deep mining to reassume its proper role in coal production. Automation of face equipment, improved haulage techniques, hydraulic coal cutting, controlled mine atmospheres, and computer monitoring of mine environments to sense and eliminate hazardous working conditions are examples of exciting research that could revitalize the underground mining industry.

SUGGESTING READINGS

Osborn, Elburt, F., 1974, Coal and the Present Energy Situation, *Science 183*, pp. 477–481.

Agony of the Northern Plains

ALVIN M. JOSEPHY, JR.   1973

In October 1971, a "Coordinating Committee," composed of the U.S. Bureau of Reclamation and 35 major private and public electric power suppliers in 14 states from Illinois to Oregon, issued a dramatic document. Innocuously titled the "North Central Power Study," it stunned environmentalists throughout the country and sent waves of horror among the ranchers, farmers, and most of the townspeople of the northern plains. Rushed through in a little over a year (the project was initiated in May, 1970 by the then Assistant Secretary of the Interior for Power and Water Development, James R. Smith), and reflecting the goals and points of view of utility interests that were in business to sell electricity, the study proposed a planned development and employment of the coal and water resources of some 250,000 square miles of Wyoming, eastern Montana, and western North and South Dakota for the generation of a vast additional power supply for the United States.

The scope of the proposal was gargan-tuan—rivaling the grand scale of the region itself. One of the most serene and least spoiled and polluted sections of the nation, it averages about 4,000 feet above sea level and stretches below the Canadian border roughly from the Badlands and Black Hills in the east to the Bighorn Mountains in the west. It is a huge, quiet land of semiarid prairies, swelling to the horizon with yellow nutritious grasses; rich river valleys, lined with irrigated farms; low mountains, buttes, and rimrock ridges dark with cedar and ponderosa pine; open, wind-swept plains covered with sagebrush, greasewood, and tumbleweed; and hundreds of meandering creeks edged with stands of cottonwoods. The rains average only 12 to 14 inches a year, the topsoil is thin and fragile, easily eroded and blown or washed away, and the vegetation in most places must struggle for life. Towns and cities are small and few and far between, and distances measured along the infrequent highways and ribbons of railroad track are great. For almost a hundred years the nat-

Alvin M. Josephy, Jr. is a leading authority on American Indian affairs and western American history and his most recent book is *Red Power: The American Indians' Fight for Freedom*. He has served as a consultant to the President, the Secretary of the Interior, the Public Land Law Review Commission, and the National Congress of American Indians. Presently he is vice-president of American Heritage Publishing Company.

Adapted from AUDUBON, July 1973, the magazine of the National Audubon Society: copyright © 1973.

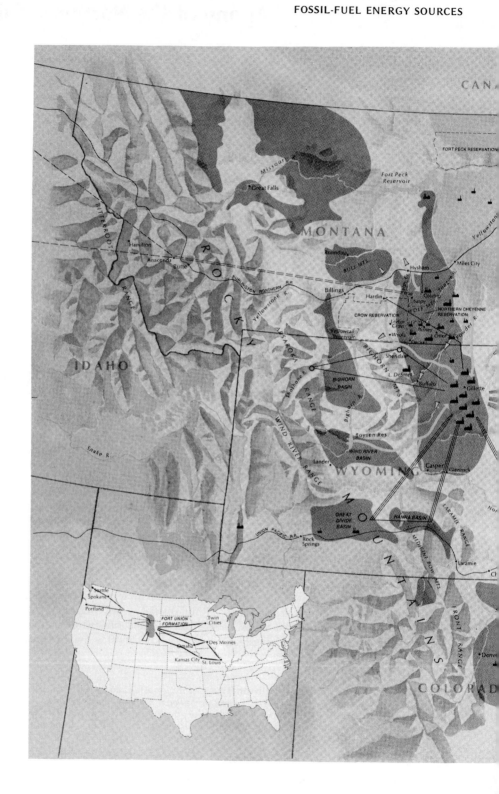

Bituminous
and sub-bituminous
coal deposits

Lignite coal deposits

Thermal generation plants

10,000 megawatts

5,000 megawatts

3,000 megawatts

1,000 megawatts

○ Pumped storage generation

◁ Proposed reservoirs

Major water conveyance
conduits

Transmission lines

Land-grant railroads

Source: North Central Power Study, Oct 1971

ural grasses and irrigated hay fields have sustained big flocks of sheep and herds of cattle, and the region has been one essentially of large, isolated ranches and farms, whose owners have fought endlessly against blizzards, drought, high winds, and grasshoppers—and have treasured their independence and the spaciousness and natural beauty of their environment.

Ominously for them, the surface of their part of the country sits atop the Fort Union Formation (in the Powder River Basin of Wyoming and Montana and in the western part of the Williston Basin of Montana and the Dakotas), containing the richest known deposits of coal in the world. There are at least 1.5 trillion tons of coal within 6,000 feet of the surface, and perhaps more than 100 billion tons so close to the surface—in seams 20 to 250 feet thick—as to be economically recoverable today by the relatively cheap modern techniques of strip mining. This is, staggeringly, 20 per cent of the world's total known coal reserves and about 40 per cent of the United States' reserves. (The total national figure would be able to supply the country for an estimated 450 to 600 years should the present use trend continue.) But perhaps even more significantly, in view of recent environmental concerns, the sulfur content of these deposits of high-quality subbituminous coal in Montana and Wyoming and lesser-grade lignite in northeastern Montana and North Dakota is low enough to meet the new air pollution standards for coal-burning power plants in urban areas.

In the past, very little of the northern plains coal has been mined, principally because of its comparatively lower BTU heat content and its distance from major markets, which made it less desirable competitively than eastern coal. But by May 1970 the need for low-sulfur coal in the cities was hurrying a change in that thinking. In addition, an energy panic was in the offing—a panic concerned more with sources of future supplies of conventional fuels than with conservation, realistic planning and pricing, dampening of demand, and the development of

alternative, nonpolluting fuels. A large-scale (though little-publicized) rush to acquire exploration permits and leases for the low-sulfur coal in the northern plains—together with plans on how to maximize short-term and long-range profits from the enormous deposits—was already stirring the energy industry. It appeared evident that national policy, guided by the industry, would inevitably encourage the exploitation of the western states' coalfields as an answer to the apparently diminishing supplies of fuels from elsewhere, the threat of growing dependency on the oil-producing nations of the Middle East, and power-plant pollution in the cities. So, strict over-all government planning and regulation were necessary if the imposition of coal-based industrialization on the traditional farming-ranching economy and environment of the north central states was not to bring disaster to the area and its people.

Viewing this as a mandate, the Department of the Interior and the 35 cooperating utilities launched their study. There were few persons in the affected region who were not already aware of the increasing attention being given to their coal; indeed, many landowners were already being subjected to the pressures of lease brokers, speculators, and coal companies. But the threat to the region as a whole was not yet visible, and the implications of the stupendous changes that the coal reserves would bring to the lives and environment of the people were not even dreamed of. The release of the North Central Power Study shattered that innocence.

Together with an accompanying document that dealt with the utilization of the region's water resources for the proposed coal development, the study suggested the employment of strip mines in Montana, Wyoming, and North Dakota to supply massive amounts of coal to fuel mine-mouth power plants, which by 1980 would produce 50,000 megawatts of power, and by the year 2000 approximately 200,000 megawatts. The power would be sent east and west over thousands of miles of 765-kilovolt transmission lines to users in urban areas. The study located sites for 42 power plants—21 in eastern Montana, 15 in Wyoming, 4 in North Dakota, and 1 each in South Dakota and Colorado. Their suggested sizes were mind boggling. No fewer than 13 of them would generate 10,000 megawatts each (about *14 times as much* as the original capacity of the Four Corners plant in New Mexico, much criticized as the world's worst polluter, and almost five times more than the 2,175 megawatts which that plant is now capable of generating). Other plants would range from a 1,000- to 5,000-megawatt capacity. In addition, 10 of the proposed giant 10,000-megawatt plants would be concentrated in a single area, 70 miles long by 30 miles wide, between Colstrip, Montana, and Gillette, Wyoming; another group, with a combined capacity of 50,000 megawatts, was targeted for another compact area close by.

To supply some 855,000 acre-feet of cooling water (an acre-foot is enough to cover one acre with one foot of water) which would be needed each year by the plants at the 50,000-megawatt level, the study proposed a huge diversion of water form the rivers of the Yellowstone Basin, requiring a large system of dams, storage reservoirs, pumping heads, and pipeline aqueducts to be built by the Bureau of Reclamation. As if that were not enough, the water resources document went further, envisaging—with great realism, as it has turned out—the construction of immense coal gasification and liquefaction plants and petrochemical complexes, located near the strip mines and power plants, and raising the need for water to at least 2,600,000 acre-feet a year.

Once they got over their shock at the stupendous dimensions of what was being proposed, environmentalists set to work dissecting the study. It was entirely oriented to the producer of electricity and dealt scarcely, or not at all, with such overwhelming problems as air, water, and noise pollution, strip mining and the reclamation of ravaged land, the diversion of major rivers and resultant conflicts over water rights in the semiarid country, the degradation of the human

and natural environments, the disruption of the region's economy, soil erosion, the destruction of fish and wildlife habitat, and the explosive influx of population with attendant social and economic strains and dislocations that would follow the carrying out of the project's individual schemes. Dr. Ernst R. Habicht, Jr. of the Environmental Defense Fund found the plan almost unbelievable, pointing out that it called for the generation of "substantially more electricity than is now produced either in Japan, Germany, or Great Britain (and would be exceeded only by the present output of the United States or the Soviet Union)." The 855,000 acre-feet of water needed annually, just for the 50,000-megawatt goal, Habicht noted, was more than half of New York City's annual water consumption, and if the need rose to the proposed 2,600,000 acre-feet, it would exceed "by 80 per cent the present municipal and industrial requirements of New York City (population 7,895,000)." Moreover, in wet years, the mammoth diversion would reduce the flow of the Yellowstone River by one-third, and in dry years by about one-half. "Water use of this order of magnitude in a semiarid region ... will have significant environmental impacts," the scientists warned. "Extreme reduction in river flows and the transfer of water from agricultural use will drastically alter existing agricultural patterns, rural life styles, and riverine ecosystems."

All of this the study had, indeed, overlooked, but there was more. Analysis showed that coal requirements for the 50,000-megawatt level in 1980 would be 210 million tons a year, consuming 10 to 30 square miles of surface annually, or 350 to 1,050 square miles over the 35-year period, which the study proposed for the life of the powerplants. At the 200,000-megawatt level, the strip mines would consume from 50,000 to 175,000 square miles of surface during the 35-year period. In addition, each coal gasification plant, producing 250 million cubic feet of gas per day, would use almost 8 million tons of coal a year, eating up more land, as well as 8,000 to 33,000 acre-feet of water (estimates vary widely) and 500 megawatts of electric power.

The astronomical figures continued. At the 50,000-megawatt level nearly 3 per cent of the tristate region would be strip-mined, an area more than half the size of Rhode Island. The transmission lines would require approximately 8,015 miles of right of way, which, with one-mile-wide multiple-use corridors, would encompass a total of 4,800 square miles, approximately the size of Connecticut. Power losses over the network of lines would exceed 3,000 megawatts, greater than the present average peak demand requirements of Manhattan, and would raise a serious problem of ozone production.

. A population influx of from 500,000 to 1,000,000 people might be expected in the tristate area. (The present population of Montana is 694,000; Wyoming, 332,000; and North Dakota, 617,000.) Half a million newcomers would mean a 500 percent increase in the present population of the coal areas and would result in new industrial towns and cities, putting added pressures on the states for public services and increased taxes. The quality of life, as well as the environment, would change drastically. At the 50,000-megawatt level, the proposed plants, even with 99.5 per cent ash removal, would fill the air with more than 100,000 tons of particulate matter per year, detrimental to visibility and health. The combustion of the coal would introduce dangerous trace elements like mercury into the atmosphere; and the plants would emit at least 2,100,000 tons of sulfur dioxide (yielding, in turn, sulfurous and sulfuric acids that would be deposited by the wind on farms, ranches, communities, and forests) and up to 1,879,000 tons of nitrogen oxides per year. Though the study ignored the prospect, living in the Colstrip-Gillette area, with ten 10,000-megawatt powerplants, not to mention an unspecified number of coal gasification plants as neighbors, could be lethal.

If the simplistic report, blithely ignoring the need for scores of impact studies, bewildered environmentalists, it sent peals of alarm among many of the people of the three states. The powerful energy companies and utilities of the country, with the encouragement of the federal government, were going to turn them into an exploited and despoiled colony, supplying power to other parts of the nation. Far from planning the orderly development of their region, the study had considered only the needs of industry and, without publicity, without public hearings, without representation from, or accountability to, those who would be affected, had shown a green light to the devastation of life on the Great Plains.

Throughout the region, individuals were soon comparing notes and discovering that a coal rush of gigantic proportions was, indeed, already under way. Lease brokers, syndicate agents, and corporate representatives—many of them from places like Louisiana, Texas, and Oklahoma, with a long experience of wheeling and dealing in gas and oil rights—had been swarming across the plains country, and more coal lands than anyone had dared imagine were already locked away in exploration permits and leases. Ranch owners found out with a start that neighbors had already signed agreements, and that a strip mine and power plant might soon be disturbing their cattle or destroying their range. Irrigation farmers learned of corporations from Pennsylvania, Ohio, and Virginia buying options on the limited supplies of water, and worried about their own water rights. The areas of busiest activity matched the study's proposed sites for development, and rumors multiplied of industrial plans and commitments being made so fast that they could not be stopped. In half a dozen districts in Montana and Wyoming that seemed most threatened, ranchers and farmers hastily organized landowners' associations, which banded together as the Northern Plains Resource Council—a loose federation based with volunteer officers and staff in Billings, Montana—to pool their information, pledge landowners to hold out against the strippers, and contest the coal interests in the courts and the state capitals.

From the start, opposition to the coal

development was hobbled by a lack of reliable knowledge of what was going on. In the first place, it soon became evident that the coal and energy companies that were buying up the land and making plans to exploit the region had rejected the proposals of the North Central Power Study even before the document had been made public, and were proceeding, instead, on a voracious, every-developer-for-himself basis. Alarming as the suggestions of the Bureau of Reclamation and the utilities had been, they had nevertheless reflected the federal government's desire to guide development according to a comprehensive and orderly plan. Even the critical environmentalist groups had recognized that, if coal development was inevitable, the study was something with which to work—a plan susceptible to detailed examination and protective actions and modifications that would ensure a minimal degradation of the human and natural environments.

Now the study was nothing but a check list of some—but far from all—of the opportunities for the fastest corporations with the most dollars. Aside from alerting the region's people to the scope of the calamity they faced, the study's effect was to draw additional attention in Wall Street and elsewhere to the possibilities of the immense coal fields and accelerate what was becoming a frantic, modern-day version of the California Gold Rush. By October 1972 the guideline aspects of the study were dead, and Secretary of the Interior Rogers C. B. Morton, aware of the concern in the region over the chaotic exploitation taking place, announced the formation of an interagency federal-state task force and the launching of a Northern Great Plains Resource Program to assess the social, economic, and environmental impacts of the coal development and, hopefully, "coordinate on-going activities and build a policy framework which might help guide resource management decisions in the future."

It was pretty much a case of locking the barn door after the horse was stolen. The 1971 study had been issued well after the coal rush had started, and the new study

group—which was criticized because it did not provide fully enough for the participation of the public—would not release its final report until December 1975, although results were expected to be "incorporated into regional planning and decision-making by the end of the first year," or October 1973. In view of the rapid developments taking place, even this seemed too late. Regional planning by then would be almost impossible.

Meanwhile, other factors were adding to the confusion. Without the over-all guidance, planning, or authority of any federal or state agency, it became difficult for anyone, including state officials, to assemble accurate and comprehensive information about who was acquiring what rights and where, and what they intended to do with them. The roster of those who were buying coal deposits read like a who's who of the energy industry: Shell Oil, Atlantic Richfield, Mobil, Exxon, Gulf, Chevron, Kerr-McGee, Carter Oil, Ashland Oil, Consolidation Coal (Continental Oil), Peabody Coal (Kennecott Copper), Westmoreland Coal, Reynolds Metals, North American Coal, Kewanee Oil, Kemmerer Coal, Concho Petroleum, Island Creek Coal (Occidental Petroleum), Cordero Mining (Sun Oil), Arch Minerals, Hunt Oil, Pacific Power & Light, Valley Camp Coal, Penn Virginia Corporation, National Gas Pipeline (Star Drilling), Farmers Union Central Exchange, Cooper Creek, and Western Standard.

They were all there, but so, also, were subsidiaries, subsidiaries of subsidiaries, fronts for bigger names, syndicates, partnerships, speculators, and lease brokers. Rights were acquired by a firm named Meadowlark Farms, suggesting to the public the bucolic image of dairy cows and buttercups rather than a coal strip mine. The company was a subsidiary of Ayrshire Coal Company, formerly Ayrshire Collieries Corporation, which with Azure Coal Company was owned by American Metal Climax's Amax Coal Company. The worldwide construction firms of Peter Kiewit Sons in Omaha, Nebraska, and

Morrison-Knudsen Company in Boise, Idaho, also held rights; the former, moving into Montana in a big way, owned the Big Horn and Rosebud Coal companies and half of Decker Coal Company, and the latter held 20 per cent of Westmoreland Resources. There were names relatively unfamiliar to the public: Temporary Corporation, Tipperary Resources, Pioneer Nuclear, J & P Corporation, Ark Land Company, Badger Service Company, Allied Nuclear Corporation, BTU Inc., as well as dozens of individuals like Violet Pavkovich, Fred C. Woodson, E. B. Leisenring Jr., Billings attorney Bruce L. Ennis, and lease brokers Jase O. Norsworthy and James Reger.

All of them, to a greater or lesser extent, were engaged competitively, and the securing of permits and leases and the making of plans and commitments for exploitation were done with great secrecy. But the necessity to conceal activities and intentions from rivals also frustrated interested officials and the public, who were kept in the dark about plans for such projects as strip mines, power plants, new railroad spurs, water purchases, and coal gasification plants—all of which would affect their environment and lives—until the companies were prepared to announce them. By that time, commitments had been made, and though clues to some of the projects—like the number of companies or the amount of capital involved, the large size of a water pipeline, or the required tonnage of coal—implied immense undertakings with serious impacts on the people and environments of large areas, questioners had to grapple for detailed and meaningful information and were at a disadvantage.

Perhaps the greatest confusion stemmed from the complex ownership rights to the coal and the land surface above it. Some of the coal is owned by the federal government and is administered by the Bureau of Land Management. Some is owned by the states; some by the Union Pacific or Burlington Northern railroads (though their legal rights to the coal, acquired originally with the railroad land grants of the last century, are being questioned by certain congressmen

and organizations); some by Indian tribes (the Crow, Northern Cheyenne, and Fort Peck reservations in Montana and the Fort Berthold reservation in North Dakota); and some by private owners. A purchaser may secure an exploration permit or lease for the coal; but to get at it, he also has to deal with the owner of the surface—which frequently produces a problem. The surface rights, again, might be owned by the federal government, the states, the railroads, the Indians, or private owners. Where the same interest owns both the surface and the coal and is willing to part with them, there is no complication. But more often than not, private ranchers own or lease land above coal that does not belong to them. In the past, they or their forebears might have gotten their land from the federal government (under the various Homestead Acts) or from the railroads, but in both cases the government and the railroads reserved the mineral rights, including the coal, for themselves. Similarly, when the Crow Indians ceded some of their land to the government in 1904 and the government opened it to white settlers, the government retained the mineral rights. But in 1947 and 1948 it returned those rights to the Crows, creating a situation of Indian tribal ownership of coal under white-owned ranches.

Strip mining was not a concern when the original homesteaders bought their lands. If the coal were ever to be mined, they and the sellers undoubtedly envisioned deep mining, which would have disturbed only a small part of the surface. A strip mine is a different matter, for it eats away the pasture, range, farmland, and buildings that constitute one's home and means of livelihood. The question of the surface owner's rights versus the rights of the purchaser of the coal beneath his land is a matter of contention and will inevitably be tested in the courts. But the necessity of acquiring separate items of coal rights and surface rights from different owners (and sometimes—when trying to create a large compact block of coal—from several different adjoining owners of both the coal and the surface) introduced bitter conflict and more confusion to the harassed region.

In Montana, a surface condemnation law that favored the coal purchasers made the situation worse. Under the influence of the Anaconda Company, which had wished to condemn land for copper mining at Butte, that state in 1961 had declared mining a "public use" and had given mineral companies the right of eminent domain. Speculators, lease brokers, and the agents of corporations acquiring coal rights—sometimes even before they had bought the coal—now abused that law. They frightened many Montana landowners into signing exploration permits and leases or selling their lands on the purchaser's terms ("better than you'll get from any court"), and threatened condemnation proceedings against those who resisted. Episodes of angry confrontation and near violence multiplied as the purchasers—nervously eyeing the progress of competitors and aware of large secret corporate plans that depended on timely acquisitions—pressured the landowners.

The unpleasantness visited on the Boyd Charter family in the Bull Mountain region north of Billings is typical of many small, human agonies. The Bull Mountain area is a particularly fragile one, a grassy parkland with irregular topography that includes rimrock walls and picturesque hills covered with dense growths of ponderosa pine. Because coal seams are exposed on the rock walls and outcrop on the hillsides, contour stripping—the most destructive of all open-cut techniques—probably would be necessary, and reclamation to restore the present natural beauty and scenic values would be virtually impossible. A Montana Coal Task Force, established by the state government in August 1972, urged that no strip mining be permitted there unless a severe national coal shortage occurred in the future (an unlikely event for half a millennium), and Montana's Senator Mike Mansfield singled out the area as one district of the state in which strip mining should be banned outright.

Nevertheless, the Bull Mountain area contains approximately 130 million tons of coal, the rights to which were quietly purchased by Consolidation Coal Company in permits and leases from Burlington Northern railroad and the State of Montana. Owned by Continental Oil Company (whose chairman, John G. McLean, also head of the National Petroleum Council, has been in the forefront of industry leaders warning of an energy crisis and advocating governmental encouragement of western coal development), Consol, as the coal company is known, plans an $11.5 million strip mine in the Bull Mountains, to be worked over a 25-year period. Its initial production would be about two million tons a year, but the figure would rise. For the present, there are no plans for a mine-mouth power plant, and there is not enough coal to sustain a coal gasification development. Most of the coal would be shipped by train to customers in the upper Mississippi Valley, and a total of some 3,500 acres of the Bull Mountains would be subject to mining for those distant users, with additional acreage being disturbed by roads, installations, and the operations of the miners.

Though Consol officials recently made verbal promises to reshape the stripped land "to a contour similar to and compatible with its virgin contour, to save and replace topsoil, to revegetate, fertilize, and continue reclamation work, with as many replantings as necessary, until reclamation is successful," the company's leases, reflecting a traditional looseness in state and federal regulations, bound them to no such obligations. For instance, a lease made with Montana on June 3, 1970, for 640 acres of state-owned coal in the Bull Mountains merely obliged Consol "so far as reasonably possible" to "restore the stripped area and spoil banks to a condition in keeping with the concept of the best beneficial use," adding vaguely that "the lessee may prescribe the steps to be taken and restoration to be made." A $1,000 bond accompanied the lease, considered hardly enough to guarantee the reclamation of one acre in that area.

In 1970 Consol set about purchasing the

surface rights necessary to make exploration drillings and mine the Bull Mountain coal. Many of the people in the nearest town, Roundup (population 2,800), welcomed the development. Small-scale deep mining had been done for many years in the Bull Mountains; Tony Boyle, the former United Mine Workers president, had come from the area; and the townspeople, without landholdings at stake, saw prosperity for themselves in Consol's promises to spend $1,400,000 each year in the region and employ 80 men, whose needs, said the company, would generate 240 other jobs. To the Boyd Charters and other ranchers, however, plans for the strip mine became a nightmare.

Originally from western Wyoming, the Charters and their three sons and a daughter ran cattle on approximately 20 sections of land, 10 of which they owned and the rest leased from the Burlington Northern. Without warning, they were visited one day by a land agent from Consol, who told them that the company had bought the coal beneath their land from the federal government and the railroad and now wanted to drill exploratory core holes preparatory to mining. He produced a form for them to sign, offering one dollar to release the company from any damages done to their property by the drilling. When the Charters refused to sign, the agent left them and made a tour of other ranches, relating, according to the word of one ranch owner, that the Charters had signed, and thus winning the agreement of a few of them.

The company thereafter began harassing the Charters. Higher officials, including a Consol regional vice-president from Denver and company attorneys, began showing up at their home, increasing the pressure on them, and gradually driving the family frantic with worry. After numerous sessions the visits stopped, and the Charters wondered if condemnation proceedings, under the Montana law, were to be instituted against them. Then, one morning, they heard a racket near their house. They ran out, discovered a Consol crew drilling core holes on a deeded part of their land, and ordered them to stop.

A fat man, according to Boyd Charter, came over to them, threatening a fist fight. "I got as much goddamn right on this land as you," he said. The infuriated Charters finally drove the crew off the property, and they have heard nothing more since then from Consol. But the company has tested all around the Charter ranch, it can get Burlington Northern to break the lease for its part of Charter's holdings, and it still intends to strip-mine the Bull Mountains in the near future. Far down the line from Continental Oil's national policy planner, John McLean, this small Montana ranching family is one of his victims.

Many similar conflicts have occurred elsewhere. Almost 150 miles by road southeast of the Bull Mountains, the Billings firm of Norsworthy & Reger helped Westmoreland Resources (a partnership of Westmoreland Coal, Kewanee Oil, Penn Virginia, Kemmerer Coal, and Morrison-Knudsen) assemble a package of rights to about one billion tons of very rich coal deposits at the head of Sarpy Creek for a huge strip mine and at least one coal gasification plant. The area, a beautiful basin under the pine-covered Wolf Mountains in southeastern Montana, encompassed land ceded by the Crow Indians. White ranchers now owned the surface, but the tribe still owned the coal. In a series of transactions, Norsworthy & Reger and E. B. Leisenring, Jr., a director of the Fidelity Bank in Pennsylvania, won permits for approximately 34,000 acres of Crow coal—apparently paying the Indians an average of $7.87 per acre and a royalty of 17.5 cents a ton for the first two years of production and 20 cents a ton for the next eight years—and then assigned their rights to Westmoreland Resources.

Surface rights still had to be won from the ranchers. Under threat of condemnation, some of them sold, but others resisted, including the family of John Redding. Westmoreland and its agents became desperate for the Reddings' signatures. The company had plans to begin stripping in March 1974; a giant 75-cubic-yard walking dragline was under construction; contracts were being made to sell 76.5 million tons of coal over a 20-year period to four Midwestern utilities (Wisconsin Power and Light, Iowa's Interstate Power Company, Wisconsin's Dairyland Power Cooperative, and Minnesota's Northern States Power Company to fuel a 1,600-megawatt generating plant near Henderson, Minnesota); and a 10-year option agreement for the delivery of a whopping 300 million tons of coal had been signed with Colorado Interstate Gas Company, which was planning to build up to four coal gasification plants in the region.

Moreover, the abundant and rich coal deposits guaranteed enormous growth potential in the value of the area. Consol was acquiring coal and surface rights nearby, with leases in which the language implied coal gasification plants and a large-scale industrialization of its own; and just to the east was still another huge developing coal-and-power center at Colstrip, where Montana Power Company was building new power-plant units, two transmission lines of which would come through the Sarpy district to Hardin, Montana. The region was going to become one of the principal new coal-based industrial centers in the northern plains, with a city of perhaps 25,000 people, and Westmoreland's plans and needs to assemble and invest capital required the combining of their package of coal and surface rights as quickly as possible.

On February 25, 1972, Billings attorney Bruce Ennis served written notice on the Reddings that unless they agreed to sell the entire, or necessary, portion of their ranch to Westmoreland at $137 an acre within one week, Westmoreland would begin condemnation proceedings against them. John Redding had come to Sarpy 56 years before, had lived in a tent, then a cabin, and finally had established a home, a family, and a 9,000-acre ranch. Through good years and lean, fighting the elements and the Depression, the Reddings had reflected the tradition of Westerners who treasured the place they lived because they could "stand tall and breathe free," and they now proved tougher than the coal company. Calling Ennis' bluff,

they stood firmly over their property with gun in hand, and the company eventually backed away. "We've gotten enough people to agree that, at least for the time being, we don't have to go the condemnation route," Westmoreland's president, Pemberton Hutchinson, announced. "We needed to settle with eight landowners, and we settled with six—and that's enough." (Actually, at last count, there were still three holdouts, including one who claimed that a Westmoreland agent had told her, "You'll be down on your knees begging to sell." The lands of the holdouts are so strategically located as to split the coal company's surface rights and limit initial operations to a comparatively small tract.)

In four instances in a different area, but one connected with the Sarpy Creek development, landowners actually had condemnation proceedings instituted against them, but by Burlington Northern railroad, which is building a 37-mile spur line from its main tracks at Hysham, Montana, up Sarpy Creek to take out coal from the new Westmoreland mine. Ranchers and other landowners opposed the railroad's demands for right-of-way easements, often through the best parts of their land, and the conflicts became angry and tense. One woman, harassed by the railroad, suffered a nervous breakdown. Another, Mrs. Montana Garverich, 67 years old, a widow with 14 grandchildren and 8 great-grandchildren, who had lived on her land since 1912 and still operated her 4,000-acre ranch with the help of some of the children, fended off attempts to take her bottomland and was hauled into a U.S. District Court by the railroad. When the court found in favor of the Burlington Northern, Mrs. Garverich announced she would appeal, and the railroad, not relishing further action and its attendant publicity, rerouted its line in several places and dropped its suits.

As might be expected, hundreds of landowners in the three states, willingly or unwillingly, have already leased or sold their surface rights. Some, getting on in years and tired of a strenuous, often harsh, existence on the plains, did so happily, taking what they could get and planning on retirement to an easier life somewhere else. Others became frightened, were cajoled, or failed to understand what was involved, and signed whatever was asked of them, while still others hired lawyers, dickered back and forth, and finally felt they had outsmarted the purchaser and had gained a good deal for themselves. On the whole, the negotiated terms differed from one lease to another, depending on how badly a company wanted a particular right and how resistant the owner was. One rancher may have given up all his rights for a dollar an acre, while his neighbor received more than $100 an acre and a small percentage royalty on each ton of coal taken from beneath his surface. The operations of the land buyers inevitably stirred up jealousies and divisions within families and among old friends and neighbors, some of whom wanted to sell out while others hoped for a united show of resistance against the purchasers. At Sarpy Creek, at Otter, and elsewhere, distrust and defensiveness soured relationships that had existed happily for decades.

A division of opinion also affected those who did not have land at stake. Like the townspeople of Roundup, many citizens in all three states regard the coal-field development as an economic boon to the region and, not sharing the torment that such a point of view visits on a Montana Garverich or a John Redding, agree with the comment of Los Angeles financier Norton Simon, a development-minded director of the Burlington Northern: "For a state like Montana to have only 700,000 people is cockeyed." But others enjoy living on the northern plains precisely because of the small population and are fearful of pollution, the degradation of the environment, higher taxes, a change in life style, and other unfavorable impacts that the development will have on their part of the country and their lives.

Meanwhile, the absence of hard information concerning exactly what the impacts will be, and when they will start to be felt, has become something of a scandal. Despite all the developments that have occurred, not a single meaningful impact study has yet been made of any one of them; nor will an in-depth study be available for the region as a whole, or for any one of the affected states, until Secretary Morton's resource program report is finished at the end of 1975. It has been estimated that more than 5.5 million acres of federal- and Indian-owned land have already been let out in coal permits and leases. More acreage has been let out by the states, the railroads, and private individuals. In Montana, the Northern Plains Resource Council, checking documents on file in many of the counties, estimates that at least 1.7 million acres, more than half of that state's surface covering economically strippable reserves, are already signed away. The figures in Wyoming and North Dakota are believed to be far greater. But such information, lacking the addition of anything but occasional and very brief and bare corporate announcements on how a certain quantity of coal at some particular locality is to be utilized, has only increased the sense of helplessness.

In Wyoming, with strippable coal reserves of 23 billion tons in seven major coal areas, only a few of the mammoth projects that are certainly in store for the use of the resource have yet been described with any detail. Near Rock Springs, the $300 million, 1,500-megawatt Jim Bridger power plant is being constructed by Pacific Power & Light and Idaho Power Company, threatening an even worse degradation of Wyoming's air quality than is already caused by Pacific Power's offending 750-megawatt Dave Johnston plant at Glenrock on the North Platte River. And near Buffalo, Reynolds Metals has proposed the organization of a consortium of companies to build and operate a uranium enrichment plant requiring, according to Reynolds, "millions of kilowatts" of power. Coal for the power plant to supply electricity to the $2.5 billion project would come from a strip mine at the site, utilizing deposits of more than two billion tons owned by Reynolds. To provide the large amount of water that would be required, Reynolds has bought nearby Lake De Smet and has dammed Piney Creek for the diversion of its water into the lake, causing fears already among ranchers and farmers in that semiarid area of limited water. The uranium plant, the first one to be privately owned, might export some of its product to Japan; similarly, coal producers are known to be shopping for customers outside the United States. This raises the question of how valid is the exploitation of Western coal as an answer to the so-called energy crisis.

The Sierra Club, the Sheridan County Action Group, and several other Wyoming citizens' bodies, together with editor Tom Bell of the crusading *High Country News* in Lander, Wyoming, have tried to ring the alarm bells in that state. Very much a specter to them is the North Central Power Study's suggestion that ten 10,000-megawatt plants could be built in the Gillette area. That possibility is made more real by the knowledge that the massive, 100-mile-long Wyodak beds, all in Campbell County, contain more than 62 billion tons of coal—the national high for a county—and that a single township contains 2.87 billion tons in spectacular seams averaging about 70 feet in thickness and lying within 500 feet of the surface. A number of energy companies have paid record prices—as high as $505 an acre— for the Campbell County coal, but although the Black Hills Power & Light Company has been stripping some 500,000 tons of coal annually from the area for years, only one new development has yet occurred. In May, 1973, American Metal Climax's Amax arm opened the Belle Ayr mine to strip 6 million tons a year from its 6,000-acre holdings. Kerr-McGee, Exxon, Atlantic Richfield, Ark Land Company, Mobil, and Cordero Mining (Sun Oil) are among the other large lease holders in the area, all capable of opening additional strip mines and building polluting complexes.

Moreover, the State of Wyoming generally, its governor and junior senator, and a majority of the members of the state legislature are development-oriented, welcoming the coal industrialization as a boost to the state's economy, and showing little appetite for conducting significant studies or enacting sufficiently strong reclamation and other

laws that would give protection to the state but, at the same time, irritate and impede the energy companies.

In Montana, where large-scale coal mining is a new fact of life, the reverse is true, and state officials and agencies have, if anything, been ahead of many of the people in evidencing genuine concern over the uncontrolled character of the coal exploitation. On March 9, 1971, the state passed an Environmental Policy Act, which among other things, created a 13-member Environmental Quality Council, headed by George Darrow, a Billings geologist and state representative who had been one of the chief architects of the act. Fletcher E. Newby, another concerned Montanan, became executive director of the council, the functions of which include watchdogging the environmental problems in the state, recommending protective actions, and furthering state environmental impact statements. On August 2, 1972, on the recommendation of the council, the state created a Coal Task Force to watch the developing coal situation, identify problems, and recommend needed legislation or other action.

Both Montana bodies have tried to gather adequate information for laws necessary to protect the state, but cooperation from the federal level has been sorely missed. Aware of the regional character and the enormity of what was just beginning, the governor and state officers, from December 1971 on, appealed to the Environmental Protection Agency and various federal officials for a coordinated federal-state study of the total regional and state impacts of the coal development, but until the launching of the Interior Department's long-range study in late 1972, they were told that reviews could only be made of impact statements on individual projects. This was ironic, in view of the fact that the regulations requiring the filing of such statements were, themselves, not being enforced.

By the fall of 1972, the every-man-for-himself development in Montana, occurring without meaningful impact statements or regulations strong enough to provide protection to the environment, was becoming alarming. A study made by Thomas J. Gill for the state Environmental Quality Council, and based on data supplied by various state agencies, pointed out that total strip-mined coal production in Montana would jump from 1.5 million tons in 1971 to 16 million tons in 1973 and to 75 to 80 million tons in 1980. At the 16 million-ton level in 1973, 275 to 520 acres of Montana land would be disturbed by the mines. Four strip mines were already in operation in the state: at Colstrip, the Rosebud Mine of Western Energy, owned by Montana Power, was producing 5.5 million tons a year and in five years would raise the figure to 11.5 to 13 million tons, disturbing 240 to 350 acres annually. Also at Colstrip, Peabody Coal Company's Big Sky Mine was producing 2 million tons a year and would double the production in five years, disturbing 100 acres a year. In addition, Peabody was writing a mining plan for a new mine at Colstrip on 4,306.5 acres leased on April 1, 1971, without a preliminary environmental impact statement, from the Bureau of Land Management. At Decker, Montana, where Decker Coal Company, owned by Peter Kiewit and Pacific Power & Light, possessed 1 billion tons of strippable coal, the company had startled long-time ranchers in the area by disrupting a large part of the peaceful countryside within a matter of months, building a 16.5-mile-long railroad spur line, rerouting the main road, and beginning operations on a huge strip mine committed to ship 4 million tons of coal annually to the Midwest. The fourth mine, a smaller one operated by Knife River Coal Company, produced about 320,000 tons a year and disturbed 20 acres annually. The state also expected the big Westmoreland mine at Sarpy, the Consol mine in the Bull Mountains, and another Peabody mine on the Northern Cheyenne Indian Reservation to begin operations within a couple of years.

Reclamation of the mined land was only one of the problems posed by the increased stripping in the state. Neither federal nor state regulations written into the leases car-

ried any guarantees that the lands would be successfully restored, and railroad, private, and Indian leases were so deficient that they almost guaranteed that there would be no reclamation. For anyone concerned about the preservation of the land, Montana Governor Thomas L. Judge pointed out to Congress early in 1973, "the lease agreements make sinister reading." One contract, for instance, gave a company the "right to use and/or destroy so much of said lands as may be reasonably necessary in carrying out such exploration and mining." Reclamation experiments were being carried out by Big Horn Coal Company and at Colstrip, but they were inconclusive. The best estimates were that it would take many years and successive replantings with much fertilizer and large amounts of water, and would cost upward of $500, perhaps as much as $5,000, per acre, before one could tell if reclamation had truly worked in that dry and fragile land of thin topsoil. Yet the leases carried no bonds, or ridiculously low ones, usually less then would be required to pay for the restoration of a single acre. A company could make a try at reclamation, then walk away, forfeiting the bond and leaving it to the state or someone else to struggle with reclamation problems.

In addition, there was little information available about water problems that would result from the strip mines. Some of them would seriously disturb patterns of drainage and surface runoff; at Decker, aquifers that lie among the coal deposits would disappear. The implications for the entire region's future water supply, especially as it felt the impact of increased demand for industry, were great, but no meaningful hydrological studies existed.

The power-plant problem in Montana, Gill's study showed, was still a relatively small cloud in the sky, but already an ominous one. On a 50-50 ownership basis with Puget Sound Power & Light Company, Montana Power was constructing two 350-megawatt units of a new plant at Colstrip, and had announced two more units of 700 megawatts each, with Puget Sound owning 75 per cent of them. The first units were to be completed in 1975 and 1976, and the next two in 1978 and 1979. An initial environmental impact study, based on data supplied by Montana Power, was submitted by the State Department of Health's Division of Environmental Sciences, but was deemed inadequate and deficient on many counts. Fears of ineffective emission controls; widespread pollution harmful to vegetation, trees, and livestock; degradation of the quality of the air; and disruption of the ecosystem of a large region all seemed justified to many of those who analyzed the study. A final, 400-page version was more complete, but failed to still the fears. Alarm was heightened, moreover, by the prospect that additional polluting power plants and other industrial installations were already being planned for the same area. In its own notice of appropriation for Yellowstone River water in 1970, Montana Power had indicated it planned to run a 31-mile-long, 60-inch pipeline, capable of conveying 250 cubic feet of water a second, from the river to Colstrip. This was more water than the power plant units would need, would divert from downstream users about one-eighth of the Yellowstone's water at low flow in an average year, and suggested a future use for something else, perhaps a coal gasification plant, at Colstrip.

As to water, the study noted that the state's total existing and potential supply from the rivers of the Yellowstone Basin was 1,735,000 acre-feet a year; yet energy companies (possibly planning gasification and liquefaction plants) had already received options from the Bureau of Reclamation for 871,000 to 1,004,000 acre-feet per year and had requested or indicated interest in another 945,000 acre-feet per year from those streams! Where this would ultimately leave farmers, ranchers, towns, Indian tribes, and others with claims on the water was not stated, but Gill suggested that "it seems safe to assume that a supply of water sufficient to accommodate the coal developments . . . would require complete development of the area's water resources," includ-

ing more dams, as well as the interbasin and interstate transportation of water via a network of aqueduct pipelines, built by the Bureau of Reclamation.

As if to underscore the pressures that were already building for water, Gill noted an intention of the HFC Oil Company of Casper, Wyoming, to construct two or more gasification plants in Dawson County, Montana; the proposed Colorado Interstate Gas plant at Sarpy, and another one near Hardin; and Consol's plan to build a complex of four of them on the Northern Cheyenne Indian Reservation in Montana. Since a three-plant complex would require 50,000 to 75,000 acre-feet a year, the total water needs, he suggested, would probably limit the number of complexes in Montana "to 12 or less," an observation that, in fact, focused on the one definitive limit (outside of the vast total coal supply) to the ultimate coal-field development of the entire region. In other words, he who gets the water can build, and after the water is all taken, there can be no more users.

Gill's study also dealt with looming problems of transmission line corridors. Much of the power generated at Colstrip would be transmitted to consumers in the Pacific Northwest, requiring corridors for new lines across central and western Montana as well as Idaho. Conflict was already breaking out with landowners over rights of way for a new 40-mile-long corridor in the Bitterroot Valley in the western part of the state, and it was only the forerunner of what was sure to be a mass of angry confrontations as more plants were built and more corridors were sought to carry power east and west to distant consumers.

The report finally mentioned problems of air pollution, the increase in population, and changes in the human environment. All were matters of pressing concern to the state, but in the absence of overall planning and controls, none of them could be discussed intelligently until plans for each project were made public. Then the impacts would have to be assessed on an *individual* project basis—a sure formula for the rapid deterioration of the human and natural environment.

Montana's growing distress over these problems was reflected when the state legislature convened early in 1973. Numerous regulatory bills were introduced, and by April several significant ones had become law. Coal was eliminated from the condemnation statute, and operators were prohibited from prospecting or mining until they had secured the permission of the owners of the surface rights. Both measures came too late to help all those who had already sold their surface under threat, but they took some of the pressure off the many Boyd Charters and John Reddings who were still holding out. Ahead, however, lay legal battles over the rights of coal purchasers versus those of the landowners. The companies, claiming that other state and federal statutes gave them rights, felt that they still had ways of getting the surface rights they needed. The legislature also passed a strong reclamation law that spelled out required reclamation procedures in detail, increased sharply the state tax on coal, set up a Resource Indemnity Trust Fund to rectify damage to the environment caused by the extraction of nonrenewable natural resources, established a centralized system for water rights, and created a power facility siting mechanism, giving the state's Department and Board of Natural Resources and Conservation authority to approve the location of generation and conversion plants, transmission lines, rail spurs, and associated installations.

Still missing, at that late date, was convincing evidence of concern or commitment on the part of agencies of the federal government. A major portion of the coal lands in the northern plains is public domain, administered by the Bureau of Land Management of the Department of the Interior. Every aspect of the bureau's practices in the granting of federal coal permits and leases has been severely criticized in Congress and by the General Accounting Office. In March 1972, the General Accounting Office focused on the question of whether the United States was receiving a fair price for its coal, and concluded that it probably was not.

In the past, the lack of competition for western coal had permitted the securing of permits and leases for bonuses and royalties so low as to constitute a virtual steal in present-day terms. But the agreements ran for twenty years before they could be adjusted, and many of them still have long periods to run before the royalty can be raised. So the "steals" on those leases continue. Moreover, even the prices paid to the government today can be questioned. Permits and leases are awarded to applicants who pay the highest bonus in competitive bidding. But the royalty rate which the applicant must pay the government for each ton of coal produced is recommended to the Bureau of Land Management by the U.S. Geological Survey and is set as a fixed term or percentage for a specified number of years. Of late, the figures have usually been 17.5 cents for subbituminous coal and 15.5 cents for lignite—considered by many critics to be too low, in view of actual market conditions. In non-Bureau of Land Management deals, for example, producers have revealed with uninhibited realism the extent of their ravenous appetite for coal lands by offering higher royalties and letting speculators who assign them their rights tack on increased tonnage royalties for themselves. Moreover, companies who have leased the coal are now asking the federal government to do research that will establish the value of the coal—something which, if done before the leasing, might have gotten the government a higher price for it.

The General Accounting Office was even more critical on other points. Speculators could buy rights cheaply, hold onto them for long periods of time with no plans to mine the coal, then sell the rights at a large profit in the rising market. Reclamation and environmental requirements were almost nonexistent in older leases, and the Bureau of Land Management was ignoring this deficiency, waiting for each lease to come up for renegotiation on the twentieth year after the lease had been made. Newer leases had stiffer requirements, but they were not being enforced. In August 1972 a second General Accounting Office report spelled out its criticisms on this score more sternly, aiming its charges also at the Bureau of Indian Affairs, which was administering the leases of coal owned by Indian tribes. Technical examinations of environmental effects were not being conducted by either agency; coal operators were permitted to proceed with exploration and mining without approved plans; compliance and performance bonds covering the requirements, including reclamation, were not being obtained—or, if in some cases they were, the amounts were insufficient to cover estimated reclamation costs; required reports were not being received from operators; and procedures did not exist for the preparation of environmental impact statements, so they were not being made.

The criticisms pinpointed numerous violations of federal laws and the code of federal regulations by both the Bureaus of Land Management and Indian Affairs. The Department of the Interior made no meaningful response, and in October and November 1972, both Russell E. Train, chairman of the Council on Environmental Quality, and William D. Ruckelshaus, then administrator of the Environmental Protection Agency, urged the department to undertake remedial actions. Train particularly recommended an environmental impact statement on the over-all coal leasing program. Except for directives to the field for a minor tightening up of enforcement procedures, silence in Washington continued, presumably because of a desire not to do anything until the President's national energy policy could be prepared and made public or the Northern Great Plains Resource Program study could issue a report.

Meanwhile, Secretary of the Interior Morton refused to uphold a resolution passed by the United States Senate on October 12, 1972, calling for a moratorium on further coal leasing of federal lands in Montana for one year or until the Senate could act on strip-mining legislation. Senators Mike Mans-

field and Lee Metcalf of Montana and Frank E. Moss of Utah wrote angrily to Morton, terming his decision "arrogance of the executive branch" and "unconscionable," and criticizing his statement that the Senate could rely on the regulations of the Interior Department to guarantee "environmentally acceptable mining."

Actually, after April 1971, the Bureau of Land Management had held up the approval of all federal coal permits and leases in the northern plains until it could assess how much coal was already under lease and ascertain the demand and need for additional coal. It was conducting a study of the coal-rich Birney-Decker area in the Tongue River Basin of southeastern Montana, where many applicants hoped to secure rights to deposits of some 11 billion tons, and it used the study as one of the excuses for the unofficial moratorium. But the study was released (angering the coal companies by proposing the mining of only a limited strip, two townships wide, just north of the Montana-Wyoming border—"leaving out the best coal and including only the poorest area," according to one operator), and still no new Bureau of Land Management leases were approved. But now, according to Secretary Morton, the department would proceed "cautiously on a case-by-case basis," suggesting to the companies that even the desired part of the Birney-Decker region would soon be opened to them.

In a Senate speech on January 12, 1973, Mansfield called attention to what the energy crisis was doing to his state, complaining that the individual landowner was being treated "shabbily," attacking the utilities and coal companies for "approaching this situation with little compassion and regard for the future of this part of our nation," and asserting that "if we cannot have orderly and reasonable development of the vast coal resources in Montana and the west, there should be no strip mining of coal."

Meanwhile, if the federal government was not protecting the non-Indian people of the region, it was actually selling out the Indians. The General Accounting Office criticisms of the Bureau of Indian Affairs merely scratched the surface of the derelictions of government trust obligations to the tribes. Indian lands in Montana contain approximately one-third of the state's total 30 billion tons of strippable coal reserves. Some of it is owned by the Fort Peck Indian Reservation in northeastern Montana, but the largest and most valuable deposits underlie the entire Crow and Northern Cheyenne reservations in the southeastern part of the state, roughly in the heart of the prized Colstrip-Gillette area. Beginning in 1966, the Bureau of Indian Affairs—which as legal protector of Indian resources must approve all tribal permits and leases—brought coal companies to the Northern Cheyenne tribal council, encouraging that body ultimately to sign a total of eleven exploratory permits for the tribe's land. Uninformed of the ramifications of strip mining and of the omissions and deficiencies of Bureau of Indian Affairs coal leases (the terms and regulations of which adhered pretty closely to those of the Bureau of Land Management), the tribal council put its trust in the Bureau of Indian Affairs, one of whose officials was quoted as saying as late as 1972, "There are indications coal will be a salable product for only a few years." Encouraged to take money while the taking seemed good (bonuses, rentals with a floor of 1 dollar an acre, and royalties of 17.5 cents a ton), the tribe let out to Peabody, Amax, Consol, Norsworthy & Reger, and Bruce Ennis a total of 243,808 acres—a startling 56 per cent of the reservation's entire acreage!

The permits were loosely worded as to reclamation and other environmental considerations; and, like Bureau of Land Management and most other permits, gave the operators the right to exercise lease options which were appended as part of the original agreements and which set forth the monetary and other terms of the leases. Thus, a permit holder could explore for the coal, discover its value, then secure it without the seller being able to negotiate for the really true value of the coal. The leases, in turn, gave the purchaser the right to use the Indian land for all manner of buildings and installations necessary for the production, processing, and transportation of the coal, opening the way for the construction of power, conversion, and petrochemical plants, railroad lines, associated industrial complexes, and new towns of non-Indians, whose numbers would submerge the approximately 2,500 Northern Cheyennes and turn the reservation quickly into an industrialized white man's domain.

Most members of the tribe were uninformed about the terms of the leases, but when Peabody and Amax exploration crews appeared, drilling among Indian burial grounds and disrupting the Indians' lives, friction and unrest developed rapidly. Fearful for the future of the reservation, their culture, and the tribe itself, a number of Indians, mostly those who held allotments of their own land on the reservation, formed the Northern Cheyenne Landowners' Association to oppose the coal development. At almost the same time, Consol entered negotiations with the tribal council for another 70,000 acres of the tribe's land (which would have brought the total acreage held by permittees to 72 per cent of the reservation). Consol's proposal, which was not made public to the tribal members, offered $35 an acre and a royalty of 25 cents a ton (7.5 cents above what the federal government was getting for Bureau of Land Management coal and what the Indians had received in all previous leases).

To the startled Indians, Consol explained that it intended to invest approximately $1.2 billion in an industrial complex that would include four coal gasification units and that implied a city of perhaps 30,000 non-Indian people on the small reservation. The company was in a rush to get the permit signed. It urged the Indians to forgo the usual practice of asking for competitive bids (it would mean "the loss of several months' " income to them), and it offered the tribe $1.5 million toward the cost of a new health center (needed badly by the Indians, but also by the non-Indian industry, the white employees of which would, according to a clause in the proposed agreement, have access to the facility—inevitably becoming the center's major users). It also tried to pressure the Indians with a threat: "If Consol cannot conclude negotiations with the Northern Cheyenne tribe at an early date, Consol will be forced to take this project elsewhere . . . this project will be lost to the Northern Cheyenne, and it may be a long time before a project of this magnitude comes again, if ever."

But the company, which had prospective customers of its own for the coal, needed the deal more than the Indians did. Word of the proposal leaked out to the Northern Cheyenne Landowners' Association, and public meetings were held, cautioning the tribal council to go slowly. The higher price offered by Consol for the coal started some new thinking. Gradually, the tribal council could recognize problems with all the permits. The exercise by Peabody of its options to lease raised the question of whether the coal company should have had to negotiate anew, treating the leases as separate documents and letting the tribe ask for a fairer price for the coal. The company's activities also were causing many resentments among the Indians; the terms of the Peabody lease were now seen to be too loose for the protection of the reservation; the enforcement of strip-mining procedures in the code of federal regulations was not being observed by the Bureau of Indian Affairs; and the possibility that corporations would erect gasification plants and other installations on Peabody's leased land posed a fearful threat to the Indians' future. The same questions were raised about Amax's permit, while in connection with a third permit, given to Bruce Ennis, the Billings lawyer, and then assigned by him to Chevron, the Indians wondered if this had been speculation with their property and if Ennis had received a royalty from Chevron on top of their own 17.5 cents—which would have been illegal.

After more public meetings and deliberations, the Northern Cheyennes called in an attorney of the Native American Rights Fund in Boulder, Colorado, for advice and to write an environmental code that would protect the reservation. Other attorneys were consulted, and on March 5, 1973, postponing further consideration of the Consol proposal, with its threat of gasification plants, the Northern Cheyennes demanded that the Bureau of Indian Affairs declare null and void all their existing coal permits and leases. At the same time, the tribe implied that if the agency refused to undertake such action, the Northern Cheyennes would consider suing the federal government for not having protected the tribe and its resources, either in the drawing up and approval of the agreements or in the observance of provisions in the code of federal regulations. The tribal council indicated, moreover, that the Indians might prefer to mine and market their own coal themselves, drawing on independent expertise and, with the advice of competent environmental scientists, protecting the reservation with proper planning, regulations, and controls.

While the tribe's demand was being pondered by solicitors of the Interior Department, the coal companies' plans went forward. On March 21, 1973, Peabody announced it would supply 500 million tons of coal from its Northern Cheyenne strip mine to the Northern Natural Gas Company of Omaha and the Cities Service Gas Company of Oklahoma City, which jointly would build four gasification plants, at a cost of $1.4 billion, presumably in the vicinity of the mine. Each plant would employ up to 600 people (meaning an influx of many more non-Indians), and construction of the first plant would start in 1976. Peabody's coal, moreover, would only fuel two of the giant plants; the gas companys would need another 500 million tons from a second mine, which the Indians guessed would be opened by one of the other permit holders.

Somewhat similar events were transpiring, meanwhile, on the Crow Indian Reservation, which abuts that of the Northern Cheyennes. The Crows had let out permits for 292,680 acres, including rights to the coal in the off-reservation Sarpy area, the surface of which the Crows no longer owned. Some of the rights to that coal had been bought from them for 17.5 cents a ton by Norsworthy & Reger, which had then assigned the rights to Westmoreland. In view of the situation on the Northern Cheyenne Reservation, the Crows began to question the 17.5 cents-a-ton price they had received, as well as a 5 cents-a-ton overriding royalty that Westmoreland had paid Norsworthy & Reger, making it clear that Westmoreland had actually been willing to pay at least 22.5 cents for the coal.

In addition, when making the original deal, Norsworthy & Reger had persuaded the Crows that they could not sell their coal unless they also handed over rights to 30,000 acre-feet of water a year (which would be needed for gasification plants). Unknowledgably, the Crows obliged, transferring one of their water options from agricultural to industrial use and turning it over to Norsworthy & Reger. Altogether, in fact, the Crows gave away to the different coal companies valuable options for 140,000 acre-feet of water per year without a penny of payment. Testimony by James Reger to the Montana Water Resources Board in Helena on May 20, 1971, relating how he had maneuvered the water from the Crows, angered the Indians when, almost two years later, it came to their attention. Again, the tribe felt that the Bureau of Indian Affairs had not offered protection, and now, as with the Northern Cheyennes, violations were noted in all the permits, and fears were raised for the people's future. Early in 1973, lease options were exercised by Gulf and Shell for reservation lands. A report was circulated that a non-Indian city of up to 200,000 people was being considered for the neighborhood of Wyola or Lodge Grass on the reservation. Sentiment for canceling all the tribe's leases spread rapidly, and the tribal chairman, meeting with attorneys and Montana environmental experts, indicated that the Crows might take actions paralleling those of the Northern Cheyennes.

The resentments of the two tribes could seriously threaten some of the major projects being planned for the heart of one of the principal coal fields. As such, they would prove a significant impediment to the federal government's encouragement of the full-scale exploitation of the western coal. But there is a greater threat inherent in the indictment that Indians, once again, were defrauded by their trustee, the Bureau of Indian Affairs, which, abetting the coal companies, opened the reservations to an exploitation marked by unfair terms, lack of protection, and deceit. Throughout the country, other Indians are coming to recognize that the massive nature of the coal developments means the end of the Crow and Northern Cheyenne reservations as they have been, and, with it, the almost certain extinction of those peoples as tribal groups. As a result, the situation has a growing significance to all Indians and bids fair to become another source of explosive confrontation between Native Americans and the federal government.

The lack of impact statements, the nonobservance of regulations, and the many violations of laws that have characterized the first years of the coal rush throughout the region have provided concerned environmentalists with opportunities for numerous lawsuits. The Natural Resources Defense Council, the Environmental Defense Fund, the Sierra Club, and other organizations, consulting with attorneys, scientists, landowners, and environmental advocates like William L. Bryan, Jr. in the region, are currently preparing a number of cases which may attack some of the worst evils, bring about tighter controls and a modicum of order, and slow the headlong exploitation. In addition, an independent committee of twelve prominent natural scientists headed by Dr. Thadis W. Box, dean of the College of Natural Resources at Utah State University in Logan, was formed in April 1973 under the auspices of the National Academy of Sciences and the National Academy of Engineering. The committee reviewed the ecological and environmental consequences of the coal and power operations, and its report was completed in July 1973. Meanwhile, each week new projects are announced, the hurried pattern of development grows more chaotic, and the threat to the northern plains increases.

In Wyoming, Tipperary Resources, holder of 1 billion tons of coal, announces that it will build a 1,200-megawatt power plant near Buffalo, using water from 58 wells in the dry country; a new Atlantic Richfield strip mine will ship 10,000 tons of coal a day to Oklahoma; Wyodak Resources Development Corporation will build a 200- to 300-megawatt plant near Gillette, using 1 to 1.5 million tons of strip-mined coal a year; and the total Wyoming coal production will jump from 10.9 million tons in 1972 to 30 million tons in 1976. In Montana, Basin Electric Power Cooperative will build a generating plant to send power to eight states; coal will be shipped to two 600-megawatt plants that will be built in Oregon; a new Montana Power transmission line is planned to run from Anaconda to Hamilton, another from Billings to Great Falls, small parts of an eventual great new network.

In North Dakota, more than 2 million acres of land are believed already leased for strip mines; companies holding rights to 1 billion tons of coal in Hettinger County will build 4 large-scale power plants; the Michigan Wisconsin Pipe Line Company, arranging for the purchase of 1.5 billion tons of strip-mined lignite from the North American Coal Corporation, asks for 375,000 acre-feet of water per year from Garrison Reservoir on the Missouri River, enough for no less than 22 gasification plants; still another company wants water for 8 more gasification plants. And so it goes.

The horrors conjured up by the North Central Power Study in 1971 are coming true even faster than that document proposed—without the focus for planning and control which its blueprint provided. Is it,

then, all over for the northern plains? Will they inevitably become another Appalachia? On the Tongue River near Birney, Montana, where strip mines, power plants, gasification plants, and other industrial installations threaten the land, air, water, and quality of life of the Irving Alderson, Jr. family, fifth-generation owners of the Bones Brothers Ranch, Mrs. Alderson gives voice to a desperate, last-ditch courage that says there is still time to save the region:

"To those of you who would exploit us, do not underestimate the people of this area. Do not make the mistake of lumping us and the land all together as 'overburden' and dispense with us as nuisances. Land is historically the central issue in any war. We are the descendants, spiritually, if not actually, of those who fought for this land once, and we are prepared to do it again. We intend to win."

SUGGESTED READINGS

Box, Thadis, W. et al., 1973, Potential for Rehabilitating Land Surfce—Mined for Coal in the Western United States: National Science Foundation—National Academy of Engineering.

Caudill, Harry M., 1963, *Night Comes to the Cumberlands*, Boston: Atlantic-Little, Brown.

Caudill, Harry M., 1973, Farming and Mining, *Atlantic Monthly*, September 13, 1973, pp. 85–90.

Conaway, James, 1973, "The Last of the West: Hell, Strip It!," *Atlantic Monthly*, September 13, 1973, pp. 91–103.

Gillette, Robert, 1973, Western Coal: Does the Debate Follow Irreversible Commitment? *Science 182*, pp. 456–458.

Strip and Surface Mine Study Policy Committee, 1967, Surface Mining and Our Environment: U.S. Dept. of Interior, Washington, D.C., 124 p.

Wolf, Anthony, 1972, Showdown at Four Corners: *Saturday Review,* June 3, 1972, pp. 29–41.

Biologic Effects of Air Pollution

ROBERT FRANK   1972

Air pollution arises chiefly from the combustion of fossil fuels. Its composition varies with the type of fuel used, the method and efficiency of combustion, and the techniques of emission control applied. Nevertheless, there are five classes of pollutants released whenever fossil fuels are consumed. They are oxides of sulfur, oxides of nitrogen, particles (sometimes referred to as aerosols or fly ash), hydrocarbons, and carbon monoxide.

In many parts of the country electric power plants are major sources of pollution. It is estimated that power plants in the United States emit by weight about 50 per cent of all oxides of sulfur, 30 per cent of all oxides of nitrogen, 20 per cent of all suspended particles, and 0.8 per cent of all carbon monoxide. These emissions, either alone or in combination, are capable of damaging health, reducing safety in the performance of certain tasks, and offending our sense of well-being. The adverse effects of pollutants on man and his environment are measured by and referred to legislatively and legally as *air quality criteria.*

The effects of urban air pollutants on health, except for those caused by a few specific elements, such as carbon monoxide, lead, and beryllium (*NAPCA,* 1969a,b, 1970a,b,c; *EPA,* 1971; *NAS* and *NAE,* 1969; Higgins and McCarroll, 1970), are neither discrete nor readily quantified. Urban pollution probably acts in combination with other forms of stress rather than independently. Its principal effect may be to worsen rather than initiate disease (Lawther, 1966). The other forms of stress that enter into this complex interplay include socioeconomic factors (crowding, poor hygiene, malnutrition), weather (coldness, dampness, and perhaps sudden, extreme changes in temperature), and most important, tobacco smoking. Smoking, a personal form of air pollution, is considered to be the greatest cause of chronic respiratory disease and cancer. Unfortunately, it has frequently confounded efforts to assess the role of pollution on

Dr. Robert Frank is professor of environmental health in the School of Public Health and Community Medicine at the University of Washington, Seattle, Washington. His research interests include the effects of pollutant gases and aerosols on the function and structure of the lung.

From "Electric Power Consumption and Human Welfare, The Social Consequences of the Environmental Effects of Electric-Power Uses," *AAAS/CEA Power Study Group,* Section 1, Chapter 2, pp. 31–49, 1972. Reprinted with author's revisions and light editing and by permission of the author.

health because of its overwhelming toxicity. In fact, studies of the impact of air pollution on children have been undertaken in part for the purpose of avoiding this complication, as well as the complication of occupational exposure.

Evidence of the biologic effects of air pollution, exclusive of damage to vegetation, derives from three main avenues of investigation.[1]

TOXICOLOGIC STUDIES ON ANIMALS

Whereas these studies have been valuable for clarifying the mechanisms of biologic response and for describing dose-response relations, they have proved to be of only limited use in providing criteria for the establishment of ambient and emission standards. The overwhelming majority have focused on acute, essentially reversible effects (if the animal survived) rather than on chronic exposures that might lead to cumulative and persistent damage to the lungs. Perhaps the principal shortcoming of such studies has been a reliance on exposures too simple in composition, unrealistically high in concentration, or both. For example, the levels of sulfur dioxide in urban atmospheres associated with increasing morbidity and mortality have virtually no effect in laboratory investigations when the gas is administered in otherwise clean air. Since there is no certainty about which individual pollutant or combinations of pollutants may be critical in urban pollution,[2] the toxicologist has been understandably reluctant to generate a complex mixture of gases and particles, particularly for use in chronic studies, that may

1. An excellent critique of these types of investigation is found in a publication of the National Institute of Environmental Health Sciences (*NAPCA*, 1970).
2. There are exceptions to this statement, as, for example, the levels of carbon monoxide found in the vicinity of heavy automotive traffic, and those of fly ash and oxides of sulfur immediately downwind to large industrial sources. But these circumstances are analogous to the acute severe episodes of air pollution rather than the more chronic, low-grade forms to which larger numbers of persons are exposed.

ultimately prove to be irrelevant or benign. Such studies are also beset with the perennial uncertainty over how readily their results can be extrapolated to man.

EXPOSURE OF HUMAN SUBJECTS TO CONTROLLED DOSES OF POLLUTANTS

Human exposures have usually lasted up to several hours. They share some of the advantages and disadvantages of the studies on animals. Though such studies have yielded information about the mechanisms of response and the probable importance of individual pollutants, they have frequently had to resort to unrealistically high concentrations of one, or at most, a combination of two pollutants to evoke a response. Recent experiments that combined the pollutant with an additional stress, such as exercise, have elicited responses at more realistic concentrations (Bates et al., 1972). Though potentially a valuable source of criteria, the studies have yielded disappointingly little information in the past. Several factors contribute to this poor record, perhaps the most important of which is the ethical dilemma posed by the use of human subjects, particularly those having pulmonary or cardiovascular disease. Though most of the epidemiologic evidence for adverse effects of air pollution on health have been obtained on such patients, there is monumental unease over subjecting sick persons to pollutants experimentally, even under controlled laboratory circumstances. Until recently, this form of applied research on man and the laboratory animals attracted relatively few competent investigators. Fortunately, there has been a recent influx of capable investigators into the field, as concern about the environment has intensified and financial support for research has grown.

EPIDEMIOLOGIC SURVEYS AND COMMUNITY HEALTH

Acute, severe episodes of air pollution in combination with specific weather patterns have been associated with increased rates of morbidity and mortality, particularly among the aged and sick. Again, however, the possible role of chronic, low-grade pollution on health remains uncertain. Though community studies have provided the bulk of our criteria, they have yielded only fragmentary information on the specific pollutants responsible, as well as their concentrations. One great difficulty has been to estimate the dose of a specific pollutant to which the individual is exposed. There may be considerable gradients in concentration between the indoors and outdoors, within a small outdoor area (vertical and horizontal gradients are not uncommon), and from one neighborhood to the next. The complexity of the problem increases enormously when one attempts to estimate doses for extended periods of time, particularly those for an increasingly mobile population.

Vulnerability to air pollution varies widely among different segments of the population. The very young and the elderly appear to be especially vulnerable. DuBos and co-workers (1968) have shown in experimental animals that the effects of pharmacologic agents on the fetus at or near term and on the newborn are exaggerated in magnitude and duration. Whether man at the same stages of development is equally vulnerable to chemical stress, including that of air pollution, is unknown at present. The difficulty of obtaining such vital information directly will be tremendous, but the problem does deserve assiduous investigation both epidemiologically and experimentally. The problem might be more amenable to solution if we increased our capacity to accumulate and analyze medical, health, and air-sampling information on a national scale and were prepared to cope with mobile populations; that is, some equivalent of large data banks would be most helpful. Pregnancy and childhood are two additional periods of life during which susceptibility to pollutants may increase. The association between air pollution and lower respiratory infection from infancy into adolescence is quite impressive (Douglas and Waller, 1966; Holland et al., 1969). Whether this vulnerability is innate or may be related, at least in children,

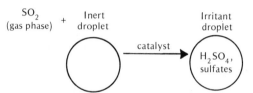

Figure 2-37. Conversion of SO_2 to a more toxic aerosol. The gas is removed by the upper airways. Irritant droplets may be deposited along the tracheo-bronchial tree or alveoli, depending on their physical characteristics. This reaction is catalyzed by humidity, heavy metals (lead, mercury, etc.), photochemical pollutants, and dust.

to increased activity outdoors is not known.[3]

The most susceptible persons are likely to be those already ill, particularly those with chronic respiratory disease, cardiovascular disease, or both. The dramatic increases in morbidity and mortality rates associated with severe episodes of air pollution have traditionally centered on these patients. But there is evidence that less marked fluctuations in urban air pollution may also aggravate their illness. A recent survey in Chicago showed that elderly patients[4] with moderate to advanced chronic bronchitis suffered an increase in symptoms during and immediately after any increase in local pollution for which sulfur dioxide and particulates (see Figure 2-37) were used as the indices (Carnow, 1970; Carnow et al., 1969). Perhaps implicit in this observation is the notion that we shall probably continue to unveil greater evidence of the harmful effects of air pollution as our methods of study become more sensitive and sophisticated, and as they are applied more extensively. The question that may confront society as our knowledge is expanded is not, "What is a safe level of pollution?" but, "What is a tolerable or acceptable level of sickness from pollution?"

3. The increased ventilatory rate and obligatory mouth-breathing characteristic of exercise and heavy work are equated with increased exposure of the lungs. The increased exposure occurs in part because the upper airways become less efficient gas scrubbers as ventilatory rates increase.

4. The elderly and chronically ill segments of our society are likely to grow in size as improved medical and community services help increase the average span of life. These groups are among the most vulnerable to the toxic effects of pollution.

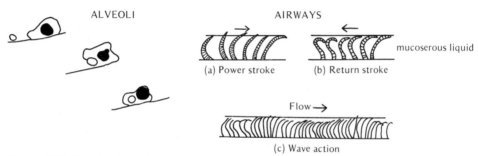

Figure 2-38. Healthy lungs are virtually sterile, "cleaning house" by two related processes. The first involves alveolar macrophages, mobile scavenger cells that engulf particles on alveolar surfaces. Their cellular enzymes kill and digest microorganisms. The second involves the mucociliary system. Cilia, hairlike projections in the mucoserous liquid, line the airways of the mouth, nose, pharynx, and tracheo-bronchial tree. Beating over 1,000 times/minute, they propel the liquid mouthward to be swallowed or expectorated. (Denton et al., 1965.)

A variety of biologic effects are imputed to air pollutants.[5] A partial list follows:

Reduced Resistance to Infection of the Lung

The occurrence and severity of respiratory infections in laboratory animals is increased, and the capacity of the animal to inactivate or kill inhaled organisms is impaired following exposure to oxides of nitrogen (Ehrlich and Henry, 1968; Henry et al., 1970) or ozone (Coffin and Blommer, 1965). The basis for these effects may be manyfold. Ozone, in particular, damages the alveolar macrophage, a cell (Figure 2-38) that functions to engulf, inactivate, and dissolve invasive organisms (Coffin et al., 1968). Both gases may also damage the epithelial lining of the airways and thereby alter mucociliary function (Figure 2-38), as the latter is vital in cleansing dust and bacteria from the lung. It is notable that the destruction of inhaled bacteria is also impeded by malnutrition, acute alcoholism, and smoking, all of which, of course, coexist in the community with pollution. These different forms of stress may be expected to reinforce each other's toxic potential. Paradoxically, though the concentrations of photochemical oxidants encountered in the urban and occupational environments can be shown to reduce the body's resistance to infection in laboratory animals, there is virtually no evidence for such an effect in humans. An equal but opposite paradox applies to sulfur dioxide and particulates—the association between the latter pollutants and respiratory infection in community studies is strong, whereas laboratory confirmation, based on realistic dosages of these agents, has been wanting.

Aging and Chronic Disease of the Lung

Collagen and elastin, the two most abundant structural proteins of the lung, contribute importantly to its elastic behavior and structural integrity. Collagen and probably elastin as well are denatured by nitric oxide and ozone (Buell et al., 1965). As a consequence, the molecular configuration is thought to have important implications for the rate at which the lung ages structurally, and may also provide a basis for degenerative processes that culminate in diseases like emphysema. There is evidence, for example, that the lifetime exposure of laboratory animals to concentrations of nitric oxide as low as 2 parts per million may cause irreversible damage to the lungs, a damage bearing resemblance to human emphysema (Freeman et al., 1968). Whether man is equally sensitive to these gases is unknown. To isolate such effects epidemiologically is probably beyond our present abilities.

5. See the publications of *NAPCA* (1969a,b; 1970a,b,c), *EPA* (1971) for detailed effects on vegetation.

Carcinoma of the Lung

Death rates from carcinoma of the lungs are higher in cities than in rural areas, but the role of air pollution in this difference is uncertain (Anderson, 1967). Aromatic hydrocarbons such as 3,4-benzopyrene, which induce carcinoma experimentally, are present in urban air (and cigarette smoke). Two important sources for the carcinogens are auto exhaust and coal-burning plants. These hydrocarbons, and perhaps other elements in urban air, probably play a secondary role rather than a primary one in the complex process leading to carcinoma.

Altered Respiratory Mechanics

The lung is designed structurally to distribute inspired air over a diffusing surface that is about 60–80 square meters in area, and comprises about 300 million alveoli.[6] The conducting airways, starting at the trachea, may undergo more than 20 dichotomous divisions or branchings before terminating at the alveoli. Despite such structural complexity, the inspired air is distributed evenly at little cost in energy. This remarkable performance requires that the frictional resistance to air flow remain slight and be approximately equal among the many parallel airways. Both requisites are met in the normal lung. Inhaled irritants may provoke either localized or widespread narrowing of the airways and so cause an increase in the work of breathing. If, as frequently happens, the constriction is nonuniform, the air that is inhaled may be unevenly distributed throughout the lung. (See Figure 2-38.) When these defects become severe, shortness of breath and inadequate exchanges of oxygen and carbon dioxide across the lung may become disabling. Sulfur dioxide, nitric oxide, ozone, certain hydrocarbons, particulates, are among the irritants capable of producing these effects. The magnitude of the

6. The alveolus is the functional unit of the lung, the site of exchange of oxygen and carbon dioxide. It has very thin walls (less than 1 micron thick) and is about 150 microns in diameter.

response will depend on the concentration of the irritant in inspired air, the mode of breathing (breathing by mouth at an increased rate, as in exercise, is equated with an increased exposure of the lower airways) (Frank et al., 1969), and the reactivity of the tracheobronchial tree. Patients with asthma or tracheobronchitis may react excessively and experience bronchospasm amid levels of pollutants that have no appreciable effect on healthy individuals. The most conspicuous laboratory evidence for the synergistic effects of mixtures of pollutants, namely, sulfur dioxide and sodium chloride aerosol, has been obtained with these measurements of airway mechanics (NAPCA, 1969a,b; Amdur, 1961).

There is uncertainty over the relation between short-term airway narrowing and either the capacity of the lung to resist infection or the occurrence and progression of chronic irreversible disease. There is little doubt however, that the superimposition of new or aggravated mechanical defects on a lung that is already diseased can be seriously disabling.

Interference with Oxygenation of Tissue

There are several ways whereby air pollutants may cause a reduction in the supply of oxygen to the tissues of the body. Mention has already been made of the impairment in the distribution of air in the lung that may attend airway narrowing. Another way is by damage to the alveolar-capillary membrane. The latter separates the air in the alveoli from the neighboring circulation. It is the path across which oxygen and carbon dioxide move to enter and leave the body. The extreme thinness of this membrane enables the lung to serve as an efficient diffusing surface. The transfer of oxygen is impaired whenever this membrane is thickened by inflammation, edema (leakage of fluid from the blood vessels), or scarring. (See Figure 2-39.) Low concentrations of ozone may damage the alveolar-capillary wall and produce edema. Perhaps the most damaging component of photochemical

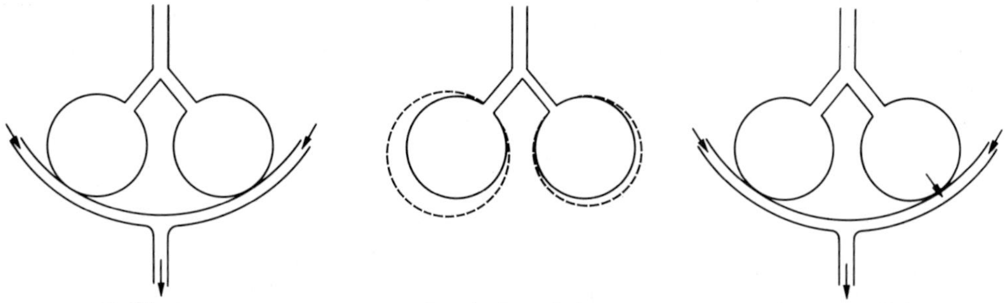

Healthy lung

Impaired ventilation

Unevenly distributed airway narrowing may in turn cause underventilation of some portions of the lung. This is a common cause of inadequate oxygenation of blood.

Impaired diffusion

Thickening of the alveolar capillary membrane interferes with diffusion of oxygen into the pulmonary circulation. At the same time, damage to the capillary wall may result in edema.

Figure 2-39. Inteference of air pollutants with tissue oxygenation.

smog—ozone—has been shown to diminish (reversibly) the diffusing capacity of the lung in healthy volunteers (Bates et al., 1972; Young et al., 1964).

A third source of interference with the supply of oxygen to tissues is carbon monoxide, an asphyxiant. Carbon monoxide has an affinity for hemoglobin over 200 times greater than that of molecular oxygen (NAPCA, 1969a,b, 1970a,b,c). Therefore, it displaces oxygen from hemoglobin, competitively. Moreover, the oxygen that does combine with the hemoglobin is less readily released to the peripheral tissues when carbon monoxide is also present as carboxyhemoglobin (CO·Hb). These two factors, decreased volume of oxygen at the tissue level, may seriously interfere with the function of peripheral tissues. The two organs most sensitive to oxygen deprivation are the brain, especially at the cortical level, and the heart. In healthy volunteers, CO·Hb levels as low as 2 to 5 per cent have reportedly been associated with changes in visual and time perception, and in certain psychomotor tests (NAPCA, 1969a,b, 1970a,b,c). Such changes might interfere with performances requiring a high degree of skill and attention. The population at greatest risk includes individuals who are already hypoxic from a variety of diseases or circumstances—anemia, chronic pulmonary or cardiac disease, and high altitude—or those who have increased oxygen needs as in pregnancy, fever, and hyperthyroidism. The association between cigarette smoking and increased mortality from coronary vascular disease may partly reflect the increase in CO·Hb typical of smokers (Goldsmith et al., 1969). The fetus may be particularly vulnerable to asphyxia from carbon monoxide. Thus, newborn babies of mothers who smoke appear to be smaller than those of nonsmoking mothers, a difference that may be related to the CO·Hb shared by both the maternal and fetal circulations (MacMahon et al., 1965).

Nondisease Effects

Air pollution afflicts us in ways that are intensely disconcerting, even dangerous, but not directly damaging to health. Particulates create haze, reduce visibility, and increase the hazards of traffic on the ground and in the air (Anderson, 1967). Quite often, pollution is malodorous. It can irritate the conjuntival lining of the eyes and the mucosal lining of the throat. Eye irritation is reportedly the most widespread symptom of the photochemical pollution in Los Angeles County; and it is experienced by about three-fourths of the population (Goldsmith, 1969).

SUMMARY

We are only just beginning to study the possible impact of air pollution on behavior (Anderson, 1967). We surmise that esthetic reactions are among those driving the more affluent segments of society out of polluted cities. We (probably most of us) are frustrated and angered by the sight of a city or a valley enshrouded in dirt.

How strong is the case that urban air pollution affects health? Most authorities would probably agree that such evidence is good and that it is being continually strengthened, though gaps and weaknesses persist. It is unlikely that we shall ever know precisely all the effects on man of individual pollutants or their mixtures as both the population and the pollution are two dynamic, too changing, and ultimately too difficult to characterize to permit unqualified answers. Perhaps, for this reason, we ought to upgrade our reliance on the secondary criteria concerned with welfare, and give more emphasis to esthetic and economic factors.[7] This suggestion is not intended to demean our concern about health, but rather to broaden the basis for our decisions and to take into fuller account the enormity of the problem.

REFERENCES

Amdur, M. O., 1961. The effect of aerosols on the response to irritant gases: C.N. Davies (ed.), *Inhaled Particles and Vapours.* Pergamon, New York.

Anderson, D. O., 1967. The effects of air contamination on health: A review. Part II. *Canad. Med. Ass. J. 97:* 585–593.

Bates, D. V., G. M. Bell, D. C. Burnham, M. Hazucha, J. Mantha, L. D. Pengelly, and F.

7. Lave and Seskin (1970) estimated that $2,080,000,000 would be saved annually by a 50 per cent reduction in levels of air pollution.

Silverman, 1972. Short-term effects of ozone on the lung. *J. Appl. Physiol. 32:* 176–181.

Buell, G. C., Y. Tokiwa, and P. K. Mueller, 1965. Potential crosslinking agents in lung tissue. *Arch. Environ. Health 10:* 213.

Carnow, B. W., 1970. Relationship of SO_2 levels to morbidity and mortality in "high risk" populations. Presented at the Air Pollution Medical Research Conference, American Medical Association, New Orleans, October 5–7, 1970.

Carnow, B. W., M. H. Lepper, R. B. Shekelle, and J. Stamler, 1969. Chicago air pollution study. SO_2 levels and acute illness in patients with chronic bronchopulmonary disease. *Arch. Environ. Health 18:* 768.

Coffin, D. L., and E. J. Blommer, 1965. The influence of cold on mortality from streptococci following ozone exposure. *J. Air Poll. Control. Adm. 15:*523.

Coffin, D. L., D. E. Gardner, R. S. Holzman, and F. J. Wolock, 1968. Influence of ozone on pulmonary cells *Arch. Environ. Health 16:* 633.

Denton, R., M. Little, and S. H. Hwang, 1965. Chemical engineering aspects of obstructive lung disease. *Chemical Engineering in Medicine 62:* 12–18.

Douglas, J. W. B., and R. E. Waller, 1966. Air pollution and respiratory infection in children. *Brit. J. Prev. Soc. Med. 20:* 1.

DuBos, R., R. W. Schaedler, and R. Costello, 1968. Lasting biological effects of early environmental influences. I. Conditioning of adult size by prenatal and postnatal nutrition. *J. Explt. Med. 127:* 783.

Ehrlich, R., and M. C. Henry, 1968. Chronic toxicity of nitrogen dioxide. I. Effect on resistance to bacterial pneumonia. *Arch. Environ. Health 17:* 860.

Environmental Protection Agency, 1971. Air quality criteria for nitrogen oxides. Washington, D.C.

Frank, N. R., R. E. Yoder, J. D. Brain, and E. Yokoyama, 1969. SO_2 (^{35}S labeled) absorption by the nose and mouth under conditions of varying concentration and flow. *Arch. Environ. Health 18:* 315.

Freeman, G., S. C. Crane, R. J. Stephens, and N. J. Furiosi, 1968. Environmental factors in emphysema and a model system with NO_2. *Yale Journal of Biol. and Med. 40:* 5, 6, 566.

Goldsmith, J. R., 1969. Nondisease effects of air pollution. *Environ. Res. 2:* 93.

Goldsmith, J. R., R. R. Beard, and B. D. Dinman, 1969. Epidemiologic appraisal of carbon monoxide effects. Committee on Effects of Atmospheric Contaminants on Human Health and Welfare (*NAS* and *NAE*): *Effects of Chronic Exposure to Low Levels of*

Carbon Monoxide on Human Health, Behavior, and Performance. Washington, D.C.

Henry, M. C., J. Findlay, J. Spangler, and R. Ehrlich, 1970. Chronic toxicity of NO_2 in squirrel monkeys. III. Effect on resistance to bacterial and viral infection. *Arch. Environ. Health 20:* 566.

Higgins, I. T. T., and J. R. McCarroll, 1970. Types, ranges, and methods for classifying human pathophysiologic changes and responses to air pollution. A. Atkisson, R. S. Gaines (eds.), *Development of Air Quality Standards.* Chas. E. Merrill, Columbus, Ohio.

Holland, W. W., T. Halil, A. E. Bennett, A. Elliott, 1969. Factors influencing the onset of chronic respiratory disease. *Brit. Med. J. 2:* 205.

Lave, L. B., and E. P. Seskin, 1970. Air pollution and human health. *Science 169:* 723.

Lawther, P. J., 1966. Air pollution, bronchitis and lung cancer. *Postgrad. Med. J. 42:* 703.

MacMahon, B., M. Alpert, and E. J. Salber, 1965. Infant weight and parental smoking habits. *Amer. J. Epidem. 82:* 247.

National Academy of Sciences and National Academy of Engineering (*NAS* and *NAE*), 1969. *Effects of Chronic Exposure to Low Levels of Carbon Monoxide on Human Health, Behavior, and Performance,* Washington, D.C.

National Air Pollution Control Administration (*NAPCA*), 1969a. Air Quality Criteria for Sulfur Oxides, Summary and Conclusions. Arlington, Va.

National Air Pollution Control Administration, 1969b. Air Quality Criteria for Particulate Matter, Summary and Conclusions. Arlington, Va.

National Air Pollution Control Administration, 1970a. Air Quality Criteria for Photochemical Oxidants, Summary and Conclusions. Washington, D.C.

National Air Pollution Control Administration, 1970b. Air Quality Criteria for Hydrocarbons, Summary and Conclusions. Washington, D.C.

National Air Pollution Control Administration, 1970c. Air Quality Criteria for Carbon Monoxide, Summary and Conclusions. Washington, D.C.

National Institute of Environmental Health Sciences, 1970. Task Force on Research Planning in Environmental Health Science. Man's Health and the Environment—Some Research Needs. Washington, D.C.

Young, W. A., D. B. Shaw, and D. V. Bates, 1964. Effect of low concentrations of ozone on pulmonary function in man. *J. Appl. Physiol. 19:* 756–768.

3

NUCLEAR FISSION
AS AN ENERGY SOURCE

The San Onofre nuclear generating station, near San Clemente, California. This station provides 430 megawatts (electric), enough to supply well over half a million users. The sphere houses a pressurized water reactor. (Photo courtesy of Southern California Edison Company.)

The control room of the San Onofre nuclear generating station. Any one of approximately 60 types of malfunction will shut down the plant automatically. (Photo courtesy of Southern California Edison Company.)

Fueling the San Onofre nuclear generating station. (Photo courtesy of the U.S. Atomic Energy Commission.)

The Experimental Breeding Reactor II at Argonne National Laboratory's Idaho site. The power plant is at the left, the reactor containment shell, center, and the fuel cycle facility is behind the stack. (Photo courtesy of Argonne National Laboratory.)

INTRODUCTION

One area of energy policy in which the federal government has taken a firm, decisive position is that involved in developing fission energy and, particularly, the breeder reactor, as rapidly as feasible. Such a position poses severe dilemmas for society, and, in the nature of all true dilemmas, there is no obvious, unanimously acceptable "correct answer." Owing to the profound social implications of this decision, serious analysis by well-informed, objective observers has led to fundamental disagreement as to the wisdom of continuing in this direction.

Those questioning the wisdom of opting for a "plutonium economy" point to uncertainties surrounding reactor safety, possible diversion of plutonium to militant nations, the problem of radioactive waste storage, and environmental pollution through heat and radioactive waste. Those supporting such a policy point to the vast resources of fuel, proven feasibility, excellent safety record of existing reactors, relative environmental advantages over fossil fuels, and our projected increasing energy needs. All who have studied the issues agree that the choice of the "plutonium economy" imposes an obligation on society for substantial, long-term modifications to assure the safe handling, storage, and disposal of all radioactive materials associated with fission power.

Since nuclear fission as an energy source poses such intrinsic and inescapable dilemmas for society, it is essential that its benefits and risks be as widely understood as possible. Public policy is determined as much by *perceptions of the truth* as by *the truth* itself. Hence, we present both informational articles on fission reactors and differing interpretations of the impact fission power will have on the social system.

Bernard I. Spinrad very forcibly presents his ideas in *The Case for Nuclear*

Energy. He sees increased energy consumption as inevitable and generally "good." After considering various alternative energy sources, he concludes that breeder reactors hold the best promise for meeting increased energy demands. Problems involved with reactor operation and waste disposal are evaluated and deemed solvable. Finally, he presents an interesting analysis of antinuclear sentiments.

In *Electric Power from Nuclear Fission,* Manson Benedict gives a detailed picture of present fission reactors and the economics of fission energy compared with fossil fuel energy. The liquid-metal fast-breeder reactor is proposed as the most promising route to utilization of the potential energy in uranium-238, which constitutes 99.3 per cent of all uranium found in nature, and which cannot be used in current light-water reactors. Professor Benedict also is optimistic that the problems in safely operating fission reactors can be solved at acceptable levels of risk.

David J. Rose provides additional perspectives on the promises and risks involved in utilizing uranium as an energy source in his comprehensive article, *Nuclear Eclectic Power.* According to him, the very elastic supply of uranium and thorium fuel for breeder reactors guarantees a long-term supply of low-cost energy. He considers many of the hazards associated with nuclear power in more detail, and compares them with those of fossil fuel. His analysis, somewhat less sanguine than the previous two, touches on the uncertainty surrounding plutonium and public health, which is investigated in more detail in a subsequent article.

Three areas of immediate concern surrounding fission energy are (1) reactor safety and the risk of catastrophic accident, (2) disposal of fission wastes, and (3) biologic effects of radioactive emissions. The next three articles address these problems.

In *Reactor Safety Study: An Assessment of Accident Risks in Commercial Nuclear Power Plants,* the U.S. Atomic Energy Commission presents the conclusions of an intensive two-year study by a group of over fifty scientists and engineers headed by Professor Norman Rasmussen of the Massachusetts Institute of Technology. Funded by the Atomic Energy Commission, this report provides an independent assessment of the risks of catastrophic accident in commercial reactors. Here, probability versus consequences curves are calculated through complex "event tree analysis" techniques, and the likelihood of a reactor accident comparable in severity to the crash of a large airliner is computed to be one chance in a million per reactor year of operation.

In *Disposal of Nuclear Wastes,* Arthur S. Kubo and David J. Rose present the various options available for keeping radioactive wastes out of the biosphere. The relative costs of each technique are estimated, and social considerations are shown to be of primary importance in selecting any final solution. The crucial necessity of separating and recycling the actinides (elements heavier than uranium) is emphasized as a method of shortening the required safe storage time from a time scale of a million years to one of less than a thousand years.

The committee on the Biological Effects of Ionizing Radiations of the National Academy of Sciences National Research Council has studied the effects of low-level ionizing radiation and produced a detailed report (*BEIR Report*), the *Summary and Recommendations* of which are presented here. The genetic and somatic effects of the previous Atomic Energy Commission 170 millirem guide lines on the population of the United States are calculated, and recommendations are made for reducing the upper limits allowed from the nuclear-power industry. This report recommends a consideration of health hazards from alternative energy sources when setting radiation standards. Though no guide lines are recommended specifically, general principles are suggested for use in setting such guide lines.

In *Plutonium and Public Health,* Donald P. Geesaman discusses the "hot particle" problem. The risk from deposition of plutonium dioxide aerosol particles in the lungs is not well understood, and he raises questions about applying maximum permissible lung burden dose limits for hot particles. Since the current power reactors "breed" about 500 pounds of plutonium per year, and this will be the primary fuel of future breeder reactors, it is essential that any questions on its health hazards be resolved.

The Clear and Present Danger: Nuclear Power Plants is a summary by the Environmental Education Group (an active environmentalist group organized in California) of many of the criticisms being raised over the decision to "go nuclear." The conflicting roles of the former Atomic Energy Commission as promoter and regulator of the nuclear industry are described. Many of the problems seen by previous authors as "engineering challenges" are viewed here as being intrinsically unsolvable or at least serious enough to invalidate the present policy of promoting fission power production.

Donald P. Geesaman and Dean E. Abrahamson present a thoughtful analysis of the impact that fission power will have on society in *The Dilemma of Fission Energy.* They think that the requirements for absolute security of fissionable materials and by-products will impose constraints on our social system that are nearly impossible to foresee. It is crucial that we understand the nature of the "Faustian bargain" into which we as a society are entering.

Finally, the Twenty-third Pugwash Conference, in *Pugwash Warns of Nuclear Power,* calls attention to the dangers involved in fission power and makes recommendations for dealing with them. This sober analysis by many of the world's leading thinkers cannot be easily dismissed as blind opposition to progress by the uninformed. The issues raised here must be kept clearly in mind by all of those in the public sector as they make decisions on nuclear energy matters.

Many of the problems surrounding fission as an energy source are dealt with directly in the selections included, and others are suggested. Additional issues deserving serious consideration include the following:

1. What will be the impact of fission energy on international relations? Since the production of fission energy depends upon abundant capital and an advanced technology, the major industrial nations will control the flow of reactors and reactor fuel to less industrialized nations. The effects are clearly problematic and suggest the possibility of a "nuclear colonialism."

2. What are, in fact, the economics of nuclear power in a period of rapid inflation and tight money? Since fission energy prices are fixed primarily by the initial costs of the plant, the cost per installed kilowatt of electricity will be a strong function of the price of money. Clearly, fission-power cost projections made when interest rates were 6 per cent per year are not valid when interest rates run to 12 per cent per year.

3. How does fission energy measure up in a comprehensive global cost-benefit analysis of all energy options? Is fission energy environmentally superior to coal as a source of central power-plant electricity? How do fission energy plants compare environmentally to solar, geothermal, or large-scale wind power facilities? How does one compare the known environmental risks of fission power to the uncertain prospects of fusion power or the possible ecological disruption of thousands of square miles of desert by utilization of solar energy? If we had spent billions of dollars on solar energy research and millions on fission (instead of vice versa), what would our present energy position be?

4. What is the social morality of burning fossil fuels when fission presents an alternative? Coal and crude oil serve as the raw material for lubricants, the plastics industry, much of the chemical industry, and some of the fertilizer industry. It been claimed that the entire world population could be fed by algae farms that would consume only a small fraction of our present crude oil production. Uranium, on the other hand, produces 2.5 million times more energy per pound than does coal and has little use other than energy production. Certainly we should call into question the wisdom of burning the vital fossil resources that provide us with clothing and many other life necessities and that are now dwindling so rapidly. The choice between fossil-fuel-powered or uranium-powered electric plants must take the relative abundance and alternative uses of these natural resources into account.

Nuclear fission is the second major source of energy for meeting present and near future needs. It is expected to overtake fossil fuels in electric energy production by 1985. And, just as uranium has orders of magnitude greater energy content than coal, the problems of its safe and efficient utilization are orders of magnitude more severe. We now examine these issues surrounding nuclear fission energy.

The Case for Nuclear Energy

BERNARD I. SPINRAD   1973

I am an advocate of nuclear energy. I believe that the technology of nuclear power reactors offers great benefits to mankind, without commensurate costs, and that we can reduce (and have reduced) the risks associated with it to an acceptable, and indeed very low level. Specifically, I assert that nuclear power is economic and can be more so; that it is produced with more dignified labor than other forms of manufactured energy; that its burden of air pollution is far smaller than energy sources with which it is competing now; that its total environmental impact is minor compared with these competitive sources; and that it is safe by the standards which we apply to our normal living, our businesses and our industries.

The case for nuclear energy is based on five propositions:

(1) The world needs more energy, and specifically more electrical power, and needs it indefinitely.

(2) The scale of this need is such that economy in procuring this power is enormously important.

(3) Nuclear power is the only proven way of getting the needed energy economically and through use of a fuel which is available indefinitely.

(4) The effluents of nuclear power pose a minor burden to humanity when measured by the scale of burdens which we currently find acceptable.

(5) The risks to the public which might arise from accidents in nuclear facilities have been generously overcompensated by cautious design, careful operational analysis, accident analysis, inspections and incorporation of multiple, independent safety systems, to the point where these risks are too low to estimate.

These propositions are elaborated in the subsequent parts of this paper.

Dr. Bernard I. Spinrad is professor of nuclear engineering at Oregon State University. He has previously worked with the International Atomic Energy Agency (Vienna) on projections of world energy needs and has served as director of the Nuclear Reactor Division of Argonne National Laboratory. His research interests have included all phases of reactor development and are presently centered on the study of "after heat" following reactor shutdown. He has given courses at several universities on the social aspects of nuclear energy.

Originally presented to the Oregon Academy of Sciences, February 24, 1973, and subsequently published in *Ambiente-Environment*, Vol. 1, No. 2, 1973. Reprinted with light editing and by permission of the author and *Ambiente-Environment*. Copyright 1973 by *Ambiente-Environment*.

ENERGY NEEDS

The most opulent world citizens are conspicuous consumers—indeed, squanderers—of energy. Another group of world citizens, whom I label "comfortable," could level off their energy consumption right now. Specifically, with regard to personal use of energy, I think that the current consumption of upper-middle-class Americans is an acceptable standard. This does not mean that our actual standard of living will or ought to remain at a level, but that there is a continuing trend toward more efficient use of energy, so that more amenity can be available for a fixed energy supply. To give just two examples, improved household insulation decreases heating and air-conditioning energy demand, and the substitution of transistors and flourescent lights for vacuum tubes and incandescent bulbs leads to further energy economy.

Nevertheless, even in America, and *a fortiori* in the world as a whole, the great majority of the people do not live by an American upper-middle-class energy standard. There is a factor of about 30, I believe, between the personal energy consumption of a "comfortable" American and an average world citizen, and I see no reason for growth in energy demand to stop, even if population is stabilized, until that gap is bridged. (I further estimate that it will take America about fifty years to achieve a level energy demand for personal living, and the world as a whole about a century.)

As it is with personal demand, so it is with commercial and industrial demand for energy. Again, one can find conspicuous overconsumption of power in places (the bright lights of Las Vegas being a glaring example), overuse of energy-intensive materials such as throw-away aluminum cans, and industrial inefficiencies both in processes (open-hearth steel) and in products (prematurely obsolete automobiles). Again it is probable that the correction of these deficiencies will be accompanied by an increase in the number of new energy-intensive products made in America and the world, the net result being a steady, upward trend in commercial and industrial demand, toward a 100-year asymptote.

A final large sector of growth in energy demand is in the field of transportation. In the United States, in spite of significant academic worrying about the consequences of it, both domestic automobile travel and airplane travel are increasing in mileage and decreasing in fuel efficiency. It makes no difference that I, personally, would favor very strong restrictions on automobile travel in cities (which is, I think, the best way to make public mass transportation attractive). It also does not seem to matter that smaller cars with intrinsically low pollution can be developed, or that an efficient combination of ground and two modes of air travel could move people by public transportation faster and with lower fuel cost than is now the case. The transportation sector continues to gulp increasing amounts of hydrocarbons, and we do not seem to be able to do anything about it.

In addition to requirements of manufactured energy for the good life, we also must consider incremental energy to recycle the raw materials of spaceship earth—fresh water, metals, and the like, for we are using up high-concentration materials (like the iron ore which used to be in the Mesabi Range) and will soon be forced to use low-concentration materials or impure ones. The process of purification is the heart of recycling; stated in thermodynamic terms, purification is the same as the local reduction of entropy, which can only be accomplished by work. (The word *work*, which means two different things in thermodynamic and popular use, is correct in the above sentence according to either usage.)

In summary, no matter what we do about improving the reasonableness and efficiency of our energy use, the world is demanding more energy. The most rapid area of demand growth is electricity (10-year doubling time worldwide, 12 to 15 years in America), both because of its convenience and because the centralization of electric power generation has permitted this industry to take full advantage of economies of scale. Moreover, if any larger fraction of the transportation sector electrifies (as is possible in the case of electric personal vehicles and buses or a rejuvenated, electrified rail system), electrical demand will take a further quantum jump. It is this rising demand that the nuclear industry is trying to supply.

ECONOMIC FACTORS

In 1970, America alone consumed about 1.5 trillion kilowatt-hours of electricity. The cost of generating this electricity averages to about 6 mills per kilowatt-hour, so that generating cost was about 9 billion dollars.

Though this is a small fraction—around 1 per cent—of the gross national product, it is not an insignificant fraction. However, the important point is the total sum involved.

By the year 2000, I have forecast a five-fold increase in United States electrical energy demand.[1] This is a very conservative estimate, amounting to a 5.35 per cent annual increase as compared with a historical 7.2 to 7.5 per cent rate. (The assumed rate of increase from 1980 to 2000 was even smaller—4.8 per cent.) In constant dollars, we are discussing an industry of about 50 billion dollars-per-year cost.

Roughly half of the generating cost is accommodated by cost of fuel. In 1970, the electrical industry spent about 4 billion dollars on fossil fuels—coal, gas, and oil. If the real cost of fuel did not change, and if the United States' pattern of electrical generation remained as it was in 1970, we would be spending 20 billion dollars per year on fuel for power plants.

However, things are not remaining the same:

(1) We have become aware that we have been paying a heavy price in order to use our cheapest fossil fuels. That price is in air pollution—grime and soot, sulfur dioxide,

and other noxious by-products. Persistent research, carried on for many years, has failed to find a cheap way of producing power from our coals without significant pollution: either the coal must be pre-processed in a costly plant to remove pollutants from the fuel, or expensive scrubbing systems must be added to the plant to remove pollutants from the exhaust gases.

We can also defer concern about air pollution by such devices as dispersing power-plant effluents throughout larger volumes of air through the use of tall exhaust stacks. I say "defer" because our pressing immediate concern is with air quality in industrial areas, which can be improved by this device; but at some time, general air pollution will become serious in America, as it already is in Europe and Japan.

We can also use "clean" fossil fuels—natural gas, Middle Eastern oil, or low-sulfur coal—but the first two have a premium value for their relative scarcity, whereas the last occurs in locations far from the point of consumption, and is expensive to transport.

In brief: having chosen to be concerned about the quality of our air, we find that the fuels we permit ourselves to use are rapidly increasing in price.

(2) The world's supply of coal, gas, and oil is very inhomogenously distributed. The United States has ample, albeit hard-to-get-at, reserves of coal, but not enough gas and oil for another generation. The Middle East has most of the good oil in the world today. Europe has coal and some gas; Japan has no fuels at all to speak of.

The requirements of the transportation sector, plus power plants already built, have already committed Europe, Japan, and the United States to purchase Middle Eastern oil for the next fifteen years, at least. The currency payments are a steady drain on the balance of trade of the industrialized countries, and have serious diplomatic consequences, as well. Most important, the sellers' market in oil is severely nonlinear; each increment in demand brings a greater price increase than the last. Thus, not so much as to balance payments for fuel as to prevent

1. B. I. Spinrad, "The Role of Nuclear Power in Meeting World Energy Needs," from "Environmental Aspects of Nuclear Power Stations", IAEA, Vienna (1971), pp. 57–82.

them from becoming a major burden, it is imperative to the industrialized countries that oil consumption not grow rapidly.

Regardless of future price increases, the price of fuel has soared even within the last years (and in constant dollars). On this evidence, high-quality organic fuels are likely to cost at least double what they used to, and the United States fuel bill for electrical power plants might be $50,000,000,000 per year by A.D. 2000. (The price of electricity would rise, on this scenario, by 50 per cent.)

In response to the likelihood, a substitution of technologies is indicated, and it is for this reason that nuclear power is now taking over a very large share of our new investment in electrical power facilities—50 to 60 per cent. It would be responsible for upward of 75 per cent of new American plant were it not for two factors that have raised the cost of nuclear plant investment by increasing the lead time between ordering and commissioning. One of these factors is the exacting standards of the nuclear industry, to which vendors and contractors are only slowly learning to conform. The other is the "antinuclear crusade," which has not been without its influence on the utilities.

THE ALTERNATIVES

Up to now, I have argued that we cannot satisfy the demand for electricity, either in America or the world, by suppressing it, and that an attempt to satisfy the demand by technologies that derive from our "conventional" steam-driven, fossil-fueled thermal plants will lead either to more massive air pollution or a sharp escalation in price. At this point, I must enter both a plea of innocence and a disclaimer of expertise.

As an engineer, I have no more (or less) right than anyone else to discuss the alternative of letting the price rise until the demand drops; or of taking other sociopolitical steps to balance demand and supply. Thus, I state my opinion that humanity will be well served by continuing to use electricity in a rational way, and indeed by using (as a whole) much more of it than now. I do not state this opinion with any special authority.

My job in society is to provide technical alternatives as best I can. Whether society uses them or abuses them is society's business. If you don't like the way mankind is using the airplane, the food additive or the bomb, don't blame the inventor, and don't blame technology: blame yourself. "I" refuse to be the scapegoat; "we" are the ones.

With this parenthesis, I proceed to discuss those technical alternatives which have been seriously offered.

Nuclear Fusion

Large-scale nuclear fusion was first demonstrated in 1952 through the destruction of Eniwetok atoll by an H-bomb. Research on tapping this same energy in a controlled and sustained way by a fusion reactor dates from before that event and has continued to the present. The prospects for a controlled thermonuclear reactor can be judged to some extent by the fact that, though progress toward it has been definite, the intervening twenty years have not yet provided a proof-of-principle experiment. Optimists[2] believe that such an experiment is still five years away, whereas a consensus of researchers in the field would probably center on ten years. (These forecasts, incidentally have hardly changed since 1960). Further grounds for optimism now exist in the addition of laser-initiated fusion of a reagent pellet to the previous schemes of confining a reacting plasma in a "magnetic bottle."

The problem, unfortunately, only begins once a proof-of-principle is achieved. It took 27 years from the time that the fission chain reaction was first demonstrated (1942) to the time of significant use of this principle in the electrical industry (1 per cent of the world's electrical generation was by nuclear power in 1969). Fusion is a much more complex system than fission. It involves high-vacuum and superconducting magnets and delicate electric windings (or precisely aimed and timed particle guns and super-efficient lasers), plus the solution of problems

2. R. F. Post, "Fusion Power—The Uncertain Certainty," from "The Energy Crisis", *Science & Public Affairs,* Chicago (1972), pp. 114–120.

in radiation, materials, and heat, which are avoided by accepting performance limitations in the present era of fission reactor development.

Therefore, even with massive expenditures, fusion reactors will not, in my opinion, be economically significant before A.D. 2010. Even then, we must reserve judgment, for much will hinge upon the cost of making components which we are not even able to specify or describe now.

Solar Power

Modern research on solar power has been going on for quite some time. Most of the research is on ways of using it at the point of consumption—for heating, cooking, and other domestic purposes—and there have been some successes. Converting this diffuse power into a more concentrated energy source has also been achieved for special purposes: for example, a solar furnace in France reflects the sun's rays from a hillside through a parabolic mirror array, at the focus of which the world's cleanest furnace (and one of the hottest) is located.

More recently, researchers have been looking at ways of achieving lesser concentrations of solar energy at reasonable cost. Haas and Haas of the University of Houston,[3] and Meinel and Meinel of the University of Arizona[4] both have evaluated schemes for doing so. The Haas's have looked at inexpensive mirror arrays, while the Meinels have developed and evaluated stacked solids which function as light-frequency converters, producing the equivalent of a greenhouse effect at the bottom of the stack.

These evaluations are roughly equivalent. In neither case would an exotic technology be required to make the schemes operable, but in both cases, even with economies of

3. G. Haas, reported in Report M72-50, "Symposium on Energy, Resources and the Environment," Fifth Meeting of the Symposium Committee (Feb. 1972) pp. 241–339; The MITRE Corp., McLean, Va.
4. A. G. Meinel and M. P. Meinel, "Physics Looks at Solar Energy," *Physics Today* 25, No. 2 (Feb. 1972), pp. 44–50.

scale, the use of components available today would result in electrical generating costs about ten times greater than our current prices. It is an open research question whether manufacturing processes to bring the costs down enough can be developed, and it should be pursued. Meanwhile, we can only describe solar power as intriguing, worthy of research, and currently uneconomic.

Hydroelectricity and Tidal Power

Although I have suggested previously that hydroelectricity is fully developed in the United States (and also in Europe), the world's good sites for hydroelectric power are less than 10 per cent developed. Unfortunately, most of the best remaining sites are in undeveloped parts of the world: the Andean slopes in Brazil, the lower Congo, Madagascar, and arctic Canada. As such, they can (and should) be of great value to selected countries. However, they are not vast enough to supply a large fraction of the world's energy needs.

There are also selected locations at which tidal energy concentration is large enough so that tidal impoundment could power a large hydroelectric plant. The Bay of Fundy is probably the world's best. Again this would be only a local solution to a continental and world problem.

Geothermal Energy

Geothermal energy is energy stored in the earth's crust. Just as hydrocarbon fuels represent fossil storage of solar energy, geothermal heat is largely the stored energy from past radioactivity of the earth. It is recoverable whenever circumstances have resulted in the retention of water in pockets deep below the surface. Then the heat is transferred to the water, and can be extracted by drilling into the pocket; live steam or hot pressurized water flows to the surface and, if it is clean, can be used directly to power a steam turbine. If the steam is impure (hydrogen sulfide is a common constituent), its heat is exchanged to clean steam at a somewhat increased plant expense.

Geothermal energy is comparable to hydropower, in being a valuable local resource in some areas, but not a major national or world energy source. The technology is not new, and if it were as promising as its strongest advocates claim, it ought to have been widely utilized by now. The lack of utilization suggests that its problems—corrosion and effluent control, and stability of the energy source over a long time period—are more severe than is usually advertised.

Nuclear Fission—Breeders

Nuclear power plants are now producing electricity at very reasonable prices. There is no reason for the costs to go up within the next fifteen or twenty years, and many reasons for the prices (in real dollars) to drop. After that period, our supply of fresh natural uranium fuel will become more expensive, but the rise in power costs (if it occurs at all) will be limited: the cost of generating power from breeder reactors is not more than double the conventional cost now, and will certainly decrease. At some time, therefore, we can switch over to breeder reactors; then our power costs will stabilize or continue to decrease, for the fuel supply for breeder reactors is cheap and, by human scales, infinite.

The two points that require emphasis here are the immediacy of nuclear power—it exists economically now—and its endurance—it could serve us forever. To bridge the gap, breeder reactors are needed, and they are under development. However, it is just as easy to exaggerate the difficulty of this development as to minimize, for example, the difficulties of controlled thermonuclear reactor or solar-power development. It is therefore worth putting the breeder development problem in perspective.

The world's first civilian power from nuclear energy was generated from sodium-cooled reactors in 1951. There are two claimants for the distinction: the Submarine

Intermediate Reactor (*SIR*) test at Knolls Atomic Power Laboratory in Schenectady, N.Y., and the Experimental Breeder Reactor No. 1 (*EBR-I*) at the Idaho site of Argonne National Laboratory. (Knolls was actually first, but its electrical delivery was intermittent, and *EBR-I* is usually credited with the official laurels.) Both ran successfully as designed, and both showed troubles when operated in experimental tests. The *SIR* was decommissioned when the Navy decided that pressurized water reactors would be preferable submarine systems (this decision followed on the heels of engineering problems with the U.S.S. Seawolf, the prototype *SIR*-powered submarine). The *EBR-I* suffered a core meltdown during a safety test experiment—a severe accident accompanied by *no personal injuries whatsoever*—was rebuilt, and ran until its experimental value was vitiated by the construction of larger experimental reactors.

On the heels of these two reactors, the United States initiated two sodium-cooled fast-breeder reactor projects: Argonne's *EBR-II*, designed for 20 megawatts of electricity, and the privately funded Enrico Fermi reactor near Detroit, Michigan, rated at 185 megawatts. The *EBR-II* is still running. The Fermi fast reactor had a variety of engineering problems, including a partial fuel meltdown (again, *unaccompanied by any personal injuries*), was repaired and rerun, and has now been shut down as not economic for power generation.

In addition to these developments, the United Kingdom, the Soviet Union, and France have built and are operating successful experimental fast reactors.

I interpret this history as meaning that the fast breeder is a sure thing. If we limit our conditions of operation to those that have already been successful, we can price power from fast reactors as two to three times more expensive than conventional power, now. If we permit ourselves to design within current technology (which is ten years more advanced than the systems running now), the price factor is certainly below two. Thus, current development needs to demonstrate only that economics of scale—which would lower the gap further—can indeed be engineered. The very large sums now being spent on this development (in the United States, $500 million for a 300-megawatt demonstration, over $150 million for a test reactor) testify not to the intrinsic difficulty of the job, but to the increased expense of the process of final product development as compared with earlier research.*

Coal Fluidization

There is one method of generating energy that has the potential of using coal—our most abundant hydrocarbon fuel—to give cheap and clean power. This is to react coal, which is fundamentally impure carbon, with water. Depending on the conditions of the reaction, one can get as products hydrogen, hydrocarbons, carbon monoxide, or carbon dioxide, or all of them. When these products are burned, the amount of heat produced is sensibly the same as was available from burning the coal in the first place.

Except for the heavier hydrocarbons, which are liquids, most of the products of the reaction are gases. The process can be adjusted to yield a liquid (Fischer-Tropsch process, used in Germany in World War I to make oil from coal) or a gas (water gas, producer gas, and the like) as a product. Most of our current interest centers on the gasification processes.

The reason for this interest stems from a number of advantages which gasified fuel has over solids:

(1) Mineral impurities are left behind in a residue. Since the conversion plant would presumably be at mine mouth, this saves money on sorting and shipping at the mine, and provides an ash-free fuel.

(2) Gases are much cheaper to transport, via pipeline, than coal, via rail and barge.

(3) Sulfur can be much more easily removed form the gas product than from

*Editors' note: In the fall of 1974, cost projections for the 300-megawatt demonstration plant were increased to $1.8 billion.

either the original solid fuel or the ultimate combustion gases.

There are a number of competitive processes for coal gasification under development. All have seen industrial application in the past or have been successfully operated at a pilot-plant scale. They are all too expensive for current application, by about a factor of two, and all have potential for reducing this expense very significantly by going to large plants with their economies of scale.

In brief, coal gasification is in the same situation as the breeder reactor—a sure-fire solution to our energy problems at some increase in the price we pay, and a good bet for development to fully competitive power costs.

Recommendations

I am now in a position to recapitulate this portion of my argument in the form of recommendations. I do so by setting aside the question of the "nuclear controversy" for the moment, and assuming that nuclear energy is in fact environmentally acceptable:

(1) Hydroelectric, tidal, geothermal, and other forms of power from earth forces require no engineering development. They should be harnessed wherever favorable economic circumstances for their use (taking into account both their side effects and their investment priority) exist. They are probably being brought to use as rapidly as is proper, right now, and cannot be relied upon as principal inputs to ultimate world power needs.

(2) Both fusion and solar energy have such dramatic ultimate potentials that they are worth serious attention. They are both at the research stage rather than the developmental one (although the natures of the researches required are quite different). Research is cheap compared to engineering development—sums of the order of $100 million per year over 30 years could be tolerated for both items, even if successful results were delayed (assuming that research forces

of several hundred to a few thousand people could be usefully deployed on the research missions). If, and when, research results are favorable, the scale of activity for engineering development should be assessed and allocated.

(3) Both the breeder reactor and coal gasification should be developed. The potential returns justify expenses of tens of billions of dollars over the next decades, and even though current programs may be costly, the most efficient development would not be cheap. Gasification offers a solution to the problems of America and Europe for a few hundred years, which gives plenty of time to develop an asymptotic power source. Breeder reactors represent one of the possible asymptotic sources.

(4) For the next generation, fission reactors of the now existing types—light- and heavy-water reactors, gas-cooled graphite reactors—should be commissioned at whatever rate is necessary to prevent further increases in our demands on the world's petroleum.

NUCLEAR POWER EFFLUENTS

Nuclear power from fission reactors produces effluents, and these effluents have been the rallying point of many of the critics of nuclear construction. The problem is often divided into two parts—one concerned with the power plant itself, and the other with the nuclear fuel cycle. Critics have contended, however, that these two aspects of nuclear power are inextricably related, and I am inclined to agree with them.

The primary dichotomy I shall make is between problems (real or imaginary) associated with the normal operation of the total system—power-plant and fuel-cycle facilities—and those concerned with (real or imaginary) accidents and malfunctions. Here, as effluents, I shall discuss waste products or side effects arising from normal operation. These problems naturally are grouped into aesthetic, ecological, and radiological categories.

Aesthetics

Under "aesthetic" considerations, one includes the aspects of the industry that contribute or detract from our sense of beauty in the world. On this basis, among thermal plants, we can immediately consider coal-fired power plants as the worst, gas power plants the best, nuclear ones next, and oil-fired ones next. The mining of coal scars both the earth and the lungs of its miners; its transportation leaves a trail of black dust; the plants where coal is burned dedicate a large plot of land to coal storage; and the burning produces a secondary mountain of ashes, which we must cover up. At the other extreme, gas-fired power plants best approach our desire for inconspicuous utilities; they are compact, ashless, and use fuel delivered by a sometimes graceful, often invisible pipeline. The delivery of gas from the well to the pipeline is done with equipment which, while bold, has a beauty of its own.

Nuclear power plants are an architect's pleasure, their containment spheres or protective reactor buildings permitting some expression of the spirit of *Homo fabricator*. The interiors of these plants are, above all, clean; they are examplary of modern industry. The volume needed for fuel storage and waste handling is so small that it is housed in the reactor building itself or in a harmoniously designed auxiliary structure. The fuel cycle plants range from the similarly hypermodern fuel assembly, enrichment and reprocessing plants, to some rather tacky, but fortunately small, ore mills. The mass of mineral that must be mined to supply a given quantity of energy is something less than 1 per cent of the equivalent coal, and its volume is several-fold smaller still, so that the mining operation is really a minor matter.

Oil is somewhere in between. Its recovery is a messier operation than is the case with gas, and so is its transportation, but it is far cleaner than coal. Its storage in enormous tanks is not beautiful, but it is acceptable in a modern setting. Oil-fired power plants are somewhat harder to accommodate to good visual design than are nuclear ones, but again they are acceptable. A unique problem with oil plants is their tendency to create foul odors, a factor which is not always removed in their design.

Ecology

Under "ecological" effluents, we consider those aspects of energy production which adversely affect nature. The key word is "adversely," and the question is often a matter of judgment—both because there are divergent views as to what end products of man's intervention with nature are desirable and undesirable, and because nature is a complex web of life that can respond to stresses in unpredictable ways. To give an example, the strip mining of coal is usually condemned as an insult to nature, and certainly, our aesthetic senses are disturbed by a gash in the hillside. Yet, in central Illinois, nature has improved on itself in accepting the challenge of reinvading strip mines, and has made of the scars some of the best bird-refuge areas in the United States.

Similarly, a trout fisherman will view as a disaster what a recreational swimmer or boater will consider a boon: the warming of natural waters beyond optimal temperature for fishing, but to the comfort range for water sports.

All this is by way of preface to one of the most artificial obstacles placed in the path of nuclear power: it contributes to "thermal pollution" of the environment.

The facts can be stated succinctly. Nuclear reactor plants based on light- and heavy-water reactors convert heat less efficiently into work than the best modern, fossil-fired steam plants. In fact, per unit electrical output, their heat rejection is between one and a half and two times as great. Their efficiency comes close to the average of all power plants currently operating. This heat can be rejected in a variety of ways: by transfer to natural waters; by transfer to artificially created bodies of waters; by

evaporation of water (wet cooling towers); and by transfer to air (dry cooling towers). The best method of doing so is clearly a matter of local conditions; for example, throughout Canada, the warming of natural water is welcomed, whereas in the southern United States, one has to be very careful. In some locations, the creation of cooling lakes opens up desirable recreational opportunities, and in others it just creates too much fog. And so judgments must be made. They involve somewhat larger quantities of heat when the plant is nuclear, but there are no qualitative differences from one thermal power plant to the next.

Therefore, "thermal pollution" is not a problem of nuclear energy. In fact, it is not a problem at all, except in the engineering sense that problems are sets of conditions which a good design must cope with. I view waste heat as that sort of problem, and therefore as an opportunity.

Radiation

There are only a few people in the world who would argue that increased nuclear radiation is good for people. At the same time, there is widespread confusion over how much increase would have noticeable effects. Therefore, a very cursory summary is in order.

Our biosphere is bathed in nuclear radiation. We receive of the order of 100 millirems (a rem is 100 ergs per gram) of radiation per year, roughly half each from cosmic rays and from the natural radioactivity of the soil. When we climb mountains or take jet airplane trips, we get more cosmic rays, and if we live on certain rocky soils, we get more from the ground; backgrounds in locations with very similar public health records vary by factors of two or more. Modern medical and dental practice is responsible for the average individual receiving several tens of millirem more each year from x-rays.

The radioactivity from nuclear weapons test fallout has rarely added more than 10 millirem per year to backgrounds. There have been a few fallout hot spots where this much radiation has been exceeded, but it tends to decrease again as fallout particles wash away and settle into the ground.

To this, nuclear power plants will be adding directly about one millirem by the year 2000, if our expectations of having 1 million megawatts of installed nuclear power at that time are fulfilled. This is about the same increment as would be added by coal, either directly or gasified, the radioactivity from coal being the result of release of trapped radon and thoron.

The increment from nuclear power would be another millirem from the release of radioactive krypton at the reprocessing plants, provided that the processes adopted for krypton retention have the disappointingly low yield of 90 per cent.

The above refers to the United States as a whole.

Gofman and Tamplin[5] have estimated as an upper limit that an incremental exposure of 200,000,000 people to 170 millirems per year would result in 30,000 extra deaths each year from cancer and leukemia. Their number has been criticized as being based on a very pessimistic model, but if I accept it, 2 millirems per year would result in 400 extra deaths per year at maximum. As a realistic estimate, many biologists would divide the Gofman-Tamplin estimate by a factor of ten[6]; this indicates the uncertainty of these estimates.

Attempts to specify this number through examination of vital statistics in areas of varying background radiation have yielded null results. It appears, as Tamplin pointed out in a definitive refutation of Sternglass[7]

5. A. R. Tamplin and J. W. Gofman, "The Radiation Effects Controversy" and L. Pauling, "Genetic and Somatic Effects of High Energy Radiation," Bull. Atomic Scientists 26, No. 7 (Sept. 1970), pp. 2–8.
6. P. J. Lindop and J. Rotblat, "Radiation Pollution of the Environment," from "The Energy Crisis," cf. Ref. 2; pp. 41–48, particularly pp. 44, 46.
7. A. R. Tamplin, Y. Ricker, and M. F. Longmate, "A Criticism of the Sternglass Article on Fetal Mortality," Report UCID-15506, Lawrence Radiation Laboratory, Livermore (1969).

(who has been repeatedly discredited), that radiation is not a major contributor to infant mortality, nor even to cancer, and any affects attributable to radiation are lost in the much more significant variations arising from socioeconomic factors.

The annual statistics for cancer mortality fluctuate by thousands from year to year; so what the number 40 (or even 400) really means, is hard to determine. It is actually a delayed risk spread over the population, and such risks are usually interpreted as shortening life expectancy. The more pessimistic number translates into a 15-minute decrease in average life expectancy. I do know that we tolerate a lot of things that are a lot worse, which I think deserve prior attention—malnutrition and poverty-induced crime, for example.

The results for genetic damage to our population are, incidentally, similar, except that genetic defects have some tendency to be selected out in the gamete or the fetus, so that increased mutation rates tend not to be accompanied by a proportionate increase in genetically damaged people.

I can conclude this section by the summary statement that nuclear energy has some favorable environmental aspects, and some unfavorable ones; but if the industry works as it is supposed to, none of the effects are overriding, or even major.

RISKS TO THE PUBLIC

Of course, all of the foregoing is minor compared to the question of public hazard brought about by nuclear power. The real issue is whether we can prevent accidents that will expose the population to radiation burdens many orders of magnitude greater than those attendant on normal system operation.

I believe we can. One of the ingredients of this belief is that the nuclear power industry has avoided accidents leading to major public releases of radioactivity from its infancy; the systems we have settled down with are safer than many with which we have successfully operated. Another ingredient is the safety consciousness of the industry, which is very probably unique among the world's industries: the nuclear engineer is trained to dream up ways in which an accident could occur, and then design to avoid it. Enforcing this consciousness is an institution of safety analysis reporting for licensing, which is also unique, and of operational reporting, which insures that good engineering solutions are found for even minor malfunctions so that they never get to be major ones.

In a word, the nuclear engineer runs scared, and in so doing, he assures the safety of the public.

The whole question of safety analysis is surely one of the most widely misunderstood processes. The point of the analysis is, as stated, to make sure that all accidents can be kept within the confines of the plant. To do this, the analyst conceives of increasingly improbable accidents, and then demonstrates how his design will cope with them. At some point, the probability of accident drops below the point of credibility, but the analyst keeps going. (We used to have the phrase "maximum credible accident," but it was dropped, primarily because it was too difficult to assign credibility to some of the major accidents which are covered by design "just in case"!) At the end, he postulates the consequences of a standard release, which nobody knows how to bring to pass, and which is required primarily to remind all concerned that this is what the design is supposed to avoid. All this is perfectly clear in the wording of the report; yet nuclear opponents have convinced many citizens that these exercises in discipline are imminent hazards.

The cautions of the nuclear industry with regard to transport, fuel processing, waste disposal, and management have been equally misrepresented. The specifications on shipping containers for spent fuel and high-level wastes are so rigid that they can survive accidents you would not believe. You also would not believe how many contingency layers of air filters a reprocessing plant has

"just in case." Many, many people who should know better have assumed that, because permanent nuclear waste disposal is not practiced, it is an unsolved problem. It is not; it is a too many-times solved one. In fact, some solutions, such as deep-sea disposal, have been forbidden, on fallacious arguments. Let me list some possibilities:

—Near-surface vaults of air-convection-cooled, chemically immobilized waste-solid containers.
—Deep-sea placement of bronze-clad, chemically immobilized containers.
—Deep burial of chemically immobilized, loose solids in geologically stable, inactive rock formations.
—Deep burial of raw wastes in dry rock formations (including salt).
—Chemical segregation of wastes, with variable disposal according to half-life and chemical properties.

I summarize this section by the conclusions that the nuclear power system is as safe as we can make it, and that this is safe enough provided only that:

(1) The nuclear industry is constantly reminded of its responsibility.
(2) The reminding is in the form of legitimate scientific or technical questions, rather than emotional or crackpot assertions.

COMMENTS AND CONCLUSIONS

The preceding material is the basis for the beliefs I stated at the beginning of this paper: the economic, social and physical well-being of the peoples of our world are best served by continuing the pace at which we are substituting inanimate for animate energy; this can be done through nuclear power at less jeopardy to the environment and less public risk than what we now accept as "standard" adjuncts of industry; indeed, a clean and abundant source of energy, which nuclear fission is, is vital to us in order to repair some of the insults of our previous, less knowledgeable industrial revolution; and, although competitive forms of energy can be foreseen, none of them combine the advantages of being immediately practical, and yet durable from an economic and resource standpoint, except nuclear power.

These arguments are not new, and I believe their validity has in fact been demonstrable for at least ten years. It is therefore an occasion for comment to speculate why (and particularly in the last five years, when the benefits of nuclear power rather than the costs of the atomic bomb were the most visible parts of the nuclear picture) so much opposition to nuclear power has developed.

I suggest that there are several components to this opposition: interest, jealousy, fear, and reaction.

Under "interest," we must consider that nuclear power is indeed competing with other forms of energy production; and to a certain extent is capturing markets which might have otherwise gone to industries that are not nuclear. Even where competitive capitalist forces are not at work, there are national and regional interests which may be adversely affected by nuclear power (such as the perceived interests of oil-exporting countries). It would be asking too much of human nature to expect anyone not to oppose a perceived threat to his interest, and it is not proper to decry the early opposition of fossil-fuel interests to nuclear power; it is only necessary to remind the public that, if nuclear power is defeated, someone is bound to profit.

Again, as is normal, the nuclear field has its own interest group: commissions, industries, and even researchers such as myself. Again, I ask the reader to discount our interest, as well as that of the opposition. But all in all, and even considering some of the wilder errors which originate in nuclear propaganda, I have to assert my opinion that, in the conflict of interests, the nuclear people are more sinned against than sinning.

Under "jealousy," I classify the opposition of many in the intellectual world. Nuclear energy has occupied a glamorous position among the various fields of learned inquiry, and has been extremely generously supported by the countries of the world.

Now that it appears that (at least in the advanced countries), we have reached the limits of growth in our total research commitment, the large "piece of the action" which nuclear energy has had is seen as an obstacle to aspirations of other fields for increased support.

The situation is aggravated by a problem internal to the nuclear field: the "bureaucratization" of the nuclear development field. The fact is that nuclear development has spawned a large management apparatus, which is responsible for grave decreases in the efficiency of the development process, and is insatiable in demanding more money. Parkinson's Law should have another corollary added: "The less justified spending is, the more money is needed by the organization whose job it is to justify the spending." And of course, the more this goes on, the more justified become the demands for diversion of spending to other fields.

Under "fear," I consider, not the uninformed hysteria of the citizen who has been frightened by the apocalyptic statements of nuclear opponents, but the motives of some of these opponents. I am personally on friendly terms with many scientists who have a more overriding fear—that a successful nuclear industry will make it difficult, or impossible, to achieve effective abolition of nuclear weapons. This article has not been addressed to that issue, and I leave it to other forums.

However, it is pertinent here to note that a great many scientists have, at minimum, subordinated their scientific passion for the truth to the end of achieving, by any means whatsoever, the goal of nuclear disarmament. Though there are only a few cases of willful deception (such as that of one eminent biologist who deliberately perpetrated a popular fallacy as to the spread of radiation-induced genetic defects), there are many who stretch the truth to the same result: encouraging popular fears so as to ensure public attention to problems of nuclear weapons.

Finally, I come to the point of "reaction," a phenomenon that I can describe, but do not understand. At least in America, there may be found large segments of opinion which seem to be yearning for the past: for lower populations, elimination of automation, less public mobility, and fewer conveniences. Thus, one can find a corps of activists to oppose any new changes. Many of these people call themselves ecologists, and all claim to be preserving the pristine world. I have never been able to understand why the rape of the soil in the southern United States due to the cultivation of cotton is to be preferred to the substitution of synthetic fibers from a small number of chemical plants; but this reasoning is their touchstone.

There is much more about the movement that I do not understand. It is counted as a leftist cause, but the archaism of its viewpoint, its preference of elite over popular tastes, its consideration that the wilderness (read: "royal") preserve is to be preferred over the public playground, and its blithe assumption that the working man ought to be happy with the conditions of the unmechanized farm or the handicraft (read "sweatshop") industry, all proclaim a fascist viewpoint. It not only puzzles me: it frightens me.

This reaction is what the nuclear industry, in its own way, directly opposes.

SUGGESTED READINGS

Cohen, Bernard L., 1974, "Perspectives on the Nuclear Debate," *Science and Public Affairs (Bulletin of the Atomic Scientists),* Vol. XXX, No. 8, pp. 35–39, October, 1974.

Jordan, Walter H., 1970, "Nuclear Energy: Benefits versus Risks," *Physics Today,* pp. 32–38, May, 1970.

Electric Power from Nuclear Fission

MANSON BENEDICT   1971

An abundant supply of electric energy generated at low cost with minimal adverse environmental effects is essential to civilized society. Generation of electricity by nuclear fission is capable of meeting all three of these requirements: abundant supply, low cost, and minimal environmental effect.

My purpose is to describe the types of nuclear reactors used for electric generation in the United States today; the advantages, deficiencies, and problems of today's reactors; and the steps that need to be taken to realize the full potential of nuclear fission as a source of electricity. However, all types of reactors under development are not described here. Attention is focused on light-water and fast-breeder reactors, the principal types in use or under development today in the United States. Heavy-water reactors and molten-salt reactors, which may also play a role in the future, are not dealt with.

The type of nuclear reactor used throughout the world today for electric power generation obtains most of its energy from slow-neutron fission of the scarce isotope of uranium, uranium-235, which occurs in natural uranium only to the extent of 1 part in 140. To obtain the full potential of nuclear energy it will be necessary to develop effective means for utilizing the abundant isotope, uranium-238, which makes up the remaining 99.3 per cent of natural uranium. The most promising type of reactor for this purpose is the fast-breeder reactor, in which uranium-238 is converted to plutonium, which then undergoes fission with fast neutrons.

The breeder reactor will provide the world with electric energy for thousands of years, far beyond the capability of all fossil fuels—coal, oil, and gas—now known or likely to be discovered. Already, nonbreeding reactors in operation in many parts of the United States and elsewhere in the world are generating electricity at a cost as low as the cost of electricity from fossil fuels in the

Dr. Manson Benedict is institute professor emeritus in the Department of Nuclear Engineering, Massachusetts Institute of Technology, Cambridge, Massachusetts.

Originally presented at the National Academy of Sciences Symposium on "Energy for the Future", April 26, 1971, and published in the *Proceedings of the National Academy of Sciences*, Vol. 68, No. 8, pp. 1923–1930, August 1971, and subsequently appeared in *Science and Public Affairs (Bulletin of the Atomic Scientists)*, Vol. XXVII, No. 7, pp. 8–16, September 1971, and in *Technology Review*, Vol. 74, No. 1, pp. 32–41, October-November 1971. Reprinted with light editing and by permission of the author and the National Academy of Sciences. Copyright 1971 by the National Academy of Sciences.

same place. Breeder reactors, under development in many countries, are likely to generate electricity at an equally low cost.

LIGHT-WATER REACTORS

Turning first to today's power reactors, the slow-neutron, nonbreeding, uranium-235-consuming kind, the reactors used predominantly in the United States are of the light-water type. In these reactors ordinary water, under pressure and at temperatures up to 600°F, is used both as coolant to transport the heat released in fission and as moderator to slow down the fast neutrons initially produced in fission. There are two principal types of light-water reactor: the pressurized water reactor, used in around 60 per cent of the light-water reactor installations, and the boiling-water reactor, used in about 40 per cent. Both of these reactor types were developed initially in the United States through the joint efforts of industry and government, and their use throughout the world today has reduced the cost of electricity and paid important dividends in enhanced United States prestige and favorable foreign trade.

Pressurized-Water Reactor

The principle of the pressurized-water nuclear power plant is illustrated in Figure 3-1. Fuel rods for this reactor consist of uranium dioxide, enriched to about 3 per cent in uranium-235, hermetically sealed in tubes of a zirconium alloy. This zirconium tubing constitutes the first barrier against escape of the highly radioactive fission products that form in the fuel as the end product of the nuclear reaction. Assemblies of these zirconium-clad fuel rods are mounted in a heavy-walled steel pressure vessel and are surrounded by flowing water entering at a temperature of around 540°F and leaving around 600°F, held at a pressure of around 2250 pounds per square inch to prevent boiling. The water slows down neutrons produced in fission and increases their probability of reacting with uranium-235 to such an extent that the uranium fuel constitutes a critical mass capable of sustaining a nuclear-fission chain reaction. To hold the chain reaction at a steady rate, a variable amount of neutron-absorbing boron is used in the reactor, partly as movable control rods and partly as boric acid dissolved in the water. Pressurized water is pumped through the reactor by the circulating pump, past the gas-cushioned pressurizer which holds the pressure constant, and through the steam generator. There heat is transferred from the primary pressurized water at 600°F and 2250 pounds per square inch to secondary water boiling at a lower pressure of around 720 pounds per square inch to make steam at around 506°F.

The steam flows through a turbine driving an electric generator and then passes to the condenser, where it is condensed at subatmospheric pressure. The condensate is returned to the steam generator by the condensate pump.

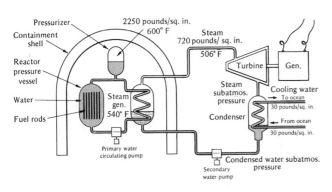

Figure 3-1. Schematic diagram of pressurized-water nuclear power plant.

In the condenser, heat from the condensing steam is transferred through cooling coils to cooling water at a pressure above atmospheric, which leaves the condenser at a temperature typically 20°F warmer than the incoming water. In some plants, this cooling water is drawn from the ocean or other natural sources; in others, it is recirculated through cooling towers. Disposal of the heat contained in this warm water without adverse effect on the environment is one of the problems of all steam-electric plants, nuclear as well as fossil. In light-water nuclear plants, however, about 50 per cent more warm water must be handled than in an efficient fossil-fuel plant, because the thermal efficiency of the water-cooled nuclear plant is only 32.5 per cent due to its relatively low steam temperature of 506°F, compared to a fossil-fueled plant with a thermal efficiency over 40 per cent obtainable from steam temperature around 1000°F. The environmental impact of warm water from a nuclear plant can be dealt with satisfactorily either by siting on a natural body of water with adequate heat-absorption capacity, such as the ocean or a large river, or by use of cooling towers. The cost of heat disposal from a water-cooled nuclear plant is, however, somewhat higher than from a fossil-fueled plant.

Radioactivity Aspect

A more significant potential environmental aspect of a nuclear power plant is the enormous amount of radioactivity contained in it. Water-cooled nuclear power reactors are provided with many barriers against escape of radioactivity, which have kept releases to insignificant levels, and can be provided with even more elaborate safeguards if these should be required. As an illustration, precautions taken to minimize radioactive releases from a pressurized water reactor will be described.

Most of the radioactivity is in the form of fission products contained in the uranium dioxide fuel. Some of the fission products are refractory oxides, insoluble in water. Others, however, are volatile or water soluble, and a small fraction of them appear in the primary water whenever zirconium tubes leak, as they sometimes do. The primary water also carries corrosion products made radioactive by neutron activation. The radioactive content of the primary water is kept low by continuous purification by filtration and ion exchange. Escape of radioactivity from the primary water is prevented by the leak-tight pressure vessel and piping system within which this water circulates.

Even if the primary water system should unexpectedly leak, there are further barriers against escape of radioactivity. The primary system is completely housed within a steel-and-concrete containment shell. This shell is tested periodically for leak tightness and will prevent escape of serious amounts of radioactivity, even if a large fraction of the fission products were to get out of the zirconium tubes and then out of the reactor. Any radioactivity leaking through the steam generator from the high-pressure primary side to the lower-pressure secondary side is prevented from further escape by the leak tightness of the secondary system. Even if the condenser leaked, no radioactivity could inadvertently get into the cooling water and into the environment, because the cooling water is at higher pressure than the steam being condensed.

When the reactor is shut down for refueling and the pressure vessel is opened, precautions as stringent as deemed necessary can be taken to concentrate, package, and confine radioactive materials present in the water or pressurized gases.

I have dwelt this long on the many barriers against escape of radioactivity because it is important to realize that, although nuclear power plants contain enormous amounts of radioactivity, release of radioactivity to the environment from them can be controlled to any degree desired, but with increasing cost. All United States nuclear power plants are monitored by the U.S. Public Health Service, and have been found to add to the environment only a minute fraction of the amount of radioactivity naturally present.[1]

Boiling-Water Reactor

The boiling-water reactor differs from the pressurized-water reactor mainly in that the primary water in the reactor is held at a lower pressure, around 1,000 pounds per square inch, and is allowed to boil in the reactor. Steam and water flowing past the fuel are separated, the water being recirculated and the steam flowing directly to the turbine, after which it is condensed and returned to the reactor. The boiling-water reactor needs no separate steam generator, as the reactor itself performs this function. A boiling-water power plant has about the same thermal efficiency as a pressurized-water plant.

Nuclear versus Fossil-Fueled Power Plants

In 1971, 17 light-water nuclear power plants, with a total generating capacity of 7,000,000 kilowatts, were in operation in the United States and 108 plants, with a total capacity of close to 90,000,000 kilowatts, were operating, under construction, or planned. The total capacity of such plants in operation by 1980 is expected to be 145,000,000 kilowatts. What are the reasons that lead to the widespread adoption of light-water reactors in competition with conventional plants?

The first group of reasons relate to the favorable environmental aspects of nuclear power plants. The traffic of fuel into a nuclear plant and waste products out is negligible compared with a fossil-fueled plant. Whereas a 1,000-megawatt fossil-fueled plant consumes over 2,000,000 tons of fuel per year, a nuclear plant of the same size needs only around 35 tons of uranium dioxide. In contrast to the constant traffic of coal into and ashes out of a coal-fired plant, a nuclear plant needs only one shipment of fuel in and (spent) fuel out per year. Whereas a coal-fired plant needs a large, unsightly reserve coal pile, with noisy coal-handling machinery, a nuclear plant requires only about one-third as much land and is quiet. A fossil-fueled plant discharges an enormous amount of effluent into the atmosphere; a 1,000-megawatt coal-burning plant emits around 10,000,000 tons of carbon dioxide per year and several hundred thousand tons of sulfur dioxide, nitrogen oxides, and ash particles, whereas a nuclear plant emits practically none. Because a nuclear plant requires no combustion air, it can be built partially or wholly underground if desired, and can be designed to be less obtrusive aesthetically than a fossil-fueled plant. In an evaluation of the environmental impact of fossil-fired, hydroelectric, and nuclear power plants, a writer in the October 1970 newsletter of the New England chapter of the Sierra Club, an organization not noticeably biased in favor of nuclear energy, concluded that the choice was "overwhelmingly in favor of nuclear power."[2]

Safety Record

Another factor that has led to widespread adoption of light-water nuclear power plants has been their excellent safety record. Despite their large inventory of radioactivity, there have been no serious accidents and no overexposure of the general public in over 100 reactor years of operation of commercial light-water reactors.* Furthermore, another 780 reactor years of operation without a reactor accident have been recorded by pressurized-water reactors in the United States Navy.

Another aspect of nuclear power that has favored its adoption is its relative invulnerability to interruptions in fuel supply, because a year's supply of fuel is stored right in

---

1. See, for instance, *Nuclear Power Reactors and the Population,* Public Health Service Report BRH/OCS 70-1, Jan. 1970, p. 14.

2. "The Power Dilemma—Our Choices and the Environmental Burden," *New England Sierran,* Oct. 1970.

*Editors' note: By fall of 1974, this safety record had been extended to nearly 250 reactor years.

the reactor. Uranium will be plentiful for at least twenty years, whereas in some places low-cost fossil fuels are already in short supply. The low cost of transporting uranium permits a nuclear plant to obtain its fuel economically from great distances, whereas a fossil-fueled plant is limited to sources from which transportation costs are low. The cost of electricity in a nuclear plant has been less affected by fuel prices than in a plant burning fossil fuel, both because the cost of fuel is a smaller fraction of the cost of electricity, and because the price of uranium has been stable, while the price of fossil fuels has nearly doubled in the past few years.

But all these advantages of nuclear power would not have been effective were it not for the fact that in many parts of the United States and elsewhere, electricity can be generated by light-water nuclear power plants at as low a cost as by other means. This is especially true where the cost of fossil fuel is high because it is not produced nearby, as in most of the United States except near low-cost coal mines or natural gas fields. Absolute values for the cost of electricity from

different kinds of plants at different times and different places are hard to interpret because of the rapid increases occurring in construction costs and the effect of local climatic and environmental conditions on costs. One example of a comparison between the cost of electricity from a pressurized-water reactor plant and a plant burning coal will illustrate how nuclear power has been able to compete economically with electricity from coal. Before deciding to build two 940,000-kilowatt-hour pressurized-water nuclear power units for operation in 1974–1975 near Fredericksburg, Virginia, the Virginia Electric and Power Co.[3] compared the cost of building either the nuclear plant or a coal-fired plant of the same size, and estimated the cost of heat from nuclear fuel and coal. They estimated that the unit cost of the nuclear plant would be $255 per kilowatt hour, for the coal plant, $202 per kilowatt-hour. More than offsetting this

3. Personal communication from Mr. Stanley Ragone, Virginia Electric and Power Co., April 22, 1971.

Table 3-1. Comparison of Cost of Electricity from Coal and Pressurized Water Reactor (two 940,000-kilowatt generating units to be operational in 1974–75 near Fredericksburg, Va.).

	Coal	Nuclear
Unit investment cost of plant (dollars per kilowatt), C	202[a]	255
Annual capital charge rate, per year, i	0.13	0.13
Kilowatt-hours generated per year per kilowatt capacity, k	5,256[b]	5,256[b]
Heat rate (million BTU per kilowatt-hours), h	0.009	0.0104
Cost of heat from fuel (cents per million BTU), f	45	18
Cost of electricity (mills per kilowatt-hours)		
Plant investment, $1,000 \times Ci/k$	5.00	6.31
Operation and maintenance	0.30	0.38
Fuel, $10 \times hf$	4.05	1.87
Total	9.35	8.56
Break even cost of heat from coal, cents per million BTU	36.2	

[a]With no allowance for sulfur dioxide removal.
[b]In actual system, coal plant would generate less electricity than nuclear.

capital-cost disadvantage of the nuclear plant is its lower fuel cost, 18 cents per million BTU compared with 45 cents per million BTU estimated for coal at that place and time. Table 3-1 summarizes a calculation of the cost of electricity for both types of plant, resulting in 8.56 mills per kilowatt-hour for the nuclear plant and 9.35 for the coal-fired plant. For the coal-fired plant to be competitive with the nuclear plant, coal would have to be available at the plant at a breakeven price of 36.2 cents per million BTU. In view of the rapid rise of coal prices, this is not likely to occur.

The economic advantages of nuclear power are even greater in places like Chicago, where local restrictions on burning high-sulfur coal have required importation of low-sulfur coal from Wyoming and Montana at costs of 50 cents per million BTU or higher.

Radioactive Wastes

This discussion of light-water reactors would not be complete without mentioning their problems and disadvantages. One problem that light-water reactors share with all other fission reactors is safe, long-term storage of their highly radioactive fission products. In this brief account I can only outline the problem and the presently favored solution. Spent fuel assemblies removed from the reactor are sealed in shielded containers, specially designed to withstand shipping accidents. Fuel assemblies are transported to a reprocessing plant where they are cut open, their contents are dissolved in acid, uranium and plutonium are recovered, and the radioactive fission products are concentrated and stored in solution for around five years in double-walled containers. At the end of this time, present Atomic Energy Commission regulations require that the solutions be evaporated to dryness and that the solid fission products be sealed in steel containers. Finally, these containers are to be shipped to a national radioactive waste repository to be located in a salt mine in central Kansas for storage 1,500 feet (457.2

meters) underground. This location is chosen because geologic evidence indicates that the radioactive wastes will remain out of contact with ground waters for far more than the 1000-year period in which they must be safeguarded before their radioactivity will decay to a harmless level.

The technology for shipping and reprocessing radioactive fuel has been developed and proved safe by a number of years of operation, although additional measures for retention of the long-lived gaseous radioactive species tritium and krypton-85 released in reprocessing will be required when the volume of fuel is greater than now.

The proposed storage of radioactive wastes in salt deposits was examined by a committee of the National Academy of Sciences[4], which reported in November 1970 that: "The use of bedded salt for the disposal of radioactive wastes is satisfactory. In addition, it is the safest choice now available, provided the wastes are in an appropriate form and the salt beds meet the necessary design and geological criteria. The site near Lyons, Kansas, selected by the AEC, is satisfactory, subject to development of certain additional confirmatory data and evaluation." Despite this favorable report, the widespread concern about this proposal that has recently arisen and the unprecedented responsibility involved in safeguarding highly toxic material for 1,000 years dictate thorough test and careful, gradual exploitation of this proposed storage method.**

Uranium Consumption of Light-Water Reactors

Particular disadvantages of light-water reactors include their low thermal efficiency, already mentioned, and much more significant, the fact that they utilize effectively

4. *Disposal of Solid Radioactive Wastes in Bedded Salt Deposits,* report by the Committee on Radioactive Waste Management, National Academy of Sciences–National Research Council, Nov. 1970. **Editors' note: Salt deposit storage is still under consideration, but the Lyons site proposal has been abandoned.

Table 3-2. United States Uranium Requirements and Resources.

	Actual (1970)	Projected 1980	Projected 2000
Electric generating capacity Million kilowatts			
Total	300	523	1,550
Nuclear	6	145	735
Total tons uranium concentrates consumed	–	200,000	1,600,000

Price of Uranium Concentrates (dollars per pound U_3O_8)	Tons of Uranium Resources at This or Lower Price[5]	Increase in Cost of Electricity from Water-Cooled Nuclear Power Plants (mills per kilowatt-hour)
8	594,000	0.0
10	940,000	0.1
15	1,450,000	0.4
30	2,240,000	1.3
50	10,000,000	2.5
100	25,000,000	5.5

only the scarce isotope uranium-235. The incomplete use of natural uranium will be a very serious disadvantage unless much greater amounts of uranium are discovered than have now been found. Table 3-2 gives the cumulative consumption of uranium ore concentrates by light-water reactors in the United States if this type were used to generate the 145,000,000 kilowatts of nuclear power expected by 1980 and the 735,000,000 kilowatts expected by 2000. The 200,000 tons of uranium concentrates consumed by 1980 and the 1,600,000 tons consumed by the year 2000 are to be compared with the total resources of uranium in known deposits and expected to be found in extensions of them as a function of uranium price. Unless more uranium is found, the 1,600,000 tons which would be consumed in light-water reactors by the year 2000 would raise the price of uranium from the present value of less than $8 per pound to over $15. Since the cost of electricity from light-water reactors increases by about 0.06 mill per kilowatt-hour per dollar per pound increase in the price of uranium, this would add more than 0.4 mill per kilowatt-hour to the cost

of electricity. Even more serious, of course, would be the complete exhaustion of all of our presently estimated resources of low-cost uranium in less than thirty years.

FAST-BREEDER REACTORS

This wasteful consumption of low-cost uranium by light-water reactors can be arrested if fast-breeder reactors now being developed in many countries prove to be reliable and economic. The fast-breeder reactor is fueled with a mixture of plutonium and abundant uranium-238. In this reactor, a coolant which does not slow neutrons down is substituted for water; the two principal candidate coolants are helium under pressure or liquid sodium near 1 atmosphere. Fission of a plutonium atom by a fast neutron produces about 2.5 neutrons for every neutron consumed. One of these neutrons continues the fission chain reaction and the remaining 1.5 are absorbed by uranium-238 to produce

5. *Potential Nuclear Power Growth Patterns,* U.S. Atomic Energy Commission, Report WASH-1098, Dec. 1970.

1.5 atoms of new plutonium. One of these replaces the plutonium consumed in fission, leaving a net gain of around half an atom of plutonium for every 1.5 atoms of uranium-238 consumed. In this way, all the abundant uranium-238 in natural uranium can be used as fuel, which would multiply our nuclear fuel resources more than a hundredfold.

But this is only part of the improvement in our nuclear fuel resources made possible by the breeder. Because a fast-breeder reactor consumes so little natural uranium, the cost of electricity from it would be practically independent of the price of uranium. Consequently, if the fast-breeder reactor proves to be economic at today's uranium price of $8 per pound, it still could be fueled economically with uranium costing $50 or $100 per pound. At $100 per pound, United States uranium resources are estimated to be 25,000,000 tons, compared with less than 1,500,000 at $15 per pound. Thus, by being able to use this high-cost uranium, the breeder reactor would extend nuclear fuel resources by another factor of over 15.

Table 3-3 compares the amount of electricity that could be generated by breeders and light-water reactors from the amount of uranium available at different prices, and shows the increase in cost of electricity caused by progressively higher uranium prices. Light-water reactors could generate 8,480,000,000 kilowatt-years of electricity from all the uranium available at $15 per pound, about the highest uranium price at which this type can compete with fossil fuel at today's prices. Fast breeders, on the other hand, could generate 19.2 million million kilowatt-years of electricity from uranium at $100 per pound, more than 2000 times as much as light-water reactors. At the present rate of electricity generation in the United States, 300,000,000-kilowatt, fast-breeder reactors fueled with United States uranium resources available at $100 per pound could provide all our electricity for 64,000 years. It is this tremendous extension of our fuel resources that makes development of the breeder reactor so challenging and important.

Table 3-3. Comparison of Uranium Requirements of Water-Cooled Reactors and Fast-Breeder Reactors.

	Water Cooled	Fast Breeder
Principal raw material	^{235}U	^{238}U
Uranium concentrate consumption, tons per million kilowatt years	171	1.3
Increase in cost of electricity caused by increase in price of uranium, mills/kW·h/$/lb	0.06	0.0

Uranium Price (dollars per pound)	U.S. Uranium Resources (tons)	Increase in Cost of Electricity (mills per kilowatt-hours)		Million Kilowatt-Years of Electricity Which Could be Generated in	
		Water Reactor	Fast Breeder	Water Reactors	Fast Breeders
8	594,000	0.0	0.0	3,470	460,000
10	940,000	0.1	0.0	5,500	720,000
15	1,450,000	0.4	0.0	8,480	1,120,000
30	2,240,000	1.3	0.0	13,100	1,720,000
50	10,000,000	2.5	0.0	58,300	7,700,000
100	25,000,000	5.5	0.0	146,000	19,200,000

Where does development of the fast breeder stand today? Its ability to compete economically with the light-water reactor is not yet proved, but the fact that it would consume so much less natural uranium is a favorable factor that should make the fuel cycle cost of the fast breeder around 1 mill per kilowatt-hour lower than the fuel cycle cost of the light-water reactor. With this fuel cost advantage, the unit capital cost of a fast breeder could be as much as $50 per kilowatt hour higher than a light-water reactor without the breeder losing its economic advantage. Reactor development authorities in this country[6] and abroad believe that a fast breeder need not cost this much more than a light-water reactor. I share this hopeful view.

The two types of breeder reactor that have the best chance of meeting this cost requirement and being proved technically sound are the liquid-metal fast-breeder reactor and the gas-cooled fast-breeder reactor. Each type would use a mixture of plutonium oxide and natural uranium oxide as fuel, would have a thermal efficiency close to 40 per cent, and would be capable of producing close to 1.5 grams of new plutonium for every gram of plutonium consumed. This high breeding ratio gives these reactor types a tremendous advantage over other types of breeders.

6. See, for instance, *Report of the EEI Reactor Assessment Panel*, Edison Electric Institute, New York, 1970.

Except for their similar fuel and breeding potential, the liquid-metal- and gas-cooled breeders have little in common, technically or historically. The liquid-metal breeder uses molten sodium at low pressure as coolant, whereas the gas-cooled breeder uses helium at 50 to 75 atmospheres.

Gas-Cooled Fast Breeder

The gas-cooled fast breeder can draw on the helium-cooled reactor technology already developed for the high-temperature gas-cooled reactor. One 45-megawatt example of this slow-neutron, nonbreeding, helium-cooled reactor is now in operation in the United States, and a second 330-megawatt unit is being built. Figure 3-2 illustrates schematically the principle of the gas-cooled fast breeder. The prestressed-concrete pressure vessel, helium circulator, and helium-heated steam generator developed for the high-temperature gas-cooled reactor could be used without major change for this type of fast reactor.

Fuel for the gas-cooled fast-breeder reactor consists of a mixture of 15 per cent plutonium dioxide and 85 per cent (uranium-238) uranium dioxide clad in stainless steel; fuel would be generally similar to fuel developed for the liquid-metal fast-breeder reactor. Helium gas at 1,250 pounds per square inch is pumped by a helium circulator over the fuel and is heated to 1,200°F. The helium then flows through the steam genera-

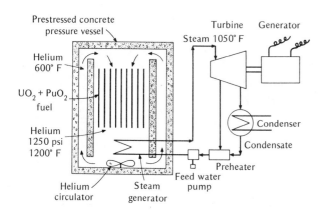

Figure 3-2. Gas-cooled fast-breeder reactor.

tor, producing steam at 1,050°F. Fuel, helium circulator, and steam generator are all contained within a prestressed concrete pressure vessel, as in the high-temperature gas-cooled reactor. Because of the high temperature at which steam is produced, the thermal efficiency of this gas-cooled fast-breeder system would be around 40 per cent, much better than water-cooled reactor systems and as high as a modern fossil-fueled plant or the liquid-metal fast-breeder system.

The major drawbacks of the gas-cooled fast-breeder reactor stem from the relatively poor heat-removal characteristics of helium gas, compared with liquid sodium. The most serious of these is the difficulty of preventing overheating of fuel in the event of loss of helium pressure. Engineering studies of solutions to this and other problems of the gas-cooled fast-breeder reactor are being conducted in this country, and in Europe, and solutions are likely to be found, but there are as yet no firm plans for building a reactor of this type. It is, nevertheless, an attractive concept because it promises high breeding ratio, high thermal efficiency, low capital cost, and freedom from the problems of handling sodium that complicate the liquid-metal breeder. The gas-cooled fast breeder should receive substantially increased development funding because of the advantages it would have over the liquid-metal breeder if emergency cooling of gas-cooled fast-breeder reactor fuel could be proved reliable.

Liquid-Metal Fast Breeder

Despite the difficulties of working with sodium, its use as a coolant for fast reactors has appealed to reactor engineers since 1950 and a number of sodium-cooled fast reactors have already been built. Sodium has excellent heat-removal characteristics—very high thermal conductivity and high volumetric heat capacity—so that overheating of the fuel during normal operation or in emergencies can be practically guaranteed not to occur. By designing the reactor as a double-walled vessel with no openings below the top of the fuel, assurance is provided that the

fuel will remain submerged in and cooled by liquid sodium even in the event of mechanical failure of the cooling system external to the reactor. Because of sodium's high boiling point and compatibility with stainless steel, sodium can be heated to 1,150°F in the reactor and can be used to generate steam at high temperatures and pressures similar to those used in conventional power plants, yielding a thermal efficiency of over 40 per cent.

But sodium has numerous drawbacks. It is opaque, so that refueling and other operations in the reactor have to be carried out blind. It is made intensely radioactive in the reactor, so that it is considered prudent to interpose a secondary, nonradioactive, sodium coolant loop between the primary-reactor sodium and the steam generator, as shown in Figure 3-3. Sodium reacts with air or water, so that it is necessary to provide argon cover gas for free sodium surfaces and to design the steam generator so that leakage of water into sodium is prevented. If local boiling of sodium should occur in the reactor, the rate of fission might increase to an unacceptable degree unless adequate precautions are taken in the design and operation of the reactor. Finally, a commercial liquid-metal breeder reactor requires the development of pumps, valves, and heat exchangers for handling sodium in unprecedently large volumes.

These difficulties with sodium have been largely overcome. The development of liquid-metal fast-breeder reactors is far advanced, and a number of power plants of this type have been built. Tabld 3-4 shows the principal liquid-metal fast-breeder power plants that have been built or are under construction.

In the United States, as long ago as 1951, the Experimental Breeder Reactor I demonstrated a breeding ratio greater than unity and was the world's first reactor to generate electricity. Later, the larger Experimental Breeder Reactor II and the Enrico Fermi Reactor have come into operation. Although other countries entered the breeder field later than the United States, Russia, En-

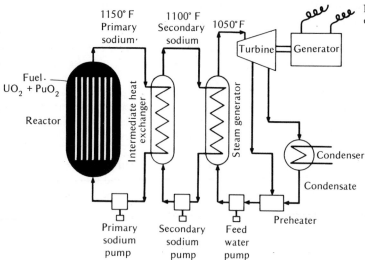

Figure 3-3. Liquid-metal fast-breeder reactor.

gland, and France have run reactors comparable to our experimental breeder reactors and are now building demonstration breeder power plants larger than any now authorized in the United States. Germany is about to authorize construction of a 300-megawatt fast-breeder power plant and Japan and Italy are making rapid progress.

Atomic Energy Commission Policies

In the United States, the policy of the Atomic Energy Commission has been to build facilities for testing the novel components of a liquid-metal fast-breeder reactor before constructing large demonstration power plants. Three industrial companies have built these testing facilities: General Electric has participated in the *SEFOR* project, which is testing the dynamic and safety characteristics of a small liquid-metal fast-breeder reactor; North American Rockwell has built facilities for engineering test of sodium components; and Westinghouse is building the Fast Test Reactor to test fuel, scheduled to start operation in 1974. Each of these firms has submitted to the Atomic Energy Commission a proposal to build a demonstration liquid-metal fast-breeder reactor power plant in the 300- to 500-megawatt

Table 3-4. Liquid-Metal Fast-Breeder Power Plants

Country	Reactor	Electricity (megawatts)	Years Operated
U.S.	EBR-I	0.3	1951–1963
U.K.	Dounreay	60	1963–
U.S.	EBR-II	20	1965–
U.S.	Fermi	70	1965–
France	Rapsodie	20	1967–
U.S.S.R.	BR-60	60	1970–
Under construction		*Scheduled to operate*	
U.S.S.R.	BR-350	150	1971–
U.K.*	PFR	250	1972–
France*	Phenix	250	1973
U.S.S.R.	BN-600	600	1973–1975

*Presently in operation (1974).

range, each of which would cost around $500,000,000.

A number of difficult decisions must now be made. Among the questions to be answered are these: Shall one, two, or three demonstration plants be built? If more than one, in which order? If less than three, how can all three companies be kept in the development to provide competitive sources of supply for commercial plants? On what time scale shall demonstration plants be built?

How shall the large sums needed for their construction be apportioned among the government, the three interested companies, and the many electric power companies of the United States?

While these questions are being debated in the United States by many government and commercial interests, foreign projects have been moving faster. In the Soviet Union, England, and France, government agencies have had full responsibility for breeder projects and have driven ahead with construction of demonstration plants without waiting for the exhaustive testing of components now characteristic of the United States program and without trying to establish several competing sources of supply. Unless these foreign projects run into major difficulties, liquid-metal fast-breeder reactors will become commercially available abroad many years earlier than in the United States. There is a definite possibility that in the 1980's United States power companies will be buying breeder reactors from France, England, or Germany instead of from Schenectady, Pittsburgh, or Los Angeles. Many United States reactor engineers, including myself, believe that the technology of the liquid-metal fast-breeder reactor is sufficient advanced to justify starting construction of at least one demonstration plant here and now before we lose the momentum of our present fast-breeder reactor program.

We cannot wait to start construction of these demonstration plants until the eventual shortage of uranium is upon us. It will take at least seven years to bring the first demonstration plant into operation, and another seven years to build the first full-scale commercial plants. Construction of the first demonstration plants for light-water reactors started 15 years ago, and these reactors are only today reaching full commercial maturity. The development period for liquid-metal-cooled breeders is apt to be even longer.

We urge Congress, the Atomic Energy Commission, and the Office of Management and Budget to agree on a national program and a timely schedule for building these demonstration plants. We urge these governmental groups, the interested manufacturers, and the power companies to agree on an equitable plan for funding their construction. One mechanism proposed for obtaining the necessary funds is a federal tax on electricity to be remitted by the federal government to groups building and operating the first fast-breeder reactor demonstration plants, somewhat as the federal gasoline tax is remitted to the states for highway construction. In this way the costs would be borne by the United States consumer of electricity, the ultimate beneficiary of this abundant source of low-cost electricity for all time to come.

CONCLUSIONS

Let me conclude by reviewing the most important points of this brief summary of power from nuclear fission.

First, I would emphasize that properly designed and operated light-water reactors have been proved to be economic, reliable, relatively unobtrusive, safe, and nonpolluting. With these reactors we have a new energy source that can serve as our principal means for generating electricity for the remainder of this century. They are the best solution we have to the current energy crisis. Light-water reactors are economically competitive with fossil fuel for generating electricity. Their use for this purpose will conserve fossil fuel for transport, home heating, chemical synthesis, and other more versatile applications for which uranium cannot be used.

Second, I would like to urge maintenance of a sensible attitude toward the low levels of radioactivity emitted by these power plants. Though it is proper to recommend that these levels be kept as low as practicable, we should bear in mind that we are surrounded on all sides by natural radiation. We therefore should not require that extreme measures be taken and excessive costs incurred to limit radiation exposure from nuclear power plants to an unnecessarily small fraction of natural levels.

Third, I would like to recommend

prompt implementation and thorough testing of the proposed long-term storage of high-level radioactive wastes in bedded salt formations. Until this or an equivalent procedure has been proved by actual operation to be completely reliable, critics of nuclear power will have a valid cause for concern.

Finally, we should move ahead aggressively with the development of breeder reactors, because today's slow-neutron, nonbreeding reactors provide only a short-term solution to our electric energy needs. One or more demonstration plants of the liquid-metal fast-breeder type should be built as soon as possible. All present indications are that the fast-breeder reactor is our best hope for providing electric energy in practically unlimited amounts for thousands of years. We cannot afford to allow other nations to pass the United States as the leader in this vital development.

SUGGESTED READINGS

Abajian, Vincent V. and Alan M. Fishman, 1973, "Supplying Enriched Uranium," *Physics Today,* pp. 23-29, August.

Bupp, Irvin C. and Jean-Claude Derian, 1974, "The Breeder Reactor in the U.S.: A New Economic Analysis," *Tech. Review,* pp. 26–36, July–August.

Creagan, R. J., 1973, "The LMFBR," *Mech. Engineering,* p. 12–16, February.

Inglis, David R., 1973, *Nuclear Energy: Its Physics and Social Challenges* (Reading, Mass.: Addison-Wesley).

Nuclear Eclectic Power

DAVID J. ROSE   1974

Enough work has been done to permit a reasonable assessment of the major issues of nuclear power. Most of the recent fluctuations in energy patterns tend to reinforce what seemed evident even several years ago: a massive switch to nuclear power for electric energy generation, and perhaps later for other purposes. The total installed electric utility generating capacity in the United States is expected to be 480,000 megawatts by the end of 1974[1]; the average generation rate in March 1974 was 212,000 megawatts.[2] The present nuclear installed capacity is about 30,000 megawatts. Serious predictions of 1,000,000 megawatts of nuclear power installed by A.D. 2000 may come true; the total cost of those nuclear plants would be more than $600 billion. The grand total, including factories to produce the equipment and facilities to enrich uranium,

process fuel, and handle wastes, may come to $1 trillion, plus the cost of transmitting and distributing the energy. Also, as alternate fuel costs rise, nuclear heat will become interesting for large-scale industrial and commercial applications. If events turn out this way, nuclear power will constitute the largest coherent technological plunge to date, with long-lasting consequences.

Any assessment of nuclear power, to be useful, must be comparative; the question is, compared to what? Until about A.D. 2000, the major choices are nuclear power, fossil fuels (of various sorts), or nothing, in varying proportions. In the twenty-first century, they are advanced nuclear power, increasingly sophisticated chemical fuels, probably derived from coal or oil shales, perhaps hydrogen (but made with nuclear power), perhaps solar power (more likely for many small-scale applications, in my opinion), or nothing. Beyond that era, resource limitations increasingly exclude fossil fuels. The benefits of nuclear power are lower production costs, vastly larger resources, and (I will try to

1. From a survey by the National Electrical Manufacturers Association, as quoted in *IEEE (Inst. Electr. Electron. Eng.) Spectrum 11,* 115 (1974).
2. Edison Electric Institute, *Elec. Output No. 49* (6 March 1974).

Dr. David J. Rose is a professor in the Nuclear Engineering Department, Massachusetts Institute of Technology, Cambridge.

From *Science*, Vol. 184, No. 4134, pp. 351–359, April 19, 1974. Reprinted by permission of the author and the American Association for the Advancement of Science. Copyright 1974 by the American Association for the Advancement of Science.

show) substantially less total adverse environmental impact compared with fossil fuels. Costs and hazards include not only the usual economic ones of the facilities themselves, but also those associated with (a) illegal diversion of nuclear fuels, (b) accidents, (c) radioactive waste storage, and (d) other environmental and societal impacts.

In this article, I will not seriously entertain the notion of opting for less electric power, believing that economic, demographic, and other social forces already present will lead to substantial increases in demand over the next several decades. The precise amount of growth turns out not to affect this discussion. Thus, while recognizing the necessity, importance, and consequences of limiting growth and of conserving energy wherever possible, one can debate the issue of nuclear power during the next fifty years separately.

NUCLEAR PLANT PROPERTIES, ECONOMIC COSTS, AND DEMAND

This is not an article on reactor principles, but a few remarks will facilitate the debate that follows.

Virtually all present-day nuclear power reactors work on the basis of fissioning the relatively rare isotope uranium-235 (0.77 per cent of natural uranium) to produce fission products (chiefly intermediate-weight elements), about 2.5 neutrons per event, and energy (200 million electron volts per event). Some neutrons go on to initiate further uranium-235 fissions; controlling the neutron fate by initial design and by adjustment of neutron-absorbing materials controls the reactor. Also, some neutrons are absorbed in the predominant uranium isotope uranium-238 to make a substantial amount of plutonium (plutonium-239). The conversion ratio (plutonium-239 formed/uranium-235 fissioned) is about 0.5 for reactors being installed today. Thus, current reactors bring with them plutonium handling and hazard problems, which are incorrectly thought by some to apply only to future

breeder reactors. Some of the plutonium-239 actually fissions in the reactor; most is removed at fuel reprocessing time, then stored for later use in the first breeder reactors, for which it is the fissionable fuel. Also, plutonium-239 can be recycled as fuel in present-day reactors, but that has not been done yet.

Turning now to more specific types, the reactors most commonly used and ordered in the United States contain fuel and ordinary (light) water in large pressure vessels, the so-called light-water reactors. They operate at 1,000 to 2,200 pounds per square inch ($\sim 15 \times 10^6$ newtons per square meter) and 315°C; the safety of this arrangement and its associated piping has been the subject of recent lively debate. The relatively low operating temperature limits the net efficiency of these plants to 32 per cent. There are two subspecies: the pressurized-water reactor, where the water does not boil, but passes into a heat exchanger that produces steam for the turbines in a separate loop; and the boiling-water reactor, where the steam from the pressure vessel passes directly through the turbines. The first of these was developed principally by the Westinghouse Electric Company, partly from experience in building smaller versions for the U.S. Navy. The second was developed solely by the General Electric Company. Importantly, both companies could afford to offer loss leaders, so to speak, in order to capture a substantial early share of the reactor market; thus, light-water reactors have proliferated, almost to the exclusion of other types. Advantages are: relatively well-developed technology and predictable performance, well-developed fuel cycle technology, best-known cost. Disadvantages are: low efficiency, public questions about reactor safety, only moderate conversion efficiency (0.5). A somewhat neutral feature is the requirement for fuel enriched to 3 per cent uranium-235, present uranium enrichment facilities will require substantial augmentation, at a cost of several billion dollars, in the 1980's.

Other reactor species need mentioning here. In the United States the General

Atomic Company has for years been developing slowly a high-temperature gas-cooled reactor, but could not afford to offer a loss leader. Thus the high-temperature gas-cooled reactor, which many imagine to be a better idea, lagged until the Gulf Corporation, and later Shell also, took it over. Advantages are: higher operating temperature and efficiency (40 per cent), possibly lower cost, higher conversion ratio (0.7, intermediate between the light-water reactor converters and a true breeder), and the absence of some mechanical failure modes discussed in relation to light-water reactors. Disadvantages are: there is only one supplier, whose fuel must be fully enriched fissionable uranium, which is weapons-grade material and thus raises the specter of illegal diversion (but design changes could permit using lower enrichment); and the fuel cycle is incompletely developed. In this last respect, a typical problem is what to do with the large amount of slightly radioactive graphite in which the fuel will be embedded. Burning it freely seems not acceptable, but I see nothing to prevent turning it into an insoluble chemical (a carbonate?) and sequestering it safely. At present, one high-temperature gas-cooled reactor is about to commence operation, six more are on order, and some analyses have shown the high-temperature gas-cooled reactor taking over a substantial share of future markets. The present fuel for the high-temperature gas-cooled reactor is uranium-235, but eventually it would work mainly on the thorium-uranium-233 cycle.

Another candidate is the Canadian CANDU reactor, of which several are in operation there, with others on order for Mexico and outside North America. The CANDU is of pressure-tube construction (thought to be a safer design than the United States light-water reactors) and permits on-line refueling. It can run on natural uranium (but does better with enriched fuel), because it is moderated and cooled with heavy water; it is an older design than most, and operates at even lower efficiency than the United States light-water reactors. It is probably more expensive than the United States re-

actors, but how much is hard to discover, because of differences between United States and Canadian costing policies. Another version is the heavy-water-moderated organic-liquid-cooled demonstration reactor at Whiteshell station, Manitoba. If the United States were starting its reactor program today, with no large commitment already made, the Canadian ideas would merit serious consideration; but under present circumstances, the need for development and prototype construction makes any such introduction to the United States market unlikely.

All modern nuclear power reactors are large, to capture economies of scale inherent in building larger components (up to a limit). The U.S. Atomic Energy Commission has limited all reactor approvals (until 1978 at least) to units that develop not more than 3,800 megawatts of nuclear heat. Thus, light-water reactors at 32 per cent efficiency will be limited to 1200 megawatts electric and high-temperature gas-cooled reactors at 40 percent to 1500 megawatts electric.

These comparisons have international impact. New electric plants in France will be light-water reactors of the United States type. The United Kingdom is finishing a complex internal debate over whether to build more of its ancient but familiar (to the British) gas-cooled Magnox reactors, to adopt light-water reactors of United States type, or to adopt a version of the Canadian reactors.

Turning now to costs, an excellent study has been made of nuclear and fossil fuel possibilities for the Northeast Utilities system by Arthur D. Little, Inc.[3] Little estimates total capital cost per kilowatt of $389 for an oil-fired plant, $588 for a coal plant with sulfur and particulate removal, and $702 for light-water reactor, all for operation in 1981. These figures are much higher than those guessed in the 1960's for several

3. "A study of base-load alternatives for the Northeast Utilities System," report to the Board of Trustees of Northeast Utilities by Arthur D. Little, Inc., Acorn Park, Cambridge, Mass., 5 July 1973. The report has had substantial circulation.

reasons: substantially higher construction costs of all kinds, need for environmental controls on fossil fuel plants, need for more experience by the nuclear industry, increasing complexity of nuclear plants, and construction delays (which add interest and inflation charges). These capital costs appear in the electric bill at about 1¢ per kilowatt-hour for each $500 (for 7,000 operating hours per year and a 14 per cent rate of return on capital before taxes). Thus, at this stage nuclear power has a disadvantage of about 0.2¢ per kilowatt-hour with respect to coal and 0.64¢ per kilowatt-hour with respect to oil.

Operation and maintenance costs (0.09¢, 0.18¢, 0.135¢ per kilowatt-hour for oil, coal, and nuclear fuel, according to Little) give a slight further advantage to oil. But the greatest single fact influencing present demand for nuclear power is that petroleum at $10 per barrel (0.16 cubic meters) represents 1.5¢ per kilowatt-hour of the cost of electricity, using the most efficient plant available (40 per cent). Total nuclear fuel cycle costs will be less than 0.3¢ per kilowatt-hour, giving an advantage over oil of 1.2¢ per kilowatt-hour, and an overall system advantage of 0.5¢ per kilowatt-hour. Much the same situation is predicted for coal, and the gap between nuclear and fossil fuels will widen with time. Petroleum prices, now about three times higher than those prevailing in early 1973, are not likely to decline very much, being supported both by a shortage of cheap oil in oil-consuming countries and increasing economic sophistication in the oil-producing ones. The likely advent of more domestic crude oil at prices of $7 to $10 a barrel, or of synthetic crude at perhaps the same price (after the middle 1980's?) still leaves nuclear power with strong economic advantage.

Studies like this show that nuclear power has become by far the most attractive large option for electric power generation, both in the United States and abroad, with few exceptions. The energy crisis of 1973–1974 is one of fossil fuels, not nuclear ones, and nuclear power tends to benefit on that ac-

count. Development of a strong conservation ethic will not have the effect of strongly decreasing the growth rate of nuclear power; the first results would be a limitation of nonnuclear plant construction or operation.

Here are some projections. The year A.D. 2000 is a good date to focus on: Plants planned today will be halfway through their operating life. The Atomic Energy Commission estimates 1,200,000 megawatts of nuclear capacity installed by that date,[4] and Dupree and West's predictions[5] amount to about 960,000 megawatts. Considering the capacity of plants now operating (30,000 megawatts), under construction, or planned for operation before 1985 (about 200,000 megawatts), the various factors mentioned above, and the traditional growth of electric power (7 per cent per year), an installed capacity of 1,000,000 megawatts nuclear seems likely to be exceeded. Beyond A.D. 2000, the rate of growth is more difficult to predict; acceptable sites will be particularly hard to find, even with use of wet cooling towers, and dry cooling towers are even larger and more expensive ($100 per kilowatt?); and the likelihood and timing of new major applications (such as making hydrogen from cheap nuclear heat or electricity, or powering personal electric vehicles) are unknown factors at present. The 1,000,000 megawatts of nuclear electric power is imagined to be perhaps two-thirds of the total electric capacity predicted for A.D. 2000, about four times the capacity of present installations. Western Europe and Japan, with less fossil fuel resources than the United States, find nuclear electric power even more attractive.

RESOURCES

Various scenarios[6] predict a total cumulative requirement of about 2.5 million tons of

4. "Nuclear power 1973 to 2000," *USAEC Rep. WASH 1139* (1972).
5. W. G. Dupree, Jr., and J. A. West, *United States Energy through the Year 2000* (Government Printing Office, Washington, D.C., 1972).
6. For example, see *Report of the Cornell Workshops on the Major Issues of a National Energy*

Table 3-5. United States Uranium Reserves[7] (The amount of U_3O_8 available comprises reasonably assured plus estimated additional reserves.

Concentration: U_3O_8 in Ore (parts per million)	U_3O_8		Electricity Producible (10^3 megawatt-years)	
	Cost (dollars per pound)	Amount Available (10^3 tons)	Light-Water Reactor	Breeder Reactor
>1600	Up to 10^b	1,127	6,600	880,000
>1000	Up to 15^b	1,630	9,500	1,270,000
>200	Up to 30^b	2,400	14,000	1,860,000
>60	Up to 50^c	8,400	49,000	6,500,000
>25	Up to 100^c	17,400	102,000	13,500,000
3^a	Several hundred	$10^6 - 10^7$		

[a]Natural crustal abundance.
[b]Includes copper leach residues and phosphates.
[c]Includes Chattanooga shale.

uranium oxide (U_3O_8) by A.D. 2000 and 4 to 4.5 million tons by A.D. 2010. The precise amount depends on the mix of reactor types, on when breeder reactors are introduced, and on the actual future electric demand.

The United States uranium resources according to the Atomic Energy Commission[7] are shown in Table 3-5. The resources available up to $10 a pound ($\sim$0.45 kilograms) would, if used in light-water reactors, generate 6,000,000 megawatt-years of electricity, a total electric supply for almost thirty years at present rates, but not enough for the growing nuclear demand to A.D. 2000. With light-water reactors, the increase in the cost of electricity would be 0.1¢ per kilowatt-hour for each increase of $17 a pound in U_3O_8. To wipe out the present nuclear cost advantage of at least 0.5¢, U_3O_8 would have to reach nearly $100 a pound, at which cost Table 3-5 shows a great deal of uranium available.

Research and Development Program (College of Engineering, Cornell University, Ithaca, N.Y., 1973), Chapter 4 by H. A. Bethe and Appendix G thereto by H. A. Bethe and C. Braun.
7. "Nuclear fuel supply," USAEC Rep. WASH 1242 (1973); "Nuclear fuel resources and requirements," USAEC Rep. WASH 1243 (1973).

The numbers in Table 3-5 provide a substantial fraction of the input for the debate pro and con nuclear power and especially for the nuclear breeder debate. The Atomic Energy Commission saw trouble ahead, from these and similar estimates made a few years ago: If used in light-water reactors, the low-cost reserves are modest, and rising U_3O_8 prices after (say) A.D. 2000 would place a new penalty on nuclear power, perhaps 0.1¢ per kilowatt-hour. Thus, it was necessary to develop the breeder reactor; by converting the common uranium-238 to fissionable plutonium, it utilizes almost all the nuclear energy of the uranium, instead of only the uranium-235 plus a small additional plutonium conversion. In effect, not only are the nuclear energy resources multiplied by a factor of 100, but (since almost all of it is used) the resource cost per energy unit drops similarly. Thus, to a good approximation, the cost of electricity from nuclear breeders becomes independent of uranium prices. Also, at least by implication, if the breeder were not developed and massively deployed in time, nuclear power might become expensive enough to drive electric utilities back to fossil fuels. Thus, a "slot" in time was imagined when breeder reactors must be introduced— late enough that plutonium would be available from existing light-water reactors to make initial fuel charges, but not so late that power from light-water reactors would have become expensive because of the rising cost of uranium. The period 1985 to 1995 was envisaged for commerical introduction.

With present oil and coal prices and environmental protection costs the slot is virtually open ended, and the nuclear advantage is unlikely to be overcome by any large-scale fossil option, except in special locations. Also, the entire debate is likely to have been mistaken, as follows. With a real market for U_3O_8 at (say) $10 a pound, prospectors seriously searched for high-grade ore, and the 1,127,000 tons of reserves shown in Table 3-5 is the result. But what of the approximately equal increment represented by the third row of Table 3-5, supposed to include all ores with a U_3O_8 concentration greater than 200 parts per million? Those lower grades were not actively sought *per se,* but were found somewhat incidentally. Thus we can explain the anomalous dip in reserves at intermediate prices and concentrations—no one seriously looked for them. The outcome of the reasoning is that if a definite offer to purchase were made at more than $10 a pound, a great deal more would easily be found, and the Atomic Energy Commission would have no scarcity argument in favor of the breeder until well into the twenty-first century. Three additional circumstances support this interpretation:

(1) In the middle and late 1950's and 1960's, when the federal government offered incentives to discover new uranium resources, much showed up at $8 to $10 a pound.

(2) The reserves at $10 a pound are enough for nearly 30 years for domestic purposes. This is an anomalously large amount in terms of the economic optimum (at private investment rates); for decades our reserves of many minerals have represented a supply for 8 to 10 years, and this has been determined by economic pressures to buy

them and economic penalties for exploring for things that will not be used for a long time.

(3) Canada and Australia, for instance, report increasingly large resources.

A generally similar debate could be constructed for thorium, which can be used in a breeder to make uranium-233; it is thought to be at least as plentiful as uranium.

DISECONOMIES AND NONMARKET COSTS

More important issues appear here than there is space properly to discuss, so I choose to develop three in subsequent sections—illegal acts, accidents, and radioactive waste disposal—and give brief mention to others that have already been well illuminated in public debate.

Present nuclear plants are less efficient than fossil fuel ones, and in addition do not exhaust any waste heat directly to the atmosphere through a chimney. Thus, a light-water nuclear plant will reject into cooling water almost twice as much waste heat as do the most efficient fossil-fuel plants, for the same electric power. Siting problems then ensue, exacerbated by the large size of the power plants. But these problems seem surmountable, perhaps at a cost of visual pollution from cooling towers or design complications arising from siting offshore; and introducing more efficient reactors will eventually ameliorate the difficulty.

Licensing is a complex issue. Two separate federal licenses are required: before construction and again before operation. At each stage, detailed studies are required, including environmental impacts, and interveners must be heard. Long delays then become possible; they are expensive—$50,000,000 a year in interest and other charges on a completed nonoperating plant. One might imagine, therefore, some reversion to fossil-fueled plants for which, alas, no such licensing is required; but cost penalties make that alternative unattractive.

ILLEGAL ACTS

The problem is vexatious, and even discussion can be dangerous. Two nonproblems are: (a) Present security arrangements at reactors make it highly unlikely that one or a few persons, even well-prepared, can cause a large disruptive accident. (b) Stealing new fuel for light-water reactors is useless, because the fuel is nonradioactive and cannot be enriched to weapons-grade uranium-235 (90 per cent +) without a technology capable of doing the whole job, starting with natural uranium.

However, a number of possibilities exist for illegal acts, against which the reactor operators and public authorities (particularly the federal government) take increasingly strict precautions, the adequacy of which has been questioned from time to time.

1. A large, organized raid on an operating reactor. The outer containment shell will resist the impact of moderate-size airplanes, and shells of reactors near airports are designed to resist the impact of the largest loaded airplanes. Thus, ingress would have to be made by direct attack on the entrances, and various security steps have been taken against this possibility. It would be logical to arrange reactor protective devices so that the reactor would shut down when any such hostile event occurred and could not be restarted except through time-consuming operations by experts; in that case attackers could cause financial mischief, but no public calamity. In addition, reactors are so complex that the active assistance of knowledgeable persons inside seems necessary; thus, to summarize this item, an economic calamity causable by an irrational employee seems the dominant danger.

2. Theft of used fuel elements. Until recently, used fuel was shipped in technically safe casks, but with almost no guard. That has now been corrected, and the used fuel is monitored during shipment in various ways. Making a bomb from it requires technological facilities and sophistication comparable to those of the Atomic Energy Commission

itself. The most likely threat would come from the fuel being ground up and used for blackmail, a strategy which would be very hazardous to the conspirators.

3. Theft of makeup fuel for high-temperature gas-cooled reactors, which is 93 per cent uranium-235. This is weapons-grade material, and must be handled as such.

4. In the future, theft of breeder fuel, which is weapons-grade plutonium. The possibilities for mischief and handling requirements are the same as for item 3 above; but the risk to conspirators is immense, because plutonium is so lethal (see below).

It seems to me that the greatest diversionary hazard is related not to civilian nuclear power, but to weapons and their components. At increasing cost, more protection can be bought, and no one—public or private—would imagine settling for less than enough. But in dealing with irrationality, how much is enough? No one knows. It would be bitter irony if civilization had to renounce its claim to that name through inability to control these aspects of nuclear power; meanwhile, illegal use is to me the most worrisome and least resolved hazard, and a prime motivation for exploring the possibilities of controlled nuclear fusion.

ACCIDENTS AND RELATED HAZARDS

An immense amount has been done; the situation is in no way as some critics of nuclear power portray it, but trouble spots persist. Before some hotly debated topics are assessed, consider Table 3-6, which summarizes a study made by Walsh[8] of the casualties associated with nuclear power. Events are normalized per unit of electric energy produced—per 1,000 megawatt electric plant year (8.76×10^9 kilowatt-hours).

Several features of Table 3-6 are notable: (a) "conventional" accidents are dominant, especially in the hazardous occupation of

8. P. Walsh, Environmental Engineer's thesis, Massachusetts Institute of Technology (1974).

Table 3-6. Summary of Health Effects of Civilian Nuclear Power, per 1000 Megawatts Electric Plant Year. (8)

Activity	Fatalities			Injuries (days off)
	Accidents (not radiation-related)	Radiation Related (cancers and genetic)	Total	
Uranium mining and milling	0.173	0.001	0.174	330.5
Fuel processing and reprocessing	0.048	0.040	0.088	5.6
Design and manufacture of reactors, instruments, and so on	0.040		0.040	24.4
Reactor operation and maintenance	0.037	0.107	0.144	158
Waste disposal		0.0003	0.0003	
Transport of nuclear fuel	0.036	0.010	0.046	
Totals	0.334	0.158	0.492	518

mining; (b) the total fatality rate is about 0.5 per reactor year, *tout compris;* (c) most of the hazards are occupational, not public. These numbers will be compared later with others for fossil fuel power.

The data of Table 3-5, gleaned from a large number of sources, are in reasonable agreement with those of other studies. Hub et al.[9] report 0.932 fatality and 373 total days off due to injuries, and Sagan[10] reports 0.390 and 1,022, respectively. The Atomic Energy Commission[11] settles on the range 0.161 to 0.364 for fatalities, and does not give an estimate of injuries. A variation of a factor of 2 should be expected because of the periods studied, assignment of casualties, and so forth. For example, of 11,870 short tons of U_3O_8 produced domestically in 1969, 4,700 were sold for electricity production and only about 350 were actually used up that year in operating reactors (1 short ton ∼ 0.9 metric ton). Also, the various investigators agree fairly well that occupational accidents unrelated to radiation domi-

nate. For example, Lave and Freeburg,[12] in an exceptionally well-documented comparison of the effects on health of electricity generation from coal, oil, and nuclear fuel, cite about 0.12 fatality per 1,000 megawatts electric plant year from mining and milling accidents, compared with Walsh's 0.173.

Table 3-6 does not seem to contain the item most hotly debated: the probability of large nuclear accidents, for example, from a pipe rupture, followed by failure of the emergency core cooling system, followed by transfer of a substantial fraction the radioactive mess to the external environment. It was just this possibility that stimulated a marathon debate between the Union of Concerned Scientists and the Atomic Energy Commission from 1972 to 1974.[13] An intensive study of hypothetical large accidents has been made by N. C. Rasmussen and co-workers for the Atomic Energy Commis-

9. K. Hub, J. G. Asbury, W. A. Buehring, P. F. Gast, R. A. Schlenker, J. T. Weills, *A Study of Social Costs for Alternate Means of Electric Power Generation for 1980 and 1990* (Argonne National Laboratory, Argonne, Ill., 1973).
10. L. Sagan, *Science 177,* 487 (1972).
11. "The safety of nuclear power reactors and related facilities," *USAEC Rep. WASH 1250* (1973).

12. L. B. Lave and L. C. Freeberg, *Nucl. Saf. 14,* 409 (1973).
13. D. F. Ford, T. C. Hollocher, H. W. Kendall, J. J. MacKenzie, L. Scheinman, A. S. Schurgin, *The Nuclear Fuel Cycle, a Survey of the Public Health, Environmental and National Security Effects of Nuclear Power* (Union of Concerned Scientists, Cambridge, Mass., 1973); see especially Chapter 3; "Emergency core cooling systems (ECCS) hearings," AEC Docket RM50-1, AEC Public Document Room, Atomic Energy Commission, Washington, D.C.

sion.* No such reactor failures have occurred, but the data base is nevertheless substantial—all the large high-technology, high-pressure, high-temperature chemical processors and vessels. According to Chairman Ray of the Atomic Energy Commission[14] the Rasmussen study can be roughly summarized by assigning a chance of about 10^{-6} per reactor year of a major accident with loss of several hundred lives, including later cancers and genetic deaths. For the sake of argument, make this 1,000 lives. The actuarial hazard would be 10^{-3} per 1,000 megawatts electric plant year, and it duly appears as a 1 per cent contribution to one of the entries in Table 3-6.

Also present in Table 3-6 are the public hazards from releasing radioactive gases—principally krypton-85 and tritium—from boiling-water reactors during operation, from pressurized-water reactors during refueling, and from fuel reprocessing plants. Radioactive xenon presents less total hazard, because it decays quickly. New reactors built with longer gas holdup permit even lower releases.

No substantial quantitative study is in disagreement with these results. The Atomic Energy Commission itself prepared a now-notorious report[15] predicting several thousand deaths and billions of dollars damage if a large amount of material from a modest-sized reactor got into the atmosphere under adverse meteorological conditions. The authors of that work never considered the probability of any sequence of events leading to the hypothetical radioactive release; this makes it somewhat like analyzing (say) the consequences of the New York World Trade Center falling over.

If Rasmussen's estimates are believed, the fatality rate per person would be about

10^{-12} per hour in A.D. 2000, about the same as the probability of being struck by a meteorite, and a thousand times less than the probability of being electrocuted.

Having seemed to bury the reactor accident bogey, let me now resurrect it. Accidents seem so remote only because of intense, persistent, and highly competent professional effort. Will that continue indefinitely, or not, and is reactor technology that good worldwide? One can easily imagine inadequate vigilance, both here and abroad, or what is just as bad, lack of social responsibility toward these matters; then disaster would surely follow.

The accident issue cannot be closed without mention of plutonium, a principal and proximate cause for the worry. An excellent and convenient review has been given by Bair and Thompson[16]; the Atomic Energy Commission has long been concerned.[17] Plutonium is an alpha-particle emitter; when introduced into the body in a soluble form, it (a) circulates as complexes in the blood in large molecules; (b) gets into the bones; (c) goes to the liver, where it tends to stay unless there is a stress on the body's iron stores; and (d) as with iron, gets caught up in the body's transport and storage system, which prevents loss and promotes reuse. When introduced as particulates (usually the oxide) through wounds or by inhalation, some of it forms local tiny hot spots, and some travels throughout the body. It is thought to be about five times as toxic as radium, and present maximum occupational body burden is now set at 40 nanocuries (0.6 microgram of plutonium-239, but only 2.3 nanograms of plutonium-238). These limits, based in part on comparisons with the effects of radium, have been questioned because of the concentration in the liver, which is not the case with radium, and the

---

*Editors' note: see summary of this study in the next article.

14. Reported by AEC Chairman Dixy Lee Ray, at the National Press Club, Washington, D.C., 21 January 1974.

15. "Theoretical possibilities and consequences of major accidents in large nuclear power plants," *USAEC Rep. WASH 740* (1957).

16. W. J. Bair and R. C. Thompson, *Science 183,* 715 (1974).

17. For example, see the report "Meeting of the advisory committee on reactor safeguards, plutonium information meeting" (Los Alamos, New Mexico, 4 and 5 January 1974), AEC Document Room, Atomic Energy Commission, Washington, D.C.

problem of hot spots. On this latter point, Tamplin and Cochran[18] pick a probability of 1/2000 that a single hot plutonium particle will cause cancer, and propose a body burden limit of two such particles. That would be 1.4×10^{-13} curie, a factor of 300,000 below the present limits. So far, the evidence indicates that plutonium is not that dangerous, and that something near the present limits will eventually be well justified. In the meantime, experimental work on animals continues, and persons inadvertently exposed in the past to plutonium are carefully monitored.

It is very unsettling that present reactors contain substantial amounts of plutonium, and breeder reactors will contain a huge quantity—close to 10^6 curies. The extreme ratio between the resource available and the allowable body burden emphasizes the necessity of vigilance, which must be presumed to exist everywhere, forever. If the Tamplin and Cochran risk estimates turn out to be correct, nuclear fission power will need to be rethought, because the consequences of even a single large accident become disastrous.

NUCLEAR WASTE DISPOSAL

The costs of proper waste disposal are higher than were originally imagined, but still small compared to the total costs of nuclear power. The main problems seem to have been failing to appreciate the importance of public concern and failing to explore the available options with enough money and imagination. Fortunately, those shortcomings in the civilian waste disposal program are being corrected.

A more comprehensive assessment of the situation has appeared,[19] and a brief summary will suffice here. The wastes fall into two categories. First, there are fission products of intermediate atomic weight, stron-

tium-90, cesium-137, and krypton-85, for example, and all the main ones have half-lives of 30 years or less; thus, in 700 years less than 10^{-7} of the waste remains, which further calculation shows is innocuous. Second, there are the so-called actinides, mainly plutonium, neptunium, curium, americium, and so on, all heavy elements made by neutron absorption in the original uranium (or thorium, if the reactor works on a thorium-232-uranium-233 breeding cycle). These typically have very long half-lives—24,600 years for plutonium-239 for example. All these elements are very toxic, because of their radioactivity, proclivity to settle in bone and other body sites, and so on. If merely stored they last a million years or more, beyond the time horizon of present rational planning.

At present, only plutonium and uranium values are extracted from the wastes, and that only to about 99.5 per cent (the limit of profitable recovery); this narrow economic optimum is clearly not the social one; an extraction of 99.9 per cent uranium, neptunium, and plutonium and 99 per cent americium and curium reduces the long-term activity by a factor of 100 compared with present practices, leaving essentially just the fission products. The extracted actinides can be recycled in the reactor at small penalty; they all turn eventually into fission products, and the million-year problem is effectively eliminated. Kubo and Rose[19] estimated an additional cost of perhaps 0.02¢ per kilowatt-hour for implementing this option.**

Now the nuclear waste problem becomes a 700-year one—a long time, but short compared to geologic eons. Sequestering the remainder, perhaps in the form of a borosilicate glass, in selected salt deposits, hard rock sites, or even near-surface repositories (with complete retrievability) makes sense; such sequestrations can be accomplished with great assurance (for example, in granite monoliths near the sea where any drainage

18. A. R. Tamplin and T. B. Cochran, *Radiation Standards for Hot Particles* (Natural Resources Defense Council, Washington, D.C., 1974).
19. A. S. Kubo and D. J. Rose, *Science 182,* 1205 (1973).

**Editors' note: This article is found in its entirety later in this nuclear fission section.

paths would lead under the continental shelf).

Some options that are not appealing at present are disposal in the ocean deeps (either buried or not), in ice sheets or continental rocks, or in space; but the possibilities should be reviewed from time to time.

Certainly, the responsibility for radioactive waste disposal must eventually lie with the government, because the time horizon of conventional economic groups cannot guarantee concern for so long; private industry can at best act as the agent of the public interest. Narrowly regional solutions are also difficult to find, because the patterns of desirable sites for nuclear reactors, fuel reprocessing plants, and waste disposal do not come anywhere near coinciding. Thus, for example, it would be highly desirable for Europe to develop an integrated nuclear waste management strategy; but a set of national decisions to accept wastes and the responsibility for them must presage workable broader agreements.[20]

BREEDER REACTORS

A few more remarks are required on breeders, to augment material in previous sections. The liquid-metal fast-breeder reactor has been the prime energy goal of the Atomic Energy Commission, which budgeted $357 million for it in fiscal year 1975 and plans a total of $2,556 million from 1975 to 1979. The 1974 allotment is 36 per cent of the entire federal budget for energy research and development; similar or higher percentages in previous years, plus the attitudes expressed by the Atomic Energy Commission, the Executive Office of the President, and the Joint Committee on Atomic Energy, have led to criticism of "all the eggs in one basket."

The liquid-metal fast-breeder reactor has substantial technical points in its favor, besides the eventual but presently slippery advantage of resource conservation. It is

nonpressurized, and in the respect more completely sealed, simpler, and safer. It will operate at higher temperatures than lightwater reactors, giving an efficiency of 41 per cent, comparable to the high-temperature gas-cooled reactors or modern fossil fuel plants. The high thermal conductivity and heat capacity of its coolant—liquid sodium—make it virtually immune to damage in case of mechanical failure of the cooling system external to the reactor. It may eventually be cheaper, but that depends on the outcome of the expensive development program presently under way. There are disadvantages too. The liquid sodium is at about $620°C$, and becomes intensely radioactive, forcing refueling and other operations on the reactor to be carried out blind, and constituting a chemical hazard in respect to failures in circulation pumps, pipes, or heat exchangers. The plutonium hazard has already been discussed. The Soviet Union and France have prototype or demonstration liquid-metal fast-breeder reactors already operating, and the United Kingdom will soon follow. The United States program has suffered from a plethora of rigid directions from the Atomic Energy Commission to its field offices and contractors, which has led to excessive delay and expense. The Fast Flux Test Reactor, a prototype for the United States demonstration reactor, has escalated in cost from an initial $80 million to an uncertain $600 million to $800 million; the first demonstration plant, planned to be built in Tennessee, is slated to cost $700 million including development costs, and will produce 300 megawatts electric.

The liquid-metal fast-breeder reactor is not the only breeder. The General Atomic Company, with some Atomic Energy Commission help, is making slow progress on a helium gas-cooled fast-breeder reactor; having no massive coolant, it promises to breed plutonium with much higher conversion ratio than the liquid-metal fast-breeder reactor; but by the same token it has very little internal thermal inertia, and overheats in seconds if the cooling gas fails to circulate. It has little internal resemblance to General

20. D. J. Rose and G. Tenaglia, *Ambio 2*, 233 (1973).

Atomic's high-temperature gas-cooled reactor other than utilizing the same sort of prestressed concrete pressure vessel and helium circulation system, but leans substantially on the Atomic Energy Commission's liquid-metal fast-breeder reactor program, for example for fuel and fuel rod development. Another possible candidate is the molten-salt breeder, where a mixture of lithium, beryllium, and uranium fluorides circulates through the reactor space, then through the heat exchangers and pumps. There is no solid fuel. The principal advantages are that it can operate completely on the thorium-uranium-233 breeder cycle, there is no need for outside fuel reprocessing (the molten salt is continuously purified on line), and it utilizes quite different technology from the other reactors. Thus, it is a possible alternate route to fission breeders, in case the other programs fail. Its main disadvantage is chemical engineering complexity, plus the fact that hot radioactive salt must flow outside the reactor. Even the light-water reactor and the high-temperature gas-cooled reactor can be technologically upgraded, to increase their conversion ratio and extend fissionable uranium resources.

The Atomic Energy Commission has supported these alternate breeder approaches only reluctantly, and sometimes not at all, through fiscal year 1974, and now plans to allocate $11 million for them in 1975. That is better, but will do little more than keep those high-technology programs alive.

When should the breeder reactor be introduced? That cannot yet be answered exactly, but the preceding discussion permits some guesses. The breeder promises to be cheaper because of its very low uranium cost per unit of energy. But fuel costs are not the dominant ones in any reactor, and the saving over light-water reactors will be about 0.1¢ per kilowatt-hour for each $17-a-pound rise in the cost of U_3O_8. To avoid offsetting penalties of expensively reprocessing the plutonium fuel too often, the burnup in the breeder must be high. Long fuel life means more fission products to absorb neutrons, plus mechanical limitations imposed because

of dimensional changes—all of which conflicts with good plutonium breeding. If goals of (say) 100,000 megawatt-days per ton of burnup and a breeding ratio of 1.24 can be achieved, a saving corresponding to about $50-a-pound U_3O_8 would accrue to the breeder, translatable into a capital advantage of $150 per kilowatt. If the burnup is less for that breeding ratio, or if the breeder costs more, or when plutonium is recycled in present-day reactors (thus introducing competition for the fuel), then the advantage shrinks. I estimate that the breeder will almost surely be attractive when U_3O_8 reaches $50 a pound in 1974 dollars. That will not happen in the first few decades of the twenty-first century (see the "resources" debate). In the meantime, nuclear power is in no danger of losing out to other fuels, and there does not need to be a crash breeder program. Economic introduction at A.D. 2000 would be a sign of technological good fortune, not of resolving an energy crisis with a time limit.

CONTROLLED FUSION

Controlled nuclear fusion may appear as a twenty-first-century option to (say) advanced fission reactors, but that is not yet assured. The United States fusion program has grown from its inception in the early 1950's, through a long level period of physics-oriented experimentation supported at $20 million to $30 million a year, to its present stage of rapid growth ($102 million for fusion via magnetic confinement, plus $66 million for laser fusion, in fiscal year 1975).

The trick is to heat up a mixture of deuterium and tritium to a temperature between 10^8 and $10^{9\,\circ}$K (10^4 to 10^5 electron volts) at a density high enough and for a time long enough that the product of the two quantities exceeds (about) 3×10^{14} per cubic centimeter per second. This is the so-called Lawson criterion. One major class of schemes depends on confining the highly ionized plasma with strong magnetic fields, say 50,000 to 100,000 gauss, at a density of

10^{14} per cubic centimeter for a few seconds. The most successful example of this technique so far is the Tokamak, wherein the plasma is confined and heated in a toroidal magnetic field.

It seems fairly clear that the scientific feasibility of controlled fusion can and will be demonstrated: A large enough well-designed magnetic structure can be made to achieve the Lawson criterion. Whether laser fusion will get there is less certain: the process depends critically on the laser pulse ablating a deutorium-tritium target pellet so fast and so evenly that the reaction forces on its surface compress it by a volumetric factor exceeding 1,000. Then it undergoes nuclear fusion in about 10^{-12} second.

Controlled fusion has, during this scientific stage, been the most challenging and difficult of all such assignments ever given to physical scientists, and they deserve credit for doing so well. But there is much more to controlled fusion than applied plasma physics, and now controlled fusion shows signs of becoming, in addition, the most difficult and challenging assignment given to technologists and engineers. Thin-section vacuum walls, operating at high temperatures, cooled by liquid lithium or gas, possibly under cyclic mechanical stress as well, and bathed in an immense flux of 14 million electron volt neutrons—what will they be made of? No one knows, or even whether materials with adequate life under those conditions are developable. Many of the fusion concepts require pulsed or cyclic operations, which introduces new complexities and constraints, further eroding the option space desired by any fusion reactor designer. Whereas having a favorable neutron balance in the fusion reactor still does not seem to be a problem, power balance may be one: in every concept, a great deal of energy must be spent to heat the fusion plasma, overcome energy losses while building up or taking down large magnetic fields, operate lasers, or do other tasks. Thus, the deuterium-tritium fusion reaction, which has some inherent disadvantages but the great advantage of 50 times the reaction rate of any other, seems mandatory. Because of all these, and a host of other problems connected with recovering tritium fuel, removing spent plasma, injecting new fuel, and assuring reasonable possibilities of repairing such a complicated device, the Atomic Energy Commission's implied goal of beneficial installation after 1995 seems optimistic. If economic attractiveness then could be assured, the fission breeder would be superfluous. But such success with fusion is still problematic; so the breeder programs which are the only assured routes to long-term nuclear energy, should not be appreciably modified now on account of fusion.

The advantages of success are substantial: (a) deuterium fuel is sufficient for 10^{10} years, and lithium (used with fusion neutrons to breed the tritium fuel component) is in somewhat uncertain supply but probably adequate for any technological age to come; (b) the only appreciable radiation hazard is from tritium, which is less hazardous than plutonium by many orders of magnitude, per unit of weight or nuclear energy content; (c) the reactor structure, while surely made radioactive by 14 million electron volt neutrons, is not liable to pose any appreciable hazard; (d) there is no reason to steal the nuclear material: hydrogen bombs are best made by other processes, and require atomic bombs to trigger them.

Thus, a strong sense of social purpose keeps driving the controlled fusion program, with $1,450 million planned for fiscal years 1975 to 1979; it is the next largest federal plan after the liquid-metal fast-breeder reactor and the newly upgraded coal programs.

NUCLEAR VERSUS FOSSIL POWER

With some reservations, the social costs of nuclear power are being measured, as has been shown earlier. The social costs of burning fossil fuels to generate electricity are very hard to determine: one must extract the electric power contribution from the general costs associated with extracting, processing, and burning fossil fuels, which are quite different for different modes (local pollution by home heating equipment differs from the effects of effluents from tall power plant stacks, for instance). In addition, the studies themselves have been on a relatively small scale.

Lave and Freeburg[12] in their general comparison of power from coal, oil, and nuclear fuel, conclude that nuclear power is substantially less hazardous than coal. First, the hazard for coal mining was judged to be 18 times as high (per unit of energy) as that for uranium mining; on that count alone, coal would rank much worse than all the nuclear power hazards of Table 3-6. In addition, Lave and Freeburg estimate mortality and morbidity arising from power plant effluents, finding that a pressurized-water reactor appears to offer at least 18,000 times less health risk than a coal-burning power plant, and a boiling-water reactor 24 times less health risk. Comparing low-sulfur oil and nuclear fuels, they are less sure: uranium mining is more hazardous than oil drilling, but power plant effluent data would again favor nuclear power.

Preliminary results of studies under way at Massachusetts Institute of Technology tend to support the view that even low-sulfur fuels are unlikely to be as benign as nuclear ones. Lave and Seskin's earlier data[21] indicate that lowered urban air quality reduces the average life span by about three years. They also show that the most hazardous pollutant is sulfur dioxide, and that most of it comes from home heating systems.[22] Even so, an appreciable fraction comes from electric power production, and analysis of their data and other data indicates a fatality rate on this account alone at least ten times the total nuclear one (per unit of energy), even after cleanup to the Environmental Protection Agency's sulfur dioxide standard of 80 micrograms per cubic meter. These analyses assume a linear relation between concentration and damage, which of course will lead to overestimates of damage if a threshold exists; the same linear

approximation is applied over many more orders of magnitude to effects of penetrating radiation, in the nuclear case.

Other workers tend to the same general conclusion: for example, Starr et al.[23] What seems increasingly clear is that the hazards of burning fossil fuels are substantially higher than those of burning nuclear ones, yet many debates have enticed the uncritical spectator to just the opposite conclusion. Several reasons can be put forward to explain this peculiar response. First, the hazards of reactors and radiation were perceived as "unknown," and hence very possibly large. Second, the public had come to accept the social cost of polluted air, not realizing (a) that much could be done (until recently) and (b) that its perception of the fossil fuel hazard was faulty. But I think a third reason dominates: over the past 20 or 30 years, the federal government has invested well over $1 billion attempting to measure the public health costs associated with nuclear power, and until recently almost nothing was done to measure similar hazards of fossil fuel power—in retrospect, a scandalous omission. Thus, even with sometimes clumsy words and bad grace, a vast amount of literature appeared about nuclear hazards, providing material for a great public debate. The absence of any appreciable parallel assessment of fossil fuels ensured that the debate would be unbalanced, and only now are semiquantitative social cost figures starting to appear. This profound issue can hardly fail to be resolved in the next few years as more data accumulate, especially on effects of fossil fuels. I conclude from the evidence to date that all the costs—economic and social—will favor nuclear power, unless the problem of illegal use of nuclear materials gets out of hand, or plutonium turns out to be as bad as its worst critics believe.

CONCLUSION

The uranium and thorium resources, the technology, and the social impacts all seem

21. L. Lave and E. P. Seskin, *Science 169,* 723 (1970).
22. L. Lave and E. P. Seskin, *Am. J. Publ. Health 62,* 909 (1972).

23. C. Starr, M. A. Greenfield, D. F. Hausknecht, *Nucl. News 15,* 37 (1972).

to presage an even sharper increase in nuclear power for electric generation than had hitherto been predicted. There are more future consequences.

The "Hydrogen Economy"

Nuclear power plants operate best at constant power and full load. Thus, a largely nuclear electric economy has the problem of utilizing substantial off-peak capacity; the additional energy generation can typically be half the normal daily demand. Thus, the option of generating hydrogen as a nonpolluting fuel receives two boosts: excess nuclear capacity to produce it, plus much higher future costs for oil and natural gas. However, the so-called "hydrogen economy" must await the excess capacity, which will not occur until the end of the century.

Nonelectric Uses

By analyses similar to those performed here, raw nuclear heat can be shown to be cheaper than heat from many other fuel sources, especially nonpolluting ones. This will be particularly true as domestic natural gas supplies become more scarce. Nuclear heat becomes attractive for industrial purposes, and even for urban district heating, provided (a) the temperature is high enough (this is no problem for district heating, but could be for industry; the high-temperature gas-cooled reactors and breeders, with 600°C or more available, have the advantage); (b) there is a market for large quantities (a heat rate of 3,800 megawatts thermal, the reactor size permitted today, will heat Boston, with some to spare); and (c) the social costs become more definitely resolved in favor of nuclear power.

Capital Requirements

Nuclear-electric installations are very capital-intensive. One trillion dollars for the plants, backup industry, and so forth is only 2 per cent of the total gross national product between 1974 and 2000, at a growth rate of 4 per cent per year. But capital accumulation tends to run at about 10 per cent of the gross national product, so the nuclear requirements make a sizable perturbation. Also increasing the electric share of energy provision means increasing electric power utilization, which has a high technological content and demands yet more capital. Thus, provision of capital is a major problem ahead, especially for electric utilities.

The Need for People

The supply of available trained technologists, environmental engineers, and so on, especially in the architect-engineer profession, is insufficient for the task ahead, especially since the same categories of people will be in demand to build up a synthetic fuels industry and do other new things.

Beyond these specific items and beyond the technological discussion, one can feel deeper currents running in this debate. Issues that started out seeming technological ended up being mainly societal: prevention of clandestine use, either by vigilance or by public spirit; a determination to maintain quality and to safeguard wastes that transcends narrow interests; a perception of social benefits and damage much more holistic than before; the need to manage programs more openly and better than before. Questions and doubts become more acute, answers and methods less sure.

Here is a final question. We have never before been given a virtually infinite resource of something we craved. So far, increasingly large amounts of energy have been used to turn resources into junk, from which activity we derive ephemeral benefit and pleasure; the track record is not too good. What will we do now?

Reactor Safety Study

An Assessment of Accident Risks in
United States Commercial Nuclear Power Plants

U.S. ATOMIC ENERGY COMMISSION   1974

INTRODUCTION AND RESULTS

The Reactor Safety Study was sponsored by the U.S. Atomic Energy Commission to estimate the public risks that could be involved in potential accidents in commercial nuclear power plants of the type now in use. It was performed under the independent direction of Professor Norman C. Rasmussen of the Massachusetts Institute of Technology. The risks had to be estimated, rather than measured, because although there are about 50 such plants now operating, there have been no nuclear accidents to date. The methods used to develop these estimates are based on those developed by the Department of Defense and the National Aeronautics and Space Administration in the last 10 years.

The objective of the study was to make a realistic estimate of these risks and to compare them with nonnuclear risks to which our society and its individuals are already exposed. This information will be of help in determining the future use of nuclear power as a source of electricity.

The basic conclusion of this study is that the risks to the public from potential accidents in nuclear power plants are very small. This is based on the following considerations:

(1) The consequences of potential reactor accidents are no larger, and in many cases, are much smaller than those of nonnuclear accidents. These consequences are smaller than people have been led to believe by previous studies that deliberately maximized risk estimates.

(2) The likelihood of reactor accidents is much smaller than many nonnuclear accidents having similar consequences. All nonnuclear accidents examined in this study, including fires, explosions, toxic chemical releases, dam failures, airplane crashes, earthquakes, hurricanes and tornadoes, are much more likely to occur and can have consequences comparable to or larger than nuclear accidents.

Figures 3-4, 3-5, and 3-6* compare the

*An example of the numerical meaning of Figures 3-4 to 3-6 can be seen by selecting a vertical consequence line and reading the likelihood that various types of accidents would cause that consequence. For instance, in Figure 3-4, 100 plants

This is the draft of the Summary Report of the Reactor Safety Study directed by Professor Norman C. Rasmussen of the Massachusetts Institute of Technology. This draft was provided through the courtesy of Saul Levine, Project Staff Director of the Reactor Safety Study, Atomic Energy Commission. The final report will be issued in 1975.

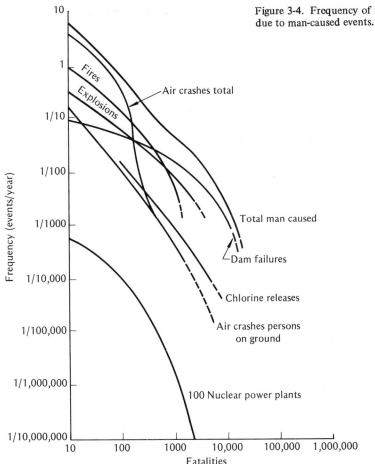

Figure 3-4. Frequency of fatalities due to man-caused events.

nuclear reactor accident risks for the 100 plants expected to be operating by about 1980 with risks from other man-made and natural phenomena. These figures indicate the following:

(1) Figure 3-4 and 3-5 show the likelihood and number of fatalities from both nuclear and a variety of non-nuclear accidents. These figures indicate that non-nuclear events are about 10,000 times more likely to produce large accidents than nuclear plants.

(2) Figure 3-6 shows the likelihood and dollar value of property damage associated with nuclear and non-nuclear accidents. Nuclear plants are about 100 to 1,000 times less likely to cause comparable large dollar value accidents than other sources. Property damage is associated with three effects: (a) the cost of temporarily moving people away from contaminated areas, (b) the denial of use of real property during the few weeks to a few months during which the radioactivity is cleaned up, and (c) the cost of assuring that people are not exposed to potential sources of radioactivity in food and water supplies. This latter cost reflects

would cause this consequence with a likelihood of one in 10,000 per year. Chlorine releases are about 100 times more likely, or about one in 100; fires are about 1,000 times more likely, or about one in 10 per year; air crashes are about 5,000 times more likely, or about one per 2 years.

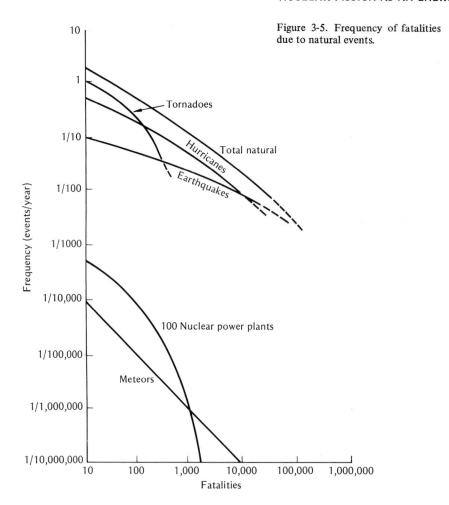

Figure 3-5. Frequency of fatalities due to natural events.

Table 3-7. Risk of Fatality by Various Causes

Accident Type	Total Number	Individual Chance per Year
Motor vehicle	55,791	1 in 4,000
Falls	17,827	1 in 10,000
Fires and hot substances	7,451	1 in 25,000
Drowning	6,181	1 in 30,000
Firearms	2,309	1 in 100,000
Air travel	1,778	1 in 100,000
Falling objects	1,271	1 in 160,000
Electrocution	1,148	1 in 160,000
Lightening	160	1 in 2,000,000
Tornadoes	91	1 in 2,500,000
Hurricanes	93	1 in 2,500,000
All accidents	111,992	1 in 1,600
Nuclear reactor accidents (100 plants)	0	1 in 300,000,000

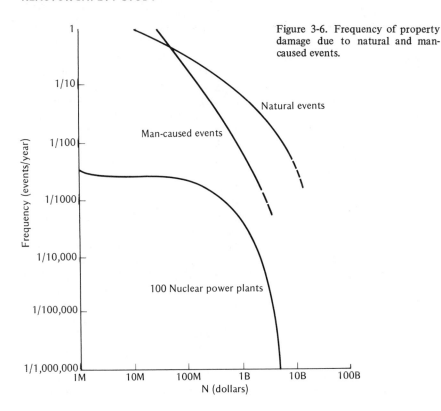

Figure 3-6. Frequency of property damage due to natural and man-caused events.

the efforts required to survey agricultural products, plus the loss of products which might be contaminated.

In addition to the over-all risk information in Figures 3-4 through 3-6, it is useful to consider the risk to individuals of being fatally injured by various types of accidents. The bulk of the information shown in Table 3-7 is taken from the 1973 U.S. Statistical Abstract and applies to the year 1969, the latest year for which this data has been tabulated. The nuclear risks are very small compared to other possible causes of fatal injuries.

In addition to fatalities and property damage, a number of other health effects can be caused by nuclear accidents. These include injuries and long-term health effects such as cancers, genetic effects, and thyroid gland illness. The injuries expected in potential accidents would be about twice as large as the fatalities shown in Figures 3-4 and

3-5; however, such injuries would be insignificant compared to the 8 million injuries caused annually by other accidents. The number of cases of genetic effects and long-term cancers are predicted to be much smaller than the normal incidence rate of these diseases. Even for a large, very unlikely accident, the small increases in these diseases would not be detected.

Thyroid illnesses that might result from a large accident are the formation of nodules on the thyroid gland that can be treated by medical procedures and rarely lead to serious consequences. For most accidents, the number of nodules caused would be small compared to their normal incidence rate. The number that might be produced in very unlikely accidents would be comparable to their normal rate of occurrence. These would be observed during a period of 10 to 20 years following the accident and would be about equal to their normal incidence in the people exposed.

While the study has presented the estimated risks from nuclear power plant accidents and compared them with other risks that exist in our society, it has made no judgment on the acceptability of nuclear risks. Though the study believes nuclear accident risks are very small, the judgment as to what level of risk society should accept is a broader one than can be made here.

QUESTIONS AND ANSWERS ABOUT THE STUDY

This section of the summary presents more information about the details of the study than was covered in the introduction. It is presented in question and answer format for ease of reference.

Who did this study and how much effort was involved?

The study was done principally at the Atomic Energy Commission headquarters by a group of scientists and engineers who had the skills needed to carry out the study's tasks. They came from a variety of organizations including the Atomic Energy Commission, the national laboratories, private laboratories, and universities. About 10 people were Atomic Energy Commission employees. The director of the study was Professor Norman C. Rasmussen of the Department of Nuclear Engineering of the Massachusetts Institute of Technology, who served as an Atomic Energy Commission consultant during the course of the study. The staff director who had day-to-day responsibility for the project was Mr. Saul Levine of the Atomic Energy Commission. The study was started in the summer of 1972 and took two years to complete. A total of 60 people, various consultants, 50 man years of effort and $3 million were involved.

What kind of nuclear power plants are covered by the study?

The study considered large power reactors of the pressurized water and boiling water type

being used in the United States today. This present generation of reactors are all water cooled and, therefore, the study limited itself to this type. Although high-temperature gas-cooled and liquid-metal fast-breeder reactor designs are now under development, no large reactors of this type are expected to operate in this decade; thus they were not considered.

Nuclear power plants produce electricity by the fissioning (or splitting) of uranium atoms. The nuclear reactor fuel in which the uranium atoms fission is in a large steel vessel. The reactor fuel consists of about 100 tons of uranium. The uranium is inside metal rods about 0.5 inch in diameter and about 12 feet long. These rods are formed into fuel bundles of about 50 to 200 rods each. Each reactor contains several hundred bundles. The vessel is filled with water which is needed both to cool the fuel and to maintain the fission chain reaction.

The heat released in the uranium by the fission process heats the water and forms steam; the steam turns a turbine to generate electricity. Similarly, coal and oil plants generate electricity using fossil fuel to boil water.

Today's nuclear power plants are very large. A typical plant has an electrical capacity of 1,000,000 kilowatts, or 1,000 megawatts. This is enough electricity for a city of about 500,000 people.

Can a nuclear power plant explode like an atom bomb?

No. It is impossible for nuclear power plants to explode like a nuclear weapon. The laws of physics do not permit this because the fuel contains only a small fraction (3-5 per cent) of the special type of uranium (uranium-235) used in weapons.

How is risk defined?

The idea of risk involves both the likelihood and consequences of an event. Thus, to estimate the risk involved in driving an automobile, one would need to know the likelihood of an accident in which, for example, an individual could be (a) injured or (b) killed. Thus, there are two different consequences, injury or fatality, each with its own likelihood. For injury, an individual's chance per year is one in 130 and for fatality, it is one in 4,000. This type of data concerns the risk to individuals and can affect attitudes and habits that individuals have toward driving.

However, from an over-all societal viewpoint, different types of data are of interest. Here, 1.5 million injuries per year and 55,000 fatalities per year due to automobile accidents represent the kind of information that might be of use in making decisions on highway and automobile safety.

The same type of logic applies to reactors. From the viewpoint of a person living in the general vicinity of a reactor, the likelihood of being killed in any one year in a reactor accident is one chance in 300,000,000 and the likelihood of being injured in any one year in a reactor accident is one chance in 150,000,000.

From a broader societal viewpoint, one individual of the 15 million people living in the vicinity of 100 reactors might be killed and 2 individuals might be injured every 25 years. This type of information might be of some use to the Congress or other decision makers in thinking about the over-all risk to society from reactor accidents.

What causes the risks associated with nuclear power plant accidents?

The risks from nuclear power plants are due to the radioactivity formed by the fission process. In normal operation nuclear power plants release only minute amounts of this radioactivity under controlled conditions. In the event of highly unlikely accidents, larger amounts of radioactivity could be released that could cause significant risks.

The fragments of the uranium atom that remain after it fissions are radioactive. These radioactive atoms are called fission products. They disintegrate further with the release of nuclear radiations. Many of them decay away quickly, in a matter of minutes or hours, to nonradioactive forms. Others decay away more slowly and require months and, in a few cases, many years to decay. The fission products accumulating in the fuel rods include both gases and solids. Included are iodine, gases like krypton and xenon, and solids like cesium and strontium.

How can radioactivity be released?

The only way that potentially large amounts of radioactivity can be released is by melting the fuel in the reactor core. The fuel removed from a reactor after use and stored at the plant site contains considerable amounts of radioactivity. However, accidental releases from such fuel were found to be very small compared to potential releases of radioactivity from the full reactor core.

The safety design of reactors includes a series of systems to prevent the overheating of fuel and to control potential releases of radioactivity from the fuel. Thus, to get an accidental release of radioactivity to the environment there must be a series of sequential failures that cause the fuel to overheat and release its radioactivity. There must also be failures in the systems designed to remove and contain the radioactivity.

The study has examined thousands of potential paths by which radioactive releases could occur and has identified those that determine the risks. This involved defining the ways in which the fuel in the core could melt and the ways in which systems to control the release of radioactivity could fail.

How might a core-melt accident occur?

It is significant that not once in some 200 reactor years of commercial operation of reactors of the type considered in the report has there ever been fuel melting. To melt the fuel requires that a failure in the cooling system occur that allows the fuel to heat up to its melting point, about 5,000°F.

To those unfamiliar with the characteristics of reactors, it might seem that all that is required to prevent fuel from overheating is a system to promptly stop, or shutdown, the fission process at the first sign of trouble. Although reactors have such systems, they alone are not enough since the radioactive decay of the fuel continues to generate heat (called *decay heat*) that must be removed even after the fission process stops. Thus, redundant decay heat-removal systems are also provided in reactors. In addition, emergency core cooling systems (ECCS) are provided to cope with a series of potential but unlikely accidents.

The Reactor Safety Study has defined two broad types of situations that might potentially lead to a melting of the reactor core: the loss of coolant accident and transients. In the event of a loss of coolant (LOCA) the normal cooling water is lost from the cooling systems and core melting would be prevented by the use of the emergency core cooling system. However, melting could potentially occur in a loss of coolant if the emergency core cooling system failed to operate.

The term *transient* refers to any one of a number of conditions that can occur in a plant that require the reactor to be shut down. Following shutdown, the decay heat removal systems operate to keep the core from overheating. Certain failures in either the shutdown or the decay heat removal systems have the potential to cause melting of the core.

What features are provided in reactors to cope with a core-melt accident?

Nuclear power plants have numerous systems to prevent core melting. Furthermore, there are inherent physical processes and additional features that remove and contain the radioactivity released from the molten fuel should core melting occur. Although there are features provided to keep the containment building from being damaged for some time after the core melts, the containment will ultimately fail, causing a release of radioactivity.

An essentially leak-tight containment building is provided to prevent the initial dispersion of the air-borne radioactivity into the environment. Although the containment will fail a number of hours after the core melts, until that time, the radioactivity released from the fuel will be deposited by natural processes on the surfaces inside the containment. In addition, plants are provided with systems to contain and trap the radioactivity released within the containment building. These systems include such things as water sprays and pools to wash radioactivity out of the building atmosphere and filters to trap radioactive particles prior to their release. Since the containment buildings are made essentially leak tight, the radioactivity is contained as long as the building remains intact. Even if the building were to have sizable leaks, large amounts of the radioactivity would be removed by the systems provided for that purpose or would be deposited on interior surfaces of the building by natural processes.

Even though the containment building would be expected to remain intact for some time following a core melt, eventually the molten mass would be expected to eat its way through the concrete floor into the ground below. Following this, most of the radioactive gases will be trapped in the soil; however, a small amount would escape to the surface and be released. Almost all of the nongaseous radioactivity would be trapped in the soil.

It is possible to postulate highly unlikely core-melt accidents in which the containment building fails by overpressurization or by missiles created by the accident. Such accidents could release a larger amount of air-borne radioactivity and have more serious consequences. The consequences of these less likely accidents have been included in the study's results shown in Figures 3-4 to 3-6.

How might the loss of coolant accident lead to a core melt?

Loss of coolant accidents are postulated to result from failures in the normal reactor cooling water system and plants are designed to cope with such failures. The water in the reactor cooling systems is at a very high pressure (between 50 to 100 times the pressure in a car tire) and if a rupture were to occur in the pipes, pumps, valves, or vessels that contain it, then a "blow out" would happen. In this case the water would flash to steam and blow out of the hole. This could be serious since the fuel could melt if additional cooling were not supplied in a rather short time.

The loss of normal cooling in the event of a loss of coolant accident would stop the chain reaction so that the amount of heat produced would drop almost instantly to a few percent of its operating level. However, after this sudden drop the amount of heat being produced would decrease much more slowly and would be controlled by the decay of the radioactivity in the fuel. Though this decrease in heat generation is helpful, it would not be enough to prevent the fuel from melting unless additional cooling were supplied. To deal with this situation, reactors have emergency core-cooling systems, the function of which is to provide cooling for just such events. These systems have pumps, pipes, valves, and water supplies capable of dealing with breaks of various sizes. They are also designed to be redundant so that if some components fail to operate, the core can still be cooled.

The study has reviewed a large number of potential sequences of events following loss of coolant accidents of various sizes. In almost all of the cases, the loss of coolant accident must be followed by multiple failures in the emergency core-cooling system for the core to melt. The principal exception to this is the massive failure of the large pressure vessel that contains the core. However, the accumulated experience with pressure vessels indicates that the chance of such a failure is indeed very small. In fact, the study found that the likelihood of pressure vessel failure is so small that it does not contribute to the overall risk from reactor accidents.

How might a reactor transient lead to a core melt?

The term *reactor transient* refers to a number of events that require the reactor to be shut down. These range from normal shutdown for such things as refueling to such unplanned but expected events as loss of power to the plant from the utility transmission lines. The reactor is designed to cope with unplanned transients by automatically shutting down. Following shutdown, cooling systems would be operated to remove the heat produced by the radioactivity in the fuel. There are several different cooling systems capable of removing this heat, but if they all should fail, the heat being produced would be sufficient to eventually boil away all the cooling water and melt the core.

In addition to the above pathway to core melt, it is also possible to postulate core melt resulting from the failure of the reactor shutdown systems following a transient event. In this case, it would be possible for the pressure to increase enough so that the normal reactor cooling system might rupture. This would create a loss of coolant accident and could lead to core melting.

How likely is a core-melt accident?

This Reactor Safety Study carefully examined the various paths leading to core melt. Using methods developed in recent years for estimating the likelihood of such accidents, a probability of occurrence was determined for each core melt accident identified. These probabilities were combined to obtain the total probability of melting the core. The value obtained was one in 17,000 per reactor per year. With 100 reactors operating, as is anticipated for the United States by about 1980, this means that one such accident would occur, on the average, every one and three-quarters centuries.

It is important to note that a melting of the core in a nuclear power plant does not necessarily involve an accident with serious public consequences. One of the major findings of the study is that only about one in

10 potential core melt accidents, occuring on the average of once every 17 centuries, might produce measurable health effects.

What is the nature of the health effects that a core-melt accident might produce?

It is possible for a core-melt accident to release enough radioactivity so that some fatalities might occur within a short time (a few weeks) after the accident. Other people may be exposed to radiation levels that would produce observable effects requiring medical attention but from which they would recover completely. In addition, some people may receive even lower exposures, which produce no noticeable effects but may increase the incidence of certain diseases over a period of many years. The observable effects that occur shortly after the accident are called *short-term* or *acute* effects.

The *delayed* or *latent* effects of radiation exposure can cause some increase in the incidence of diseases such as cancer, genetic effects, and thyroid gland illnesses in the exposed population. These effects would appear as an increase in these diseases over a 10- to 20-year period following the exposure. Such effects would be difficult to notice because the increase is usually small compared with the normal incidence rate of these diseases.

The study has conservatively estimated the increased incidence of potentially fatal cancers over the 20 years following an accident. This has been done by following a procedure that estimates the number by extrapolating data from high dose rates to low dose rates. It is generally believed that this procedure probably overestimates the effect considerably, but it is not possible to do experiments with large enough populations to determine these very small effects. The number of latent cancers are predicted to be very small compared with the normal incidence of cancer. Thyroid illness refers to small lumps on the thyroid gland that can be felt by an examining physician; they are treated by medical procedures that some-

times involve simple surgery and rarely lead to serious consequences. For very large potential reactor accidents, the increase in nodules would be about equal to their normal incidence rate.

Radiation is recognized as one of the factors that can produce genetic effects which appear as defects in a subsequent generation. From the total population exposure caused by the accident, the expected increase in congenital defects in subsequent generations can be estimated. These effects are predicted to be very small compared with their normal incidence rate.

What are the most likely consequences of a core-melt accident?

The most likely core-melt accident would occur on the average of one every 17,000 years per plant. The size of the consequences of such an accident are given in Table 3-8.

How does the annual risk from nuclear accidents compare to other common risks?

Considering the 15 million people who live within 20 miles of current or planned United States reactor sites, and based on current accident rates in the United States, the annual number of fatalities and injuries expected from various sources are shown in Table 3-9.

What is the number of fatalities and injuries expected as a result of a core-melt accident?

A core-melt accident is similar to many other types of major accidents such as fires,

Table 3-8. Consequences of the Most Likely Core-Melt Accident

	Consequences
Fatalities	<1
Injuries	<1
Latent fatalities	<1
Thyroid nodules	~4
Genetic defects	<1
Property damage[a]	$100,000

[a]This does not include damage that might occur to the plant.

Table 3-9. Annual Fatalities and Injuries Expected among the 15 Million People Living within 20 Miles of United States Reactor Sites

Accident Type	Fatalities	Injuries
Automobile	4,200	375,000
Falls	1,500	75,000
Fire	560	22,000
Electrocution	90	–
Lightning	8	–
Reactors (100 plants)	0.3	6

explosions, dam failures, and the like, in that a wide range of consequences is possible depending on the exact conditions under which the accident occurs. In the case of a core melt, the consequences depend mainly on three factors; the amount of radioactivity released, the way it is dispersed by the prevailing weather conditions, and the number of people exposed to the radiation. With these three factors known it is possible to make a reasonable estimate of the consequences.

The study calculated the health effects and the probability of occurrence for 4,800 possible combinations of radioactive release magnitude, weather type, and population exposed. The probability of a given release was determined from a careful examination of

the likelihood of various reactor-system failures. The probability of various weather conditions was obtained from weather data collected at many reactor sites. The probability of various numbers of people being exposed was obtained from United States census data for current and planned United States reactor sites. These thousands of calculations were carried out with the aid of a large digital computer.

These calculations showed that the probability of accidents having 10 or more fatalities is predicted to be about 1 in 250,000 per plant per year. The probability of 100 or more fatalities is predicted to be about 1 in 1,000,000 and for 1,000 or more, 1 in 100,000,000. The largest calculated value was 2,300 fatalities with a probability of about one in a billion.

The estimates given above are based on the assumption that evacuation procedures would be used to move most of the people out of the path of the airborne radioactivity. Experience has shown that evacuations have been successfully carried out in a large number of nonnuclear accident situations. Since nuclear power plants have evacuation plans prepared and since there is warning time before radioactivity would be released to the environment, it seems highly likely that evacuation would be effective in the case of nuclear accidents.

Table 3-10. Probability of Major Man-Caused and Natural Events

Type of Events	Probability of 100 or More Fatalities	Probability of 1,000 or More Fatalities
Man-caused		
Airplane crash	1 in 2 years	1 in 2,000 years
Fire	1 in 7 years	1 in 200 years
Explosion	1 in 16 years	1 in 120 years
Toxic gas	1 in 100 years	1 in 1,000 years
Natural		
Tornado	1 in 5 years	very small
Hurricanes	1 in 5 years	1 in 25 years
Earthquake	1 in 20 years	1 in 50 years
Meteorite impact	1 in 100,000 years	1 in 1,000,000 years
Reactors		
100 plants	1 in 10,000 years	1 in 1,000,000 years

If we consider a group of 100 similar plants then the chance of an accident causing 10 or more fatalities is 1 in 2,500 per year or, on the average, one such accident every 25 centuries. For accidents involving 1,000 or more fatalities the number is 1 in 1,000,000 or once in a million years. Interestingly, this is just the probability that a meteor would strike a United States population center and cause 1000 fatalities.

Table 3-10 can be used to compare the likelihood of a nuclear accident to nonnuclear accidents that could cause the same consequences. These include man-caused as well as natural events. Many of these probabilities are obtained from historical records but others are so small that no such event has ever been observed. In the latter cases the probability has been calculated using techniques similar to those used for the nuclear plant.

With regard to injuries from potential nuclear power plant accidents, the number of injuries that could require medical attention shortly after an accident is about two times larger than the number of fatalities predicted.

What is the magnitude of the latent or long-term health effects?

As with the short term effects the magnitude of latent cancers, treatable latent thyroid illness, and genetic effects vary with the exact accident conditions. Table 3-11 illustrates the potential size of such events. The first column shows the consequences that would be produced by core-melt accidents, the most likely of which has one chance in 17,000 per plant per year of occurring. The second column shows the consequences for an accident that has a chance of 1 in million of occurring. The third column shows the normal incidence rate.

In these accidents, only the production of thyroid nodules would be observed and this only in the case of an exceedingly unlikely accident. These nodules are easily diagnosed and treatable by medical or surgical procedures. The other effects are too small to be discernable above the high normal incidence of these two diseases.

What type of property damage might a core-melt accident produce?

A serious nuclear accident would cause no physical damage to property beyond the plant site but may contaminate it with radioactivity. At high levels of contamination, people would have to be moved temporarily from their homes until the radioactivity either decayed away or was removed. At levels lower than this, but involving a larger area, people might take simple actions to reduce possible contamination, but would continue being able to live in the area. The principal concern in this larger area would be to monitor farm produce to keep the amount of radioactivity ingested through the food chain small. Farms in this area would have to have their produce monitored and any produce above a safe level could not be used.

Table 3-11. Magnitude of Latent Health Effects Expected in a 20-Year Period for an Accident That Produces 100 Fatalities

Effect	Chance per Plant per Year		Normal[a] Incidence Rate
	One in 17,000	One in 1,000,000	
Latent cancers	<1	450	64,000
Thyroid illness	4	12,000	20,000
Genetic effects	<1	450	100,000

[a]This is the normal incidence that would be expected for people in the vicinity of any one reactor.

The most likely core-melt accident, having a likelihood of one in 17,000 per plant per year, would result in little or no contamination. The probability of an accident that requires temporary evacuation of 20 square miles is one in 170,000 per reactor per year. Ninety per cent of all core-melt accidents would be expected to be less severe than this. The largest accident might require temporary evacuation from 400 square miles. In an accident such as this, agricultural products, particularly milk, would have to be monitored for a month or two over an area about 100 times as large until the iodine decayed away. After that, the area requiring monitoring would be very much smaller.

What would be the cost of a core-melt accident?

As with the other consequences, the cost will depend upon the exact circumstances of the accident. The cost calculated by the Reactor Safety Study included the cost of moving and housing the people that were evacuated, the cost caused by denial of land use and the cost associated with the denial of use of reproducible assets such as dwellings and factories. The most likely core-melt accident, having a likelihood of one in 17,000 per plant per year, would cause property damage of about $100,000. The chance of an accident causing $100,000,000 damage would be about one in 50,000 per plant per year. Such an accident would be expected on the average to occur once every 5 centuries for 100 operating reactors. The probability would be about one in 1,000,000 per plant per year of causing damage of about $2 to 3 billion. The maximum value would be predicted to be about $4 to 6 billion with a probability of about one in 1,000,000,000 per plant per year.

This property damage risk from nuclear accidents can be compared to other risks in several ways. The largest man-caused events that have occurred are fires. In recent years there have been an average of three fires with damage in excess of $10 million every year. About once every two years there is a fire with damage in the $50 to $100-million range. There have been four hurricanes in the last 10 years which caused damage in the range of $0.5 to $5 billion. Recent earthquake estimates suggest a $1 billion earthquake can be expected in the United States about once every 50 years.

A comparison of the preceding costs shows that, though a severe reactor accident would be very costly, it would not be significantly larger than a number of serious accidents with which our society deals quite often, and the probability of such a nuclear accident is, of course, estimated to be much smaller than the other events.

What will be the chance of a reactor melt down in the year 2000 if we have 1,000 reactors operating?

One might be tempted to take the per plant probability of a particular reactor accident and multiply it by 1,000 to estimate the chance of an accident in the year 2000. This is not a valid calculation, however, because it assumes that the reactors to be built during the next 25 years will be the same as those being built today. Experience with other technologies such as automobiles and aircraft show that as more units are built and more experience is gained the overall safety record in terms of the probability of accidents per unit decreases. There are already changes in plants now being constructed that appear to be improvements over the plants analyzed in the study.

How do we know that the study has included all accidents in the analysis?

The study devoted a large amount of its effort to ensuring that it covered all potential accidents important in determining the public risk. It relied heavily on over 20 years of experience that exists in the identification and analysis of potential reactor accidents. It also went considerably beyond earlier analyses that have been performed by considering a large number of potential failures that had never before been analyzed. For example, failure of reactor systems that can lead to core melt and the failure of systems that affect the consequences of core melt have been analyzed. The consequences of the failure of the massive steel reactor vessel were considered for the first time. The likelihood that various external forces such as earthquakes, floods, and tornadoes could cause accidents were also analyzed.

In addition there are further factors that give a high degree of confidence that all significant accidents have been included. These are: (a) the identification of all significant sources of radioactivity located at nuclear power plants, (b) the fact that a large release of radioactivity can occur only if reactor fuel melts, and (c) knowledge of the factors that can cause fuel to melt. This type of approach led to the screening of thousands of potential accident paths to identify those that would determine the public risk.

Whereas there is no way of proving that all possible accident sequences which contribute to public risk have been considered in this study, the systematic approach used in identifying possible accident sequences make it very unlikely that an accident which would contribute to the overall risk was overlooked.

How do your calculations of reactor accidents compare with those of earlier studies that predicted much larger consequences?

The principal earlier study of reactor accidents (WASH-740) was published by the Atomic Energy Commission in 1957, before any commercial nuclear power plants were operating. Thus, this study was necessarily vague about the engineering details of reactor accidents. The purpose of that study was to essentially maximize the consequences that could occur in an accident. This was done because it was to serve as a basis for the Congress to use in establishing adequate indemnification of the public in the event that an accident occurred. Thus, WASH-740 served as the basis for the Price-Anderson Act which provides such indemnification.

The reactor used for the WASH-740 study was one that generated 500 million watts (megawatts) of thermal energy as opposed to today's reactor of about 3,200 megawatts. To compare the earlier estimates with the more realistic approach used in this study, calculations were made for a 500 megawatt reactor using the reactor safety study model. The results are presented in Table 3-12. The differences between these two sets of results can in large part be explained as follows:

(1) This study used actual population data from the census bureau for the areas in the vicinity of actual reactor sites. The WASH-740 study used an estimated population that was much higher.

(2) The WASH-740 study assumed that 50 per cent of all the core radioactivity would be released to the environment. This study, using available experimental data, finds it

Table 3-12. Comparison of Consequences from Accidents in a 500-Megawatt Reactor as Calculated in WASH-740 and as Predicted by WASH-1400

Parameter	WASH-740 Peak	WASH-1400 Peak	WASH-1400 Average
Acute deaths	3,400	92	0.05
Acute illness	43,000	200	0.01
Total dollar damage (billions)	7[a]	1.7[b]	0.51[b]
Approximate chance per reactor year		One in a billion	One in ten thousand

[a]This is the value predicted in 1957 dollars.
[b]The values shown are in 1973 dollars. In 1957 dollars, these values should be about two-thirds of that shown.

physically impossible to attain total core releases as large as those used in WASH-740.

(3) The WASH-740 calculation made no provisions for the evacuation of people. Experience shows that evacuation is highly likely and would significantly reduce the consequences of an accident should it occur.

(4) The radioactivity released in a potential reactor accident would be in the form of a plume such as can be seen from smoke stacks. The radioactivity has sufficient heat associated with it to cause the plume to rise, thus reducing the concentration of radioactivity near the ground. This has some effect in reducing consequences. The calculations of the WASH-740 study did not include this effect.

What techniques were used in performing the study?

The latest methodologies, developed over the past ten years by the Department of Defense and National Aeronautics and Space Administration, were used in this study. These techniques are called *event trees* and *fault trees* and help to define potential accident paths and their likelihood of occurrence.

An event tree defines an initial failure within the plant. It then examines the course of events which follow as determined by the operation or failure of various systems that are provided to prevent the core from melting and to prevent the release of radioactivity to the environment. Event trees were used in this study to define thousands of potential accident paths that were examined to determine their likelihood of occurrence and the amount of radioactivity that they might release.

Fault trees were used to determine the likelihood of failure of the various systems identified in the event tree accident paths. A fault tree starts with the definition of an undesired event, such as failure of a system to operate, and then determines, using engineering and mathematical logic, the ways in which the system can fail. Using data covering (a) the failure of components such as pumps, pipes and valves, (b) the likelihood of operator errors, and (c) the likelihood of maintenance errors, it is possible to estimate the likelihood of system failure, even where no data on total system failure exist.

The likelihood and the size of radioactive releases from potential accident paths were used in combination with the likelihood of various weather conditions and population distributions in the vicinity of the reactor to calculate the consequences of the various potential accidents.

How will the results of the study affect safety decision-making?

This study, using an over-all methodology directed toward risk assessment, has developed new insights that contribute to a better understanding of reactor safety. However, many of the techniques used were developed and used only for the purpose of over-all risk assessment and are not directly applicable for optimizing safety designs or evaluating the acceptability of specific designs or reactor site locations. Though the techniques developed in the Study may someday be useful for such purposes, considerable additional development is needed before they can assist effectively in safety decision-making.

Decision-making processes in many fields, and especially in safety, are quite complex and should not lightly be changed. This is especially true where a good safety record has already been obtained, as is so far true for nuclear power plants. The use of quantitative techniques in decision-making associated with risk is still in its early stages and is highly formative. It appears that for the near future considerable additional development is needed in quantitative techniques before they can be used effectively in safety decision-making processes.

SUGGESTED READINGS

Primack, Joel and Frank Von Hippel, 1974, "Nuclear Reactor Safety," *Science and Public Affairs (Bulletin of the Atomic Scientists) XXX:* 5–12, October.

Disposal of Nuclear Wastes

ARTHUR S. KUBO and DAVID J. ROSE   1973

Disposing of radioactive wastes from nuclear fission reactors has been much debated, both in public and in private. The assessments and discussions require rebalancing from time to time, and we attempt this here.

Do the perceived difficulties of the waste disposal problem arise from severe scientific or technological limitations, or from lack of understanding and institutional restrictions? Both contribute, but we think that the latter are dominant: we find several attractive technological options that have been given little consideration, and institutional arrangements that have contributed to premature narrowness of thought. For instance, until late 1971, both the United States Atomic Energy Commission and the public debated the merits of disposing of nuclear wastes in a salt mine near Lyons, Kansas, almost as if that were the sole option.

Misassessment has led to a nuclear waste disposal program too small to match our needs and has contributed to public debate unworthy of the topic. We believe that a greater but still modest effort should be successful, and that the cost of an adequate program will remain small compared with the overall cost of nuclear power. Since the cost is small an extended range of options can be explored.

HUMANISTIC AND TECHNOLOGICAL ASPECTS

Societal problems involving technology, such as radioactive waste management, have features related to present and future costs, perceived benefits, and time scales of concern. Here, for illustration, is one basic dilemma that arises in our problem, somewhat fictionalized for emphasis. Suppose we know that our social structure will persist unperturbed and that we will remain fully responsible up to a time T many years in the future; then society will collapse completely, and we will revert to savages. Suppose also

Dr. Arthur S. Kubo is a lieutenant colonel in the United States Army Corps of Engineers and is an assistant professor in the Department of Engineering, United States Military Academy, West Point, New York.
Dr. David J. Rose is a professor in the Nuclear Engineering Department, Massachusetts Institute of Technology, Cambridge, Massachusetts.

From *Science*, Vol. 182, No. 4118, pp. 1205–1211, December 21, 1973. Reprinted by permission of the authors and the American Association for the Advancement of Science. Copyright 1973 by the American Association for the Advancement of Science.

that we feel as responsible toward the savages as toward ourselves, and that no reasonable technological option is too expensive for us to afford. (In this hypothetical example, we exclude disposal in space and some other option-terminating stratagems.) Under those circumstances we might choose, until nearly time T, to store irreducible nuclear wastes in surface mausolea built to withstand natural disasters, and to watch them assiduously—each radioactive waste container on its own plinth in a gallery, so to speak. We gain the advantage of preserving technological options, in case anything unforeseen goes awry. But shortly before time T we would transfer those wastes to some subterranean place, chosen, if possible, so that the geologic strata themselves were no special attraction, and there seal them forever as best we could.

The two stratagems entail very different features and trade-offs: the first retains options and provides more safety against both error and disasters, but at a cost of full societal responsibility; the second guards against an irresponsible society at the cost of increased environmental risk. One stresses complete retrievability, the other stresses irretrievability. Deciding which to choose was not a technological matter for that imagined society; it depended on two uncertain things: (a) present assessment of future societal stability and (b) depth of responsibility toward future people. Our own choices depend on these two humanistic issues, just as in the caricature above.

Of course, technology enters more than that. What mausoleum, and what subterranean disposal? Can the radioactive inventory be reduced or separated into more easily handled materials? Do safe, inaccessible places exist? These questions and more influence the major decisions, and all require study. This article is substantially technological; we will try to present the physical options and estimate the costs and benefits as well as we can. Such an assessment should show the best options in each major category and put matters in the best order for public discussion. These considerations, together with the humanistic ones, help us to decide not only whether we have the "correct" tangible assets for the cost, but also what are the definitions of assets and costs.

BACKGROUND

A few numbers will help put the following discussion in perspective. First, consider the cost of radioactive waste management schemes compared with that of the nuclear reactor installations themselves. A nuclear capacity of at least 900,000 megawatts has been predicted for the United States by A.D. 2000[1]; the nuclear components alone will cost more than $100 billion (at present costs) and the plants over $300 billion. In contradistinction, every nuclear waste disposal scheme which has been discussed (except perhaps shooting the wastes into the sun) costs only a small fraction of that—often less than 1 per cent. Thus, public debate on this topic should be viewed somewhat less as bearing on absolute limitations to nuclear acceptance, and more as speculation on the very costly consequences of technical failure of too-cheap disposal schemes.

A second set of orienting numbers concerns categories. The major one in terms of short-term radioactivity comprises the fission products, that is, atoms of medium atomic weight formed by fission of uranium or plutonium. Strontium-90, cesium-137, and to a lesser extent krypton-85 are the main culprits; zirconium-93, tellurium-99, iodine-129, cesium-135, and others are much less important. The main ones have half-lives not greater than 30 years.* In 700 years less than one ten-millionth remains, and for this discussion we take 700 years as the end point of practical concern for this category of radioactive wastes.

1. "Potential nuclear power growth patterns," *USAEC Rep. WASH 1098* (December 1970). *Excepting ^{129}I, which has a half-life of 16 million years and is therefore very weakly radioactive. Eventually (after many centuries) its buildup in the environment could become a problem; but the amount to be released in the present technological age is insignificant.

The other category of radioactive wastes consists of the actinides (the elements actinium, thorium, uranium, neptunium, plutonium, and so on), which are formed not by fission but by neutron absorption into the original uranium (or thorium) fuel. All are very toxic and most have long half-lives—for example, about 25,000 years for plutonium-239, the most abundant transuranium actinide, which is formed either in conventional light-water reactors or in proposed liquid-metal fast-breeder reactors. The actinides cause waste management difficulties at two distinct points in nuclear fuel cycle. Some are carried over with the fission products during nuclear fuel reprocessing, with which this article is concerned, but also some highly dilute plutonium wastes will appear from fuel manufacturing plants.** Thus, at the entrance to the waste facility we find a mix of many different transuranic actinides intimately combined with the shorter-lived and temporarily more hazardous fission products. The important things to notice about this category of wastes are that: (a) the offending actinides are relatively toxic,† and (b) although initially far less radioactive than the fission products,

**We do not discuss here the fuel manufacturing wastes and other solid objects contaminated with plutonium, commonly referred to as alpha wastes. This does not imply that alpha wastes are less important, for about as much plutonium is lost there as is sent to waste at the fuel reprocessing facility. But some of our option analyses can apply to those wastes too.

†Relative toxicity is an important concept which, while still unofficial, finds increasing use; it is used here to compare hazards of nuclear wastes with those of naturally occurring substances. For each hazardous species, a maximum permissible concentration (MPC) in water has been defined by federal regulations [Code of Federal Regulations, Title 10 (Government Printing Office, Washington, D.C., 1973), regulation 20 for the substances discussed here, for unrestricted discharges]. Thus, for any initial concentration, a water dilution factor can be specified, to dilute the substance to MPC. We define relative toxicity as the ratio: (volume of water required to dilute a mass of solidified waste to MPC)/(volume of water required to dilute an equal mass of uranium ore to MPC). The ore is assumed to contain 1.4 per cent U_3O_8 by weight; the solidified fission wastes are assumed to be concentrated to 4.14 kilograms per 1000 megawatt-day (thermal) burnup.

they become dominant at 500 years because of their much longer half-lives.

This disparity between the categories, graphically presented in Figure 3-7, suggests that it would be advantageous to separate them chemically and adopt different strategies for each kind.

Now we turn to what has been done about the problem. Our present policy, established almost two decades ago, aims for deep underground burial in selected geologic formations; that option seems to have been a logical follow on from waste management practices commenced in the 1940's. At present three major locations are envisaged: the Atomic Energy Commission production facilities at Hanford, Washington, in local basalt; the commercial wastes in the salt beds of Kansas; and the salt beds of southern

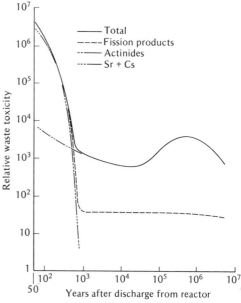

Figure 3-7. Toxicity of wastes from light water reactors, for an equilibrium fuel cycle, with 99.5 per cent removal of uranium and plutonium. Each metric ton of fuel is assumed to deliver a total thermal energy of 33,000 megawatts × days during its operating lifetime. The turn up at 10^6 years arises from growth of daughter products not present in the original material, which is not in decay equilibrium.

New Mexico, which underlie some largely mined-out potash deposits.†† But the underground disposal schemes are opposed by concerned scientists, politicians, and laymen, and the Atomic Energy Commission is now reconsidering other options, such as storage in vaults, disposal in space, and so forth.

The present state of waste management policy is based on requirements determined early in the development of commerical nuclear power: (a) safety beyond any reasonable doubt, and (b) reasonable cost—that is, it should not hinder appreciably the development of commercial nuclear energy. These criteria, in themselves unexceptionable, receive various interpretations.

Aware of complications, the congressional Joint Committee on Atomic Energy asked the National Academy of Sciences on several occasions (and the General Accounting Office as well) to help assess the situation. The National Academy of Sciences concluded in 1957 that, for the near term, salt mine burial appeared most attractive; this assessment was based on the 700-year toxicity of strontium-90 and cesium-137. A broader assessment of different disposal schemes has never formally been made, and indeed since 1957 the National Academy of Sciences seems to have been concerned with increasingly fine points of an already-made decision. The concept of long-term storage in vaults near the surface was dropped early as a safe but temporizing solution requiring active surveillance, although a recent Atomic Energy Commission announcement[2] indicates a return to this concept. Disposal in the oceans seemed unsafe for lack of adequate knowledge about all the consequences of failure—a situation that still obtains. No complete study of disposal in space has been made to date, because of apprehension about the consequences of shortfall. How-

ever, there is a growing interest in space disposal. Success for this project depends on the space shuttle to make it economically feasible, and a sophisticated container to survive possible shortfall; both requirements await the successful outcome of the shuttle program, so that prognosis is difficult and likely to be biased by attitudes for or against the space program.

Selective waste management of heat-producing or highly toxic isotopes has been suggested, but never much analyzed because of the costs and added complications of chemical separation. Thus, through the 1960's and into 1971, at a rate of about $5 million a year, research and development focused on deep disposal, including the intermediate step of solidification.[3] Of that annual commitment, about $500,000 has been applied to developing the salt mine disposal concept, mostly by the Oak Ridge National Laboratory (Oak Ridge, Tennessee); about $2 million has been applied to developing waste solidification processes; and the biggest portion of the remainder has been used to develop a deep underground disposal scheme (now abandoned or at least substantially delayed) for the Atomic Energy Commission's Savannah River wastes. This $5-million commitment must be compared to: (a) the total waste management budget of the Atomic Energy Commission, over $40 million a year in the 1970's and (b) the anticipated scale of nuclear operations, as outlined earlier. A waste disposal

---

3. The discussion of AEC budgetary actions is based on various hearings on AEC authorizing legislation, particularly: U.S. Congress, Joint Committee on Atomic Energy, *Hearings before the Subcommittee on Legislation* (88th Congress, 1st session, 9, 10 April and 2 May 1963) on the 1964 budget; *Hearings before the Subcommittee on Legislation* (88th Congress, 2nd session, 22, 23, 27 January and 3, 4, 7 February 1964), part 1, on the 1965 budget. Also, *Nucl. News 15* (No. 3), 26 (1972) on the 1973 budget; H. F. Soule, private communication. Additional material on the program for fiscal years 1968 to 1971 is contained in: Comptroller General of the United States, *Progress and Problems in Programs for Managing High Level Radioactive Wastes* (Document G 164052, General Accounting Office, Washington, D.C., 28 January 1971).

---

††Disposal in gneiss at the AEC facility on the Savannah River, South Carolina, was considered, but on 17 November 1972 the AEC announced that the Savannah River Bedrock Project was indefinitely postponed.
2. This announcement appeared in *Nucl. News 15* (No. 7), 55 (1972).

research and development budget several times $5 million a year seems more appropriate.

TAXONOMY OF OPTIONS

Figure 3-8 shows what we think are the major options to be considered. Of course, variations exist, and some will be mentioned later. We draw several routes on the map of Figure 3-8, so to speak. All start with very radioactive liquid wastes from reprocessing fuels based on either uranium, plutonium, or thorium.

Route 1 is the scheme (until recently in favor) of solidifying the wastes, including whatever actinides were present, and transporting them to a salt mine for permanent disposal. But other types of mines could be used, as shown in route 1A near the bottom of Figure 3-8. In routes 2 and 2A various wastes are separated, which ameliorates the disposal problem. With any of these schemes, a temporary visit may be made to near-surface storage facilities with full retrieval capability, which we call mausolea. Several other disposal options of possible interest can follow from the upper segments

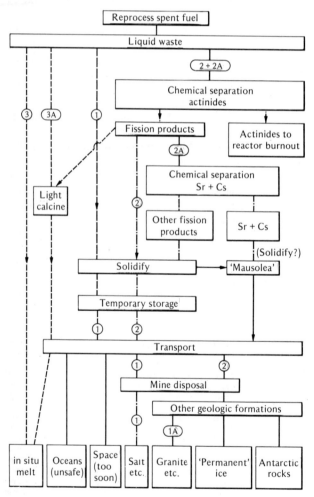

Figure 3-8. Taxonomy of nuclear waste disposal options.

of routes 1 and 2, particularly the latter. We rule out dumping in the oceans and in space, for the time being, for reasons already given. We are skeptical about the two remaining options, "permanent" ice and Antarctic rocks, for reasons given later.

Routes 3 and 3A, *in situ* melting, are quite different; the wastes, upon being inserted into a selected underground site, fuse themselves into a permanent glassy mass.

We now discuss each of these options in more detail, giving the advantages, disadvantages, and costs as we see them. The costs and some other details are not always easy to ascertain, and the degree of present uncertainty in our figures varies substantially between options. In this article we present a summary; more detailed justifications can be found elsewhere.[4]

SIMPLE MINE DISPOSAL AND SALT VAULTS (ROUTE 1)

The Kansas salt vault project, until recently the Atomic Energy Commission's sole commerical option, has proceeded far enough that the major technical and economic uncertainties have been resolved.[5] The cost would range between 0.045 and 0.055 mill per kilowatt-hour electric for disposal (after 10 years of temporary storage) of solidified, unaltered, high-level radioactive wastes. That is about 0.5 per cent of the cost of generating the nuclear power, a negligible increase, small compared even to annual inflation. Retrieval of these wastes from the repository, if it were ever required, grows increasingly problematic as the project passes from demonstration to operation, and finally to a

decommissioned state with sealed shafts and backfilled corridors.

The long-term safety of the project depends on preventing the intrusion of water into the salt beds by any means. This could occur by natural means such as erosion, failure of overlying or underlying shale beds, boundary dissolution, and by man-induced means such as well borings. The Lyons site had several chiefly man-induced flaws.

The concept has some advantages: salt is easy to mine, it will in time flow plastically to seal the whole midden, and surely the very presence of the salt guarantees that no water was present in the geologic past. But these advantages are two-sided, for the very fragility (vulnerability to water) of the geologic structure is used as an argument in its favor, and the demonstrated stability refers only to past time, and not to the future, when conditions will likely be different. We may mistake an indicator of past quality for a substantive future property.

Arguments like these, related to future uncertainty, now appear in the scientific and public literature. From those discussions, we note that: (a) the prognosis is likely to be better in some other salt deposits, and (b) similar disposal is possible in other geologic structures—other evaporites or granite monoliths, for example—with some advantages and disadvantages (that is, route 1A in Figure 3-8).

The extensive beds of salt and some potash of southern New Mexico are now being viewed hopefully by the Atomic Energy Commission. The advantages of the site are remoteness from present occupation and a more favorable political climate [due to large Atomic Energy Commission commitments to the New Mexican economy at Sandia Base (near Albuquerque) and Los Alamos Scientific Laboratory]. The disadvantages are similar to many discussed for Lyons.

Route 1A leads to disposal in hard rock, whose advantages counter some of the shortcomings associated with salt: the rock is insoluble, ubiquitous, and normally not associated with valuable mineral resources. But

4. A. S. Kubo, thesis, Massachusetts Institute of Technology (1973).

5. There are many salt mine references, but the basic ideas are covered in: "Siting of fuel reprocessing plants and waste management facilities," *Report ORNL-4451* (July 1970); U.S. Congress, Joint Committee on Atomic Energy, *AEC Authorizing Legislation, Fiscal Year 1972* (92nd Congress, 1st session, 9, 16, 17 March 1971), part 3; R. L. Bradshaw and W. C. McClain, Eds., *Report ORNL-4555* (April 1971); A. Boch, *Report ORNL-4751* (December 1971).

there are disadvantages: the rock is brittle and unhealing, and it may be leached by groundwater. The mining cost (less than $1 per cubic foot) almost certainly will be higher, but not prohibitively so. Also, we calculate that the extra expense of transporting unwanted salt or other evaporites to the ocean can equal the extra cost of mining hard rock, which needs no environmental treatment and may actually be salable. At the very worst, the mining cost would be doubled in hard rock.[6] But the total cost of disposal in hard rock would be only about 25 per cent more than for the salt mine repository. This increase is so modest because of the small fraction of the total cost apportioned to the mine facility (either salt or hard rock) compared to the interim storage, solidification, and transportation costs (for burial of 10-year-old wastes in formations where their heat generation limits the concentration; this is the usual case).

FURTHER CHEMICAL SEPARATIONS (ROUTES 2 AND 2A)

Whatever the final means of disposal may be, using chemical separations to alter the character of the wastes has considerable merit.[7] As discussed above, removing the actinides turns a million-year problem into a 700-year one, because we envisage burning out the actinides in a reactor; the technology is available now and can be implemented; and the method is not limited to countries with specific geologic formations. Also, at much greater expense, one can remove the principal heat-producing isotopes, strontium-90 and cesium-137. Against these advantages we find, as usual, some disadvantages: higher cost, more complex operations, and a re-

versal of waste management policy that will cause economic dislocations for commercial fuel reprocessors now in operation. Also, disposal of the strontium-90 and cesium-137 is a problem.

Extract Actinides Only (Route 2)

The extraction of actinides reduces the long-term toxicity (beyond 1000 years) of the wastes by two to four orders of magnitude (see Figure 3-9). At present and in the projected future, only plutonium and uranium values are to be extracted from the spent fuels from light-water reactors and liquid-metal fast-breeder reactors, and that to only a moderate extent (about 99.5 per cent). Even if the extraction were improved to 99.99 per cent, the extra reduction in long-term toxicity would be very small, because of other actinides are causing the trouble. Curve 1 of Figure 3-9 shows that both the 99.5 per cent and 99.99 per cent extractions of uranium and plutonium are essentially congruent. Thus, extreme extraction of the "usable" values is unhelpful for waste management purposes. Today the technological limit (as opposed to the economic optimum) to the extraction of actinides appears to be

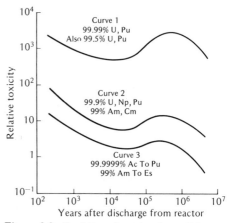

Figure 3-9. Toxicity of wastes from light water reactors, for different degrees of actinide extraction. The fuel conditions are the same as for Figure 3-7.

6. J. J. Perona, R. L. Bradshaw, J. O. Blomeke, *Report ORNL-TM-664* (October 1963).
7. D. Ferguson and H. C. Claybourne, private communications. For Cs and Sr removal: D. E. Larson and P. W. Smith, paper presented at the 66th national meeting of the American Institute of Chemical Engineers, Portland, Oregon, 24–27 August 1969; H. L. Caudill, J. R. LaRiviere, H. P. Shaw, *Report ARH-SA-41* (Atlantic Richfield Hanford Co., Richland, Washington, undated).

99.9999 per cent for actinum through pluto-nium, and 99 per cent for americium to einsteinium. If this extreme extraction were accomplished, the wastes would closely ap-proach the "nontoxic" level in 1,000 years; at that time the toxicity would be some three to four orders of magnitude less than with the current extraction goals (compare curves 1 and 3 in Figure 3-9). A more mod-est extraction of 99.9 per cent of the ura-nium, neptunium, and plutonium, and 99 per cent of the americium and curium (Fig-ure 3-9, curve 2) yields substantial benefits compared to the standard extraction, and can be accomplished more cheaply.

We propose that the troublesome ex-tracted actinides are to be recycled through a reactor,[8] which we consider in this article to be a light-water reactor. This is the most disadvantageous reactor for such a task, hav-ing a deficiency (for this purpose) of high energy neutrons, but there are no data at present for evaluating recycling through a fast breeder reactor. Thus, the economic fig-ures are pessimistic. The remaining wastes would be processed for "conventional" dis-posal (for example, in a mine or a mauso-leum).

The anticipated costs for using route 2 in this way are given in Table 3-13; these are more uncertain than the route 1 costs. The

8. An analysis of this possibility is given by H. C. Claiborne, *Report ORNL-TM-3964* (1973).

Table 3-13. Cost of Separating and Re-cycling Acinides in Wastes from a Light-Water Reactor. (The three cost categories represent 1.1, 1.5, and 2.0 times the current light-water reactor reprocessing costs.)

Item	Cost (mills per kilowatt-hour electric)		
	Optimistic	Median	Extreme
Salt mine[a]	0.045	0.045	0.045
Recycle actinides	.119	.189	.276
Total	0.164	0.234	0.321

[a]The fission products still require disposal.

three categories—optimistic, median, and ex-treme—are based on reprocessing costs 1.1, 1.5, and 2.0 times the current light-water reactor reprocessing costs. In all cases the reactor fuel was slightly and appropriately enriched to compensate for the actinides, and the fuel manufacturing costs were in-creased also.

For a liquid-metal fast-breeder reactor, the added costs should be much lower, for several reasons: (a) a less pure actinide prod-uct should be recyclable without degrading the reactor's neutron economy, which would reduce the need for extreme separations of chemical groups; (b) there would be a smaller fuel manufacturing penalty, since the whole system is full of highly toxic gamma-emitting plutonium already; and (c) more actinides are naturally present in an oper-ating liquid-metal fast-breeder reactor, so the addition of more actinides affects it less. We estimate informally that the recycle cost would be 0.020 mill per kilowatt-hour elec-tric if the actinide extraction cost could be reduced to 110 per cent of the currently anticipated cost, and that the overall dis-posal cost would be about 0.065 mill per kilowatt-hour electric (if the fission products go to a salt mine).

Admittedly, these estimates are prema-ture, but they do indicate that nuclear trans-mutation of the actinides would cost between 0.065 and 0.320 mill per kilowatt-hour electric. This is a nontrivial fraction of the total fuel cycle cost (about 8–9 mill per kilowatt-hour electric) of nuclear electric energy, but in our opinion it is promising enough to be worth further study.

In shortening the period of concern about the waste repository by a factor of at least 1000, extraction of the actinides represents a real safety improvement.

Extract Actinides, Strontium, and Cesium (Route 2A)

Removing these key isotopes reduces the long-term toxicity, as in the previous case, and also the waste thermal power. Strontium and cesium account for the major portion of

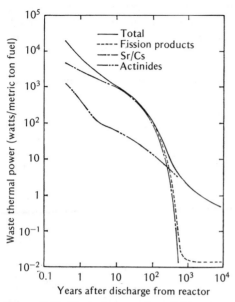

Figure 3-10. Thermal power from light-water re-actor wastes. The conditions are the same as for Figure 3-7.

the waste thermal power during the interim period, 1 year to about 100 years after dis-charge from the reactor, the precise fraction depending on reactor type. The thermal con-tributions of key groups of elements are shown in Figure 3-10.

Waste thermal power is the prime disrupt-ing force that jeopardizes the storage of wastes underground. Reducing the waste thermal power by removing strontium and cesium has two advantages. With a reduced heat load, the waste containers can be packed closer together, which reduces the mining cost by about a factor of 30. More important, it decreases the thermal stresses that work against the safety of (say) the salt disposal project, so that the remaining wastes can be buried in salt or similar struc-tures within the year after fuel reprocessing. The cesium and strontium would be stored in a mausoleum until the space disposal scheme is operational or some future dis-posal scheme is developed, say in 50 years. The stripped actinides would be recycled to a liquid-metal fast-breeder reactor as before (cost estimates based on a light-water reactor are given in the preceding section). The dis-advantages of the scheme are that no method of disposal is available yet for stron-tium and cesium and that the *in situ* melt option is precluded (because of marginal energy in the residual wastes). The costs for route 2A are estimated in Table 3-14 for the following assumptions: (a) the actinides are recycled as before; (b) strontium and cesium are extracted at $0.01, $0.05, and $0.10 per curie for the three cases; (c) strontium and

Table 3-14. Cost of Separating Strontium, Cesium, and Actinides and Recycling Actinides. (The cost of recycling actinides is the light-water reactor estimate.)

Item	Cost (mills per kilowatt-hour electric)		
	Optimistic	Median	Extreme
Extract Strontium and Cesium	0.071	0.142	0.710
Store Strontium and Cesium	.020	.025	.050
Recycle actinides	.119	.189	.276
Salt disposal	.025	.025	.025
Subtotal	0.235	0.381	1.061
Disposal of Strontium and Cesium	?	?	?
Total	?	?	?

cesium are stored at 0.8, 1.0, and 2.0 times the expected cost of long-term storage of wastes in a mausoleum.

Considering the problem of what to do with the strontium and cesium, the option would be attractive only if it promised a substantial societal advantage.

ENGINEERED NEAR-SURFACE STRUCTURES (MAUSOLEA)

Using a mausoleum or storage vault is really storage with the option, if not the explicit intent, of future retrieval. Present estimates of the cost of long-term storage are inaccurate and scanty, because the waste management policy has been oriented toward disposal. However, research at Oak Ridge National Laboratory on interim storage for periods up to thirty years indicates a cost of about 0.015 mill per kilowatt-hour electric. For a fifty-year storage period, a cost of approximately 0.025 mill per kilowatt-electric seems reasonable. The use of mausolea (if storage for future recovery of isotopes is neglected) is a temporizing measure. All countries currently using this concept intend to use some other method of disposal when new technology is developed or the wastes are more manageable.

The optional flexibility of near-surface storage is acquired at the expense of extensive surveillance, and could leave the wastes vulnerable to extremes of nature (such as earthquakes) and the irrationality of man (for example, war, sabotage, and neglect).

Such a temporizing solution would make sense if new technology is likely to be available later. If the present-worth concept of money is employed, at even as low a discount rate as 5 per cent per year, 8.7 cents today would purchase services worth $1 at the end of a fifty-year storage period. This might be attractive if expensive technological solutions, such as space disposal, were imagined to be the ultimate choice. Of course, if nothing turns up such schemes appear as procrastination.

ANTARCTIC ROCKS AND "PERMANENT" ICE

These options have features in common and both start from solidified waste.[9] First, we present the favorable points of view.

Two difficulties that exist with conventional hard rock disposal are the possibility that ground water might leach out the wastes, and the possibility that people might come across the material in some future age when markings have vanished. These difficulties can be circumvented, at least in large part, by disposal at great depths. But both would be overcome by burial at modest depth in Antarctic rocks. To a depth of about 1 kilometer, all ground water is frozen in the Antarctic; thus, insertion of the wastes might be arranged to cause only warm inclusions in the totally frozen surround. Also, none but scientifically well-prepared civilizations are likely to come upon the area.

Another scheme might apply to wastes from which the actinides had been extracted. The residual wastes could be suitably contained (in stainless steel, perhaps) to last for thousands of years in fresh water. Their activity ceases for all practical purposes after a few thousand years. It might be possible to place the residuals in deep holes in one of the long-lasting ice sheets, such as Greenland or Antarctica. The volume of ice that could be melted by all the wastes generated between now and A.D. 2000 is about 0.04 cubic kilometers (0.01 cubic mile) which is not very much. The bottom of the Greenland ice sheet is bowl shaped and below sea level. Thus, it appears that nuclear wastes with a 700-year half-life would be secure there.

One of the major difficulties is that frozen ground—either permafrost or Antarctic rocks—is not really cold enough at depth.

9. These ideas have been promoted in, for example, E. J. Zeller, D. F. Saunders, E. E. Argino, *International Radionuclide Depository (INTER RAD) Report 72-1* (Center for Research, Inc., University of Kansas, Lawrence, 1972).

The temperature ranges from $-5°C$ at -150 meters to $0°C$ at -300 meters, which leaves little room to maneuver. Rocks beneath the Anarctic ice cap are colder but inaccessible.

Ice-cap disposal has several drawbacks. First, wastes still containing actinides require extremely long periods of storage; for example, if the original concentration of plutonium-239 is 10^6 times the permissible concentration in drinking water and no credit is allowed for insolubility, dilution, or adsorption, the required period of isolation is 500,000 years, and the ice may not be that permanent. Even if the actinides were removed, an area problem remains: to preclude appreciable heating at the ground-ice interface (and hence increased ice flow), the heat generated from the wastes must be a small fraction of that appearing via the geothermal gradient—1 per cent would be 63 kilowatts per square kilometer. Wastes from the United States aged 10 years before burial, if accumulated and spread out in Antarctica to give that heat load, would cover 10^6 square kilometers by A.D. 2025, that is, 25 per cent of the area that has ice with an anticipated lifetime exceeding 10,000 years.

Finally, transportation and working conditions in the Antarctic are difficult and hazardous, and at present the Antarctic is kept free of nuclear wastes by international treaty.

IN SITU MELT (ROUTES 3 AND 3A)

This scheme was originally proposed at Lawrence Livermore Laboratory.[10] A hole is bored beneath the waste processing plant, and a nuclear bomb is set off in the hole. Then the radioactive waste is poured into the subterranean cavity so formed, over a 25-year filling period. The wastes heat up through their own activity, boil dry, and eventually melt themselves and some surrounding rock into a glassy ball. The cost is quite uncertain but was judged to be extremely attractive—0.011 to 0.016 mill per kilowatt-hour electric if government financ-

10. J. J. Cohen, A. E. Lewis, R. L. Braun, *Nucl. Techno. 13* (No. 1), 76 (1972).

ing is used. But there are substantial technical difficulties with the scheme in its present stage of planning. (a) During the boiling phase isotopes could migrate from the disposal site. (b) Stress reversal will occur as the transient temperature field moves radially from the disposal site. (c) Faulting or earthquakes might shear the feed and steam lines which join the disposal site to the surface facility. (d) Groundwater might eventually leach out the wastes. (e) An excavation procedure should be developed which does not involve the originally proposed nuclear explosive, and which can make a cavity at greater depth.

According to the original proposal, the fuel reprocessing plant would be at the waste disposal site. This is a severe restriction that protends an undesirable proliferation of disposal sites. But the concept can be modified so that lightly calcined wastes are transported from fuel reprocessing facilities to a federally controlled central repository for *in situ* melt (this is alternate route 3A in Figure 3-8). The wastes are slurried and pumped down to the prepared cavity. The waste boiling period would be reduced from the proposed 25 years to less than 1 year if adequate wastes were stored on site and charged in one operation. Solidifying and transporting the wastes would introduce an added cost, which we estimate at 0.020 mill per kilowatt-hour electric, and would bring the total to roughly 0.031 to 0.036 mill per kilowatt-hour electric. This is less expensive than the salt mine concept, and if it can be shown to be technically safe the project might be practicable. Of course, retrieval is impossible for any of these variations; chemical separation of actinides (route 2) can be incorporated, probably to advantage.

In situ melt suffers from lack of any detailed assessment, but the modified idea seems worthy of analysis comparable to that given the salt project.

SUMMARY

For the present and the foreseeable future the following options appear to be either

Table 3-15. Summary of Waste Disposal Options. (The routes refer to the diagram in Figure 308.)

Route	Option	Cost (mills per kilowatt-hour electric)	Advantages	Disadvantages
1	Salt mine	0.045–0.050	Most technical work to date; plastic media with good thermal properties occur in seismically stable regions	Corrosive media; highly susceptible to water; normally associated with other valuable minerals; difficult to monitor and retrieve wastes
1A	Granite	0.050–0.055	Crystalline rock; low porosity if sound; comparable to salt in thermal properties; retrievable wastes	Nonplastic media; presence of ground water; difficult to monitor
2	Further chemical separation; recycle actinides	0.065–0.320	Reduced long-term toxicity; technology feasible; increases future options	Additional handling and processing; more toxic materials in fuel inventory; waste dilution due to processing; fission products remain
2A	Further chemical separation; remove Strontium and Cesium; recycle actinides	0.140–1.100	Reduced long-term toxicity; reduced short-term thermal power; some reduction of fission product toxicity; increases future options	Additional handling and processing; more toxic materials in fuel inventory; waste dilution due to processing; storage and disposal of Strontium and Cesium extract; fission products remain
3	Melt *in situ*	0.011–0.016	*In situ* creation of insoluble rock-waste matrix; no transportation; reduced handling	Highly mobile wastes during 25-year boiling phase; presence of ground water; irretrievable wastes; proliferation of disposal sites; difficult to monitor
3A	Melt *in situ*, central repository	0.031–0.036	*In situ* creation of insoluble rock-waste matrix; short boiling period; no proliferation of sites	Presence of ground water; irretrievable wastes; difficult to monitor
2	Antarctic rocks		Immobile water	Very narrow temperature limits; not a permanent geologic feature; difficult environment
2	Continental ice sheets		Immobile water	Cannot dispose of actinides; limited amount of ice; not a permanent geologic feature; difficult environment

usable or worth further exploration: mausolea; disposal in mines of various sorts, and perhaps in ice; *in situ* melt; and further chemical separations. The options are interdependent.

It is too early to assess disposal in space, and disposal in the oceans remains unsafe for lack of adequate knowledge. Table 3-15 is a summary of the main ideas for which we have worked out (sometimes uncertain) costs.

For the short-term, ultimate disposal in deep mines is the best-developed plan. However, the related concept of *in situ* melt has significant advantages and should be realistically appraised. Further chemical separation with subsequent recycling of the actinides in a liquid-metal fast-breeder reactor should be investigated and implemented, for it would be universally beneficial; on the other hand, additional removal of strontium and cesium does not seem attractive. Thus, for the near future we make the following recommendations:

(1) Provide temporary storage facilities to ensure that the projected commercial high-level wastes do not become a public hazard. The Atomic Energy Commission adopts this view, and has stated an intention to construct such facilities. But, because of the capriciousness of man and nature, a workable ultimate disposal scheme must be developed soon.

(2) Fund other ultimate disposal schemes at the same rate as the salt mine project—say $1 million a year or more—to sharpen the

technological issues, so that a decision can be reached in the next few years. The schemes should include: (a) *in situ* melt, and the variation with a central repository; (b) burial in mines other than salt mines (including Antarctic rocks and permanent ice); (c) further chemical separation of actinides and recycling actinides in a liquid-metal fast-breeder reactor.

(3) Maintain liaison with the developing space shuttle technology to insure that no opportunity is lost.

The Atomic Energy Commission has a commitment to hold safety foremost in its waste management program, but budget considerations and management priorities have downgraded the program. Past funding levels and management emphasis have yet to produce, after a decade and a half, one operational long-term storage facility—a sign of both commendable caution and inadequate work. If nuclear power is to resolve our energy needs in the coming decades, its benefits should not be delayed for lack of a viable management program for high-level wastes.

SUGGESTED READINGS

Blomeke, John O., Jere P. Nichols, and William C. McClain, 1973, "Managing Radioactive Wastes," *Physics Today,* pp. 36–42, August.

Micklin, Philip P., 1974, "Environmental Hazards of Nuclear Wastes," *Science and Public Affairs (Bulletin of the Atomic Scientists),* pp. 36–42, April.

Report on Biological Effects of Ionizing Radiations

Summary and Recommendations

THE BEIR COMMITTEE   1972

In anticipation of the widespread increased use of nuclear energy, it is time to think anew about radiation protection. We need standards for the major categories of radiation exposure, based insofar as possible on risk estimates and on cost-benefit analyses which compare the activity involving radiation with the alternative options. Such analyses, crude though they must be at this time, are needed to provide a better public understanding of the issues and a sound basis for decision. These analyses should seek to clarify such matters as: (a) the environmental and biological risks of given developments, (b) a comparison of these risks with the benefits to be gained, (c) the feasibility and worth of reducing these environmental and biological risks, (d) the net benefit to society of a given development as compared to the alternative options.

In the foreseeable future, the major contributors to radiation exposure of the population will continue to be natural background with an average whole-body dose of about 100 millirems per year, and medical applications which now contribute comparable exposures to various tissues of the body. Medical exposures are not under control or guidance by regulation or law at present. The use of ionizing radiation in medicine is of tremendous value, but it is essential to reduce exposures since this can be accomplished without loss of benefit and at relatively low cost. The aim is not only to reduce the radiation exposure to the individual but also to have procedures carried out with maximum efficiency so that there can be a continuing increase in medical benefits accompanied by a minimum radiation exposure.

Concern about the nuclear power industry arises because of its potential magnitude and widespread distribution. Based on experience to date and present engineering judg-

Editors' note: The purpose of this study was to critically review the scientific information in this field and report to the Radiation Office of the Environmental Protection Agency the state of science against which the adequacy of Federal Radiation guides may be weighed.

This article is taken from the *Summary and Recommendations of the Report on Biological Effects of Ionizing Radiations*, Committee on the Biological Effects of Ionizing Radiations (BEIR Committee), National Academy of Sciences—National Research Council, Washington, D.C., 1972. Permission to use this excerpt was granted by Dr. Albert W. Hilberg of the National Academy of Sciences. This *Summary and Recommendations* was subsequently published in *Science and Public Affairs (Bulletin of the Atomic Scientists)*, Vol. XXIX, No. 3, pp. 47–49, March 1973.

ment, the contribution to radiation exposure averaged over the United States population from the developing nuclear power industry can remain less than about 1 millirem per year (about 1 per cent of natural background) and the exposure of any individual kept to a small fraction of background provided that there is: (a) attainment and long-term maintenance of anticipated engineering performance, (b) adequate management of radioactive wastes, (c) control of sabotage and diversion of fissionable material, (d) avoidance of catastrophic accidents.

The present Radiation Protection Guide for the general population was based on genetic considerations and conforms to the BEIR Committee recommendations that the average individual exposure be less than 10 roentgens before the mean age of reproduction (30 years). The Federal Radiation Council (FRD) did not include medical radiation in its limits and set 5 rems as the 30-year limit (0.17 rem per year).

EXPRESSION OF GENETIC RISK

Present estimates of genetic risk are expressed in four ways:

(1) *Risk relative to natural background radiation.* Exposure to manmade radiation below the level of background radiation will produce additional effects that are less in quantity and no different in kind from those which man has experienced and has been able to tolerate throughout his history.

(2) *Risk estimates for specific genetic conditions.* The expected effect of radiation can be compared with current incidence of genetic effects by use of the concept of doubling dose (the dose required to produce a number of mutations equal to those which occur naturally). Based mainly on experimental studies in the mouse and *Drosophila* and with some support from observations of human populations in Hiroshima and Nagasaki, the doubling dose for chronic radiation in man is estimated to fall in the range of 20 to 200 rem. It is calculated that the effect of 170 millirems per year (or 5 rems per 30-year reproduction generation) would cause

in the first generation between 100 and 1,800 cases of serious, dominant or X-linked diseases and defects per year (assuming 3.6 million births annually in the United States). This is an incidence of 0.05 per cent. At equilibrium (approached after several generations) these numbers would be about fivefold larger. Added to these would be a smaller number caused by chromosomal defects and recessive diseases.

(3) *Risk relative to current prevalence of serious disabilities.* In addition to those in (2) caused by single-gene defects and chromosome aberrations are congenital abnormalities and constitutional diseases which are partly genetic. It is estimated that the *total* incidence from all these including those in (2) above, would be between 1,100 and 27,000 per year at equilibrium (again, based on 3.6 million births). This would be about 0.75 per cent at equilibrium, or 0.1 per cent in the first generation.

(4) *The risk in terms of over-all ill health.* The most tangible measure of total genetic damage is probably "ill health" which includes but is not limited to the above categories. It is thought that between 5 and 50 per cent of ill health is proportional to the mutation rate. Using a value of 20 per cent and a doubling dose of 20 rems, we can calculate that 5 rem per generation would eventually lead to an increase of 5 per cent in the ill health of the population. Using estimates of the financial costs of ill health, such effects can be measured in dollars if this is needed for cost-benefit analysis.

Until recently, it has been taken for granted that genetic risks from exposure of populations to ionizing radiation near background levels were of much greater import than were somatic risks. However, this assumption can no longer be made if linear nonthreshold relationships are accepted as a basis for estimating cancer risks. Based on knowledge of mechanisms (admittedly incomplete) it must be stated that tumor induction as a result of radiation injury to one or a few cells of the body cannot be excluded. Risk estimates have been made based on this premise and using linear extrapolation from the data from the A-bomb survivors of Hiroshima and Nagasaki, from certain groups of patients irradiated therapeutically, and from groups occupationally exposed. Such calculations based on these data from irradiated humans lead to the prediction that additional exposure of the United States population of 5 rems per 30 years could cause from roughly 3,000 to 15,000 cancer deaths annually, depending on the assumptions used in the calculations. The Committee considers the most likely estimate to be approximately 6,000 cancer deaths annually, an increase of about 2 per cent in the spontaneous cancer death rate which is an increase of about 0.3 per cent in the over-all death rate from all causes.

Given the estimates for genetic and somatic risk, the question arises as to how this information can be used as a basis for radiation protection guidance. Logically the guidance or standards should be related to risk. Whether we regard a risk as acceptable or not depends on how avoidable it is, and, to the extent not avoidable, how it compares with the risks of alternative options and those normally accepted by society.

DOSE COMMITMENTS

There is reason to expect that over the next few decades, the dose commitments for all man-made sources of radiation except medical should not exceed more than a few millirems average annual dose to the entire United States population. The present guides of 170 millirems per year grew out of an effort to balance societal needs against genetic risks. It appears that these needs can be met with far lower average exposures and lower genetic and somatic risk than permitted by the current Radiation Protection Guide. To this extent, the current Guide is unnecessarily high.

The exposures from medical and dental uses should be subject to the same rationale. To the extent that such exposures can be reduced without impairing benefits, they are also unnecessarily high.

GENERAL PRINCIPLES

It is not within the scope of this Committee to propose numerical limits of radiation exposure. It is apparent that sound decisions require technical, economic and sociological considerations of a complex nature. However, we can state some general principles, many of which are well recognized and in use, and some of which may represent a departure from present practice.

(1) No exposure to ionizing radiation should be permitted without the expectation of a commensurate benefit.

(2) The public must be protected from radiation but not to the extent that the degree of protection provided results in the substitution of a worse hazard for the radiation avoided. Additionally there should not be attempted the reduction of small risks even further at the cost of large sums of money that spent otherwise, would clearly produce greater benefit.

(3) There should be an upper limit of man-made nonmedical exposure for individuals in the general population such that the risk of serious injury from somatic effects in such individuals is very small relative to risks that are normally accepted. Exceptions to this limit in specific cases should be allowable only if it can be demonstrated that meeting it would cause individuals to be exposed to other risks greater than those from the radiation avoided.

(4) There should be an upper limit of man-made nonmedical exposure for the general population. The average exposure permitted for the population should be considerably lower than the upper limit permitted for individuals.

(5) Medical radiation exposure can and should be reduced considerably by limiting its use to clinically indicated procedures utilizing efficient exposure techniques and optimal operation of radiation equipment. Consideration should be given to the following:

(a) Restriction of the use of radiation for public health survey purposes, unless there is a reasonable probability of significant detection of disease.

(b) Inspection and licensing of radiation and ancillary equipment.

(c) Appropriate training and certification of involved personnel. Gonad shielding (especially shielding the testis) is strongly recommended as a simple and highly efficient way to reduce the Genetically Significant Dose.

(6) Guidance for the nuclear power industry should be established on the basis of cost-benefit analysis, particularly taking into account the total biological and environmental risks of the various options available and the cost effectiveness of reducing these risks. The quantifying of the "as low as practicable" concept and consideration of the net effect on the welfare of society should be encouraged.

(7) In addition to normal operating conditions in the nuclear power industry, careful consideration should be given to the probabilities and estimated effects of uncontrolled releases. It has been estimated that a catastrophic accident leading to melting of the core of a large nuclear reactor could result in mortality comparable to that of a severe natural disaster. Hence, extraordinary efforts to minimize this risk are clearly called for.

(8) Occupational and emergency exposure limits have not been specifically considered but should be based on those sections of the report relating to somatic risk to the individual.

(9) In regard to possible effects of radiation on the environment, it is felt that if the guide lines and standards are accepted as adequate for man then it is highly unlikely that populations of other living organisms would be perceptibly harmed. Nevertheless, ecological studies should be improved and strengthened and programs put in force to answer the following questions about release of radioactivity to the environment: (a) how much, where, and what type of radioactivity is released; (b) how are these materials moved through the environment; (c) where are they concentrated in natural systems; (d) how long might it take for them to move through these systems to a position of contact with man; (e) what is their effect on the environment itself; (f) how can this information be used as an early warning system to prevent potential problems from developing?

(10) Every effort should be made to assure accurate estimates and predictions of radiation equivalent dosages from all existing and planned sources. This requires use of present knowledge on transport in the environment, on metabolism, and on relative biological efficiencies of radiation as well as further research on many aspects.

Plutonium and Public Health

DONALD P. GEESAMAN   1972

AUTHOR'S NOTE—JUNE 1972

On May 11, 1969 a major fire occurred at the large Rocky Flats plutonium facility located northwest of Denver, Colorado, and operated for the Atomic Energy Commission by the Dow Chemical Company.

Consequent to this fire E. A. Martell and S. E. Poet conducted a pilot study on the plutonium contamination of surface soils in the Rocky Flats environs. Their results suggested an off-site contamination orders of magnitude larger than that which would have been expected from the measured plutonium releases in the air effluent of the facility.

In a letter of January 13, 1970 to Glenn Seaborg, then chairman of the Atomic Energy Commission, and in a press release of February 24, 1970 by the Colorado Committee on Environmental Information, Martell and co-workers called attention to this anomalous contamination and expressed concern over its uncertain origin and over its significance to public health. In response the Atomic Energy Commission fixed the probable origin of the off-site contamination as wind dispersal of plutonium leaking from rusted barrels of contaminated cutting oil, and denied that cause existed for concern over hazards to public health.

It was my conviction that the Atomic Energy Commission response provided a distorted and inadequate representation of the possible hazards associated with the observed off-site contamination, and that the imminent large-scale commercial introduction of plutonium gave this situation a precedential significance much greater than the already considerable significance of the situation itself.

In April 1970 a representative of the Division of Biology and Medicine of the Atomic Energy Commission and I were invited to present our views at the University of Colorado. "Plutonium and Public Health"

Dr. Donald P. Geesaman is an associate professor in the School of Public Affairs of the University of Minnesota, Minneapolis, Minnesota. His professional interest is in the relationship between technology and political institutions. He spent thirteen years in the Theoretical Physics Division and the Biomedical Division of the Lawrence Radiation Laboratory, University of California, where much of his research was concerned with nuclear technologies and their implications concerning society.

From *Underground Uses of Nuclear Energy, Part 2, Hearings before the Subcommittee on Public Works,* United States Senate, August 5, 1970 and subsequently submitted as a working paper to the AAAS/CEA Electric Power Study Group (1972). Reprinted by permission of the author.

derives from the preceding history and should be so interpreted. The presentation was to a lay audience and was made with that expectation. Adequate referencing was added to the written text prior to its inclusion in *Underground Uses of Nuclear Energy, Part 2, Hearings before the Subcommittee on Air and Water Pollution of the Committee on Public Works,* United States Senate, August 5, 1970.

As it stands the paper still represents a legitimate critique, and the recent emphasis on plutonium as a major energy source increases the relevance of the discussion. An updating would involve only incremental changes, and would generally supplement rather than disturb the substantive arguments of the original paper. Hence, whereas such an updating is desirable, it is also of sufficient marginal value that it can be properly deferred at my discretion.

For those who are interested in reading the traditional Atomic Energy Commission position on the subject I would suggest "Appendix 24—Safety Considerations in the Operations of the Rocky Flats Plutonium Processing Plant," from *AEC Authorizing Legislation Fiscal Year 1971—Hearings before the Joint Committee on Atomic Energy, Part 4,* March 19, 1970.

Times have changed since May 1969. Then plutonium was regarded as a military substance and was accordingly given little public attention. Now it is much publicized as the energy source of the not too distant future. April 1970 was a time of transition, and I felt the strong presence of the earlier tradition, and the decision to speak was not an easy one for me. I have had no regrets.

D.P.G.

For the sake of completeness let me give you some background on plutonium—an element that is virtually nonexistent in the earth's natural crust. Plutonium was first produced and isolated by Dr. Seaborg and colleagues in the early 1940's—Dr. Seaborg is a former chairman of the Atomic Energy Commission. This element has several isotopes, the most important being plutonium-239, which, because of its fissionable properties and ease of production, is potentially the best of the three fission fuels, which is why this isotope is of interest. Aside from its fissionable properties, plutonium-239 is a radioactive isotope of relatively long half-life (24,000 years). Hence, its radioactivity is undiminished within human time scales. Plutonium, when it decays, emits a helium nucleus of substantial energy. Because of its physical characteristics, a helium nucleus interacts strongly with the material along its path, and as a consequence deposits its energy in a relatively short distance—about 0.04 millimeter in solid tissue. For comparison, a typical cell dimension is about one-fourth to one-tenth of that. A cell with a nucleus intercepted by the path of such a particle suffers sufficient injury that its capacity for cell division is usually lost (Bardendson, 1962; Bloom, 1959).

The cancer-inducing potential of plutonium is well known. One-millionth of a gram injected intradermally in mice has caused cancer (Lisco et al., 1947); a similar amount injected into the blood system of dogs has induced a substantial incidence of bone cancer (Mays et al., 1947), because of the tendency of plutonium to seek bone tissue. Fortunately the body maintains a relatively effective barrier against the entry of plutonium into the blood system. Also, because of the short range of the emitted helium nuclei, the radiation from plutonium deposited on the surface of human skin does not usually reach any relevant tissue. Unfortunately the lung is more vulnerable.

Before I describe the reason for this, I would like to say something about the characteristics of an aerosol. An aerosol is physically similar to cigarette smoke, or fog, or cement dust. The particles comprising an aerosol, because of their small size, remain suspended in air for long periods of time. If an aerosol is inhaled, then, depending on its physical characteristics, it may be deposited at different sites in the respiratory tree

(*Health Physics,* 1966). The larger aerosol particles usually are removed by turbulence in the nose; particles deposited in the bronchial tree are cleared upward in hours by the ciliated mucus blanket that covers the structure. This clearance system, however, does not penetrate into the deep respiratory structures, the alveoli, where the basic oxygen-carbon dioxide exchange of the lung takes place. Smaller particles tend to be deposited here by gravitational settling, and if they are insoluble, these particles may remain in the alveoli for a considerable time. The problem is this: under a number of conditions (Anderson et al., 1967; Fraser, 1967; Kirchner, 1966; Mann et al., 1967; Stewart, 1963; Wilson et al., 1967) plutonium tends to form aerosols of a size that are preferentially deposited in deep lung tissue. Plutonium dioxide, which is a principal offender, is insoluble and may be immobilized in the lung for hundreds of days before being cleared to the throat or to the lymph nodes around the lungs (*Health Physics,* 1966).

An aerosol is composed of particles of many different sizes, the radioactivity of which may differ by factors of thousands or even more. I will simplify the argument by saying that there is a class of these particles, the largest of which are deposited in the deep lung tissue, that can be expected to have a different potential of cancer induction than the particles of the smaller class. This is because they are sufficiently radioactive to disrupt cell populations in the volume of cell tissue that they expose (Geesaman, 1968a). An example might be a particle that emits 5,000 helium nuclei per day. It would subject between 1 and 20 alveoli to intense radiation, sufficient to inflict substantial cell death and tissue disruption. For reference, the alveoli are the basic structural units of the deep lung. They are shaped and bunched roughly like hollow grapes 0.3 millimeter in diameter. Their walls are thin, a few thousandths of a millimeter, and they are a highly structured tissue with many cell types. Intense exposure of local tissue by a radioactive particle is referred to as the *hot-particle problem.* The question is this: Does such a particle have an enhanced potential for cancer? No one knows. One can argue that cancer cannot evolve from dead cells; hence, a depleted cell population must be less carcinogenic. This is believable, and must be true on occasion. The facts are, however, that intense local doses of radiation are extremely effective carcinogens, much more so than if the energy were averaged over a larger tissue mass (Geesaman, 1968b). Furthermore, this can take place at high doses of radiation where only one cell in ten thousand has retained its capacity to divide. The cancer susceptibility of lung tissue to radiation has been demonstrated in many species; one can say in general that the lung is more susceptible to inhomogeneous exposures from particles and implants than it is to diffuse uniform radiation. Some very careful skin experiments of Albert and co-workers have indicated that tissue disruption is a very likely pathway of radioactive induction of cancer after intense exposure (Albert et al., 1967a, b, c; 1969). The experiments show that the most severe tissue injury is not necessary, nor even optimal, for the induction of cancer. When these notions are applied to a hot particle in the lung, the possibility of one cancer from 10,000 disruptive particles is realistic. This is disturbing because an appreciable portion of the total radioactivity in a plutonium aerosol is usually in the large particle component.

Let me demonstrate what I mean. Suppose a man received a maximum permissible lung burden for plutonium, and suppose roughly 10 per cent of the mass of the burden was associated with the most active class of particles deposited (that is, those emitting several thousand helium nuclei per day), which is reasonable. There would be something like a thousand of these particles, and each would chronically expose 1 to 20 alveoli to intense radiation. If the risk of cancer is like 1 in 10,000 for one disruptive particle, then the total risk in this situation is one in ten—one man in ten would develop lung cancer.

Put another way, about 1 cubic centimeter of the lung is receiving high doses of radiation. It would not be surprising if intense exposure of such a localized volume led to a cancer one time in ten. The question is this: If the individual volumes are separated from each other, is substantial protection afforded? No one knows. It is much easier to find two cancers using 50 exposures of 1 cubic centimeter each, than it is to find a couple of cancers in 50,000 single particle exposures. Certainly the length scales of injury are long enough that a disruptive carcinogenic pathway cannot be disregarded for isolated hot particles (Geesaman, 1968b).

One can look to the relevant experience for reassurance. In an experiment done at Hanford by Dr. Bair and his colleagues, beagle dogs were given plutonium oxide ($Pu^{239}O_2$) lung burdens of a few hundred thousandths of a gram (Bair et al., 1966; Ross, 1967). At 9 years post exposure, or after roughly half of an adult beagle life span, 22 of 24 deaths involved lung cancer, usually of multiple origin. Five dogs remain alive. For comparison, these exposures are about 100 times larger than the present maximum permissible burdens in man.*

There are two unsatisfactory aspects to this experiment. First, because all of the dogs are developing cancer, it is impossible to infer what would happen at lower exposures; simple proportionality does, however, suggest that present human standards are too lax by at least a factor of ten. Second, because the radiation dose is large, with tissue injury almost killing the dogs, and because large numbers of particles are involved (often acting in conjunction), it is improbable that the risk from disruptive particles can be inferred. After all, this is what we need to know, since almost all human exposures will involve hot particles acting independently. And if there is a risk from these particles, it

*Editors' note: A summary of recent experimental data on this subject, and plutonium toxicology generally, is given in W. J. Bair and R. C. Thompson, 1974, "Plutonium: Biomedical Research," *Science 183:* 715–722, February 22, 1974, and in "A Radiobiological Assessment of the Spatial Distribution of Radiation Dose from Inhaled Plutonium," USAEC Rep. WASH 1320 (September 1974).

will be additive throughout the population—there will be no question of a threshold burden, and there will be a possibility that a man with an undetectable burden of a few particles will develop a cancer as a consequence. For the exposures of concern, 1,000 people with 100 disruptive particles each will suffer as many total cancers as 10,000 people with 10 particles each, or as 100 people with 1,000 particles each.

Human experience does not give us the answer either. Plutonium has been around for 25 years, and people have been exposed. In 1964 through 1966 contractors indicated an average total of 21 people per year with over 25 per cent of a maximum permissible burden of plutonium (Ross, 1968). Three out of four of these exposures derived from inhalation. To be reasonably useful, the documentation of exposure must go back more than 15 years, because of the latent period for radiation-induced cancer. In recent years documentation has improved greatly, but from early days there is pitifully little of relevance to the hot-particle problem in the lung.

The official maximum permissible lung burden has been established by equilibrating the exposure from the deposited radioactive aerosol with that of an acceptable uniform dose of X-rays. The International Commission on Radiological Protection indicates this may be greatly in error, and specifically states in *Publication 9:* "In the meantime there is no clear evidence to show whether, with a given mean absorbed dose, the biological risk associated with a nonhomogeneous distribution is greater or less than the risk resulting from a more diffuse distribution of that dose in the lung" (ICRP, 1966). They are effectively saying that there is no guidance as to the risk for nonhomogeneous exposure in the lung; hence, the maximum permissible lung burden is meaningless in terms of plutonium particles, as are the maximum permissible air concentrations which derive from it.

Thus, there is a hot-particle problem with plutonium in the lung, and this hot-particle problem is not understood, and there is no guidance as to the risk. I do not think there is any controversy about that. Let me quote Dr. K. Z. Morgan, one of the United States' two members to the main Committee of the International Commission on Radiological Protection who has been a member of the committee longer than anyone, and is director of Health Physics Division at Oak Ridge National Laboratory. Dr. Morgan's testimony in January 1970 before the Joint Committee on Atomic Energy, United States Congress (Morgan, 1970), states:

There are many things about radiation exposure we do not understand, and there will continue to be uncertainties until health physics can provide a coherent theory of radiation damage. This is why some of the basic research studies of the USAEC are so important. D. P. Geesaman and A. R. Tamplin have pointed out recently the problems of plutonium-239 particles and the uncertainty of the risk to a man who carries such a particle of high specific activity in his lungs.

At the same hearing, in response to the committee's inquiry about priorities in basic research on the biological effects of radiation, Dr. M. Eisenbud, then Director of the New York City Environmental Protection Administration, in part replied:

For some reason or other the particle problem has not come upon us in quite a little while, but it probably will one of these days. We are not much further along on the basic question of whether a given amount of energy delivered to a progressively smaller and smaller volume of tissue is better or worse for the recipient. This is another way of asking the question of how you calculate the dose when you inhale a single particle (Eisenbud, 1970).

He was correct—the problem has come up again.

In the context of his comment it is interesting to refer to the National Academy of Sciences-National Research Council 1961 report on the effects of inhaled radioactive particles (U.S. NAS-NRC, 1961). The first sentence reads: "The potential hazard due to airborne radioactive particulates is probably the least understood of the hazards associated with atomic weapons tests, production of radioelements, and the expanding use of nuclear energy for power production." A decade later that statement is still valid. Finally let me quote Drs. Sanders, Thompson, and Bair from a paper given by them in October 1969 (Sanders, 1970). Dr. Bair and his colleagues have done the most relevant plutonium oxide inhalation experiments and state:

Nonuniform irradiation of the lung from deposited radioactive particulates is clearly more carcinogenic than uniform exposure (on a total-lung dose basis), and alpha-irradiation is more carcinogenic than beta-irradiation. The doses required for a substantial tumor incidence, are very high, however, if measured in proximity to the particle; and, again, there are no data to establish the low-incidence end of a dose-effect curve. And there is no general theory, or data on which to base a theory, which would permit extrapolation of the high incidence portion of the curve into the low incidence region.

I agree, and I suggest that in such a circumstance it is appropriate to view the standards with extreme caution.

There is another hazardous aspect of the particulate problem in which substantial uncertainty exists. In case of an aerosol depositing on a surface, the material may be resuspended in the air. This process is crudely described by a quantity called a resuspension factor which is remarkable in that it seems generally known only to within a factor of billions (Kathren, 1968). Undoubtedly it can be pinpointed somewhat better than this for plutonium oxide, but the handiest way to dispatch the problem is to say there is some evidence that plutonium particles become attached to larger particles and are therefore no longer potential aerosols. Unfortunately, there is also evidence that large particles generate aerodynamic turbulence, and are hence blown about more readily, and on being redeposited tend to knock small particles free. In relation to this, I would like to give you a little subjective feeling for the hazard. There is no official guidance on surface contamination by plutonium. Two years ago, in an effort to determine some indication of the opinions of knowledgeable persons with respect to environmental contamination by plutonium, a brief questionaire was administered to 38 selected Lawrence Radiation Laboratory employees (Kathren, private communication). All were persons well acquainted with the hazards of plutonium. The group consisted of 16 Hazards Control personnel, primarily health physicists and senior radiation monitors. The remainder were professional personnel from Biomedical Division, Chemistry, and Military Applications, who had extensive experience with plutonium. I had nothing to do with the survey, nor was I one of the members who was queried. The conjectured situation was that their neighborhood had been contaminated by plutonium oxide to levels of 0.4 microcuries per square meter. For reference, this value is roughly ten times the highest concentration Dr. Martell found east of the Rocky Flats Dow Chemical facility (Martell, 1970)—and bear in mind that a factor of ten is a small difference relative to the large uncertainties associated with the hazards from plutonium contamination. Several questions were asked. One was, Would you allow your children to play in it? Eighty-six per cent said "no." Should these levels be decontaminated? Eighty-nine per cent said "yes." And to what level should the area be cleaned? Fifty per cent said to background, zero, minimum, or by a reduction of at least a factor of 40. This has no profound scientific significance, but indicates that many people conversant with the hazard are not blasé about the levels of contamination encountered east of Rocky Flats.

Finally I would like to describe the problem in a larger context. Plutonium-239 has been conjectured to be a major energy source by the year 2000. Commercial production is projected at 30 tons per year by 1980, in excess of 100 tons per year by the year 2000. Plutonium contamination is not an academic question. Unless fusion reactor feasibility is demonstrated in the near future, the commitment will be made to liquid-metal fast-breeder reactors fueled by plutonium. Since fusion reactors are presently speculative, the decision for liquid-metal fast-breeder reactors should be anticipated, and plutonium should be considered as a major pollutant of remarkable toxicity and persistence. Considering the enormous economic inertia involved in the commitment, it is imperative that public health aspects be carefully and honestly defined prior to active promotion of the industry. To live sanely with plutonium one must appreciate the potential magnitude of the risk, and be able to monitor against all significant hazards.

An indeterminate amount of plutonium has gone off site at a major facility 10 miles upwind from a metropolitan area. The loss was unnoticed. The origin is somewhat speculative, as is the ultimate deposition.

The health and safety of public and workers are protected by a set of standards for plutonium acknowledged to be meaningless.

Such things make a travesty of public health, and raise serious questions about a hurried acceptance of nuclear energy.

REFERENCES

Albert, R. E., F. J. Burns, and R. D. Heimbach, 1967a. The effect of penetration depth of electron radiation on skin tumor formation in the rat. *Radiation Res. 30:* 515–524.

Albert, R. E., F. J. Burns, and R. D. Heimbach, 1967b. Skin damage and tumor formation from grid and sieve patterns of electron and beta radiation in the rat. *Radiation Res. 30:* 525–540.

Albert, R. E., F. J. Burns, and R. D. Heimbach, 1967c. The association between chronic radiation damage of the hair follicles and tumor formation in the rat. *Radiation Res. 30:* 590–599.

Albert, R. E., F. J. Burns, and R. D. Heimbach, 1969. An evaluation by alpha-particle Bragg peak radiation of the critical depth in the rat skin for tumor induction. *Radiation Res. 39:* 332–344.

Anderson, B. V., and I. C. Nelson, 1967. Plutonium air concentrations and particle size relationship in Hanford facilities. BNWL-495, December 1967.

Bair, W. J., J. F. Park, and W. J. Clarke, 1966. Long-term study of inhaled plutonium in dogs. Battelle Memorial Institute Technical Report, AFWL-TR-65-214.

Barendson, G. W., 1962. Dose-survival curves of human cells in tissue culture irradiated with alpha-, beta-, 20-kV x- and 200-kV x-radiation. *Nature 193:* 1153–1155.

Bloom, W., 1959. Cellular responses. *Rev. Modern Phys. 31:* 21–29.

Eisenbud, M. Panel discussion, 1970. In *Environmental Effects of Producing Electrical Power, Phase 2.* Testimony presented at Hearings before the Joint Committee on Atomic Energy, 91st Cong., 1970. Washington, D.C., U.S. Govt. Print. Off.

Fraser, D. C., 1967. Health physics problems associated with the production of experimental reactor fuels containing PuO_2. *Health Phys. 13:* 1133–1143.

Geesaman, D. P., 1968a. An analysis of the carcinogenic risk from an insoluble alpha-emitting aerosol deposited in deep respiratory tissue. University of California Radiation Laboratory, Livermore, UCRL-50387.

Geesaman, D. P., 1968b. An analysis of the carcinogenic risk from an insoluble alpha-emitting aerosol deposited in deep respiratory tissue; Addendum, University of California Radiation Laboratory, Livermore, UCRL-50387, Addendum.

ICRP, 1966. *Recommendations of the International Commission on Radiological Protection (Adopted September 17, 1965), ICRP Publication 9.* Oxford, Pergamon Press. ICRP-PUBL-9.

Kathren, R. L., 1968. Towards interim acceptable surface contamination levels for environmental PuO_2. Battelle Northwest Laboratory, Richland, Washington, BNWL-SA-1510.

Kathren, R. L. Battelle Northwest (private communication).

Kirchner, R. A., 1966. A plutonium particle size study in production areas at Rocky Flats. *Am. Ind. Hygiene Assoc. J. 27:* 396–401.

Lisco, H., M. P. Finkel, and A. M. Brues, 1947. Carcinogenic properties of radioactive fission products and of plutonium. *Radiology 49:* 361–363.

Mann, J. R., and R. A. Kirchner, 1967. Evaluation of lung burden following acute inhalation exposure to highly insoluble PuO_2. *Health Phys. 13:* 877–882.

Martell, E. A., P. D. Goldan, J. J. Kraushaar, D. W. Shea, and R. H. Williams, 1970. Report on the Dow Rocky Flats fire: Implications of plutonium releases to the public health and safety. Colorado Committee for Environmental Information, Subcommittee on Rocky Flats, Boulder, Colorado, January 13, 1970. (Personal communication to Dr. Glenn T. Seaborg, Chairman, Atomic Energy Commission).

Mays, C. W., et al., 1969. Radiation-induced bone cancer in beagles. In Mays, et al., (eds.), *Delayed Effects of Bone-Seeking Radionuclides.* Salt Lake City, University of Utah Press.

Morgan, K. Z., Radiation standards for reactor siting. In *Environmental Effects of Producing Electrical Power, Phase 2.* Testimony presented at Hearings before the Joint Committee on Atomic Energy, 91st Cong., 1970. Washington, D.C., U.S. Govt. Print. Office.

Morrow, Paul E. (Task Group Chairman), 1966. Deposition and retention models for internal dosimetry of the human respiratory tract. *Health Physics 12:* 173–207.

Park, J. F., et al., 1970. Chronic effects of inhaled $^{239}PuO_2$ in beagles. Battelle Northwest Laboratory, Richland, Washington, BNWL-1050, Part 1: 3.3–3.5.

Ross, D. M., 1968. A statistical summary of United States Atomic Energy Commission contractors' internal exposure experience, 1957–1966. In Kornberg, H. A., and W. D. Norwood (eds.), *Diagnosis and Treatment of Deposited Radionuclides.* Proceedings of a Symposium held at Richland, Washington, 15-17 May 1967. N.Y., Excerpta Medica Foundation, 1968. pp. 427–434. (CONF-670521).

Sanders, C. L., R. C. Thompson, and W. J. Bair, 1970. Lung cancer: Dose response studies with radionuclides. In *Inhalation Carcinogenesis.* Proceedings of a Biology Division, Oak National Laboratory, Conference held in Gatlinburg, Tenn. October 8–11, 1969. M. G. Hanna, Jr., P. Nettesheim, and J. R. Gilhert, (eds.), U.S. Atomic Energy Commission Symposium Series 18, 1970. pp. 285–303, (CONF-691001).

Stewart, K., 1963. The particulate material formed by the oxidation of plutonium. In *Technology, Engineering and Safety,* C. Nichols, (ed.). New York, The Macmillan Company 5: 535–579.

U.S. NAS-NRC-SUBCOMM, 1961. *Effects of Inhaled Radioactive Particles.* Report of the Subcommittee on Inhalation Hazards. Committee on Pathologic Effects of Atomic Radiation. National Academy of Sciences-National Research Council, Washington, D.C., Publication 848. NAS-NRC/PUB-848.

Wilson, R. H., and J. L. Terry, 1967. Biological studies associated with a field release of plutonium. In *Inhaled Particles and Vapours II,* C. Davies, (ed.). Oxford, Pergamon Press, 273–290.

SUGGESTED READINGS

Bair, W. J., and R. C. Thompson, 1974, "Plutonium: Biomedical Research," *Science 183:* 715–722, February 22, 1974.

Speth, J. Gustave, Arthur R. Tamplin, Thomas B. Cochran, 1974, "Plutonium Recycle: The Fateful Step," *Science and Public Affairs (Bulletin of the Atomic Scientists) XXX:* 15–22, November.

The Clear and Present Danger: Nuclear Power Plants

ENVIRONMENTAL EDUCATION GROUP/
ENVIRONMENTAL ALERT GROUP   1974

HISTORY

At one time, the world was led to believe that the peaceful use of the atom was indeed a safe and practical answer to solving the energy problems of the developed nations and that the commercial use of nuclear energy was the humanistic harnassing of the incredible power locked in the atom—the bypassing of the terrors of nuclear weapons. Recently, a great deal of information, much of which was formerly suppressed from public view, has brought startling awareness of inherent difficulties, and the real and potential hazards, that have accompanied the proliferation of nuclear-engendered power. And what is even more frightening is the fact that the further development of nuclear plants is dependent upon the proliferation of an even more hazardous nuclear facility, the breeder, because nuclear fuels are becoming ever more scarce. Before we

discuss in depth the nature of the hazards, it is important to discuss the reasons for the growth of nuclear power and our commitment to what has now proven to be a seriously short-sighted program in commercial and moral terms.[1]

Thirty years ago, when scientists first perceived the possibilities of practical large-scale uses for nuclear energy, they based their hopes on two facts: first, the enormous energy content of uranium compared to that of an equivalent amount of coal (almost three million times) and, second, the known chemical abundance of uranium in the earth's crust. Enthusiasm at the time led to well publicized, but premature, visions of an "abundant energy resource" which would produce "costless electricity." In the com-

1. George L. Weil, *Nuclear energy: Promises, promises* (Washington, D.C.: G. L. Weil, 1972), 40 p.

*Editors' note: This article is reprinted from "Energy Options," Part II, titled "Nuclear Energy in the Form of Fission Power," pp. 30–35, which is abridged from "The Clear and Present Danger: A Public Report on Nuclear Power Plants," a 47-page document presented to the United States Congress and international agencies in May 1973. Both reports were produced under a grant from the Environmental Alert Group, 1543 North Martel Avenue, Los Angeles, California, 90046, to the Environmental Education Group and are available upon request. The latter report was copyrighted 1973 by the Environmental Education/Environmental Alert Group.

From "Energy Options," pp. 30–35, 1974. Reprinted with light editing and by permission of the Environmental Education Group and the Environmental Alert Group.

mon phrase of that era, nuclear energy was going "to make the deserts bloom." Unfortunately, technology has not caught up with the vision.

The Atomic Energy Commission was set up in 1946 primarily to maintain United States nuclear superiority and thereby protect our national security. At the same time, the Atomic Energy Commission was commissioned to encourage the development of peaceful uses for nuclear energy. Many hoped that dramatic uses of nuclear energy in peace time would expiate in some way for the atomic bombs that ended World War II so efficiently but with such a terrifying instantaneous devastation of human life and property.

The Joint Committee on Atomic Energy was established as the Congressional watchdog over Atomic Energy Commission programs and given unchallenged authority because of cold war secrecy at the time. Of course, the need for blanket maximum security has since changed, but the power of the Joint Committee on Atomic Energy has not. Congress still seldom digs into "esoteric" nuclear legislation, or questions the sacrosanct authority of either the Atomic Energy Commission or the Joint Committee on Atomic Energy.

Our current nuclear power plants are a direct outgrowth of United States development of nuclear propulsion plants for submarines. The success of the Navy's program awakened the interest of industry in the possibilities of commercial nuclear power, an interest that was quickened with fear that if private industry did not step in, nuclear power plants might become a government monopoly. Thus, industry, encouraged by the Atomic Energy Commission, chose their basic nuclear power plant designs on the basis of existing military technology rather than on the basis of the most efficient use of nuclear fuel for commercial power. It was a short-term solution to a long-term problem. In response to industry, Atomic Energy Commission and Joint Committee on Atomic Energy pressure, Congress passed the 1954 Atomic Energy Act which opened up the field of nuclear energy to private industry. This legislation allowed private enterprise, under Atomic Energy Commission licensing and regulation, to own and operate nuclear plants and such auxiliary facilities as nuclear fuel reprocessing plants. But apparently the 1954 Atomic Energy Act was not enough to persuade utilities to build nuclear plants. By this time, the Joint Committee on Atomic Energy, the Congressional "watchdog" committee, had become as much a promoter of commercial nuclear power as the Atomic Energy Commission. In 1955, with Joint Committee on Atomic Energy authorization, the Atomic Energy Commission proposed attractive subsidies to the utilities—such relief from paying normal commercial interest charged on publicly owned valuable fuel, for research contracts, and use of Atomic Energy Commission laboratories at no charge. In 1956, the Atomic Energy Commission established guaranteed "buy" back prices for by-product plutonium produced in commercial nuclear fuel. Still not enough to get nuclear power off the drawing boards! One year later, the Atomic Energy Act was amended to provide federally funded public liability insurance for operators of nuclear facilities. Without these subsidies, few utilities, if any, would have "gone nuclear."

Westinghouse and General Electric—the two companies with experience derived from the Navy's nuclear-powered submarine program—at first had the nuclear manufacturing field all to themselves. They accepted large lead losses estimated at 20 to 30 million dollars per plant in order to create a market for their new, but inherently inefficient technology. Even today, almost ten years later, plant manufacturers are reportedly not yet earning a profit on their nuclear business, although they are now pricing plants on a more realistic cost basis.

As the nuclear industry grew, the utilities learned to use the threat of nuclear power to drive down the delivered coal prices. This tactic drove marginal producers out of business, put marginal mines of large coal companies out of operation, and encouraged these companies to develop foreign markets. Today, when utilities are pressed to turn back to coal-fired plants to compensate for shortages of generating capacity because nuclear plants are behind schedule in construction and licensing, they discover that not only is there not enough coal available for their needs, but also that prices have gone up substantially. Their tactic seriously backfired at the expense of the public welfare and accounts in part for the present so-called coal shortage. Thus, today, we face the prospect of several years of inadequate power-generating facilities and predictable brown outs.

Nuclear power plants have never lacked for enthusiastic supporters among the Atomic Energy Commission, Joint Committee on Atomic Energy, and the nuclear industry. These promoters tend to dismiss their critics as "kooks," "professional alarmists," "misguided zealots," and argue with remarkable self-assurance that the benefits of our current low-performance nuclear plants far outweigh both the risks and the huge publicly financed developmental costs.

Their three most popular and catchy arguments have been—first, nuclear energy is cheap and will more than repay, development costs through lower electric rates to the public; second, nuclear energy is urgently needed because of our "dwindling" coal resources; and third, nuclear plants are clean—do not pollute the environment.

Today's large operating nuclear plants furnish no conclusive evidence that the production of nuclear energy, even under the most favorable circumstances and most advanced technology, is appreciably less expensive than the energy production from large fossil-fuel plants. Not only is nuclear energy far from "costless," but the production cost differential between it and energy produced from fossil fuels may be extremely small, depending on factors such as the size of the plant, its location, cost of money, reliability of operation, and so on—and some of these factors cannot be determined for thirty years or more—the expected operating life of

a plant. Past history clearly indicates that nuclear plants are at least more than 20 per cent costlier to construct than conventional plants of equal generating capacity. And nuclear plants are, because of many questions concerning safety, often plagued, in their construction, with "unforeseen" delays involving achieving standards of safety, beyond Atomic Energy Commission regulations, in order to drastically reduce such dangers as the routine release of radioactivity and the release of thermal heat. And because of the misguided zeal of the Atomic Energy Commission to promote and protect the nuclear industry, especially manifest in the Commission's disregard for specific legislation dealing with environmental impact statements, it has foisted upon industry the burden of increased costs, of reopened hearings for license, and of costly delays due to the adherence to new rulings. All these added requirements and difficulties have expanded greatly the initial production cost estimates of nuclear plants.

Whether we urgently need nuclear energy regardless of cost will depend upon development of the many abundant and far more clean and trouble-free alternative energy sources.

It is important to reiterate that, despite their unique energy source, nuclear plants utilize the same relatively inefficient method of converting heat energy to electrical energy as conventional plants have since the days of wood service as fuel. Current reactor designs, many of which use water as a coolant, impose lower-temperature limitations on the extremely high-temperature potential of nuclear power. Therefore, an unduly large percentage of energy is forever lost and wasted. In fact, today's "advanced" nuclear plants, and tomorrow's, cannot match even the efficiencies of conventional plants operating over ten years ago. These inefficiencies also give rise to the spectre of thermal pollution—producing about 50 per cent more waste heat than that produced by conventional plants, for disposal into our waters and atmosphere.

ENVIRONMENTAL IMPACT

With all these limitations it would appear that the only advantage of nuclear-engendered power would be that it is virtually free of pollution. Nothing could be further from the truth. The truth of the matter is that the environmental hazards of nuclear power are extraordinarily serious and manifold. The distinct handicaps associated with the production of nuclear energy fall into the following categories: the *possibility of catastrophic disasters due to accident;* the fear of *sabotage* and *diversion* of nuclear materials for use in the *materialization of nuclear weapons,* the reality of *thermal pollution,* the *release of radioactive substances* into the environment, the *hazards of transportation* of nuclear materials and the *long-term* handling and storage of *radioactive waste.*

Radioactivity

There is no large disagreement about the biological harmfulness of radiation. A single "curie" of lethal strontium-90, with a radioactive half-life of about thirty years, will spit out 37 billion high-speed particles per second, and each emitted particle has enough energy to smash about a quarter of a million chemical bonds in human tissue. Both cancer and genetic defects can start with radiation injury to a *single* cell, if additional variables are present. Radioactivity is the ultimate pollutant.

There is also no disagreement about the quantities of radioactive poison produced by so-called "clean" nuclear plants. In one year, *one* large plant produces as much long-lived radioactivity as the explosion of about 1,000 Hiroshima bombs (including a few million "curies" of strontium-90), plus about 600 pounds of plutonium-239 with a radioactive half-life of 24,000 years. A single pound of plutonium, escaping into the environment and inhaled eventually by humans, could cause several billion cases of lung cancer.

The Atomic Energy Commission plans to license about 1,000 large nuclear power plants in the next 30 years—a plan that would produce the radioactive equivalent of 1,000,000 Hiroshima bombs every year plus 600,000 pounds of plutonium, annually.[2]

Types of Waste

Although a nuclear reactor does not create smoke, fly ash, or sulfur dioxide, it produces three kinds of radioactive pollutants: solid, liquid, and gaseous. Solid wastes may consist of such items as clothing, reactor parts, and tools, which may be highly radioactive depending upon their use and which are customarily buried in cement drums either in trenches on land or at sea.

Some liquid wastes possess low-level radioactivity, resulting from isotopes formed when impurities in the coolant water and corrosion products from the coolant pipes are bombarded with neutrons escaping from the core area. Although demineralizing the coolant water before it enters the heat exchange area can control this process somewhat, there will always be some radioisotopes generated from this source. Because of the low radiation level of these wastes (in the form of cobalt-58, chromium-52, and others), they are usually discharged into the environment.

The fuel elements themselves are an even more dangerous source of both liquid and gaseous wastes. Although clad in either stainless steel or zirconium alloy, carefully fabricated to minimize leakage, complete sealing of the fuel elements is apparently impossible to attain or to sustain.

Minute cracks allow radioactive fission products to escape into the primary coolant. In some instances, the high radiation level causes the cladding to flake or weaken, allowing still more leakage. When the steam is condensed to return to the pile, the gaseous fission products are separated and vented through the stack. One of these gases is krypton-83. Others are xenon, argon-41,

2. Environmental Protection Agency.

iodine-131, and nitrogen-13. Not all of these are fission products; some are results of impurities of air and water in the reactor.

The biggest problem lies in the disposing of the fission products. Sooner or later, usually in one to three years, fission products accumulate to the point where the chain reaction stops or is poisoned, that is, the neutrons are being absorbed at greater rates by the fission products, leaving fewer to sustain the chain reaction. At this point the fuel elements, now extremely radioactive, are removed and shipped in specially cooled and shielded containers to a fuel reprocessing plant. Here they are chopped up and placed in concentrated nitric acid. The fission products and unreacted fuel are dissolved and separated. The unreacted fuel can be used again, but the fission products, concentrated somewhat by evaporation, are stored as liquid in underground stainless steel tanks. With a radioactivity that is usually between 100 and 1,000 curies per gallon, enough heat accumulates to keep the tanks boiling like teakettles.

Assuming the present level of efficiency (25 per cent), almost one and a half tons per year of fission products are produced per 1,000 megawatts of power. Already hundreds of millions of gallons of high-level wastes are in storage. Since these wastes will remain highly radioactive for hundreds of years while the tank life is measured in decades, underground tanks represent at best a temporary solution to the waste disposal problem.

Some of those tanks are known to be leaking their poisons ... One tank drained 60,000 gallons into the ground before discovery, the Atomic Energy Commission has acknowledged. Two more have been found to be leaking. Others are suspect.

"There are 145 storage tanks at Hanford, Washington, on the Columbia River. This was where the 60,000 gallons escaped. One tank, in use only five years, leaked from a 2-foot bulge near the bottom. These containers store from 300,000 to 1,000,000 gallons of such perilous substances as strontium-90, which continues to radiate for over a thousand years. . . ."[3]

3. Richard H. Wagner, *Environment and Man* (New York: W. W. Norton and Co., Inc., 1971), pp. 213–214.

It should be added that the United States currently is the waste depository for several other countries, including Canada, Japan, and Italy, increasing the potential problems of waste disposal, storage, and transportation. (See Figure 3-11).

Mine Tailings

"Radioactive wastes involve more than the reactor and its by-products: uranium must be mined, purified, and processed into fuel elements, and as we have seen, when the elements are exhausted they must be reprocessed. Each of these steps produces radioactive wastes. One example will suffice. Scattered around the Colorado River Basin are huge piles of uranium tailings. Ground to a sandlike consistency to remove the uranium, these tailings contain radium-226 with a half-life of 80,000 years. Radium and thorium, like strontium-90, are absorbed by the bones. As radioactive dust from the piles blows into rivers and ultimately into Lake Powell and Lake Mead, the radium-226 level has increased in places to twice the maximum permissible level suggested for human consumption. Near Durango, Colorado, a pile of over a million tons containing radium-226 sits near the Animas River. As it erodes, this material will certainly increase the radium-226 content of the river. Crops grown on irrigation water from the Animas River already contain twice as much radium-226 as crops growing above the tailings. Since radium is concentrated in hay and alfalfa that is eaten by cattle and in turn by man, the possibility exists of yet another food chain being contaminated by radioactivity. Of course, the tailings should be stabilized by grading and planting, which would greatly reduce erosion into surface drainage. This approach has been followed to some extent by state and federal agencies, but apparently not at a rate that has kept up with continuing production.

Further problems have arisen from the tailing sand being used for children's sandboxes and by contractors as a base for concrete slabs or in the backfilling of basements.

Nuclear Wastes —"Limits to Growth," Potomac Associates

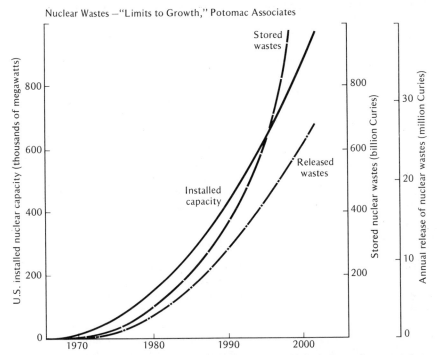

Figure 3-11. Installed nuclear generating capacity in the United States is expected to grow from 11 thousand megawatts in 1970 to more than 900,000 megawatts in the year 2000. Total amount of stored nuclear wastes, radioactive by-products of the energy production, will probably exceed one thousand billion curies by that year. Annual release of nuclear wastes, mostly in the form of krypton gas and tritium in cooling water, will reach 25 million curies, if present release standards are still in effect.

Sources: Installed capacity to 1985 from U.S. Atomic Energy Commission, *Forecast of Growth of Nuclear Power* (Washington, D.C.: Government Printing Office, 1971). Installed capacity to 2000 from Chauncey Starr, "Energy and Power," *Scientific American,* September 1971. Stored nuclear wastes from J. A. Snow, "Radioactive Waste from Reactors," *Scientist and Citizen,* 1967. Annual release of nuclear wastes calculated from specifications for 1.6 thousand-megawatt plant in Calvert Cliffs, Maryland.

As the radium-226 decays, radon, a radio-active gas, is given off. Although radon has a short half-life, it has been established as the prime cause of lung cancer in uranium mine workers. As a result of the use of uranium tailings in construction, many people have been exposed in their homes to levels of radiation many times higher than the maximum dose allowed miners in a uranium mine. In some instances houses have had to be abandoned.

Once again, as with strip mine spoil banks, no one seems interested in the tailings once plants cease operation. The 12 million tons of tailings in the Colorado River Basin may contain as much as 8,000 grams of radium-226. When it has been carried by the rivers of the basin and deposited into the bottom sediments of Lake Powell and Lake Mead, it may well be put out of circulation. But radium-226 may become a much larger problem than it is now as increasing use is made of river water in this arid region."[4]

Thermal Wastes

It is predicted that total thermal electric production, both nuclear and fossil fuel, in

4. Richard H. Wagner, *Environment and Man* (New York: W. W. Norton and Co., Inc., 1971), pp. 215–216.

the United States will amount to 2,000 billion kilowatts per hour by 1980. Such production will require 200 billion gallons of water a day, of which 94 per cent will be used for cooling. This amount of water must come from an annual runoff of 1,200 billion gallons per day. In other words, one-sixth of the total amount of available fresh water will be necessary for cooling the steam-electric power-producing plants. During the two-thirds of the year when the flood flows are generally lacking, about one-half of the total fresh water runoff will be required for cooling power plants. In certain heavily populated and industrialized watersheds, such as the Northeast, up to 100 per cent of available flows may be passed through the various power generating stations within the watersheds during low flow periods.

Temperature affects nearly every physical property of concern in water quality management including density, viscosity, vapor pressure, surface tension, gas solubility, and gas diffusion. Extreme temperature kills. Even within the zone of tolerance, heat may produce sublethal responses. Heat causes changes in metabolism, respiration, behavior, distribution, and migration, feeding rate, growth, and reproduction as well as in parasites and diseases in fishes.

Transportation

Transportation of radioactive materials is yet another area in which the critics of nuclear plants express fears. The route taken by uranium and its fission products before reaching final disposal (or dispersal) is a long one, extending from the mine to the refining mill to the fuel fabrication assembly plants to the reactor vessel to the reprocessing facility (where unused fuel and economically recoverable radioisotopes are extracted) and finally to disposal points.

David Lilienthal, former chairman of the Atomic Energy Commission, is among those who have expressed doubts on the subject:

These huge quantities of radioactive wastes must somehow be removed from the reactors, must—without mishap—be put into

containers that will never rupture; then these vast quantities of poisonous stuff must be moved either to a burial ground or to reprocessing and concentration plants, handled again, and disposed of, by burial or otherwise, with a risk of human error at every step.

And accidents in transportation have occurred. Trucks bearing radioactive materials have been involved in accidents, and in one instance a train carrying radioactive materials derailed. Containers bearing wastes have broken open while in transit, and one such accident necessitated decontamination procedures at a cost of about $27,500.

Accidents

Unique and very substantial hazards are associated with nuclear power reactors. They contain enormous quantities of radioactive materials, the "ashes" from the fission (splitting) of uranium, the accidental release of which into the environment would be a catastrophe. Great reliance is placed on engineered safety systems to prevent or mitigate the consequences of such accidents. Foremost among safety systems are the emergency fuel-core cooling systems, which, should normal cooling systems accidentally fail, are designed to prevent an overheating and melting of the reactor fuel and subsequent release of lethal radioactivity into the environment.

If the emergency cooling system did not function at all, the (fuel) core would melt and the molten mass of zircaloy and UO_2 (uranium oxide) would collapse and probably melt through the pressure vessel in 30 minutes to one hour.

It seems certain that melt-through will be a catastrophic event in that large quantities of molten material will be discharged suddenly.

If through equipment malfunction or failure, human error, or an external event such as sabotage or severe earthquake, one of the major cooling lines to a large reactor core were ruptured, the water circulating through the primary cooling system would be discharged from the system through the rupture and the reactor core would be without coolant. In the absence of water, and with operation of emergency systems, the power generation by fission would terminate.

There is, however, a source of heat that could not be turned off by reactor shutdown—the heat generated by the intense radioactivity of the "ashes" of nuclear fission, the fission products in the fuel. In a reactor generating 650,000 kilowatts of electricity (a small plant by today's standards), the heating provided by fission products 3 seconds after shutdown amounts to approximately 200,000 kilowatts; after 1 hour, 30,000 kilowatts; after a day 12,000 kilowatts; and would still be appreciable for some months.

Under normal reactor operating conditions, the sheathing of the fuel is at a temperature of about 660°F, whereas the interiors of the fuel pellets are typically at 4,000°F, near the melting point of the material. After loss of cooling water (primary coolant) in an accident, the fuel surfaces would begin to heat rapidly, both from the higher temperatures inside and from the continued heating by the fission products. In 10 to 15 seconds the fuel sheathing would begin to fail, and within one minute would melt, and the fuel itself would begin to melt. If emergency cooling were not effective within this first minute, the entire reactor core, fuel, and supporting structure would then begin to melt down and slump to the bottom of the reactor vessel. Emergency cooling water injected at this stage might well amplify the disaster because the now-molten metals would react violently with water, generating large quantities of heat, releasing steam and hydrogen in amounts and at pressures that could themselves burst the containment vessels. Approximately 20 per cent of the fission products would be gaseous, and the meltdown would release them entirely from the now-fluid core. If the containment vessels did not burst, the molten mass of fuel and entrained supporting structure would continue to melt downward, fed by the heat generated by fission-product

radioactivity. At this point in the accident there would be no technology adequate to halt the melt down—it would be out of control. How far down the core would sink into the earth and in what manner the material would ultimately dissipate itself are not entirely understood, but it is probable that virtually all the gaseous fission products and some fraction of the volatile and nonvolatile products would be released to the atmosphere.

The worst possible accident, the Atomic Energy Commission believes, would occur if the pipe that carries water to the fuel broke. With cooling water lost, the fuel would quickly overheat, and the core would melt, releasing radioactivity and endangering large numbers of people. The emergency core-cooling system is designed to flood the system in such an accident. (See Figure 3-12.)

Concern over emergency core-cooling system began in 1965 with the sale of increasingly larger nuclear reactors. When initial tests were run by Aerojet Nuclear Company at the National Reactor Testing Station in Idaho, mechanical failures occurred. In the winter of 1970-71, Aerojet ran a significant series of tests using a 9-inch-diameter model reactor core and electric instead of nuclear heating as fuel. All six tests of the model emergency core-cooling system failed. The reactor community was stunned.

Later, experiments at Oak Ridge indicated that the zircaloy-clad fuel rods of the light-water reactors may swell, rupture and block the cooling channel in the reactor, preventing emergency cooling water from reaching the reactor core. Fuel-rod swelling began in these tests above 1,400°F and at 1,800°F the coolant channels into the hot core were blocked from 50 to 100 per cent. Such blockage, holding back the emergency water, could be catastrophic.

The Possibility of Catastrophe

The basic issue revolves around the possibility of a catastrophic reactor accident with the consequent release of large amounts of

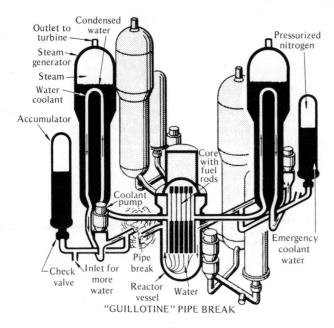

Outlet to turbine
Condensed water
Steam generator
Steam
Water coolant
Accumulator
Pressurized nitrogen
Core with fuel rods
Coolant pump
Emergency coolant water
Check valve
Inlet for more water
Pipe break
Reactor vessel
Water

"GUILLOTINE" PIPE BREAK

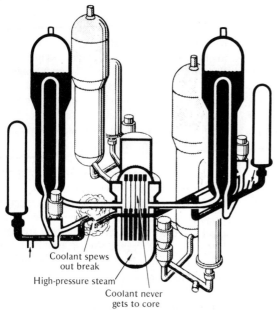

Coolant spews out break
High-pressure steam
Coolant never gets to core

CRITICS' SCENARIO: ECCS FAILS

Figure 3-12. Loss-of-coolant accident is the most feared possibility for nuclear plant designers. Here a pressurized water reactor is shown with its emergency core-cooling system. The emergency core-cooling system is designed to prevent catastrophic overheating of the reactor core in the event the core suddenly loses its coolant water because of the worst accident conceivable, "guillotine break" of the largest pipe in the coolant water loop. Just after the pipe break, water is gushing out of the breach from both the reactor and the coolant system. Instantly, control rods slam home in the core, damping the fission process, and the

radioactivity to the environment. Reactors accumulate prodigious quantities of radioactive materials as they continue in operation: A large power reactor can contain fission products equivalent to the fallout from many dozens of nuclear weapons in the megaton range. The uncontrolled release of even 5 or 10 per cent of this inventory could bring death to persons from 60 up to 100 miles from the reactor. Persons hundreds of miles distant could suffer radiation sickness, genetic damage, and increased incidence of many diseases including cancer. Large-scale and long-lasting contamination of the world's oceans could result from the accidental release of fission products from proposed offshore nuclear power plants. Several reactors now operating or under construction are sited close enough to large urban areas so that each could put more than ten million persons within range of a lethal plume of radioactivity. One reactor, proposed for a site near Philadelphia, will, if built, be within *one mile* of a planned community of 25,000 persons. The stage is set for an accident of devastating proportions.

Sabotage, Diversion, and Bombs—Nuclear Hijackings

"In addition to the hazards of serious accident are the possibilities of sabotage and the theft of nuclear materials, which, like the problems of waste disposal and accident prevention, cannot be completely removed, although the risks can be reduced. The current plague of aircraft hijackings has made clear that society is highly vulnerable to determined efforts at sabotage and that these are extraordinarily difficult to prevent. The disturbing possibilities were brought into focus by a recent incident, in which the hijackers of a commercial airliner threatened to crash their captive aircraft into the nuclear facilities at Oak Ridge National Laboratory unless a ransom were delivered, and by another report that a commercial aircraft diverted to Cuba carried on board, unknown to the hijackers, fissionable materials from which a nuclear explosive could have been constructed.

It is clearly not beyond possibility that a nuclear power plant could be held hostage for financial gain or for political purposes as aircraft now are so frequently. Power plants are so massively protected against accidents of various kinds that sabotage which released large amounts of radiation to the air or water would be difficult, but there is a real possibility that a knowledgeable individual, or an employee, might make such an attempt with the aim of obtaining a ransom, the release of political prisoners, or some other objective. It would be fairly easy to do extensive damage to a power plant, but for the general public to be endangered the reactor itself or its primary cooling system, which are massively protected, would have to be damaged from the outside. The threat of complete loss of a power plant is likely to suffice for most easily imagined saboteurs."[5]

In our time we have seen two world wars and vast revolutionary upheavals. With the increasing social tensions that are bound to accompany the growth of populations, the depletion of natural resources, and the present widening economic gap between the rich and the poor nations, it would seem prudent

5. Sheldon Novick, "Toward a nuclear power precipice," *Environment*, Vol. 15, No. 2, March 1973, p. 37.

emergency core-cooling system goes into action, releasing cool water stored in accumulators for just such an emergency. In the Atomic Energy Commission scenario (a), the system works: The check valves open, allowing pressurized nitrogen gas (at 600 pounds per square inch) to force water out of the accumulators and into the core through the unbroken pipe lines. This water fills the bottom half of the reactor and cools off the core. But in the critics' scenario (b), by the time emergency coolant water gets to the core, the temperature has risen so high that the water turns to high-pressure steam, either blocking more coolant or forcing it to exit via the pipe break, so that the core overheats disastrously. Result: the breakout of a molten glob of radioactive material (drawings by Howard Koslow). (Reprinted by permission from *Popular Science Magazine*, 1973, Popular Science Publishing Co.)

to assume that such upheavals may be even more intense in the coming years. Nuclear fission plants will be enormously attractive objects for sabotage and blackmail. A well-placed charge of explosives, in the midst of one of these huge concentrations of radioactive material, could blow into the air enough radioactivity to be carried by the winds over thousands of square miles, and perhaps render large areas uninhabitable for decades. The twisted minds, and the savage emotions, that could lead to such acts, seem utterly alien to most of us; but only a tiny minority of such people need exist to imperil all the rest of us. I agree with the view recently expressed by Hannes Alfven, in a most thoughtful article, that ". . . fission energy does not represent an acceptable solution to the energy problem. It would place an unendurable burden on the safety and health of future generations."[6]

Professor Mason Willrich, director of the Center for the Study of Science, Technology, and Public Policy, of the University of Virginia, is one expert who voices this very concern. "Most experts," he says, "consider the design and manufacture of a crude nuclear explosive device without previous access to classified data to be no longer an extremely difficult task technically." Willrich points out that "a very small amount of [nuclear] material—for example, a few kilograms [several pounds] of plutonium—is enough for a nuclear explosive capable of mass destruction, and the manufacture of such an explosive is *within the capability of many groups.*" Five kilograms (11 pounds) of plutonium could make a bomb comparable to the one used on Nagasaki.

6. Biological Laboratories, Harvard University.

CONCLUSION: THE CONCLUSION OF THE CRITICS

As Dr. Hannes Alfven, Nobel Laureate in Physics, has profoundly stated:

Fission energy is safe only if a number of critical devices work as they should, if a number of people in key positions follow all their instructions, if there is no sabotage, no hijacking of the transports, if no reactor fuel processing plant or reprocessing plant or repository anywhere in the world is situated in a region of riots or guerilla activity, and no revolution or war—even a "conventional one"—takes place in these regions. The enormous quantities of extremely dangerous material must not get into the hands of ignorant people or desperados. No acts of God can be permitted.[7]

7. Hannes Alfven, "Energy and environment," *Science and Public Affairs (Bulletin of the Atomic Scientists),* Vol. XXVIII, No. 5, May 1972, p. 6.

SUGGESTED READINGS

Ford, Daniel F., and Henry W. Kendall, 1972, "Nuclear Safety," *Environment 14:* 2–9, 48, September.

Gillette, Robert, 1973, "Radiation Spill at Hanford: The Anatomy of an Accident," *Science 181:* 728–730, August 24, 1973.

Gofman, John, and Arthur Tamplin, 1971, *Poisoned Power* (Emmaus, Pennsylvania: Rodale Press).

Holdren, John P., 1974, "Hazards of the Nuclear Fuel Cycle," *Science and Public Affairs (Bulletin of the Atomic Scientists) XXX:* 14–23, October.

Novick, Sheldon, 1973, "Toward a Nuclear Power Precipice," *Environment 15:* 32–40, March.

Novick, Sheldon, 1974, "Nuclear Breeders," *Environment 16:* 6–15, July–August.

Schleimer, Joseph D., 1974, "The Day They Blew Up San Onofre," *Science and Public Affairs (Bulletin of the Atomic Scientists) XXX:* 24–27, October.

Willrich, Mason, and Theodore B. Taylor, 1974. *Nuclear Theft: Risk and Safeguards* (Cambridge, Massachusetts: Ballinger).

The Dilemma of Fission Energy

DONALD P. GEESAMAN and DEAN E. ABRAHAMSON   1974

Man is a 100-watt thermal engine. Twenty watts of this power go to work a unique brain. Because of that brain, man has come to use external sources of energy to supplement his function. A half a million years ago Peking Man had domesticated fire and had acquired the evolutionary advantages associated with that primitive energy supplement. Somewhere in the intervening centuries *Homo sapiens* evolved, and he has amplified this energy subsidy until in an industrialized nation such as the United States, the average individual is sustained by 10,000 watts of power, a factor of one hundred greater than natural energy consumption. This growth has largely come in the past hundred years, and has occurred primarily as a consequence of exploiting fossil fuel reserves.

Fossil-fuel reserves have been a cache of latent chemical energy held and accumulated by the earth for the past 500 million years. These reserves are in the form of petroleum, natural gas, coal, and the less economically accessible tar sands and oil shales. Of these, the world reserves of natural gas and petroleum contain similar total energy, and coal contains some twenty times more than their combination. The oil shales represent a potentially large energy reserve, but they are of speculative importance. It is conservatively projected that present oil and gas reserves will be 80 per cent depleted during the next sixty years, and within half that time in the United States. Coal reserves are conjectured to last only two or three centuries more. By then an appreciable fraction of the energy stored during 500 million

Dr. Donald P. Geesaman is an associate professor in the School of Public Affairs at the University of Minnesota, Minneapolis, Minnesota. His professional interest is in the relationship between technology and political institutions. He spent thirteen years in the Theoretical Physics Division and the Biomedical Division of the Lawrence Radiation Laboratory, University of California, where much of his research was concerned with nuclear technologies and their implications to society.

Dr. Dean E. Abrahamson is a professor in the School of Public Affairs at the University of Minnesota, Minneapolis, Minnesota. His professional efforts center on energy policy, environmental affairs, and public interest law. He holds degrees in physics, medicine, and neuroanatomy, and his history includes both industrial experience as a reactor physicist and participation in the public and government forums that have grown out of the controversy over nuclear energy.

From *Science and Public Affairs (Bulletin of the Atomic Scientists)*, Vol. XXX, No. 9, pp. 37–41, November 1974. Reprinted by permission of the authors and *Bulletin of the Atomic Scientists*. Copyright © 1974 by the Educational Foundation for Nuclear Science.

years by incomplete oxidation of organic material will have been utilized in order to sustain the growth and collective functions of a few centuries of Westernized man.

If we are to continue to indulge ourselves in this ordered, pervasive pattern of life we will require a large energy subsidy. In the not too distant future we must look for sources other than fossil fuels. Nuclear energy has been promoted as such a source. In the next generation we may be passing to this new energy source—a source that has a fuel base potential unreckoned times larger than fossil fuel reserves, a source with a latent energy per unit volume of a million times that of chemical energy sources, and a source with fuels and combustion products toxic and persistent beyond human experience.

In the past many technologies have been promoted and accepted with few questions asked, and the sum of all consequent change has not been quite the paradise that was anticipated. Now some small challenge is put to new technologies—an effort at appreciating their worth in a framework other than that of the institutional promoter. The nuclear industry is being subjected to such a challenge.

In recent years the Atomic Energy Commission and the nuclear industry have presented the bright profile of their technology. The present nuclear controversy has begun to focus attention on the defects of the technology, and on the perspectives in which they can be regarded. Here one such perspective is described.

The ionization potentials of atomic species are measured in electron volts; the binding energies of the nucleons in atomic nuclei are some million times larger. Chemical and nuclear energies are related in magnitude as an hour relates to a century. A quantitative difference of a million denotes a qualitative difference. In the case of nuclear energy this qualitative difference is of profound importance.

Because the energies of the nucleus are so large, the amounts of materials involved in the nuclear fuel cycle are small compared with chemical fuels; grams of uranium or plutonium are equivalent to tons of fossil fuels. A typically sized electrical generating facility consumes either a ton of nuclear fuel each year in fission, or some million tons of coal, oil, or gas in combustion. In each case—fission or combustion—the mass of the waste material is similar to the mass of the original fuels. Therefore, chemical combustion wastes are approximately a million times larger than fission wastes for the same amount of electrical energy generated.

Consider the ton of fission wastes produced during a year of operation by a 1,000-megawatt electric nuclear power plant. Of this total inventory, ignore all but one fission product isotope, cesium-137. It is sufficiently unpleasant that if all the other fission products appeared as common ash, cesium-137 would itself constitute a major problem of radioactive waste. About 60 pounds, or 3 per cent, of the yearly reactor fission products appear as this waste material. As reference values, the official guidance for a maximum permissible body burden is about one-trillionth of that amount, and local surface contamination of one-thousandth of that amount per square mile would give an ambient exposure to ionizing radiation of about 1,000 times the usual natural background of exposure. The half-life for this isotope is 30 years, and hence over a social time interval, such as a generation, the radiological activity of cesium-137 wastes would diminish by only a half. The 60 pounds of cesium-137 produced by operating a nuclear power plant for a year have the potential of excluding humans from hundreds of square miles for decades. It is projected that the accumulated activities of cesium-137 in the United States during the 1980's will represent thousands of reactor-years of operation. Fuel reprocessing plants, such as the one under construction at Barnwell, South Carolina, might have in localized surface storage the inventory produced by 50 such reactors during ten years of operation, that is, almost 500 times the annual contribution of one typical reactor.

Cesium-137 is representative of the significant social characteristics associated with fission wastes: the amounts are physically small, the times involved may be politically long, and the potential for damage is very large. Society can be protected only if the material flows are totally contained, and the isolation of the ultimate waste is permanently guaranteed.

The problems of fission fuels are similarly grave. Neutrons produced in the fissioning of uranium-235, the only naturally occurring fission fuel, are used to breed new fission fuels, of which plutonium-239 is representative. The bred fuels are the dominant nuclear fuels in the projected breeder reactor economy of the future, and they are a vital part of the fuel cycle economics of the present generation of reactors. These bred fuels have two prejudicial characteristics: they are radiologically thousands of times more toxic than uranium-235 (with similar chemical and radiological toxicities), and they are suitable materials for the construction of nuclear explosives.

A 1,000-megawatt-electric light-water reactor produces 400 to 600 pounds of plutonium yearly, worth 2 to 3 million dollars, and having a total volume of about 1 cubic foot. That the official guidance for a maximum permissible lung burden of plutonium is a hundred-trillionth of this amount, and that the air concentration standard for plutonium is one part per million-billion is a commentary on its acknowledged potential for injury. Plutonium-239 has a radiological half-life of 24,000 years, so, unless it is consumed as fuel, it persists for geological times.

POLITICAL SIGNIFICANCE

Plutonium's suitability as a nuclear explosive is politically more significant than its toxicity. A mystique of scientific accomplishment surrounded the development of nuclear weapons during World War II, but that mystique has become illusory. The main practical impasse to weapon manufacture was the development of the expensive technologies for producing fissionable materials. During World War II both gaseous diffusion

enrichment of uranium and reactor breeding of fissionable materials were major industrial accomplishments. Today those processes are inherent in the commercial nuclear industry.

Compared with producing fissionable materials, the design and fabrication of these materials into crude nuclear explosives is a low-technology program. Moreover, the information necessary for building a bomb is available in our libraries and in the numerous nuclear technologists of our society. The amount of fissionable plutonium needed to build a nuclear explosive is cited in the open literature as from 10 to 20 pounds. Therefore, a substantial weapons potential can be inferred for a reactor producing 500 pounds of plutonium yearly. By the 1980's the commercial plutonium production is projected to exceed 50 tons annually, with inventories many times that.

If a small nuclear explosion of a kiloton is conjectured, devastation over an area of 1 square mile can be expected. Population concentrations of up to 300,000 people occur on that scale in this country, as well as concentrations of political and economic power that are necessary for our function.

The intrinsic qualities of the nuclear fission fuel cycle can be summarized: by society's standards, the quantities of materials involved are small, the times are long, and the potential for societal disruption is monstrous. In point, nuclear fuel simply is not an ultra-high-grade of fuel oil; it is something quite unique and extraordinary. Few people understand this better than Alvin Weinberg, former Director of Oak Ridge National Laboratory. Weinberg reflected:

We nuclear people have made a Faustian bargain with society. On the one hand, we offer—in the catalytic burner—an inexhaustible source of energy. . . .

But the price we demand of society for this magical energy source is both a vigilance and a longevity of our social institutions that we are quite unaccustomed to.[1]

Another person who has reservedly assessed the implications of a nuclear econ-

1. Alvin Weinberg, "Social Institutions and Nuclear Energy, *Science 177*: 27–34, July 7, 1972.

omy is Hannes Alfven, Swedish Nobel Laureate in Physics, who wrote:

Fission energy is safe only if a number of critical devices work as they should, if a number of people in key positions follow all of their instructions, if there is no sabotage, no hijacking of transports, if no reactor fuel processing plant or waste repository anywhere in the world is situated in a region of riots or guerilla activity, and no revolution or war—even a "conventional one"— takes place in these regions. The enormous quantities of extremely dangerous material must not get into the hands of ignorant people or desperados. No acts of God can be permitted.[2]

The nuclear fuel cycle then bears these constraints: the absolute containment and permanent isolation of fission products and bred fission fuels from the accessible physical environment and from any unstable sectors of society itself. The small quantities of material involved simplify the technology in and management for the strict physical containment of the material flows. Conversely, the small quantities and careful localization of the materials make them uniquely vulnerable to small-scale disruptive events.

ONE MAJOR INCIDENT

One major reactor accident could change the economics and managerial structure of the electrical generating industry. One major incident of sabotage could lock a most basic industry into a paramilitary administration. One clandestine nuclear explosive could disrupt the total structure of our institutions. Because of world-wide social instability, it is realistic to anticipate such contingencies. Hijackers of a Southern Airways flight have already threatened a national nuclear facility. Scottish nationalists have threatened an English reactor. Political terrorism at the Olympics, at the Tel Aviv airport, and in Khartoum, Ireland, and Argentina are but part of a repeating pattern in recent history.

The motive seems to be well established for terrorism on the grand scale—opportunity is ever present, and now the means are becoming available. The delicate relationship between strength and weakness is changing.

The problem casts its shadow across future global politics. Economist Robert Heilbroner has recently offered this gloomy speculation:

Thus there seems little doubt that some nuclear capability will be in the hands of the major underdeveloped nations, certainly within the next few decades and perhaps much sooner. The difficult question that must then be faced is to what use these nations might be tempted to put this weaponry. I will suggest that it may be used as an instrument of blackmail to force the developed world to transfer large amounts of wealth to the poverty stricken world. I do not raise the specter of nuclear blackmail to indulge in the dubious sport of shocking the reader. It must be evident that competition for resources may also lead to aggression in the other, "normal" direction—that is, aggression by the rich nations against the poor. Yet two considerations give a new credibility to nuclear terrorism: Modern weaponry for the first time makes such action possible; and "wars of redistribution" may be the only way by which the poor nations can hope to remedy their condition.[3]

As never before the powerful may stand in dread of the few. When this is true, suspicion and intolerance will be in their heyday. Think back on the Watergate affair and to the nearly adopted Huston plan for domestic intelligence, and remember that during John Dean's testimony before the Senate Committee the executive agency reports that formed some basis for that plan were turned over to the Committee by Judge Sirica, who in turn had received them from Dean; and remember that the material deriving from the Atomic Energy Commission was carefully classified and was kept in the custody of the Joint Committee on Atomic Energy,

2. Hannes Alfven, "Energy and Environment," *Bulletin of the Atomic Scientists,* pp. 5—7, May 1972.

3. Robert L. Heilbroner, *An Inquiry Into the Human Prospect,* W. W. Norton and Co., New York, 1974.

and was brought to the Senate Committee hearings only for Dean's identification. Why was nuclear energy significant to the people at the White House who were interested in civil disturbances? Each can provide his own answers, and at the same time he can consider how the Watergate affair would be regarded today if there had been a case of plutonium theft with subsequent political disruption or blackmail, or if domestic hijacking of aircraft had had political involvement. Would the ends then seem to justify the means?

A complex and sophisticated society must bear the burdens of vulnerability and constraint that are inherent in its technologies. The Atomic Energy Act of 1954 and subsequent legislation have brought nuclear technology into the purview of the commercial sector. Under the civilian nuclear energy program, nuclear materials will become vital elements in human commerce. They will be very special elements of commerce, and they will warp some of our existing institutions beyond recognition. This process is already incipient in existing perceptions. In the words of Atomic Energy Commissioner Larsen:

> There is no way to insure against loss of special nuclear materials. You can pay the owner of the materials the cash value of the loss, but you cannot insure against special nuclear material coming back to haunt you as a weapon.[4]

Our commercial institutions have had no experience with acute industrial calamity of this potential scale. There would be no substantive commercial reactor industry today without the 1957 Price-Anderson Act, which limits the liability of industry and government in case of a major nuclear accident. That Act is eloquent testimony to the dimension of calamity implicit in a nuclear power economy. In our society commercial interactions are mediated by economics and civil law, but the traditional concepts of

insurability and liability have been inadequate to deal with this particular commercial enterprise, and hence the intercession of the Price-Anderson Act.

ULTIMATE DILEMMA

Society must be constrained by the nature of the nuclear technology with its intrinsic hazards. Auxiliary technology can offer much protection, but, ultimately, the protection lies in the stability of society. *This is the ultimate dilemma of fission energy: The problem of guaranteeing the necessary social stability without being forced to engineer society itself.* Nuclear energy will place absolute constraints on our society, for there are certain things that absolutely must not happen. When men talk in this way, environmental operants, and manipulative drugs, and images of a paramilitary priesthood are at the edge of their awareness. No one knows from whose accounts the price of security would be paid, but we all know there would be no free lunch. The alternative to the occasional devastation of a city may be a garrison state.

Economists are the dominant theoreticians of our society. In deference to conventional wisdom the closing remarks here are prefaced by a quote from energy economist Alan Kneese, excerpted from the recent report, "Benefit-Cost Analysis and Unscheduled Events in the Nuclear Fuel Cycle." Kneese states:

> It is my belief that benefit-cost analysis cannot answer the most important policy questions associated with the desirability of developing a large-scale, fission-based economy. To expect it to do so is to ask it to bear a burden it cannot sustain. *This is so because these questions are of a deep ethical character* (emphasis added). Benefit-cost analyses certainly cannot solve such questions and may well obscure them.[5]

Accepting this statement, then much depends on our perception of man's estate.

4. U.S. Atomic Energy Commission, "Nuclear Materials Safeguards," WASH-1147, 1969.

5. Allen V. Kneese, "Benefit-Cost Analysis and Unscheduled Events in Nuclear Fuel Cycle," *Resources,* September 1973.

Man has lived until now by the chemical sublimation of the energies of sunlight. His ancient evolution has been a biochemical process. Life is mediated by collective stabilities in a space dominated by chemical energies. Persistence stems from a biochemical wisdom that is far more pervasive than conscious thought.

Nuclear energy on earth is a result of human intelligence. Human intelligence is a meaning in itself, a violent life effort to transcend the old evolution and squeeze history down into a scrap of time. Man's efforts at science and politics are a tentative and insecure facet of that evolution. "Brighter than 10,000 suns" was a prophetic description of the Trinity event, for the sun's primeval fire had burned briefly on the surface of the earth, and the former stabilities of life, sorted out by the ages in a chemical world, suddenly seemed inadequate.

We have lived consciously with nuclear energy for a generation. In that time, we have pondered it cautiously as if it were a sleeping giant. In our wars, we have been careful to leave it undisturbed. In our peaceful pursuits we are growing bolder. Now, when our society is fragmented and alienated, when it is threatened by a diminished stability and a fading sense of self-duplication, is it the proper time to try our hand at nuclear energy?

In this enterprise, the old reserves of evolutionary stabilities may be useless; and we must persist "as on a darkling plain" by the working of infallible intelligence alone.

SUGGESTED READINGS

Comey, David D. 1974, "Will Idle Capacity Kill Nuclear Power?" *Science and Public Affairs (Bulletin of the Atomic Scientists) XXX:* 23–28, November.

Gillette, Robert, 1974, "Breeder Reactor Debate: The Sun Also Rises," *Science 184:* 650–651, May.

Lovins, Amory B., 1973, "The Case Against The Fast Breeder Reactor," *Science and Public Affairs (Bulletin of the Atomic Scientists),* pp. 29–35, March.

McPhee, John, 1973, "Profiles: (Theodore B. Taylor) The Curve of Binding Energy," *The New Yorker,* (3 part series); (I) pp. 54–145, December 3, 1973; (II) pp. 50–108, December 10, 1973; (III) pp. 60–97, December 17, 1973.

Nelkin, Dorothy, 1974, "The Role of Experts in a Nuclear Siting Controversy," *Science and Public Affairs (Bulletin of the Atomic Scientists) XXX:* 29–36, November.

Pugwash Warns on Nuclear Power

WORKING GROUP FIVE ON RADIOACTIVE POLLUTION,
TWENTY-THIRD PUGWASH CONFERENCE   1973

FISSION ENERGY PROBLEMS AND ALTERNATIVES

Need for Fission Energy and Assessment of Its Ultimate Potential

A variety of factors influence the degree of the world's need for fission as an energy source and the ultimate potential of this technology. These factors include future growth rates of energy consumption, the forms in which energy will be needed, desires for national self-sufficiency in energy, and the characteristics of technological alternatives to fission: potential magnitude, economic and technical feasibility, environmental consequences in comparison to those of fission, and the time horizon on which they can be made available.

Growth Rates

In recent years, energy use in both rich countries and poor countries has been growing at an average rate of about 5 per cent per

Editors' note: The twenty-third Pugwash Conference on Science and World Affairs was held in Aulanko, Finland, from August 30 to September 4, 1973. (Since the 1950's Pugwash has brought together about one hundred of the world's most distinguished scientists to consider problems such as peace keeping and security, population, environment and resources, and the social responsibility of scientists: the conferences are unofficial—an assembly of individuals not representing anyone but themselves—and working sessions are private so that participants can exchange views with a minimum of constraint.) Scientists from 29 countries and 6 international organizations attended the 1973 Pugwash Conference. Working Group Five on Radioactive Pollution, with 21 members from 14 countries, addressed itself to the topic of Radioactive Pollution of the Environment in the context of the Energy Problem. Usually, a Statement from the Continuing Committee is the conference's only report. But this year, the conference also published the report of Group Five. Members of this group included

J. Holdren (U.S.A.)	M. Nalecz (Poland)	Kirstin Hanzon (Sweden)
H. Alfven (Sweden)	A. J. Polliart (IAEA)	P. L. Kapitza (U.S.S.R.)
S. Bjornholm (Denmark)	P. B. Smith (Netherlands)	E. Leibnitz (G.D.R.)
B. T. Feld (U.S.A.)	H. Wergeland (Norway)	M. A. Markov (U.S.S.R.)
P. Jauho (Finland)	V. Zoubek (Czechoslovakia)	I. F. Pochitalin (U.S.S.R.)
V. F. Kuleshov (U.S.S.R.)	E. Bauer (France)	J. Rotblat (U.K.)
S. Lifson (Israel)	E. Broda (Austria)	A. Vuorinen (Finland)

Excerpt from the Report of Working Group Five, "Radioactive Pollution of the Environment in the Context of the Energy Problem" as found in the *Congressional Record—Senate*, Vol. 119, October 8, 1973, pp. S18727-S18730.

year (doubling time 14 years). Electricity, now accounting for about 25 per cent of all energy consumption, has been growing much faster: 7 to 10 per cent per year in much of the world (doubling time 10 to 7 years). Continuation of these rates would give a fourfold increase in world energy consumption by the year 2000 and an eight- to sixteenfold increase in electricity generation. Projections provided by the International Atomic Energy Agency indicate about 1,000 electrical gigawatts of nuclear capacity world-wide by 1990 and 3,000 electrical gigawatts by 2000. The actual expansion, however, will depend in part on how energy growth is distributed. The average rate of consumption of energy in all forms in the United States is 11 kilowatts (thermal kilowatts) per person, that in India about 50 times lower, and the world average is 2 kilowatts per person. Growth of energy use is evidently much more badly needed in poor countries than in rich ones, and a more sensible goal than equally rapid growth in all countries would be distributing growth with the ultimate goal of achieving a roughly uniform level of per capita consumption in all countries. A figure of 5 to 10 kilowatts per person was mentioned, but some group members thought that the genuine needs are not so high. It was also noted that limits on investment capital, other resources, and social development rates are likely to prevent attainment of such figures in much of the world in the foreseeable future. These very important questions could not be thoroughly addressed by the present Working Group. In any case, if it is conceded that the most urgent needs for energy growth are in the poor countries, nuclear fission is at a disadvantage because large nuclear plants are not well suited to the presently small and dispersed needs in these countries.

Energy Potential of Fission

The potential of nuclear fission as an energy source will not be limited in magnitude in the foreseeable future by shortage of fuel, even in the event that breeder reactors are not deployed. The cost of electricity generated in light-water reactors and gas reactors is already so insensitive to the cost of raw uranium that very abundant low-grade ores could be utilized without increasing the price of electricity drastically. If all else were equal, breeder reactors would be preferable because they exploit 30 to 60 times more of the potential energy contained in the uranium. But the impact of the breeder on the safety question and on the issue of plutonium diversion should be very carefully assessed before their large-scale deployment can be supported, and the world uranium supply situation over at least the next 50 years is such that from this point of view haste is unnecessary.

Environment

How soon the use of fission energy or other forms will be limited by environmental factors is not clear. Climate on a hemispheric or global scale could conceivably be appreciably influenced by CO_2 and particulate matter from fossil fuel combustion early in the next century, and the heat from energy consumption in all forms could influence climate over regions of tens to hundreds of thousands of square kilometers by the year 2000. Understanding of global meteorological processes is inadequate to permit prediction of the level of energy use at which climatic disruption could become drastic, although some scientists have suggested that an increase of energy use by a factor of 50 might be too much. Use of solar energy, which already takes part in the global energy balance in its natural form, could alleviate the thermal limit to some degree.

Alternate Energy Sources

Wide differences of opinion exist on the potential of alternatives to nuclear fission, principally fusion, solar energy, geothermal energy, and cleaner use of fossil fuels. On the one hand, there is no obvious barrier to development of fusion, large-scale solar energy, or geothermal energy on a time scale of

20 to 50 years, and these sources evidently have many environmental advantages. On the other hand, the full difficulties and environmental liabilities of these approaches may remain to be discovered, whereas those of fission are already well known. The Working Group was in complete agreement that greatly expanded research programs on the potential alternatives to fission—and especially the apparently "clean" and thermally advantageous solar and geothermal possibilities—are urgently needed and fully justified.

Current Regulations and Practices Governing the Release of Radioactivity

A distinction must be made between routine releases of radioactivity in the everyday operations related to commercial nuclear power, on the one hand, and the much larger, unplanned and uncontrolled releases that could result from accidents, natural disasters, sabotage, or acts of war, on the other.

Routine Emissions

First, with respect to routine emissions at nuclear installations other than waste repositories (which are discussed below), it was agreed that it is technically feasible and desirable to hold the radiation exposure to members of the public below levels of one per cent or less of the average "natural background" radiation. Although some long-term genetic damage and induction of cancer presumably will be caused even at these low doses, the available data appear to justify the conclusion that the expected number of deaths from these routine emissions is significantly smaller than the expected number of deaths resulting from air pollution from fossil fuel plants generating the same amount of electricity. Such a comparison of risks of course entails the implicit assumption that the benefits of additional power generation by one means or another justify the costs. This assumption deserves separate examination, . . . as does the issue of alternative energy sources cleaner than present fossil fuel

technology or fission. Two further points must be emphasized: first, better data for comparing the health effects of fission and fossil fuels ought to be obtained, especially on the fossil fuel side—comparisons should be continually updated as new data become available; second, great technical and regulatory vigilance will be required to see that the theoretical *potential* for routine emissions far below 1 per cent of the natural background is actually achieved in practice around the world.

Radioactive Wastes

With respect to the management of long-lived radioactive wastes, strong uncertainties still exist. The principal difficulty is that the material remains highly toxic for periods measured in thousands of years; even over shorter spans, predictions about the stability and continuity of human society are impossible, and, over the longer term, significant geological change is possible in some circumstances. It is not surprising that even the "experts" cannot yet agree on what methods can guarantee the isolation of wastes over such periods, in spite of the existence of a variety of proposals (ranging from disposal in the earth's core to sending the wastes into space). Disposal in deep salt beds appears attractive, and is already being practiced in the Federal German Republic, but the viability of the method depends on the geological details of the particular salt deposit. Much more money could be spent on nuclear waste management than has yet been contemplated, without greatly increasing the price of nuclear generated electricity, but money alone does not guarantee a solution. It is impossible to be complacent about expansion of the use of nuclear power without having a solution actually in hand.

Major Releases

The possibility of a major release of radioactivity from a reactor or a fuel reprocessing plant, which theoretically could result from an accident, a natural disaster (for example, earthquake, tsunami), sabotage, or an act of war, has justifiably created deep concern. A major release could conceivably involve thousands of millions of curies of radioactivity, hundreds of thousands to millions of casualties if it occurred near a population center, and many thousands of millions of dollars in property damage. In theory the probability of such an event is quite small, but perhaps not small enough. This means that reactors should not be sited in earthquake or tsunami zones. Experts have published estimates of the probability of a major reactor accident (excluding sabotage and war, and perhaps some kinds of natural disasters) ranging from 10^{-4} to 10^{-12} per reactor year. If the latter number is true, we can be reassured, but if the former is true we cannot—faced with projections of 1,000 large reactors operating in 1990 and 3,000 in the year 2000. Better estimates of accident probability would be desirable. In the absence of agreement or proof, it is only prudent to assume the worst (the highest probability). Siting near large centers of population should be avoided. It is possible, although disputed, that siting reactors 100 meters or more underground in solid rock may provide further insurance against the consequences of major accidents. This possibility should be researched vigorously.

It is possible that fuel reprocessing plants may be more vulnerable to accidents than reactors. They are safer than reactors in that the potential for internal energy release is much smaller, but more dangerous in that some of the barriers of the reactor against release of fission products are no longer present (fuel element cladding, reactor vessel) and in that the fuel-reprocessing plant necessarily covers a much larger area.

The question of sabotage of nuclear reactors, waste shipments, or reprocessing plants generates especially grave concerns, because this possibility renders all the theoretical failure probabilities meaningless. This may be an additional reason to place reactors and reprocessing plants deep underground, if research confirms any real accident-containment advantages for this approach. Other measures against sabotage discussed by the Working Group included very careful guarding of the installations themselves, perhaps facilitated by clustering the various facilities at one location. Unfortunately, it is difficult to believe that even these measures can be 100 per cent effective.

Diversion of Fissionable Materials

The use of nuclear reactors to generate electricity necessarily involves the production and handling of fissionable materials of a character suitable for the manufacture of nuclear weapons. Light-water and heavy-water reactors yield 200 to 400 grams of plutonium per electric megawatt-year, net. Breeder reactors yield between 300 and 700 grams of plutonium per electric megawatt-year, net. High-temperature gas-cooled reactors of United States design employ fuel fully enriched in the fissile uranium isotope, uranium-235, and yield in the order of 50 grams of fissile uranium-233 net per electric megawatt-year. In each case mentioned, the materials require at most chemical separation (as opposed to more difficult and expensive isotopic separation) to be rendered usable for the production of weapons. It is to be emphasized that the high plutonium-240 content of plutonium produced in light-water reactors does not make bomb manufacture impossible, as is often supposed, but only complicates the design and makes the yield both smaller and less predictable.

There are two principal classes of possibilities for the misuse of the fissionable material from the nuclear fuel cycle. First, a government without nuclear weapons may divert material from its own nonmilitary nuclear program in order to produce weapons; second, material may be stolen from the nuclear power programs of either weapons or nonweapons nations by individuals or groups with a variety of motives—the production of a weapon for sabotage, terrorism, or blackmail, or the sale of the material to another group or even a nation intent on weapons production.

Relevant to the first class of possibilities is the Nonproliferation Treaty and the corresponding set of safeguards enforced by the International Atomic Energy Agency. Though the Working Group was in unanimous agreement that the International Atomic Energy Association is doing an excellent job of implementing safeguards within the limits imposed by the provisions of the Nonproliferation Treaty, there was some difference of opinion concerning the adequacy of those provisions and the possibility of improving them. Part of the Group felt that a nation wishing to divert material from its own reactor program might do so without detection by the safeguards permitted under the present Nonproliferation Treaty; these individuals felt that the Nonproliferation Treaty is primarily a formal expression of good intentions, and it is on these that we must rely. The majority felt that an attempt to further strengthen safeguards against governmental diversion should be made, within the framework of the present Nonproliferation Treaty, and that the International Atomic Energy Association and its safeguard activities should be strengthened and supported, as it is useful to reduce the *probability* of undetected diversion even if this cannot be made impossible. A serious gap in the Nonproliferation Treaty, of course, is that not all nations have signed and ratified it.

The possibility that material stolen from the reactor program of a weapons state could find its way into the hands of nonweapons governments represents a potential form of proliferation that seems not to be adequately deterred by the specific provisions of the Nonproliferation Treaty in its present form (although such proliferation is, of course, a direct violation of the intent and spirit of the treaty). The gap here is that the weapons states, which to date also have the largest commercial reactor programs, are not subject under the Nonproliferation Treaty to the same strict accounting procedures and safeguards as are nonweapons states. One would like to assume that considerations of internal security and sound economic and environmental practice have led the weapons states to establish strict control over their own reactor-related fissionable materials. Some members of the Working Group felt that this assumption is open to question, especially inasmuch as even a very small fractional loss of the fissionable material in the reactor program of, say, the United States or the Soviet Union, represents enough for several bombs. The Working Group was unable to reach agreement as to whether or not the Nonproliferation Treaty should ultimately be modified, or other measures instituted to provide the international community with assurance of strict national control of fissionable material in the reactors programmes of weapons states.

The problem of theft of nuclear material by internal groups or individuals intent on sabotage, terrorism, or blackmail was agreed to be a very serious one, although there was some sentiment expressed that the possibility of such activity was much smaller in the socialist states. In any case, the problem cannot be avoided simply by abandoning the breeder reactor, because, as noted above, all other reactor types also involve the use of materials suitable for weapons manufacture. It is difficult to see how the theft of such material can be made impossible in a world characterized by human failings, but measures to make such theft more difficult should be carefully studied and the best ones implemented as soon as possible. For example, relatively unsophisticated clandestine weapons manufacture might be deterred by maximizing the plutonium-240 content of reactor plutonium, adulterating material to be shipped with other neutron emitters or neutron poisons, or adulteration with hard gamma emitter to aid detection. (These measures could not stop determined and sophisticated groups.) Another possibility is to minimize shipments by concentrating enrichment, fuel fabrication, and fuel reprocessing facilities at single sites together with several reactors, and guarding the entire area with a degree of elaborateness and thoroughness hitherto reserved for strategic weapons.

The general view of the Working Group

regarding radiological terrorism using plutonium stolen from reactor programs is that
this problem is definitely of secondary importance to that of clandestine nuclear
weapons, in part because a variety of other
extremely toxic substances could be obtained by terrorists with less difficulty.

The as yet unsolved problem of radioactive waste management, and the possibly
unsolvable problems of catastrophic releases
of radioactivity or diversion of bomb-grade
material, combine to create grave misgivings
in the Working Group about the vast increase in the use of nuclear power that has
been widely forecast. Although it is evidently impossible to abandon nuclear fission altogether in the near future, we believe that
every effort must be made to minimize
reliance on fission by investigating and developing alternative, cleaner energy sources. In
the meantime, of course, every effort must
be made to minimize by technical and regulatory means the considerable hazards of
fission.

Breeder Reactors

None of the problems described in the foregoing sections is diminished significantly in
the case of breeder reactors, most especially
the liquid-metal fast-breeder now favored by
all countries, and some are significantly
aggravated. The relatively low fraction of
plutonium-240, in plutonium produced by
breeder reactors, in contrast to that from
light-water reactors, worsens the problems of
theft and diversion, as does the greater quantity of plutonium produced by the usual
breeder cycle. Additionally, some aspects of
the reactor safety problem would appear to
be compounded in the case of the breeder.
On the basis of available data, most Group
members felt that the breeder is *not* necessary in the next fifty years on grounds of
uranium supply and, therefore, that there is
no need to consider large-scale deployment
of such reactors unless and until the questions of diversion and safety are fully resolved. Some members, however, disagreed
with this view.

SUMMARY OF RECOMMENDATIONS

(1) Owing to potentially grave and as yet
unresolved problems related to waste management, diversion of fissionable material,
and major radioactivity releases arising from
accidents, natural disasters, sabotage, or acts
of war, the wisdom of a commitment to
nuclear fission as a principal energy source
for mankind must be seriously questioned at
the present time.

(2) Accordingly, research and development on alternative energy sources—particularly solar, geothermal, and fusion energy,
and cleaner technologies for fossil fuels—
should be greatly accelerated.

(3) Broadly based studies aimed at the
assessment of the relation between genuine
and sustainable energy needs, as opposed to
projected demands, are required.

(4) The hazards of fission should in the
meantime be minimized by every means
available, specifically:

(a) The greatest technical and regulatory vigilance to achieve the lowest
feasible routine emissions at all stages
of the fuel cycle, everywhere in the
world, including the establishment of
a worldwide network of radioactivity
monitoring stations.

(b) Accelerated efforts to find
technical solutions for the management of long-lived radioactive wastes.

(c) Thorough investigation of the
potential of placing underground or
clustering nuclear facilities as a means
of reducing the probability and/or
consequences of disruptions and accidents.

(d) Tightening of surveillance of
nuclear facilities.

(e) Exclusion of reactors and reprocessing plants from zones of high
seismic activity.

(f) Avoidance of siting reactors
and reprocessing plants in densely
populated regions.

(g) The safeguards authorities of
the International Atomic Energy
Agency should be strengthened and

supported, within the context of the present Nonproliferation Treaty; further studies should include research and development in order to raise the effectiveness of national and international safeguards to the highest possible level, and examination of whether the codification of uniform standards for national control of reactor-related fissile material in the weapons states would be useful.

(h) Research on technical means to render more difficult the use of reactor-related fissile materials for the construction of bombs.

(i) All nations should sign and ratify the Nonproliferation Treaty.

(5) Taking into account the fact that some problems associated with the breeder reactor are far from clear, large-scale deployment of breeder reactors should depend on the results of a thorough reexamination of these problems.

(6) Pugwash should examine the need for an International Energy Institute, or, alternatively, the need for changes in the scope and structure of existing international organizations, by reviewing the major efforts now under way in the energy field. We note with satisfaction that a Pugwash-sponsored symposium is to be convened for this purpose in the near future.

(7) The Working Group on energy at the 1974 Pugwash meeting at Baden, Austria, in 1974 should have on its agenda, in addition to the report of the symposium suggested above the two topics:

(a) International consequences of national energy demands.

(b) Energy use and well-being.

4

ALTERNATIVE ENERGY SOURCES

Scyllac, the Atomic Energy Commission's latest experiment in fusion. The goal of the Controlled Thermonuclear Research Program at the Los Alamos, New Mexico, Scientific Laboratory is to develop this process as a cheap, clean source of energy, particularly of electricity. (Photo courtesy of the Los Alamos Scientific Laboratory.)

The Solar Residence Laboratory, Colorado State University, Ft. Collins, Colorado. Top, a solar-heated, three-bedroom house, used as office space. Bottom, data-recording apparatus and the plumbing for solar hearing in the basement. (Photo courtesy of the Solar Energy Applications Laboratory.)

The Geysers Steam Field, Sonoma County, California. This is the largest geothermal power development in the world, generating 396,000 kilowatts (electric) from 10 power plant units. (Photo courtesy of C. C. Newton, Public Information Department, Pacific Gas and Electric Company.)

A wood sawing mill in the Zaanse Schans (zaan district). (Photo courtesy of the Netherlands National Tourist Office, New York; copyright VVV-Amsterdam.)

The great Canadian Oil Sands, Ltd., LMG bucketwheel excavator, the Athabasca Oil Sands, Fort McMurray, Alberta, Canada. This 4.6 million-dollar excavator can dig 54,000 tons of sand daily; currently, 50,000–55,000 barrels of oil are produced daily. (Photo courtesy of J. Perehinec, Public Relations Department, Great Canadian Oil Sands, Ltd.)

Feed lot of the Nebraska Feeding Company, Omaha, Nebraska, in 1968. About 6,000 cattle can be fed here at one time. Such feed lots may be prime locations for facilities that convert certain organic wastes, i.e., cattle manure, into methane gas to be used as a source of energy. (Photo courtesy of the U.S. Department of Agriculture, Office of Information; USDA photo.)

INTRODUCTION

Environmentalists who criticize current fossil- and fission-fuel energy systems frequently cite the promises of clean solar, fusion, geothermal, or wind energy as obvious, plentiful, and environmentally superior alternative sources of energy. It is certainly true that the promises of such energy alternatives outshine the realities of present energy sources just as fission energy promised to relegate coal-burning plants to the "ash heap of history" before the first reactors were built.

The answers to the questions "Why not solar?" or "Why not fusion?" or any of the other alternatives generally fall into one or both of the categories: economics or feasibility. Though the fuel for solar, geothermal, wind, and tidal energy is "free," the capital equipment and operating costs result in electricity prices that are generally higher than those from conventional sources. In such applications as solar heating and cooling, costs are beginning to become competitive with conventional fuels; they will certainly begin to cut into the market for these applications. The theme running through the discussion of most energy alternatives is this: *The alternative is presently somewhat more expensive than conventional fuels, but as the cost of such fuels rise and as mass production is applied to the alternative, it will surely become competitive.*

Such analyses undoubtedly have some validity and certainly justify continued or expanded research efforts aimed at making them commercially feasible. The technical feasibility of all the alternatives except fusion has been demonstrated, and only economics prevents their use on a large scale. Fusion, as a special case, has such tremendous potential as an unlimited source of energy that a continued large-scale effort is justified to verify the "proof of principle." The engineering problems of extracting fusion power are so awesome, however, that even the optimists do not expect fusion to make a significant impact on the energy picture for 30 to 50 years. Until such engineering problems can be better specified, the cost estimates of fusion energy appear highly speculative.

The scientific, economic, historical, and political forces shaping past energy policies and leading to the present priorities in energy research are extremely complex. Though certain energy alternatives have not received appropriate research funding support, we see the reasons for such an imbalance as more of a historical accident than any devious conspiracy to suppress "free" sources of energy in favor of "expensive" conventional fuels. Those seeing some sinister plot to keep solar energy from "seeing the light of day" might consider the old barroom slogan "If you're so smart, why aren't you rich?" In Part 5 we examine some of the historical influences that led to the promotion of fission energy—perhaps at the expense of research in other promising areas.

The role of economics in comparing energy alternatives with present energy sources is very difficult to unravel, particularly in a mixed, government-regulated energy industry. How does one take into account, for example, the enormous governmental research expenditures for the weapons programs that resulted in the civilian reactor program? Do present uranium prices correctly reflect actual costs involved in uranium processing and reprocessing? Does the price of coal really incorporate costs of reclaiming strip mined land? What level will the price of natural gas reach without government regulation? How can society determine the true, global costs associated with the burning of conventional fuels that will incorporate subtle environmental damage, unknown biological effects, future competitive uses, and other "fragile values"? All these questions affect the prices of current energy fuels, against which new energy costs must be measured. Generally, such present prices are artificially low owing to governmental support or regulation, or to the refusal of industry to pay the full environmental costs involved. Once the true costs of present fuels are known, all alternatives presented below will become more competitive.

Another important distinction should be made in studying energy alternatives—that between *new primary energy sources themselves* (for example, solar power and fusion) and *improvements in energy utilization and interconversion techniques* (for example, fuel cells and coal gasification). Both solar and fusion techniques promise virtually unlimited supplies of clean energy. The interconversion alternatives promise improvements in efficient utilization of present nonrenewable fuels and reduction of environmental damage associated with their use. This both prolongs the supply of such fuels and makes their use more socially acceptable.

In *The Quest for Fusion Power*, Lawrence M. Lidsky presents the basic plasma physics problems and incredible engineering difficulties that must be overcome before fusion can become an important source of energy. The "Lawson criterion," which must be satisfied before energy output exceeds energy input, is explained in detail. Several experimental approaches toward achievement of controlled thermonuclear fusion are described as are the accomplishments to date. Just as harnessing nuclear fission required much more sophisticated technology than coal, so will nuclear fusion require even more advanced technology than fission. The prospects of essentially unlimited energy supply (literally "oceans-full of fuel"), however, spurs continued research in this important area.

Next Gerald L. Kulcinski, in *Fusion Power—An Assessment of Its Potential Impact in the United States,* examines several possible effects that fusion power may have on the commercial energy market. He discusses in detail the detrimental environmental impact associated with fusion power. Though the supply of fusion fuel (deuterium and, in the first stage, lithium) is shown to present no problem, thermonuclear reactors, owing to rigid engineering specifications, will require huge amounts of such scarce metals as the superconductor niobium. The impact of such demands is difficult to predict, and it is encouraging that detailed engineering studies of both reactor feasibility and environmental impact are under way.

The prospects for utilizing energy from the sun are presented by Walter E. Morrow, Jr., in *Solar Energy: Its Time is Near.* Though the supply of solar energy is vast, its diffuse nature is seen as the main problem in tapping it for energy uses. The studies presented indicate that for heating and cooling applications, solar energy is already cheaper than electric energy and competitive with oil and gas at 1970 prices. For the production of electricity, solar energy, using solar collectors and conventional steam cycles, is still about four times more expensive than fossil-fuel-produced electricity. The much simpler and more direct method of generating electricity with solar cells, such as those used in the space program, is still about one-hundred times too expensive to be commercially competitive, though mass production would lower costs substantially.

Geothermal, tidal, and wind energy are three other renewable sources of energy that have been harnessed to a greater or lesser extent and have shown promise for alleviating energy shortages, at least on a local basis. The first, geothermal energy, is already under active commercial development and is described in *Geothermal Energy* by L. J. P. Muffler and D. E. White. The natural conditions giving rise to geothermal reservoirs are shown to be localized in geologically active regions. Though geothermal energy was produced as early as 1904, there is still no consensus among geologists of the potential impact it could have on the energy budget. Estimates presented here indicate it could supply up to 10 per cent of the electrical power demand.

Tides as a source of power, according to M. King Hubbert in *Tidal Power*, is much more limited in potential and actual development than geothermal energy owing to the limited geographical areas suitable for its production. An interesting installation at La Rance, France, is described. Tidal power, however, even if exploited completely, can never supply more than a small fraction of 1 per cent of the world's energy needs.

Finally, wind energy is boosted by William E. Heronemus in *Wind Power: A*

Near-term Solution to the Energy Crisis. This, the oldest form of inanimate energy known to man, gradually slipped into disuse, primarily because of the random nature of the wind. Professor Heronemus suggests that, with appropriate energy storage systems, wind generators could become competitive again, particularly in view of the rapidly escalating costs of fossil fuels and capital for fission reactors.

Next we turn to two sources of energy that may extend our supplies of petroleum considerably, but, until recently, have not been economically feasible. These are the oil shales, primarily those of the famous Green River Formation in the western United States and the oil sands from the Athabasca Formation of Alberta, Canada.

In *Oil Shale and the Energy Crisis,* Gerald U. Dinneen and Glenn L. Cook outline the nature of the Green River reserves and the two main techniques under investigation for extracting oil from shale. An interesting problem in both oil-shale and oil-sand mining is that the mine "tailings" take more volume than the original "ore." The disposal of such tailings raises serious environmental questions. Since the mining, crushing, and retorting of the shale constitute 60 per cent of the expense of extracting oil, *in situ* processing techniques are being given considerable attention. No oil shale is being mined commercially at present, but results of pilot projects in Colorado, Utah, and Wyoming will be used in the near future to decide on the optimum system for tapping this vast source of future oil. Early indications were that it would sell for about $4.00 per barrel, though some predict prices as high as $10.00 to $12.00 per barrel.

A. R. Allen tells of the struggle of his company, the Sun Oil Company, in *Coping with the Oil Sands.* This operation, the Great Canadian Oil Sands, now produces over 50,000 barrels of crude oil per day. Many of the problems unique to oil sands, such as handling vast quantities of this wet, sticky, quick-freezing "ore," however, have been met and apparently overcome. The complex nature of the oil-sand particles themselves, the large temperature range over which mining takes place, and the mining location (a muskeg swamp) all complicate this economically marginal operation. It is obvious after reading this article why "synthetic crude" has not made larger inroads on the oil market before now. But with rapidly rising foreign oil prices, the future of such operations seems assured, and two more ventures are planned in the near future.

Up to this point in Part 4 we have discussed new sources of energy—both renewable and nonrenewable. We now change direction and look at alternative methods of using present resources that are more efficient and environmentally acceptable. All of these techniques have been proved, in principle, and many are in operation, at least at the pilot-plant level. As public awareness of true environmental costs associate with burning the fossil fuels increases, pressures will build for switching to these new techniques. It should be emphasized, however, that all of these proposed energy interconversion technologies burn existing supplies of fossil or fission fuel—their only advantages are that they do it more efficiently or safely.

Since the United States is blessed with hundreds of years' supply of coal, the production of synthetic gas and oil from coal will become one of the most important future industries, even though our domestic production of natural gas and oil has already reached its peak. Harry Perry describes many of the established techniques for coal gasification and liquefaction in *Coal Conversion Technology*. A large number of pilot plants are now in operation, funded by both private coal, oil, and public utility companies and the federal government. It should be noted that the over-all thermal efficiencies of synthetic clean-burning, low-sulfur gas and oil range from 38 to 77 per cent. The price we pay, then, for using coal in a more environmentally acceptable way is that we use it much faster—another energy dilemma.

The synthetic gas produced from coal may be the basic fuel for fuel cells, described by Terri Aaronson in *The Black Box*. The remarkably high efficiencies and simplicity of operation make such systems extremely attractive, and they are widely used in the space program. Possible applications of fuel cells range all the way from central power-plant operations down to internal heart pacemakers. The weight-to-horsepower ratio of fuel cells is actually lower than that of the internal-combustion engine. If, through research, the requirements for noble metal catalysts can be reduced, we should be seeing much wider application of this clean energy conversion device in the near future.

Another very efficient energy conversion system, magnetohydrodynamics, is discussed by J. B. Dicks in *Magnetohydrodynamic Central Power: A Status Report*. Magnetohydrodynamic devices produce electricity directly by magnetically separating charges from a stream of hot plasma produced by burning conventional fossil fuels. Total thermal efficiencies reach 50 to 60 per cent and, when used as a topping cycle in conventional power plants, would greatly reduce present thermal pollution and extend the life of present fossil-fuel supplies. Other countries, particularly the Soviet Union, are investing large efforts in magnetohydrodynamics; the United States lags far behind in research in this promising area. Professor Dicks calls for an expanded effort in magnetohydrodynamic research.

Dr. Derek P. Gregory of the Institute of Gas Technology proposes a very interesting concept for modifying energy flow patterns in *The Hydrogen Economy*. Hydrogen is the ultimate "ecology fuel," the only by-product of which, upon burning with oxygen, is pure water, intrinsically valuable itself. Dr. Gregory visualizes hydrogen produced by electrolysis in fission reactors (located offshore for reasons of safety), the hydrogen being transported by presently existing gas pipelines to local industrial or domestic sites where it can be burned with very minimal environmental damage. Any source of energy may be used to produce the hydrogen, which then serves primarily as a storage and energy transport medium. Transporting energy via hydrogen pipelines is more efficient than high-tension electric cable transmission for distances greater than about 500 miles, and the visual pollution of unsightly overhead lines is eliminated. The hydrogen economy would even eliminate carbon dioxide pollution, which still occurs in the three previously discussed energy conversion schemes.

In *Power from Trash,* William Kasper offers partial answers to two major problems—solid waste management and energy shortages—describing several pilot projects and proved systems for recycling solid wastes. Municipal refuse contains approximately 53 per cent paper products by weight. Such refuse resembles low-sulfur, low-grade coal and may either be burned to produce electricity or recycled to produce "Garboil," which resembles fuel oil. Presently operational techniques are described for removing ferrous metals, aluminum, and glass, and these techniques, when combined with combustion, reduce the volume of the original garbage by 95 per cent. Several cities presently operate such recycling systems.

Though the energy alternatives presented do not completely exhaust the list of proposals that have been made, they certainly include the most promising alternatives and those with the greatest capacity for sizable energy production.

The Quest for Fusion Power

LAWRENCE M. LIDSKY   1972

The search for economical controlled fusion power is a scientific hunt for the Lost Dutchman Mine. Only a few true believers are *absolutely* certain that the goal exists, but the search takes place over interesting terrain and the rewards for success are overwhelming. In the case of fusion power, the potential long-term societal rewards are so enticing and the possibility of success so high that a major, truly international, research effort has developed over the last two decades. The United States has allocated over $400 million for research in controlled thermonuclear reactors to date and the Soviet Union more than twice that amount.

There seems little question that eventually either fusion energy or solar energy will be called upon to deliver the enormous quantities of "environmentally gentle" power that man will need. We should be able within five years to say whether significant amounts of energy can be obtained from controlled fusion within the next twenty to thirty years (before an irreversible commitment to an economy based on fission

breeder reactors) or whether fusion power development will have to wait much longer until the technological and economic considerations are even more favorable than they are at this moment.

My plan in the following pages is to enumerate the potentialities of controlled fusion power, to describe the physical conditions we have designated as significant milestones, to describe experiments now or soon to be in operation, and to discuss the engineering problems of controlled fusion—which may be far more difficult of solution than the physics problems. Then, finally, I shall describe a possible scheme for relaxing both the physics and engineering constraints of an economical system.

THE POTENTIAL OF FUSION POWER

The character of the atomic nucleus is such that the individual nuclear particles are most tightly bound in elements of intermediate atomic number. Thus, when we seek energy, we focus our attention on the more loosely

Dr. Lawrence M. Lidsky is an associate professor of Nuclear Engineering at the Massachusetts Institute of Technology. His current research interests center on experimental plasma physics and the emerging fields of fusion reactor engineering and fusion power economics.

From *Technology Review*, Vol. 74, No. 3, pp. 10–21, January, 1972. Reprinted with light editing and by permission of the author and *Technology Review*. Copyright 1972 by the Alumni Association of the Massachusetts Institute of Technology.

assembled elements, releasing energy by splitting (fissioning) the heavy isotopes and proposing to do so by joining (fusing) the lighter ones. There is less energy release per fusion reaction than there is per fission reaction, but the reactants are more plentiful and easier to handle.

In general, a particular fusion reaction is interesting if the power produced can be great enough to offset the power consumed in generating and maintaining the reacting medium, and if the relevant rates can be fast enough so that economically interesting regimes are accessible to modern technology.

There are, in fact, over thirty such reactions possible. The most interesting of the fusion reactions as possible routes to fusion energy are those which involve the heavy hydrogen isotopes deuterium (H_1^2 or D) and tritium (H_1^3 or T). These tend to have the largest fusion reaction probability (cross section) at the lowest energies. Deuterium is an abundant, naturally occurring isotope in wide use now as D_2O in heavy-water-moderated reactors. Tritium is a radioactive isotope with a 12.3-year half-life (it emits an electron and decays to stable helium-3) that does not occur in nature.

The deuterium (D-D) reaction chain is

$$D + D \rightarrow He^3 + n + 3.2 \text{ Mev}$$
$$D + D \rightarrow T + p + 4.0 \text{ Mev}$$
$$D + T \rightarrow He^4 + n + 17.6 \text{ Mev}$$
$$D + He^3 \rightarrow He^4 + p + 18.3 \text{ Mev}$$
$$\overline{6D \rightarrow 2 He^4 + 2p + 2n + 43.1 \text{ Mev}}$$

The first two equations represent the fact that the deuterium-deuterium reaction can follow either of two paths, producing tritium and one proton or helium-3 and one neutron, with equal probability. The products of the first two reactions form the fuel for the third and fourth reactions and are burned with additional deuterium. The net reaction consists of the conversion of six deuterium nuclei into two helium nuclei, two hydrogen nuclei, and two neutrons along with a net energy release of 43.1 million electron volts. The reaction products—

helium, hydrogen, and neutrons—are patently harmless (in contrast to some of the myriad fission products in a fission reactor), and the neutrons may in fact be used in a variety of ways. One very simple possibility uses their absorption in sodium to produce an additional 25 mega electron volts per cycle. Therefore, the deuterium-deuterium reaction produces at least 7 mega electron volts per deuterium atom (deuteron) and with absorption in sodium more than 10 mega electron volts per fuel atom.

The peak reaction rate coefficient of the deuterium-deuterium reaction is considerably less than that of the deuterium-tritium (D-T) reaction occurring within the deuterium-deuterium cycle; thus attention tends to focus on the latter. However, because tritium does not occur naturally, the reaction must be supplemented by one using lithium to reproduce the tritium fuel:

$$D + T \rightarrow He^4 + n + 17.6 \text{ Mev}$$
$$n + Li^6 \rightarrow He^4 + T + 4.8 \text{ Mev}$$
$$\overline{D + Li^6 \rightarrow 2He^4 + 22.4 \text{ Mev}}$$

This reaction is tritium regenerating and produces only helium as a reaction product.

The deuterium-tritium reactor is technologically more complicated than the deuterium-deuterium reactor because of the need to facilitate the second reaction (which takes place outside the plasma) and because very energetic neutrons must be slowed down to allow the reaction with lithium to take place. Nonetheless, the conditions needed to achieve net power output are much less demanding than for the deuterium-deuterium fuel reactor.

Deuterium occurs naturally in sea water in a ratio such that one out of every 6,500 hydrogen atoms in H_2O is in fact the heavy isotope. Isotopic separation is relatively straightforward because the mass difference between H_2O and HDO exceeds 5 per cent. At 7 million electron volts per deuterium atom, the energy attainable by fusion of all of the deuterium in a cubic meter of sea water is 12×10^{12} joules. This value is easier to visualize when compared to the energy content of a barrel of crude oil—6 × 10^9 joules. A cubic meter of sea water, then, corresponds to 2 × 10^3 barrels of oil and a cubic kilometer to 2,000 × 10^9 barrels. This last number is coincidentally nearly equal to recent estimates of the earth's total oil reserves. The oceanic volume is approximately 1.5 × 10^9 cubic kilometers, so there is more than one billion times the energy content of the world's oil reserve available to us through deuterium fusion. There are many interesting ways to illustrate the magnitude of this number (for example, it could support a world population of 7 × 10^9 people at 20 kilowatts per capita—ten times the current United States rate—for 3 × 10^9 years), but such exercises are, in fact, just exercises. It is far simpler and just as accurate to say that *fusion of deuterium represents an essentially inexhaustable supply of energy.*

The deuterium-tritium reaction will probably be exploited first, but its use will be limited by the availability of lithium. For such a light element, lithium turns out to be surprisingly rare. Recent estimates of lithium reserves place them at several times 10^7 metric tons, which limits the total energy release through the deuterium-tritium-lithium-6 cycle to approximately 5 × 10^{23} joules. The total energy content of the world's fossil fuel is 2.6 × 10^{23} joules.

It is easy to dispense with detailed computation of energy utilization trends here also. All reliable estimates of energy use conclude that we have sufficient fossil fuels to last for from 150 to 350 years. The lithium reserves will thus last for at least that long, and it is hard to believe that we will not have gained sufficient expertise in plasma physics during that time to enable us to tap the energy content of the deuterium-deuterium reaction.

FUSION HAZARDS

The end products of the fusion reaction—helium, hydrogen, and neutrons—can hardly be better chosen from the point of view of easing environmental pollution. It has even been proposed that the neutrons be used to clean the environment by transmuting certain particularly dangerous long-lived radioactive by-products of fission reactors to harmless stable nuclei. There still remain two environmental hazards to be considered, however. There is the tritium that occurs as an intermediate reactant in both the deuterium-deuterium and deuterium-tritium-lithium-6 cycles, and there is the problem of radiological activation of the fusion plant structure by neutron bombardment.

Tritium must be recycled in both fusion chains considered above and is by no means a waste product. Instead, economical design for a fusion plant demands that the inventory of tritium in the plant complex be kept as small as possible and be circulated through the plasma reaction zone as rapidly as possible. Several studies of the inventory problem have resulted in estimates of the total inventory for a 5000-megawatt thermal plant ranging from 2 to 10 kilograms. Most of the inventory is tritium dissolved in the circulating coolant and the metallic structure of the reactor. The larger value above corresponds to a radioactive burden in tritium of 10^8 curies, which is 40 per cent of the iodine-131 activity predicted for a similarly-sized fast breeder reactor. However, the inventories cannot be compared on so simple a basis. The maximum permissible concentration of tritium in the environment is 2 × 10^{-13} curies per cubic centimeter, whereas the allowable concentration of iodine is 10^{-16} curies per cubic centimeter. The relative biological hazard attributable to escape of the volatile inventory in a fusion reactor is nearly four orders of magnitude smaller than in a comparably sized fission reactor.

Because a single reactor is thus relatively safe from the standpoint of catastrophic tritium release, the tritium problem resolves itself to the minimization of leakage from all the reactors in a hypothetical fusion-power-based economy. For example, if all the world's population were using fusion-produced electricity at the current United States consumption rate (2 kilowatts per capita), then an inventory leakage of 0.01 per cent per year would contribute less than 0.1 per cent increase to the naturally occurring radioactive burden. This level of containment is easily achieved.

The other radiological hazard, activation of the reactor structure itself, is less easily quantifiable. The afterheat and residual activity depend critically on the construction materials. In the worst case studied to date, that of a niobium structure, the decay power is 10 per cent and the activity is 100 per cent of the power and after heat of comparable pressurized-water-reactor plants. In the best case studied to date, a vanadium structure, the decay power is 1 per cent and the activity only 0.1 per cent that of a comparable pressurized-water reactor plant. But, as in the case of tritium activity, a simple comparison is misleading; the activity in a fusion reactor is confined to the few activation products of a single material, whereas that of a fission reactor resides in a spectrum of fission products. Thus, it appears that the structural activity of a fusion plant will certainly be no greater than and probably much less than that of a comparable fission plant.

Rejected waste heat has more recently become appreciated as a significant environmental burden imposed by electrical power generation. The ratio of energy rejected to the environment to energy transmitted as electrical power is a sensitive function of the power plant efficiency. An increase in efficiency from 32 per cent (typical of existing fission reactors) to 48 per cent (typical of proposed fusion-reactor designs) would reduce the waste heat rejected to the environment by a factor of two. This difference exists in large measure because the fusion reactor, with separate reaction and heat removal zones, allows higher operating temperature than does the single-zone fission reactor. This return to the apparently primitive concept of "boiler and firebox" permits the design of liquid-metal-cooled systems with exit temperatures as high as 850° to 950°C.

Efficiencies much higher than 48 per cent

can be contemplated for some fuel cycles by direct conversion of the high-energy charged reaction products to electricity, but the deuterium-tritium reaction that will be the bulwark of the supply in the near future does not lend itself to direct conversion.

SCIENTIFIC FEASIBILITY

Fusion reactions can take place only when the nuclei of the fuel atoms are brought into close enough conjunction. The nuclei are, of course, positively charged and so repel each other. This repulsion is equivalent to an energy barrier that can be penetrated with reasonable efficiency only if the reacting nuclei have kinetic energy comparable to the barrier height. The level of kinetic energy required depends on the particular reaction and the desired reaction rate, but in general, plasmas of interest for use in controlled thermonuclear reactors have average energy per particle in excess of 5 kilo electron volts.

A collection of particles with average energy 5 kilo electron volts has an effective temperature of nearly 40 million degrees Kelvin (4×10^7°K). At these temperatures, the gas is completely dissociated into its constituent positively charged nuclei and free electrons. The electrical charge density is such that the behavior of the collection of particles is completely dominated by electrostatic and electromagnetic phenomena. Such a charge dominated collection of ionized matter is known as a plasma. Plasma at high temperatures cannot be confined by material walls but does respond to electromagnetic forces. The central problem in the search for controlled fusion energy has been the design of a magnetic field configuration that would allow containment of plasma in more or less stable equilibrium.

As in any gas of energetic particles, the ions and electrons in the plasma are continually undergoing collisions with each other. The collision rate is much higher than the fusion rate at reasonable energies, so the plasma particles in any magnetic trap undergo many collisions and rearrange their energy in a statistical fashion. Accordingly, there will be present in a magnetically confined plasma a spectrum of particle energies reaching up to very high values with a finite probability of fusion reactions occurring even in relatively low-temperature, low-density plasma. The occurrence of fusion reactions therefore furnishes no milestone in itself, and other more meaningful criteria must be sought. This is in contrast to the case of fission reactors, wherein the achievement of a multiplication coefficient greater than one is a clear dividing line between success and failure.

If the existence of fusion reactions in a contained plasma is not particularly noteworthy, the resultant neutrons do furnish a very useful means of describing the plasma energy distribution. These neutrons are surprisingly copious. For example, the Russian T-3 tokamak emits more than 10^{11} neutrons per second from its 500 electron volt plasma whereas the hotter, much denser Scylla IV plasma at Los Alamos Research Laboratory actually produces 40 to 50 watts of deuterium-deuterium fusion energy. If fueled with a deuterium-tritium mixture, this device would generate nearly 200 kilowatts of fusion power for a three-microsecond pulse period.

What plasma regimes must be reached to yield usable power? The so-called "Lawson criterion," one attempt to define the milestone, is developed by considering the energy balance per unit volume of plasma. (See Figure 4-1.) One assumes that the plasma is brought together at temperature T, held there in equilibrium for time τ, and then allowed to disperse. The energy of this expanding plasma is recovered. If the operation is considered to occur in cyclic fashion, the energy balance equation becomes

$$3nkT + W_r\tau = \eta\epsilon(W_n\tau + W_r\tau + 3nkT)$$

where n is the density, W_r is the radiation emission per unit volume, and W_n is the fusion energy generation rate per unit volume. If η is the efficiency, then ϵ is the fraction of the power recovered that is needed to achieve the steady state. If $\epsilon=1$,

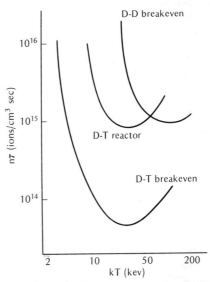

Figure 4-1. The numerical values for the "Lawson criterion," described in the text as the boundary at plasma regime, which might be expected to yield usable power, are here plotted for two values corresponding to a plasma system in which the thermal efficiency is 33 per cent and either all the energy is returned to the system ($\eta\epsilon = 0.33$) or only 10 per cent is utilized to maintain the plasma ($\eta\epsilon = 0.033$). From the plot it is clear that the minimum values of $n\tau$ are 4.5×10^{13} seconds per cubic centimeter for power break-even and 8.3×10^{14} seconds per cubic centimeter for a system of possible economic interest. (Data: Culham Lab. Report CLM-R85.)

then all the energy is required just to keep the reaction going. If ϵ is less than one, then some power is left over which presumably can be considered as output. The left side of this equation represents the energy needed to assemble the plasma ions and electrons [2 $\times$ 3/2(nkT)] and maintain the plasma in the face of radiation losses. The right side represents the energy recovered, assuming that all forms of energy are handled with the same efficiency. When the proper functional forms are inserted for the radiation and fusion rates, the equation can be solved in terms of the $n\tau$ product, as shown in the adjacent diagram. The lowest $n\tau$ system is not necessarily the lowest cost system and, in fact, economic studies of toroidal reactors predict optimum operation for those devices at temperatures much closer to 10 kilo electron volts than the 25 kilo electron volts, which corresponds to the $n\tau$ minimum.

Once the confinement geometry is determined, one can separate the terms in the $n\tau$ product by an appeal to engineering judgment. In steady-state devices the requirement that the over-all power density in the reactor be in the range of several watts per cubic centimeter leads to $n = 5 \times 10^{14}$ per cubic centimeter and $\tau = 0.5$ sec. But the fast "pinch" device would operate near $n = 5 \times 10^{16}$ per cubic centimeter, $\tau = 10^{-2}$ second. Yet another class of reactors, "stabilized mirrors," would necessarily operate far to the right of the minimum point; such reactors, if they are economical at all, will be so only if they utilize direct conversion.

There is clearly not a single goal whose attainment guarantees success but rather a set of goals corresponding to different confinement schemes. It is the object of controlled fusion researchers to predict on experimental and theoretical grounds which, if any, path is likely to reach its individual goal.

Success from the standpoint of plasma conditions alone is not sufficient, because the question of whether the particular confinement scheme is compatible with economic and engineering requirements remains to be answered. However, it is useful to remember as a touchstone that reasonable goals for a steady-state reactor are: plasma density $n = 5 \times 10^{14}$ per cubic centimeter; pulse duration $\tau = 0.5$ second; and plasma temperature $T_i = 8$ to 10 kilo electron volts. The experiments now being carried out are not aimed at these values, but rather at producing plasma conditions that allow all relevant interactions.

EXPERIMENTS IN PLASMA CONFINEMENT

The experiments now in operation or just being completed are probably the last exploratory plasma *physics* devices that will be constructed for some years. Experiments

now entering the design phase will be much more ambitious; indeed, they will aim at proof of "scientific feasibility" by generating plasma conditions so similar to those of a prototype reactor that the results can be extrapolated with reasonable confidence. The three concepts with the greatest potential are the tokamak, the stabilized mirror, and the theta pinch. These three devices are clearly differentiated from each other with respect to the dominant plasma processes, and they all scale differently with variations in plasma density, temperature, and, most importantly, with the size of the plasma.

There are many ways proposed to tap controlled fusion energy in addition to the ones described here. Most-discussed among these is laser-induced fusion, which requires no confinement whatever. In the simplest form of laser-induced fusion, a focused, very energetic laser beam is brought to bear on a small deuterium-tritium fuel pellet. If the laser pulse is energetic enough and the energy is delivered in a short enough time, the pellet can be heated to fusion temperatures. The fusion energy is released while the particle is in the process of rapid, uncontrolled expansion. The energy release is explosive, and there is a limit to the maximum amount of energy that can be released in a single pulse. When this limiting value is compared to the capital cost of the laser required to ignite the pellet, it soon appears that this system is further from economic feasibility than from scientific feasibility. Therefore, although this technique might be useful in some aspects of fusion research, it does not seem promising from the standpoint of power production.*

Tokamaks

The T-3 device in operation at the Kurchatov Institute of Technology since 1962 is the

---

*Editors' note: A sizable section of the fusion community believes that laser-induced fusion may be practicable. Recent laser experiments indicating the production of thermonuclear neutrons have provided encouragement for workers in this area (*Time*, May 27, 1974, p. 98).

prototypical example of toroidal confinement. The magnetic field lines in such closed geometries are constrained to follow toroidal surfaces, and the plasma particles (to first approximation) spiral along the field lines. But simple toroidal fields cannot confine a plasma in equilibrium, and one stabilizing scheme or another must be employed. The tokamaks, of which T-3 is an example, supply equilibrium by means of a large circulating current induced around the torus. This current also serves to heat the plasma by resistive (I^2R) heating.

The other commonly explored toroidal devices supply the required equilibrium by means of externally imposed twisted multipole magnetic fields. This class is represented by the stellarator design developed at the Princeton Plasma Physics Laboratory. Although one theoretically expects many equivalences between tokamaks and stellarators, the tokamaks to date have yielded far better results.

The induced plasma current in the tokamak generates a magnetic field that loops the minor axis of the torus; the field lines form helices along the toroidal surface, and the plasma must cross the lines to escape. It does so through the cumulative action of many random displacements caused by interparticle collisions, in effect, diffusing across the field lines and out of the system. Thermal energy is transported by much the same process.

Particle orbits in toroidal fields are exceedingly complicated. Because of the spatially varying field and the acceleration experienced in moving along the curved field lines, the particles drift away from and return to the original magnetic field lines. These excursions are quite large and the particle and energy diffusion are enhanced by large factors. The principal goal of the tokamak program has been to measure particle and energy confinement times under many conditions and to predict how these relationships may change as the size of the torus is increased. It is also apparent that the simple resistive heating of T-3 will not scale up to allow ignition of a toroidal reactor,

and another major goal of the tokamak program is the development of an additional heating scheme.

Stabilized Mirrors

If some way can be found to stop the charged particles of a plasma in their motion along the magnetic field lines, the magnetic field lines need not be closed within the plasma region and many problems associated with toroidal systems can be avoided. The "magnetic mirror" is based on a basically simple phenomenon: charged particles will reflect from a region of increasing magnetic field if their encounter with the field lines is close enough to perpendicular, and they will not be reflected if they are moving nearly parallel to the lines of force. "Mirror" devices take advantage of this by trapping the plasma in a bulge of the field, where the lines are spread. When the field increases, the lines reapproach each other and a reflecting region is formed. (See Figure 4-2.)

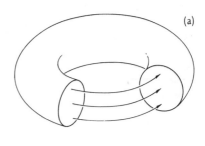

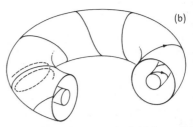

Figure 4-2. The magnetic field lines of the simple torus (a) are complicated when a current-carrying plasma is added (b). The twisted field lines result in very complex particle orbits; a typical trajectory is shown by the dashed line.

The 2X mirror experiment at the Lawrence Radiation Laboratory uses a carefully shaped set of pulsed magnetic field coils. The plasma is injected and then trapped by these coils when the magnetic field is weak; it is subsequently heated by compression as the field grows with time. The containment region has the typical stabilized mirror shape and a volume of about 16 liters; the resulting plasma has density $n = 5 \times 10^{13}$ per cubic centimeters, ion temperature $T_i = 8$ kilo electron volts, and confinement time of 0.2 to 0.4 milliseconds. This is a very interesting hot, dense plasma with confinement time

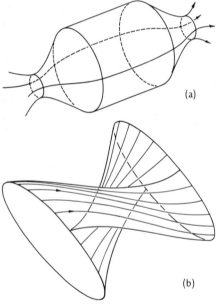

Figure 4-3. The top diagram (a) shows the basic concept of the "mirror" device: plasma is kept within the central region because the particles are repelled by the two areas of relatively greater magnetic field at the ends. In other words, where magnetic lines of force crowd together, a plasma-reflecting "mirror" is formed, and a pair of these mirrors contain the plasma. However, plasma researchers soon found that the shape of the magnetic field around the central region—that is, bulging outward—caused the plasma to be unstable. The answer, now generally accepted, is to make the magnetic field lines bulge inward toward the plasma, as in the second diagram (b), which shows the central region of one type of "stabilized mirror" device.

only a factor of 3 to 5 below the theoretically expected value. (See Figure 4-3.)

But the success of a laboratory-scale experiment does not guarantee success of a reactor concept based upon that experiment. There are very strong theoretical predictions that the mirror reactor will suffer increasingly severe plasma instabilities as the device is made larger. One ordinarily could tolerate a moderate level of instability or turbulence in a plasma because the effect is simply to increase to some extent the rate of particle loss from the trap. Unfortunately, the mirror reactor has very little margin for error.

Particles escaping from a toroidal reactor must move across magnetic field lines; but in a mirror device, it is required only that a collision send a particle in a direction nearly aligned with the magnetic field in order for it to escape. This can happen with reasonable probability in a single collision. The mirror device is thus inherently more leaky than the various closed-line systems, and this leakage manifests itself in particles streaming out the ends of the trap. If this escaping particle stream could be recaptured with high efficiency, the energy could be reinjected with minimal effect, and the development of electrostatic direct conversion was an effort to accomplish this purpose. Economic estimates show that this is sensible if the reactor performs up to its theoretical limit, but that even a small increase in the particle loss rate would doom the concept. The 2X experimental results, though not conclusive, appear to show that the mirror reactor does not perform quite up to its theoretical limit.

Theta Pinch: Scylla IV

The theta pinch experiments have been exploring yet another regime of plasma parameters—very high density. The object in these experiments is the extremely rapid compression of an existing low-density, low-temperature plasma. If the heating pulse is fast enough, there can follow a two-step process: the plasma is heated first by the shock wave generated by the rapidly rising field and then heated further by adiabatic compression as the field continues to grow more slowly in time.

The technology needed to perform these experiments is impressive indeed. Capacitors storing nearly a megajoule of energy at 50 kilovolts are discharged through hundreds of parallel paths (to reduce the inductance) into a massive single-turn compression coil to generate fields in excess of 100 kilogauss. In the Scylla-IV experiment, for example, the current rises to 8.6 million amperes in only 3.7 microseconds. The resulting plasma has density $n = 5 \times 10^{16}$ per cubic centimeter and ion temperature $T_i = 3.2$ kilo electron volts. The plasma lifetime is very short, because the plasma particles simply stream out the ends of the device, though the plasma is in stable equilibrium occupying a thin cylinder in the center of the coil during its residence there.

Plasmas achieved in the theta pinch process are so hot and dense that the kinetic pressure is several hundreds of atmospheres and the plasma energy density very nearly equal to the magnetic field energy density. Plasma behavior under these conditions is not yet theoretically understood, and we are very dependent on experimental observations.

It is obvious that the way to avoid end losses in a theta-pinch device is to close the ends upon themselves—to generate a toroidal theta pinch. This immediately reintroduces the toroidal loss-of-equilibrium problem, but there are indications that this dilemma may be resolved through a proper combination of external fields which interact strongly with the high-density theta-pinch plasma.

The Atomic Energy Commission is funding a very large research program in toroidal theta-pinch apparatus (Scyllac) at the Los Alamos Laboratory. Because the experiment is so expensive, and because the theory is not very highly developed, an initial experiment is underway to provide "proof of principle." A torus has been constructed and operated at reduced power. In this configuration, the stored energy is 3.5 megajoules and the peak current is 34.5 million amperes. Preliminary results (November, 1974) indicate that the external fields can, in fact, supply at least some degree of toroidal equilibrium, but full equilibrium has not been achieved.

The toroidal theta pinch may well provide the first proof that the appropriate "Lawson criterion" can, in fact, be achieved, but this method is possibly incapable of being scaled up to an economical power-producing reactor. The economic difficulty arises because of the great capital cost of the requisite energy storage and because the pinch tends to produce power in bursts, with relatively long periods of time between bursts for cooling, pumping out impurities, and recharging the capacitor banks.

EXPERIMENTAL SUMMARY

Table 4-1 summarizes the plasma conditions achieved in the experiments described above and their comparison with the "Lawson criterion" values. Several points are worthy of particular note. First, it is obvious that, at least in terms of density and temperature, it is *now* possible to reach the lower fringes of the plasma regimes of thermonuclear interest. Certainly we are close enough to these regimes so that almost all of the dangerous instabilities have manifested themselves. The Scylla-IV containment time looks disconcertingly small, but this is inherent to the experiment and does not denote instability. In any event, the proper "Lawson criterion" for such pinch-like devices calls for confinement time of only 2 to 10 milliseconds.

The various routes to proof of scientific feasibility are clear. The tokamak concept must be tested at large radius to see if the plasma loss scales inversely with the square of the plasma radius. It if does, then the development of an appropriate energy input scheme would lead to the goal. Several very large tokamaks have been proposed, and many heating methods are being tried on smaller scale experiments.

As for the theta pinch, it must be closed upon itself. If external coils for feedback can provide equilibrium, then nothing further

Table 4-1. Experimental Plasma Conditions versus the "Lawson Criterion."[a]

	Tokamak (T-3)	Mirror (2X)	Theta Pinch (Scylla IV)	"Lawson Criterion"
Plasma density (number per cubic centimeter)	2×10^{13}	5×10^{13}	5×10^{16}	5×10^{14}
Maximum plasma temperature T_1 (kiloelectron volts)	0.6	8	3.2	10
Confinement time τ (milliseconds)	25	0.4	0.01	200

[a]The fusion process is such that there is no sharp dividing line in a fusion reactor—as there is in a fission reactor—between "go" and "no go." Instead, there is simply the question of achieving a high enough steady-state fusion reaction rate to provide a sufficient excess of energy for power generation. This condition is roughly expressed by the so-called "Lawson criterion." As the chart shows, it is now possible—at least in terms of density and temperature—to reach what the author calls plasmas at "the lower fringes of thermonuclear interest." He suggests that five more years of research will reveal whether the goal is in fact attainable soon.

will be needed. The experiment now in progress at Los Alamos will go a long way to answering this question.

Mirror traps will have to be shown to be less prone to instability and enhanced losses than is currently predicted by theory. Experience shows that the opposite is usually true—that plasmas are, in fact, more unstable than expected. However, a mirror reactor would have a much larger ratio of plasma pressure to magnetic field pressure than the experiments carried out to date, and this regime is not amenable to theoretical analysis. Proponents of mirror-reactor systems argue that appropriate experiments should be carried out in this regime. Such experiments would be extremely expensive, and there can be no certainty that they will be funded in the near future.

For both the tokamak and the toroidal pinch, five years should be sufficient time either to demonstrate scientific feasibility or to find that unsuspected obstacles stand in our way. The distance remaining to the goal is relatively short, so the unexpected has a small domain in which to lurk; the chances for success are therefore relatively good.

ENGINEERING PROBLEMS OF THE CONTROLLED THERMONUCLEAR REACTOR

Even if and when we discover how the "Lawson criterion" can be fulfilled, there will remain the task of engineering large-scale, economic systems. And this is an issue of which the resolution is by no means certain.

We already know that in conventional power plants the greatest economic advantage can be obtained when materials are used near the limit of their capabilities. This will be true particularly of advanced power-producing schemes such as fast-breeder reactors and controlled fusion reactors, because highly stressed mechanical elements allow the designer to relax requirements on the reactor core. In the case of the fusion reactor, for example, higher heat loads at the interface between plasma and container would permit a lower ratio of container-to-plasma volume and thereby lower the unit capital cost of the plant.

There are important problem areas common to fission and fusion reactors. These include radiation damage, limiting heat fluxes and temperatures, and induced radioactive afterheat. The first of these will apparently be far more severe in fusion reactors; the other two will be comparatively less severe than in fission plants. (Other engineering problems of the fusion reactor that are specific to the particular design—tokamak, for example—will be ignored here because they are somewhat less fundamental and appear solvable by extensions of existing techniques.)

Many engineering features are surprisingly insensitive to details of the nature of the plasma. Consider, for a concrete example, a steady-state, deuterium-tritium-fueled toroidal reactor, and recall that 80 per cent of the energy release is in 14.1 million electron volt neutrons. Elementary economic analyses of steady-state reactors show that the magnetic field must be generated by superconducting magnetic coils, because the power loss in normally conducting materials is prohibitively high. On this basis one arrives very quickly at a conceptual model of the main features of the cylindrical blanket that surrounds the plasma column.

The shell immediately surrounding the plasma must be of a refractory material to withstand the enormous flux of neutrons and short-wavelength electromagnetic radiation. This first wall must be cooled very well because—although most of the energy passes through this wall—an appreciable quantity of energy (the amount depending on the thickness of the wall) will be deposited within it and must be removed. On the other hand, a well-chosen material will generate appreciable amounts of additional neutrons, and the final design thickness will be a compromise between neutron production and allowable thermal loads. The shell surrounding this one—the first wall coolant—will be devoted primarily to heat removal from the first wall, but proper material choice will allow neutron and tritium generation in this region. The third and subsequent shells will moderate the neutron flux, remove the neutron energy in the form of heat, generate tritium

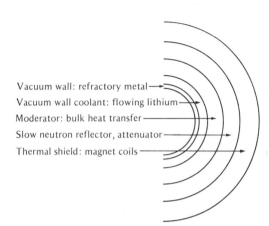

Vacuum wall: refractory metal
Vacuum wall coolant: flowing lithium
Moderator: bulk heat transfer
Slow neutron reflector, attenuator
Thermal shield: magnet coils

Figure 4-4. Cross-sectional structure of a hypothetical fusion power generator. Around the plasma there must be coils to create the magnetic field which contains the plasma, coolant fluids to remove the heat generated, and radiation shielding. In the lithium (in this particular scheme), a nuclear reaction generates tritium and neutrons, which are needed by the plasma in order to maintain the fusion reaction.

by the lithium-6 absorption reaction, and shield the magnetic field coils from the neutron and gamma ray flux. (See Figure 4-4.)

The bulk of the energy removal from the blanket occurs in the thermalization and tritium generation region, but it is not as highly stressed as the first wall because of the cylindrical geometry of the system. The radial size of the blanket is determined by the need to use with high efficiency every neutron emitted from the plasma, because at least one triton must be generated for each neutron emitted. Almost all blanket designs call for a radial thickness of 1 to 1.5 meters. This requirement impinges, through economic considerations, on the plasma physics, because the plasma radial size must be comparable to the blanket thickness or the power density of the system will be far too low.

Another strong constraint on the blanket thickness will be the necessity for using superconducting coils to generate the intense magnetic fields required. The coils operate near $4.2°K$, but the energy deposited in them by neutrons and gamma radiation must be rejected at the temperature of the environment. When the thermodynamic Carnot factor and refrigeration efficiencies are considered, it becomes clear that only 0.0001 of the total reactor power can be allowed to reach the magnetic field coils. Thus, it appears impossible to use a blanket much smaller in radial extent than 1.5 meters.

Several groups (most notably at Oak Ridge National Laboratory and the United Kingdom's Culham Laboratory) have published the results of careful studies of conceptually reasonable blankets.

A. Fraas and D. Steiner of the Oak Ridge National Laboratory have shown that published intensive studies show that the problem of material damage by radiation will be more severe for fusion reactors than for fission breeder reactors. The radiation damage effects, particularly swelling, are the limiting design considerations in breeder reactors. They may not be quite so important in fusion reactors because fission-reactor cores are characterized by a large number of narrow cooling passages, whereas the blankets of fusion reactors can be designed with much larger coolant passages and mechanical tolerances. Although the swelling is certainly a major concern, there is a paucity of information regarding the effect of very high doses of 14 million electron volt neutrons and no experimental facility is presently capable of measuring the effect. A key piece of engineering design data is simply unknown.

There is yet another problem inherent in the blanket design. The ideal material for use as heat transfer medium for the first wall is liquid lithium, because of its low vapor pressure at high operating temperatures and because lithium in this location takes full advantage of the high-energy neutrons for the lithium-7 reaction. However, this means that an electrically conducting fluid must be moved through the steady magnetic field. One consequence of this is that considerable power must be available for pumping the coolant, and the cooling passages must be designed to minimize cross-field flow. A

possibly more important consequence arises because the magnetic field suppresses the turbulence in the fluid and so reduces the heat transfer coefficient. Experiments are now in progress to measure the magnitude of this effect, but the analysis is not complete.

To summarize: many severe engineering problems, radiation damage chief among them, will limit the allowable heat flux at the blanket-plasma interface. The heat flux will determine the capital cost per unit of power, and low heat flux might possibly raise the cost to economically uninteresting levels. It may well take longer to determine the allowable engineering parameters than it will take to prove the "scientific feasibility" of fusion power. It is in belated realization of this that the Atomic Energy Commission is now starting large-scale funding of engineering research in problems of controlled thermonuclear reactors.

FISSION-FUSION SYMBIOSIS

Those who attempt to assess the prospects of generating appreciable fusion power in the near future generally assume that the deuterium-tritium-lithium-6 cycle will be used. Recall that this cycle demands that the tritium be regenerated by various reactions with lithium in the blanket. Engineering studies of such tritium regenerating blankets have been carried out at many laboratories, and all of them show that it is a relatively simple matter to breed more tritium than is needed. A typical calculation, for example, shows that 1.3 tritons are generated for each neutron incident upon the blanket. In other words, the cycle becomes

$$D + Li^6 \rightarrow 2\,He^4 + 22.4\,Mev + 0.3n$$

The neutron excess offers many intriguing possibilities. Of course, some of the excess neutrons will be needed to generate more tritium than is consumed in the reactor to allow the start up of additional fusion plants. However, the reaction rate per unit of power is very high, and a neutron excess will clearly be available. This raises the intriguing question of how to exploit the resulting change from a "neutron-poor" to a "neutron-rich" economy.

One of the more provocative possibilities involves the absorption of the excess neutrons in either thorium-232 or uranium-238 to yield the fissionable isotopes uranium-233 or plutonium-239. Each of the fissionable nuclei that might be produced would represent at least 200 million electron volts when burned in a fission reactor and could result in a net yield of more than 1,000 million electron volts in a reactor with reasonably high conversion efficiency. Thus, the excess neutrons from a fusion reactor could in fact represent far more energy than is produced by fusion reactions in the core itself. Indeed, a new "generalized Lawson criterion" can be defined in terms of this symbiotic fission-fusion scheme. Fusion reactors that would not be economically interesting by themselves become economically viable in the symbiotic system. From the viewpoint of plasma physics, the $n\tau$ curves are shifted to the lower values of the product.

In general, the technological characteristics of fast fission reactors and conceptually reasonable fusion reactors form a complementary set. Fission reactors produce low-cost power at high power density, but they can be made to breed only at the price of compromising either safety or economics. Fusion reactors may well have low power density and concomitant high costs, but because of the neutron excess, they will breed new fuel at a prodigious rate. A combined system in which fuel is bred primarily in the fusion reactor and power generated primarily in the fission reactor achieves properties attainable by neither alone—economical, safe power generation with almost arbitrarily short doubling times. Furthermore, the loosening of design constraints would allow other benefits. For example, to ease problems of thermal pollution, the fission reactor could be optimized for safe high-temperature operation, and any resulting penalty in neutron economy could be compensated for in the fusion portion of the system. In terms of such over-all considerations, the distance

between "scientific feasibility" and "economic feasibility" is materially reduced.

FINDING THE LOST DUTCHMAN MINE

The focus of the search for controlled fusion power has changed during the past several years. We have learned to confine plasmas at high temperature and high density for long periods of time. The densities and temperatures have nearly attained values of reactor interest, and energy confinement times have been stretched by many orders of magnitude from the several microseconds typical of the early, unstable devices to several tens of milliseconds in tokamaks.

The question of whether we achieve fusion power in the near future is now cast into quantitative rather than qualitative terms. The full evaluation will depend at least as much on engineering limitations of the system as it will on the plasma physics, but there is some hope for relaxing these constraints in combined fission-fusion systems.

In terms of the original metaphor, we think we have found the Lost Dutchman Mine. The hoped-for proof of "scientific feasibility" will be needed to demonstrate its existence conclusively, and we are still unsure if exploiting the mine will make economic sense. Though this is the subject of intense physics and engineering research—and far from solution—it seems highly unlikely that so glorious a resource as the vast oceans full of clean-burning deuterium will remain unavailable for use.

SUGGESTED READINGS

Gough, William C., and B. J. Eastlund, "Prospects of Fusion Power," *Scientific American,* February, pp. 50–64.

Post, R. F., and F. L. Ribe, "Fusion Reactors as Future Energy Sources," *Science 186:* 397–407, November 1, 1974.

Postma, Herman, "Engineering and Environmental Aspects of Fusion Power Reactors," *Nuclear News,* April, pp. 57–62.

Rose, David J. "Controlled Nuclear Fusion: Status and Outlook," *Science 172:* 797–808, May 21, 1971.

Fusion Power

An Assessment of Its Potential Impact in the United States

GERALD L. KULCINSKI   1974

We are now hearing less of the phrase "... *if* thermonuclear fusion can be controlled ..." and more and more scientific papers contain the statement "... *when* thermonuclear fusion power is controlled. ..." Granted that such optimism is justified, what are the implications for society in the twenty-first century? Will such a source of power be cheaper, cleaner, safer and environmentally more acceptable than the more conventional fossil fuels or the relatively new fission fuels? This article attempts to address some of the above questions, at least with respect to how they might be answered in the United States. Such an assessment, by its nature, requires a great deal of speculation on the course that the world will take in the next thirty years. The author claims no special faculty for predicting that course and therefore cautions the reader that the scenario painted in this article represents a perspective from a single vantage point. Nevertheless, it can be projected, on the basis of past experience, and what is now known about plasma physics, what type of impact fusion power might make if we pro-

ceed in the direction that we are presently heading.*

The history of controlled thermonuclear research is not a very old one. It started in the early 1950's as classified research and was declassified in 1958. The initial opti-

*This analysis will take as its basic premise that fusion feasibility can be demonstrated, that we can successfully build electrical generating stations powered by fusion reactions and that these power plants will be economically competitive. For technical details of fusion power the reader is referred to: D. J. Rose and M. Clark, Jr., *Plasma and Controlled Fusion* (New York, M.I.T. Press, Wiley, 1961); R. F. Post, *Annual Review of the Nuclear Society*, Vol. 20, p. 509 (1970); Proceedings of the British Nuclear Energy Society, Conference on Nuclear Fusion Reactors, held at *UKAFA* Culham Laboratory, September 17–19, 1969; Proceedings of the International Working Sessions on Fusion Reactor Technology, June 28–July 2, 1971, CONF-710624; Proceedings of the Texas Symposium on the Technology of Controlled Thermonuclear Fusion Experiments and Engineering Aspects of Fusion Reactors, Austin, Texas, November 1972, CONF-721111; International Conference on Nuclear Solutions to World Energy Problems, American Nuclear Society, Washington, D.C., November 13–17, 1972. See papers by R. Hancox, p. 209, R. L. Hirsch, p. 216, F. L. Ribe, p. 226, G. L. Kulcinski, p. 240, A. P. Fraas, p. 261.

Dr. Gerald L. Kulcinski is a professor of nuclear engineering and director of the Fusion Feasibility Study Project at the University of Wisconsin-Madison.

From *Energy Policy Journal*, Vol. 2, No. 2, pp. 104–125, June 1974. Reprinted by permission of the author and the *Energy Policy Journal* published by the IPC Science and Technology Press Ltd.

mism for producing large amounts of low-cost electrical power was shattered in the early 1960's when several formidable problems of plasma stability and confinement were encountered. For the next five to ten years scientists went back to "write the book" for a new field of science called plasma physics. The interest of all but the most dedicated tended to subside during this time while another form of nuclear energy, that released from the fissioning of uranium and plutonium, was developed into a commercial reality. Major advances by Russian and United States scientists in the late 1960's rekindled those early dreams and by the early 1970's the quest for fusion power was joined not only by plasma physicists but by engineers, economists, and environmentalists as well.

Despite progress made in the past few years, the basic problem of controlled thermonuclear reactors still remains, that is, to successfully contain a gas of charged particles (plasma) at ~100,000,000°K long enough to release a favorable amount of energy. This containment must be accomplished while keeping the plasma isolated from the solid structural components.

It was known from the earliest days of plasma physics that certain isotopes of hydrogen could be joined, or fused together, with a tremendous release of energy. Table 4-2 lists three of the most popular reactions of deuterium (D), tritium (T) and helium-3 (He-3). It should be noted that the deuterium-tritium reaction requires the lowest energy temperature (~10 kilo electron volts) for

reactor operation and that it releases almost 1,800 times as much energy as it takes to initiate it! The other reactions listed in Table 4-2, deuterium-deuterium and deuterium-helium-3, have certain advantages but they require higher ignition temperatures and return proportionally less of the energy required for their initiation. Therefore, the deuterium-tritium reaction is the most favorable and will be assumed to the fuel for the "fire" in the controlled thermonuclear reactors considered here.

Six major points of the deuterium-tritium reaction should be noted:

(1) The products of the deuterium-tritium reaction are *not* radioactive and therefore present no long-term disposal problems.

(2) Tritium *is* radioactive and does not occur in nature; it must be bred in the reactor by neutron reactions with lithium. The real fuel for the deuterium-tritium system is deuterium and lithium, *not* deuterium and tritium.

(3) The absorption of neutrons by structural components *will* produce radioactive isotopes which will have to be disposed of after the plant is closed down. The nature of these isotopes is somewhat different from that of fission reactors and represents a smaller, but not completely negligible problem.

(4) Most of the energy (80 per cent) from the reaction is carried away by the neutron. This kinetic energy must be converted to heat, and the heat to electricity.

(5) The energy of the fusion neutrons is about 14 million electron volts compared to about 2 million electron volts for fission neutrons. The higher-energy neutrons cause more and somewhat different types of damage in structural materials than is found in fission reactors. They also require more massive structures for complete thermalization and removal of the neutrons.

(6) Approximately four times as many neutrons must be produced per unit of energy in fusion reactors than in fission reactors. Roughly speaking, 20 million electron volts are released per fusion neutron, whereas 80

Table 4-2. Potential Fusion Reactions for Controlled Thermonuclear Reactors.

Reaction	$\dfrac{energy}{reaction}$ Kilo electron volts	
	Input[a]	Released
$D + T \rightarrow {}^4He + n$	10	17,600
$D + D \big\langle \begin{smallmatrix} {}^3He + n \\ T + p \end{smallmatrix}$	50	~3,300
$D + {}^3He \rightarrow {}^4He + p$	100	18,300

[a]For reactor-grade plasma.

million electron volts are released per fission neutron.

The confinement of the deuterium and tritium ions can be accomplished in two ways: by magnetic fields, or by the inertial confinement of the atoms themselves. The magnetic approach is the most deeply studied and understood process while the latter has only become of interest in the last five to ten years with the advent of laser-induced fusion.[1]

Simply stated, a magnetic field can contain a high-temperature mixture of deuterium and tritium ions in space because charged particles are constrained to rotate around the magnetic field lines. The magnetic fields are shaped so that the majority of charged particles are confined at distances of 50 to 100 cm from the nearest solid member of the reactor. The laser systems rely on a very different approach. A solid (frozen) pellet of deuterium and tritium atoms is simultaneously irradiated from several directions with high-intensity laser beams. The particle is rapidly heated to ignition, and enough reactions take place before the pellet flies apart to result in a net energy output. The confinement here is then due mainly to the inertia of the fuel atoms.[1]

How long does the plasma need to be confined to produce more power than required to heat and contain it? Lawson[2] has found that for the deuterium-tritium system operating above the ignition temperature (T_i) at moderate efficiencies (~33 per cent), the product of the ion density (n) and confinement time (τ) must be

$$n\tau \sim 10^{14} \quad \text{seconds per cubic}$$
$$\text{centimeter} \quad \text{at } T > T_i.$$

The problem is currently to get n, τ, and T to the proper values so that the power output may equal the power input. Some devices currently operate at high plasma

1. J. Nuckolls, John Emmett, and Lowell Wood, *Physics Today*, August 1973, p. 46.
2. J. D. Lawson, *Proceedings of the Physical Society*, Vol. 70, p. 6 (1957).

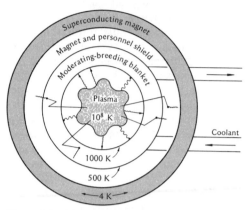

Figure 4-5. Schematic of controlled thermonuclear reactor based on the deuterium-tritium cycle.

densities and temperatures and have relatively short confinement times; other devices work at long confinement times but at insufficient plasma densities or ion temperatures, and so forth.

How then does one go about building a reactor once the plasma has been contained? The steps are outlined in Figure 4-5 for a magnetically confined deuterium-tritium plasma. First, one starts with a reacting plasma which is emitting energy in the form of neutrons, charged particles and various forms of photons. The next step is to surround the plasma with a solid wall which absorbs the charged particles and photons as well as providing a vacuum for the plasma to ignite in a magnetically confined system. This wall will absorb about 20 per cent of the energy from the plasma and must be cooled. Typical diameters of this first wall will be 5 to 10 meters for toroidal reactors such as the tokamak.*

A third step is to surround the vacuum wall with a moderator to slow down the

*The tokamak (to—toroidal, ka—chamber, mak—magnetic) reactor consists of a metal torus surrounded by a large transformer that induces an axial current to flow through and heat the plasma. The torus is surrounded by solenoidal coils which produce a second magnetic field, stronger than that developed as a result of the plasma current. Thus, the combined fields constitute a helical magnetic field which prevents leakage.

neutrons, a reflector to reduce the leakage of neutrons and a coolant to carry the heat away. This region should also contain a tritium breeding material so that the deuterium-tritium reaction can be continued. One material that would satisfy all three of the above requirements is lithium. However, other moderators such as beryllium, graphite, or even iron could be used and other coolants such as liquid metals or helium would be satisfactory. Approximately 1 meter of blanket and first wall is required to absorb about 97 per cent of the heat produced from the plasma. Unfortunately, some neutrons and gamma rays will escape, and the magnets (or lasers) must be protected from these sources of irradiation. This protection is accomplished by surrounding the blanket with a shield that completes the moderation of those neutrons that escape, and absorbs the gamma rays emitted from the blanket. This shield also serves as final radiation protection for personnel in the plant, and might consist of various combinations of iron and lead to absorb the gamma rays and boron to absorb the neutrons.

Outside the shield will be located the magnets (or laser), fueling equipment, heat exchangers, tritium-removal devices, and other equipment associated with the operation of the plant. It is expected that the sheer size of a fusion plant may be a problem from the standpoint of materials, construction, and cost because the average energy density of a deuterium-tritium reactor is one to two orders of magnitude smaller than in fission reactor cases.

UNITED STATES PLAN FOR COMMERCIAL FUSION REACTORS

A recent study conducted by the Atomic Energy Commission has revealed how the United States hopes to demonstrate commercial fusion reactors by the year 2000. The plans are summarized in Figure 4-6.[3] The first step is obviously to demonstrate that more power can be obtained from a plasma than goes into producing it; therefore feasibility experiments are required.

Research in this area is currently being conducted for the United States government at four major laboratories*: Princeton

3. R. L. Hirsch, *Proceedings of the International Conference on Nuclear Solutions to World Energy Problems,* American Nuclear Society, Washington D.C., 13–17 November 1972, p. 216.
*And a private industrial research laboratory, General Atomics.

Figure 4-6. Bar chart indicating phasing of major steps in a fusion-reactor development program.

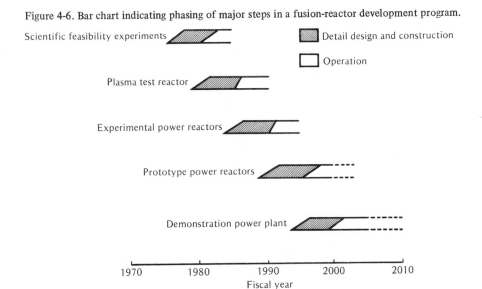

Plasma Physics Laboratory, Los Alamos Scientific Laboratory, Oak Ridge National Laboratory, and Lawrence Livermore Laboratory. Several smaller, but equally important projects are being conducted at the University of Wisconsin, University of Texas, and Massachusetts Institute of Technology. Each of these laboratories is studying one or more specific approaches to fusion.

After feasibility of the fusion had been demonstrated in the late 1970's or early 1980's, a plasma test reactor would be built to test the validity of the plasma scaling laws. The next step would be to build a reactor demonstrating the breeding of tritium and test materials under severe radiation damage conditions. Such a facility would be built around 1990. In the mid-1990's a prototype reactor of several hundred megawatts electric would be built that would probably produce electricity, although not economically. Finally, it is hoped that a reactor would be built around the year 2000 that would be in the 1,000-megawatt electric range and demonstrate that fusion power can compete economically with other sources of power.

The money required to achieve the rather ambitious program outlined above has not been officially established but estimates of $5000 to $6000 million between 1974 and 2000 are probably reasonable. Recent estimates[4] by the Atomic Energy Commission show that approximately $500 million have been spent on fusion research up to 1974 and that another $1.35 billion would be required for the fiscal years 1975 to 1979.

PENETRATION OF FUSION REACTORS INTO THE UNITED STATES ELECTRICAL GENERATING MARKET

Perhaps the best way to estimate this penetration factor is to use the method already established or predicted for fission reactors.[5]

4. D. L. Ray, "The Nation's Energy Future," WASH-1281, December, 1973.
5. "Nuclear Power 1973-2000," WASH-1139(72), 1 December, 1972.

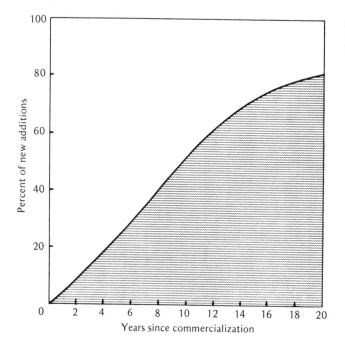

Figure 4-7. Projected penetration of fusion reactor into the United States electrical generating market.

Figure 4-7 shows the percentage of the installed electrical generating plants anticipated to be nuclear in the time period from 1966 to 1986. Note that twenty years after the introduction of the first "commercial" fission reactors, over 80 per cent of the new additions will be nuclear plants. We will assume that fusion reactors will have the same penetration factor as fission plants.

One could transfer these figures to the case of fusion power by first noting what the projected mix of electrical generating stations might be in the years 2000 to 2020. We have chosen to use the Associated Universities Report[6] to estimate these figures. Table 4-3 shows that by the year 2000 fission reactors are expected to account for about 51 per cent of United States electrical capacity, fossil-fueled plants about 45 per cent and hydroelectric about 4 per cent. The corresponding figures for the year 2020 are

6. "Reference Energy Systems and Resource Data for Use in the Assessment of Energy Technologies," AET-8, April, 1972.

60 per cent nuclear fission reactors, about 38 per cent fossil-fueled plants, and about 2 per cent hydroelectric. This information is plotted in Figure 4-8 and labeled Plan A. We shall now proceed to see how fusion might modify those figures assuming the total demand for electricity does not change.

First (Plan B in Table 4-3 and Figure 4-8), both fossil-fueled and fission-reactor capacity could be reduced proportionately by the fraction of electrical generating capacity taken by the fusion reactors. Obviously, very little effect would be seen in the year 2000, but by 2010, fusion could account for almost 10 per cent of the total generating capacity, fission reactors 49 per cent, fossil-fueled plants 38 per cent, and hydroelectric 3 per cent. By 2020, the controlled thermonuclear reactor share could rise to 29 per cent, whereas fission reactors supply 40 per cent, fossil-fueled plants 38 per cent, and hydroelectric 2 per cent.

It is entirely possible that by the year 2000 the installation of controlled thermonuclear reactors may be predominantly in

Table 4-3. Projected Mix of Electrical Generating Units (10^3 megawatts electric).[6]

Year	1977	2000	2010	2020	Year	1977	2000	2010	2020
A. Base plan					B. With fusion affecting fossil and fission system proportionate to their new additions.				
Hydro	53	72	79	86		53	72	79	86
Gas turbine and internal combustion	36	86	113	142		36	86	110	118
Gas-steam	81	90	88	86		81	90	88	86
Oil-steam	61	112	89	62		61	112	89	62
Coal-steam	203	447	757	1,070		203	446	669	762
Light-water reactor	90	407	432	443		90	407	425	429
Liquid-metal fast-breeder reactor	0	422	926	1,739		0	422	782	1,022
Controlled thermonuclear reactor						0	~1	242	1,063
Total	524	1,636	2,484	3,628		524	1,636	2,484	3,628
C. With fusion affecting fossil plants only					D. With fusion affecting fission plants only				
Hydro	53	72	79	86		53	72	79	86
Gas turbine and internal combustion	36	86	100	37		36	86	113	142
Gas-steam	81	90	88	32		81	90	88	86
Oil-steam	61	112	89	33		61	112	89	62
Coal-steam	203	446	528	195		203	446	757	1,070
Light-water reactor	90	407	432	443		90	407	421	421
Liquid-metal fast-breeder reactor	0	422	926	1,739		0	422	695	698
Controlled thermonuclear reactor	0	~1	242	1,063		0	~1	242	1,063
Total	524	1,636	2,484	3,628		524	1,636	2,484	3,628

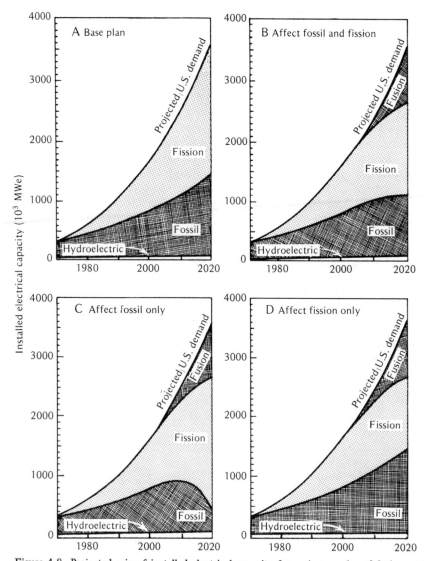

Figure 4-8. Projected mix of installed electrical capacity for various modes of fusion power introduction.

the place of fossil-fueled plants, both from the standpoint of air pollution and conservation of fossil fuels for other types of energy (for example, automobiles, home heating, chemicals). If fusion plants were to be built entirely at the expense of fossil plants (Plan C in Table 4-3 and Figure 4-8), we would see that by the year 2020, 89 per cent of the power would be generated by nuclear reac-

tors (both fusion and fission), 9 per cent by fossil fuels and 2 per cent by hydroelectric. If Plan C were to take place, the fossil-fuel requirements for electrical generating capacity by the year 2020 would be about 78 per cent of that anticipated for 1977 and 92 per cent of that required in 1973.

The final option considered here, Plan D, assumes that fusion will enter the electric-generating market at the expense of fission plants only. All the previous figures are the same until the year 2010, where now 55 per cent of the power is generated by nuclear reactions, 42 per cent by fossil fuels, and 3 per cent by hydroelectric sources. By the year 2020, 60 per cent of the total electrical generating capacity will be nuclear, with 49 per cent of that supplied by fusion and 51 per cent by fission.

The most important point to draw from this simple exercise is not the exact figures but the realization that even *if* we can achieve controlled nuclear fusion by 1980, and even *if* we can successfully bring it to commercialization by the year 2000, and even *if* we assume a rather optimistic penetration rate compared with other energy sources, fusion power will not account for more than 30 per cent of the generating capacity in the United States in the year 2020, some fifty years from now. On the more positive side, replacing the fossil-fueled plants with controlled thermonuclear reactors would reduce the demand for those sources of energy in the year 2020 to approximately the present levels of consumption. Without fusion, the United States demand on fossil fuels in the year 2020 for generation of electricity will be as much as four times our present rate.

EFFECT OF FUSION ON ENVIRONMENT IN THE EARLY TWENTY-FIRST CENTURY

Fuel Requirements and Reserves

The earlier analysis revealed that the basic fuels for a deuterium-tritium-based fusion economy are deuterium and lithium. The procurement of deuterium from the oceans, where it occurs as one part in every 6,500 parts of hydrogen, should be relatively easy, and the water can be returned virtually unchanged to the oceans. The present cost of deuterium is 20¢ per gram, and this amount, when burned with tritium, would contribute only 6×10^{-3} mill* per kilowatt-hour to the

*1 mill = 0.001 United States dollars.

electrical generating costs—an almost completely negligible cost. The amount of deuterium available is truly enormous and there is essentially no danger of ever running out of it. For example, if we were to extract the energy content from only 1 per cent of all the deuterium available in the oceans, it would supply the total energy needs of the world in the year 2000 for almost 100 million years.[7]

It requires roughly 80 grams of deuterium (in combination with tritium) to provide 1 megawatt-year electric of power assuming a 40 per cent efficiency for converting heat into electricity. This means that in the year 2010, controlled thermonuclear reactors would require 19 tons of deuterium per year. This would most probably come from heavy water (D_2O) plants so that 96 tons of heavy water would be required per year in 2010. This number rises to 425 tons of D_2O in the year 2020. Such production rates are not uncommon even today. Canada's Bruce plant[8] currently produces 800 tons of heavy water per year so that it is conceivable that *one* plant could provide the entire deuterium requirements for the United States controlled thermonuclear reactor program until the year 2020.

A slightly more limiting feature of the deuterium-tritium fuel cycle is the availability of lithium. It has been estimated that "known and inferred" United States reserves of lithium at 2¢ per gram is 6×10^6 tons,[9] and 80 per cent of this amount exists in heavy brines in Nevada and Canada. The energy content of natural lithium is 25 megawatt-hours thermal per gram,[9] and therefore the 6×10^6 tons of lithium could produce almost 6×10^9 megawatt-years electric years of energy. Such an energy supply could furnish the total United States electrical generating capacity for over 10,000 years at the 1977 rate. If that much

7. W. C. Gough and B. J. Eastlund, *Scientific American*, Vol. 224, No. 2, p. 50 (1971).
8. E. La Surf, Chalk River Laboratory, Canada, private communication.
9. J. P. Holdren, UCID-15953, December 8 (1971).

energy is not sufficient, we could always extract lithium from sea water. It has been estimated that this amount of lithium could supply the total world electric generating capacity in 1977 for almost 300,000 years.[9]

Calculation of the total lithium requirements for fuel for our projected fusion economy reveals that the minimum yearly requirement for lithium is 255 tons in the year 2010 and about 1,100 tons in 2020.

It is also possible that future controlled thermonuclear reactors will use lithium, either in elemental form or in combination with other elements, as a coolant. We shall see later that typically, 1,000 kilograms of lithium is required per megawatt electric for cooling. Hence, about 10^6 tons of lithium would be required for this purpose until the year 2020—more than 100 times the cumulative amount required for producing tritium alone. However, even this amount is well within the known reserves of just one deposit in Silver Peak, Nevada.[10]

It is not sufficient merely to state that we have the reserves but we should also be sure that there is no environmental problem resulting from extraction of those reserves from the earth such as occurs in strip mining for coal. The volume of ore (or most probably brine) which must be processed can be estimated by noting that the lithium concentration in its ores is about 5 per cent and in brines it is present at 40 to 300 parts per million.[10] Hence, if the lithium is obtained from brine, 5×10^9 gallons of liquid must be processed and reinserted into the ground, to produce the lithium required up to the year 2020. All this lithium could come from one 72 square mile deposit in Nevada without any appreciable upset in the environment. If the lithium is processed from ore and if the amount of gangue (for tunnels or overload) is three times the volume of ore removed, then approximately 40×10^6 cubic yards of earth must be disturbed. Assuming a mining depth for lithium of 10

Table 4-4. Cumulative Land Required to Provide Fuels for Electrical Generation.

Strip Mining of Coal	Square Miles[a]			
	1977	2000	2010	2020
Plan A[b]	100	7,060	11,500	18,300
Plan B	100	7,060	10,600	15,100
Plan C	100	7,060	10,100	12,400
Plan D	100	7,060	11,500	18,300
Lithium[b]	–	~0	0.35	1.5

[a]Cumulative from 1969.
[b]To a depth of 30 feet, lithium for cooling.

yards, this means that 1.5 square miles must be dug up to provide lithium for the cooling of the controlled thermonuclear reactors.

The significance of this last comparison can be more fully appreciated by noting how much land must be strip mined to produce coal for the coal fired electrical generating plants. Depending on adoption of Plan A, B, or C, Table 4-4 shows that anywhere from 12,400 to 18,300 square miles of land could be used up in the production of coal by strip mining between 1968 and 2020. The higher figure is equal in area to the states of Vermont and New Hampshire. Note particularly that almost half the strip mining for coal will be done before the year 2000, when fusion might first be introduced assuming optimistic major scientific and technological advances, as well as generous funding. Nevertheless, it can also be seen that the introduction of fusion at that time could save almost 6,000 square miles of land from strip mining by the year 2020. It is worthwhile to emphasize here that we are just talking about fuel procurement and we will treat structural metals in more detail later.*

10. D. A. Brobst and W. P. Pratt, eds., "United States Mineral Resources," Geol. Survey Paper 820, Washington, U.S. Government Printing Office, 1973.

*In summary, fuel availability is truly unlimited for fusion reactors and the procurement of this fuel would present a negligible impact on the environment. If lithium is used as a coolant as well as a breeder, then one might be limited to several hundred years of deuterium-tritium fusion reactors if we rely on land based reserves and several thousands of years if the lithium is extracted from the ocean.

Construction Materials Requirements

Resource Requirements*

The situation with respect to the procurement of structural materials for controlled thermonuclear reactors is not so optimistic as for the procurement of fuels. The problem stems from two facts. First, more exotic and less abundant elements are required to harness fusion as opposed to those required in fossil-fueled and fission plants. Higher operating temperatures favor the use of refractory metals (V, Nb, or Mo) or at the very least, high temperature, high strength iron alloys. If the controlled thermonuclear reactor uses magnetic confinement of the plasma, the iron alloys should not be magnetic, which, in turn requires significant amounts of alloying elements such as niobium or manganese to stabilize the austenitic (nonmagnetic) phase of iron. Elements such as chromium are also required for corrosion resistance. The magnets themselves require enormous amounts of low-electric-resistivity metals such as copper or aluminum and significant amounts of superconducting materials. Currently, Nb-Ti alloys, Nb-Sn, or V-Ga intermetallic compounds are the most suitable materials for the superconducting coils. Another problem arises if any scarce elements are used in the controlled thermonuclear reactor coolants. For example, lithium has already been shown to be a suitable coolant, and sodium will also be adequate. However, if more complex salts such as $Li_2 BeF_4$ are used for coolants (for reasons not outlined here), then the availability of beryllium and

fluorine becomes quite important.

The second major reason for large materials demands of controlled thermonuclear reactors is their sheer size, which is partly due to the fact that the power generation density of fusion plants (1–10 megawatt thermal per cubic meter of plasma) is much lower than that of fission plants (100 megawatt thermal per cubic meter of core). The higher energy and larger number of neutrons also requires increased shielding to protect vital reactor components as well as personnel. This shielding is accomplished most often with boron, which absorbs neutrons after they have been slowed down, and heavy elements such as lead, which absorb gamma rays emitted from elements made radioactive on the absorption of neutrons.

Before we proceed further in this comparison and assess the materials requirements of controlled thermonuclear reactors, we must list the current estimates of various materials required in controlled thermonuclear reactors. Table 4-5 lists the range of structural materials required per electric megawatt for four systems analysed thus far.[11-14] Representative values for tokamak reactors were taken from studies at Oak Ridge National Laboratory,[11] Princeton Plasma Physics Laboratory,[12] and the University of Wisconsin.[13] The Los Alamos Scientific Laboratory study[14] of pulsed systems was also included. Such a consideration of many systems necessarily produces a wide range of values because of different design philosophies and the fact that some systems are constructed with niobium-zirconium alloys while others prefer stainless steels. Nevertheless, using the worst case for all the materials ensures that we will not underestimate the materials requirement regardless of which system eventually proves to be the most economic in electricity generation.

*The elements are identified as they occur in the text of the following sections:

V, vanadium	Be, beryllium
Nb, niobium	F, fluorine
Mo, molybdenum	Zr, zirconium
Cr, chromium	Al, aluminum
Ni, nickel	B, boron
Mn, manganese	Cu, copper
Ti, titanium	Fe, iron
Sn, tin	K, potassium
Ga, gallium	He, helium
Na, sodium	Pb, lead

11. A. P. Fraas, USAEC Report ORNL-TM-3096, May 1973.
12. R. P. Mills, private communication.
13. "Tokamak Reactor Design," Vol. I, UWFDM-68, Nov. 1973, Vol. II, Aug. 1974.
14. S. C. Burnett, W. R. Ellis, T. A. Oliphant, and F. L. Ribe, LA-DC-72-234A, 1972.

Table 4-5. Estimated Maximum Materials Requirements for the "Nuclear Island" Portion of Future Fusion Reactors Which Rely on a Magnetically Confined Deuterium-Tritium Reaction.

Element	Metric Ton Required per Installed Megawatt electric	Millions of Metric Tons		
		Required for 1,063 × 10³ Megawatts electric	U.S. Reserves[e]	World Reserves[e]
Aluminium	0.6[d]	0.6	13	3,000
Boron	0.8[c]	0.8	33	66
Beryllium	0.12[b]	0.1	0.018[h]	0.38
Graphite	2[a]	2	10	large
Chromium	2[c]	2	~0	370
Copper	2[d]	2	74	310
Fluorine	1[b]	1	9	62
Iron	10[c]	11	8,500[h]	180,000[h]
Helium	0.3[d]	0.3	1.2	1.2
Potassium	0.02[a]	0.02	42	30,000
Lithium	0.95[c]	1	6	180
Manganese	0.2[c]	0.2	0	590
Molybdenum	0.2[c]	0.2	2.5[h]	6.5[h]
Niobuim	0.8[d]	0.8	0.005[h]	7.8[h]
Nickel	1.5[c]	1.6	~0.14[h]	24[h]
Lead	11[c]	12	39	85[h]
Tin	0.2[b]	0.2	0.009	6[h]
Titanium	0.8[a]	0.9	23[f]	134[f]
Vanadium	0.5[ag]	0.5	0.1	26[h]
Zirconium	0.002[a]	0.002	0.06[h]	25

[a]Ref. 11.
[b]Ref. 12.
[c]Ref. 13.
[d]Ref. 14.

[e]Ref. 15 and except where noted at present prices.
[f]Ref. 10.
[g]Use V as substitute for Nb structure.
[h]Ref. 16.

Table 4-5 shows that the major metallic structural requirements for controlled thermonuclear reactors range from 0.002 to 12 tons per megawatt electric of installed capacity. Translating these figures into the amount of finished product required to provide 1063 × 10³ megawatts electric in the year 2020 reveals that the order of 0.002 million to several million tons of some elements would be required by that time. The estimated United States reserves of these elements is also listed in Table 4-5 along with comparable numbers for word reserves. [15,16] This information can be readily classified into three categories based on *current* re-serves, (that is, no allowance for depletion between now and 2020):

A. Those elements for which there appear to be abundant United States reserves which could be used to meet controlled thermonuclear reactor demands (Al, B, Cu, Fe, K, Mo, Ti, and Zr).

B. Those elements which could be supplied by United States resources barring any great demand on these elements by other products before 2020 (F, graphite, He, and Pb).

C. Those elements which could not be supplied by the United States and reliance on foreign markets would be expected (Be, Mn, Cr, Ni, Sn, Ti, Nb, and V). However, these elements might not be able to be supplied even by world resources if other uses are found for the elements between now and the year 2020.

15. First Annual Report of the Secretary of the Interior under the Mining and Mineral Policy Act of 1970 (P.L.91-631), March, 1972.
16. E. N. Cameron, University of Wisconsin Report, UWFDM-68, Vol. 2 (1974).

In Category B, the exhaustion of helium from underground sources does not mean that the helium would be lost; it will only be dispersed in the atmosphere and would have to be extracted at perhaps 100 times its present cost.[17] The use of lead as a biological shield in controlled thermonuclear reactors could present problems if no future supplies were found. For example, approximately 1,200,000 tons of lead were consumed in the United States in 1970, an amount equal to 100 controlled thermonuclear reactors at 1,000 megawatts electric rating. However, it is anticipated that large amounts of lead could be processed at costs substantially above today's prices thus alleviating the problem of resources (but not economics).[16]

A more serious problem exists with those elements in Category C. Three of the elements (Cr, Ni, Mn) are essential ingredients in austenitic (nonmagnetic) stainless steel which, even if not used as the structural material of the controlled thermonuclear reactor blanket, would probably be used as structural support and reinforcing for the magnets because of the low temperature strength and ductility of such steel. In the Wisconsin design[13] the blanket structural metal is less than 5 per cent of the total steel required in the reactor. Tin would only be necessary if superconducting magnets capable of producing more than 125 kilogauss were required. Niobium may be required in all fusion reactors which rely on superconducting magnets. Currently niobium-titanium alloys are the most promising superconducting materials for large magnets and hence could cause a large demand on the niobium reserves. It is estimated that about 0.3 ton per electric megawatt of niobium would be required in the magnet of a controlled thermonuclear reactor, regardless of what structural material was used in the blanket. If niobium were used as a blanket structure material then an additional 0.5 ton per megawatt electric would be required.

17. G. L. Kulcinski, University of Wisconsin Report, UWFDM-83, August, 1973.

The use of vanadium as a substitute for niobium structural materials could also be made if the situation warrants this, but vanadium is also a metal which would have to be imported.

The projected average annual United States requirement for niobium in controlled thermonuclear reactors for the period 2000 to 2020 is over three times the world consumption in 1970 and over 17 times that used in the United States during 1970!

Procurement of manganese, a vital component of not only austenitic stainless steels but also for ferritic structural steel, may be a severe problem. The requirement of 200,000 tons of manganese to build the projected controlled thermonuclear reactors up to the year 2020 is only a small fraction of the total world supply. However, it is almost certain that a large amount of this element will be used for other purposes in the next 50 years. It has been estimated[15] that the United States alone will use 10 per cent of the world's known reserves of Mn between now and the year 2000 and probably another 10 per cent of that value in the period 2000 to 2020 for non-fusion purposes. If the rest of the world continues to use seven times as much manganese as the United States does, then there could be a serious shortage of manganese in the early twenty-first century with obvious implications for fields other than fusion.

Finally, the use of beryllium either as a neutron multiplier (because of its high n,2n cross section) or as a component of a lithium salt for cooling must be very closely examined. There are modest tonnage requirements for some controlled thermonuclear reactors (120,000 tons of beryllium for 10^6 megawatts electric) but the reserves are quite limited, especially in the United States. For example, the above amount of beryllium corresponds to about 300 times the 1970 United States consumption rate and five times the total projected cumulative United States consumption between 1970 and 2000. Beryllium is not essential to most fusion reactors and it could be deleted without seriously affecting the breeding ratio. Pure lithium could also be used in place of Li_2 BeF_4 salts. Therefore, unless large reserves of beryllium are found in the world, it is expected that beryllium will see only limited use in controlled thermonuclear reactors.

In summary, the use of refractory metals niobium or vanadium for high temperature structural materials in controlled thermonuclear reactors means that the United States will be essentially dependent on foreign sources for these elements. This dependence also will follow from the use of niobium-titanium or Nb_3Sn superconducting magnet materials. The unique requirement of non-magnetic low temperature high strength stainless steels will also place large demands on foreign suppliers for chromium, niobium, and manganese. It appears to be very difficult to build a fusion economy on a system that uses large amounts of beryllium. On the brighter side, the United States has adequate reserves of iron, copper, aluminum, boron, potassium, manganese, and zirconium to meet the projected needs.

Land Despoilment

There will be a certain amount of land despoilment due to the procurement of the

elements listed in Table 4-5. This will also occur for other types of power sources but it is worth demonstrating the level of disruption in view of similar effects for the procurement of fossil fuel. Table 4-6 lists the amount of ore which must be mined to get the controlled thermonuclear reactor nuclear island materials for the period 2000 to 2020. Using the assumptions given in Table 4-6, about 60 square miles (160 square kilometers) need to be disrupted over the twenty-year period prior to 2020 to provide the necessary ore. To put this figure in perspective, it is a factor of about 70 less than the land required *only* to provide strip-mined coal for the equivalent electrical generating capacity.

Three of the required metals stand out, which necessitate the largest amount of ore to be processed per plant. These are in increasing order: nickel, copper, lead. Obviously, if all other things were equal, one would like to substitute other elements with higher ore yields for these materials.

Balance of Payments Problems

One consequence of relying on foreign sources for raw materials lies with the balance of payments problem created by the

Table 4-6. Nonfuel Ore Requirement for Projected Controlled Thermonuclear Reactors from A.D. 2000 to 2020.

Element	For 1063 × 10³ megawatts electric 10⁶ tons	Approx. Yield of Metal from Ore[15]	Total Ore Requirements (10⁶ tons)
Aluminium	0.6	10	6
Beryllium	0.1	2	5
Chromium	2	5	40
Copper	3	0.9	220
Iron	13	45	29
Molybdenum	0.2	2	10
Niobium	0.8	2	50
Nickel	1	1	150
Lead	12	1.5	800
Tin	0.2	10	2
Vanadium[a]	0.5	5	10
			1351
	To account for overburden tunnels, etc.		× 4
			5404 × 10⁶ tons[b]

[a]Substitute for niobium structural material.
[b]160 square kilometers to a depth of 10 meters, assuming 0.3 cubic meters per ton of ore.

Table 4-7. Financial Requirements of Importing Metals to Supply Projected Controlled Thermonuclear Reactor from Capacity A.D. 2000 to 2020.

	Element	Required from Foreign Sources (10^6 tons)	1970 Cost ($/ton[15])	Total Cost (10^9)
Definitely	Beryllium	0.1	$132,000	13.2
must be	Chromium	2	~3,000	6
imported	Manganese	0.2	1,320	1.3
even now.	Niobium	0.8	3,630	2.9
	Nickel	1.5	2,816	4.2
	Tin	0.2	3,828	0.77
	Titanium	0.9	2,904	2.6
	Vanadium	0.5	9,614	4.7
				~36[a]
Quite	Aluminium	0.6	638	0.38
probably	Copper	2	3,630	7.3
imported	Fluorine	1	113	0.11
for economic	Iron	13	136	1.8
reasons	Lead	12	352	4.3
before 2000.				~14[a]

[a]Rounded off.

purchase of the elements. The effect of importing these elements (Be, Mn, Cr, Ni, Sn, Ti, Nb, and V) at constant (1970) prices to provide $1,063 \times 10^3$ megawatts electric of controlled thermonuclear reactor capacity is approximately $36,000 million (Table 4-7). Other elements such as aluminum, copper, fluorine, iron, and lead may also have to be imported after the year 2000 because of price considerations even though the United States may have sufficient reserves to cover the demand in the event of a national emergency. If these elements are imported, then the total costs are almost $50,000 million. Certainly, some of these problems would also be encountered by building other electrical generating stations, especially with chromium, manganese, nickel, aluminum, copper, and iron, but the costs for imported elements in the fossil systems would be lower by perhaps a factor of three or more.

Such large balance of payments charges will require serious study for other reasons such as international politics and national security. The switch to an energy source which requires large amounts of rare materials may be, at least from the United States standpoint, somewhat unattractive.

Electrical Distribution Requirements

There are two areas to consider here: the actual land used by the power plant itself (which includes building, cooling arrangements, fuel storage and exclusion areas), and the transmission line requirements. It is expected that the actual site usage for controlled thermonuclear reactors will be very much like that for fission plants. Recent estimates of the land requirements for electrical generating stations are given in Table 4-8. It can be seen that if Plan B in our scenario were to take effect, by the year 2020, 342 square miles would be saved by the introduction of controlled thermonuclear reactors. This saving will rise to about 950 square miles for Plan C where fusion plants were envisaged to take the place of newly installed fossil-fueled plants (mainly coal).

Another area in which land saving might be effected by the introduction of fusion is that of transmission requirements. If fusion reactors, which have no chemical emissions, no potential for a nuclear accident, and minimal radioactivity problems, can be sited inside big cities, substantial savings can be

Table 4-8. Land Requirements for Power Plant Sites.[6]

	Square Miles per 1000 Megawatts electric	Comment
Coal	1.60	On-site coal storage and ash disposal
Oil	0.40	On-site fuel storage
Gas	0.24	
Fission	0.47	Exclusion area required
Controlled thermonuclear reactor	0.47	Assumed to be same as fission for this study

made in the area of land required for transmission of electricity to the load centers.

For example, it has been estimated[4] that 19 square miles of land will be required per 1,000 megawatts electric for transmission line rights-of-way in 1990. Presumably, this could be reduced to 10 square miles per 1,000 megawatts electric if the plants were located in the cities. Placing all the controlled thermonuclear reactors in the cities for the years 2000 to 2020 would then save over 9,000 square miles of land for transmission lines. Other benefits might accrue from urban siting of plants such as use of waste heat for residential and industrial heating, industrial processing or sewage distillation.

Emissions to the Biosphere

Conventional Chemical Pollutants

Fusion reactors do not emit chemical pollutants such as SO_2, NO_x, or CO. Therefore,

the introduction of fusion into the economy should have a beneficial effect on the quality of the air around electrical generating plants. Table 4-9 lists the projected emissions from electrical generating plants in 1973 and the year 2020. We have also included the effect of introducing fusion according to our previous Plans A, B, and C and have included the effects of more stringent air quality standards.[4]

The effect of fusion on the amount of CO_2, CO, SO_2, NO_x, particulates, hydrocarbons and aldehydes is rather dramatic with Plan C. Without exception, the total pollutants emitted by the electrical generating stations in the year 2020 were actually less than they were expected to be in 1973! It is hard to put a dollar value on this reduction in pollution because the cost in human life and property damage is still uncertain. However, lest we deceive ourselves, we must recognize that when the pollutants from all

Table 4-9. Effect of Fusion on the Air Pollutants Emitted during the Generation of Electricity in the Year 2020 (units: 10^9 pounds per year).

Pollutant	Plan A[a]	Plan B	Plan C	Estimated 1973
CO_2	10,850	7950	2200	2750
CO	1.75	1.25	0.32	0.34
SO_2	55.2	40.1	11.0	25
NO_x	32.4	24.5	6.3	9.4
Particulates	8.97	6.45	1.7	7.5
Hydrocarbons	0.71	0.54	0.16	0.28
Aldehydes	0.099	0.04	0.018	0.029

[a]Base Plan, Ref. 6.

Table 4-10. Effect of Various Rates of Controlled Thermonuclear Reactor Penetration into the Electrical Generating Market on the Overall Pollution of the Air from All Sources in the United States in the Year 2020 (units: 10^9 pounds per year).

Pollutant	Plan A^a	Plan B	Plan C	Estimated 1973
CO_2	30,400	27,800	22,000	10,900
CO	53	53	52	180
SO_2	87	72	43	49
NO_x	81	73	55	38
Particulates	44	46	36	22
Hydrocarbons	19	19	19	31
Aldehydes	1.1	1	1	0.41

^{a}Base Plan, Ref. 6.

forms of energy (for example, cars, planes) is added up, the impact of fusion is less evident (Table 4-10). Therefore, those who look forward to fusion to produce a pristine environment may be disappointed with the differences between Plans A and C in Table 4-10, because fusion can only make a fractional impact on a part of the economy, which in turn only emits a *fraction* of the total pollutants.

Potential Hazards from Fusion Power

Conventional Sources

Probably the largest amount of stored energy in a controlled thermonuclear reactor lies in the coolant system. It was mentioned previously that either high pressure gases such as helium or liquid metals like lithium are the most likely coolants. A quick investigation shows that the most serious problem comes from the possibility of a liquid-lithium fire and the amount of heat that could be released in such a reaction.

Liquid-metal fires require oxygen and are even more spectacular in the presence of water as any chemistry student knows. If we were to imagine the maximum possible conventional accident that a controlled thermonuclear reactor plant could sustain, we might then envision a rupture of the entire controlled thermonuclear reactor confinement vessel, which results in all of the lithium (for

cooling and breeding) flowing out into the reactor building. Coupled with this release of lithium might be the rupture of cooling water pipes (if allowed in the building) or addition of air and water into the building by means of a severe storm. If all the lithium were to burn up in a 1,000-megawatt electric plant (about 1,000 tons of lithium) then the energy released would be equivalent to that released by burning about a million gallons of fuel oil. This amount of energy would not be released instantaneously, and it would take perhaps hours or days for all the lithium to burn up. If it took a day, this would be the equivalent of the amount of thermal energy released inside the boilers of a 1,000-megawatt electric oil-fired plant in the same time period. Certainly, severe damage might be incurred inside the plant, but there is little potential for damage to the surrounding populace.

Another accident that might be envisaged is the release of the energy in the superconducting magnets operating at about 100 kilogauss. The complete failure (transition from superconducting state to normal resistive state) of the magnets could release energy equivalent to that in about 1,500 gallons of fuel oil.[18] Again such an accident would be hard on the reactor and surrounding building, but would not pose any severe problem for the public.

18. "Fusion Power," WASH-1239, February 1973.

Finally, if all the fuel in a 1,000-megawatt electric fusion reactor were to burn up instantaneously (a process that present-day scientists would dearly love to happen) then the energy equivalent of about 4,000 gallons of fuel oil would be released. Such an amount of energy could hardly affect the large controlled thermonuclear reactor plants we have been discussing.

Radioactive Effluents

The only radioactive effluent from controlled thermonuclear reactors during normal operation should be tritium (half-life about 12.3 years). An analysis of a proposed controlled thermonuclear reactor has been conducted by Daley and Greenberg.[19] They have concluded that leak rates as small as 6 curies per day can be maintained in 1,000-megawatt electric controlled thermonuclear reactor plants. This number is also consistent with a value of 10 curies per day for a 1,500-megawatt electric plant.[13] Such a leak rate amounts to about 2 megacuries of tritium released per year for all the controlled thermonuclear reactors in the year 2020. Table 4-11 was prepared to compare the release rates to those expected from fission reactor and fossil fueled plants.

Martin et al.[20] have calculated that the average 1,000-megawatt electric coal fired plant releases about 48 millicuries per year of radium-226, radium-228, and thorium-232 isotopes. Hence, all the fossil plants now release about 8 curies per year and this is expected to increase to about 50 curies per year in 2020. The release of tritium and krypton-85 from fission reactors (and reprocessing facilities) is presently about 17 megacuries per year.[6] This is expected to rise to 36 megacuries per year for tritium

and 270 megacuries per year for krypton-85 in 2020.

The above figures might be put into a little more perspective by noting that tritium is produced at the rate of 4 megacuries per year in the upper atmosphere by cosmic ray action on nitrogen.[21] Since this has been happening for some time, there is an equilibrium inventory of about 70 megacuries in the stratosphere from that source. Furthermore, the amount of tritium in the atmosphere due to nuclear weapons testing is about 700 megacuries, excluding the recent French and Chinese tests.

The data in Table 4-11 show that the present gaseous radioactivity already in the atmosphere may be considerably larger than that released in the year 2020 and comparable to the total inventory of radioactivity over the period 2000 to 2020 if radioactive decay is taken into account. The introduction of fusion into the electrical generating market of 2000 to 2020 has a relatively minor (about 15 per cent) effect of radioactivity due to man-made sources if Plan B is followed, and has essentially no effect if Plan C is maintained. The implication of Plan D is that the total radioactivity released in the year 2020 could be reduced by about 40 per cent if controlled thermonuclear reactors are introduced at the expense of fission reactors. However, lest we lost our perspective, the radiation levels from these effluents are negligible when compared to those arising from cosmic rays, ultraviolet rays, natural radioactive elements, X-ray machines, weapons testing, etc.

A comparison of total radioactivity in curies is not the best way of calculating the biological impact of emission from power plants. It is more instructive to take account of the maximum permissible concentration for each isotope released. The bottom half of Table 4-11 contains the information on how many cubic kilometres of air are required to dilute specific radioactive emis-

19. J. E. Draley and S. Greenberg, in *Proceedings of the Texas Symposium on the Technology of Controlled Thermonuclear Fusion Experiments and Engineering Aspects of Fusion Reactors,* Austin, Texas, November 1972.

20. J. E. Martin, E. D. Harward, D. T. Oakley, J. M. Smith, and P. H. Bedrosian, Paper SM-146/19 in IAEA symposium on safety of nuclear power plants, Vienna, 1971.

21. D. G. Jacobs, "Sources of tritium and its behavior upon release to the environment," TID-24635, USAEC, 1968.

Table 4-11. Projected Gaseous Radioactivity Released into the Atmosphere in the Year 2020 Due to the Generation of Electricity and That Present from Selected Sources.

Source	Isotope	Plan A	Plan B	Plan C	Plan D	Estimated 1973
		(megacuries)				
Fossil fuel[a]	(Radium-228 Thorium-232)	0.00005	0.00004	0.00001	0.00005	0.000008
Fission reactors	Tritium	36[b]	23	36	18	0.6
	Krypton-85	270[b]	220	270	140	16
Controlled thermonuclear reactors	Tritium[c]	0	2	2	2	0
Weapons testing[c]	Tritium	40	40	40	40	700
Cosmic rays[d]	Tritium	70	70	70	70	70
Total (rounded)		*420*	*360*	*420*	*270*	*790*
		10³ cubic kilometers of air required to dilute to maximum permissible concentration				
Fossil fuel	(Radium-228 Thorium-232)	50	40	9	50	8
Fission reactors	Tritium	180	115	180	90	3
	Krypton-85	900	730	900	460	53
Controlled thermonuclear reactors	Tritium[c]	0	10	10	10	0
Weapons testing[c]	Tritium	200	200	200	200	3,500
Cosmic rays	Tritium	350	350	350	350	350
Total (rounded)		*1,680*	*1,450*	*1,650*	*1,160*	*3,900*

[a]Ref. 21.
[b]Ref. 6.
[c]Already present assuing no addition from 1968 onward and allowing for natural decay.
[d]Equilibrium value.

sions to acceptable standards.[22] Whereas radioactive emissions (measured in curies) from coal fired plants are negligible when compared to nuclear facilities, it is seen from this table that they could amount to 1 to 8 per cent of the total biological hazard potential of all power plants in the time period 1973 to 2020. The introduction of fusion at the expense of coal-fired plants (Plan C) would not increase the total biological hazard potential and in fact would reduce it by 2 per cent. On the other hand, if fusion is introduced at the expense of fission reactors (Plans B and D), then reductions in the year 2020 of 20 to 45 per cent in biological hazard potential could be achieved. Again, it is noted that the presence of tritium from weapons testing is a dominating feature of the present picture.

Radioactivity Inventory

The two major considerations in this category are:

(1) The release of radioisotopes in the event of an accident;
(2) The long-term storage of radioisotopes.

It is very difficult, and sometimes misleading, to discuss the release of radioisotopes during accidents because such an analysis automatically assumes that all the safety devices on a reactor will fail at the same time. This is an unrealistic assumption and must be recognized as such before we can assess the maximum possible damage that could be done by the release of all radioisotopes in a reactor.

Table 4-12 lists the major isotopes produced by the fuel of a fission reactor and in the first wall and blanket region of a controlled thermonuclear reactor. Data on four different controlled thermonuclear reactor structural materials are given: 316 stainless steel, a Nb-1Zr alloy, a V-20Ti alloy and aluminum.[23,24] Information is given for the number of curies generated per kilowatt-hour thermal and for the biological hazard potential of various isotopes in terms of the volume of air (per cubic kilometer) required to dilute the isotopes to the maximum permissible concentration.

After ten years of operating controlled thermonuclear reactors with 316 stainless steel, Nb-1Zr, Al, or V-20Ti as structural material, the total biological hazard potential of the reactor is greater than that of the entire tritium inventory by one to three orders of magnitude. This observation is particularly striking for systems using Nb-1Zr alloys.

An interesting comparison can be drawn between fusion and fission. It can be seen in Table 4-12 that fusion systems have biological hazard potential values of one to four orders of magnitude lower than fission reactors. The most critical isotopes in the time period shortly after reactor shutdown are iodine-131 and cesium-137. It is important to note that iodine or cesium are volatile elements and could conceivably escape the reactor in the event of a severe accident while most of the radioactivity in fusion systems is in the form of nonvolatile metallic elements. Stated another way, the 0.94 curies per kilowatt-hour thermal of cesium-137 in Table 4-12 would probably present more of a hazard than the 152 curies per kilowatt-hour thermal of niobium-92m because the niobium is tightly bound in the metallic structure with a very low vapor pressure at reactor temperatures, whereas the cesium-137 is in a vapor state at typical fuel element temperatures.

A brief analysis of Table 4-12 reveals that the replacement of fission reactors with fusion reactors will lower the biological hazard potential of our total electrical gener-

22. 25-FR-10914, Part 20, December 1968.
23. W. F. Vogelsang, G. L. Kulcinski, R. G. Lott,
and T. Sung, 1974, Nuclear Technology, Vol. 22, p. 379.
24. J. F. Powell, F. T. Miles, A. Aronso, W. E. Winsche, and P. Bezler, Brookhaven National Laboratory Report, BNL-18439, November 1973.
25. D. Steiner and A. P. Fraas, Nuclear Safety, Vol. 13, 1972, No. 5, p. 353.
26. T. J. Burnett, Health Physics, Vol. 18, 1970, p. 73.

Table 4-12. Major Long-Lived Radioactive Isotopes in Various Controlled Thermonuclear Reactor First Wall Materials and in Fuels of Advanced Fission Reactors[a]

System	Isotope	Half-Life	Activity (curie per kilowatt-hour thermal)	Maximum Permissible Concentration (microcurie per cubic centimeter)	Biological Hazard Potential (cubic kilometer of air per kilowatt-hour thermal)
Fusion-all	Helium-3	12.3 years	60	2×10^{-7}	0.30
	Vanadium-49	331 days	0.67	1×10^{-10}	6.7
316 Stainless Steel first wall only[b]	Iron-55	2.94 years	140	3×10^{-8}	4.6
	Cobalt-58	72 days	29	2×10^{-9}	14.5
	Nickel-57	36 hours	1.1	1×10^{-10}	11
	Manganese-54	310 days	24	1×10^{-9}	24
	Cobalt-60	5.25 years	4.7	3×10^{-10}	15.6
				Total	~77
Nb-1Zr[b]	Niobium-92m	10.1 days	152	1×10^{-10}	1,520
	Niobium-95m	3.75 days	50	1×10^{-10}	500
	Niobium-95	35 days	43	3×10^{-9}	14
	Strontium-89	51 days	38	3×10^{-10}	126
				Total	~2,200
V-20Ti[b]	Scandium-48	1.81 days	12.1	5×10^{-9}	2.5
	Calcium-45	165 days	2.6	1×10^{-9}	2.6
	Scandium-46	84 days	1.87	8×10^{-10}	2.3
				Total	7.5
Al[c]	Sodium-24	1.5 hours	630	4×10^{-8}	15.8
	Aluminum-26	7.5×10^5 years	0.004	1×10^{-10}	0.04
				Total	~15.8
Fission	Iodine-131[d]	8.04 days	31.6	1×10^{-10}	330
	Plutonium-239[e]	24,100 years	0.06	6×10^{-14}	1,000
	All plutonium isotopes		18.2		8,300
	Strontium-90[e]	25 years	0.64	3×10^{-11}	21
	Cesium-137[e]	33 years	0.94	5×10^{-10}	2
All other fission products		>1 day	-	-	18,000

[a]Neglect all isotopes with half-lives greater than 12 hours.
[b]Ref. 23 for 10-year exposure.
[c]Ref. 24.
[d]Ref. 25.
[e]Ref. 26.

Table 4-13. Long-Term Radioactive Wastes from Fusion and Fission Systems after 100 Years Decay.

System	Isotope	Half-Life	Activity (curie per kilowatt-hour thermal)	Maximum Permissible Concentration (microcurie per cubic centimeters)	Biological Hazard Potential (cubic centimeters of air per kilowatt-hour thermal)
Fusion					
316 Stainless Steel	Cobalt-60	5.25 years	8.7×10^{-6}	3×10^{-10}	3×10^{-5}
	Nickel-63	85 years	1.3×10^{-4}	2×10^{-9}	0.6×10^{-5}
Niobium-1Zr	Niobium-94	50,000 years	0.008	1×10^{-10}	0.8
	Zirconium-93	5,000,000 years	0.006	4×10^{-9}	0.0015
Vanadium-20Ti					Negligible
Aluminum	Aluminum-26	75,000 years	0.004	1×10^{-10}	0.04
Fission	Plutonium-239[a]	24,100 years	0.0006	6×10^{-14}	10
	Strontium-90	25 years	0.04	3×10^{-11}	1.3
	Cesium-137	33 years	0.115	5×10^{-10}	0.23

[a]Assume 1 percent in fission product waste.

Table 4-14. Comparison of Known Electrical Generating Costs from Fossil-Fueled and Fission Reactors to Projected Costs for Fusion Reactors.

	Coal[a]	Fission[b]	Fusion[d,e]
Capital investment (dollars per kilowatt electric)	120–220	120–220	600–800
Cost of electricity (mills per kilowatt-hour electric)[c]			
Plant investment	4–10	6–15	10–17
Operation maintenance	1–3	1–2	1–2
Fuel	3–5	2–3	<0.01
Total	8–18	9–20	11–19

[a]Reference 27, placed on line in 1972.
[b]Reference 28, placed on line in 1970–1972.
[c]In 1974 dollars.
[d]Reference 29.
[e]Reference 30.

ating system in almost direct proportion to the controlled thermonuclear reaction fraction. Hence, fusion can reduce future potential hazards from large-scale release of radioactivity in the unlikely event of a severe nuclear accident.

The long-term storage problem of fission reactor fuel stems from the actinides, plutonium-239, which is the predominant example, and the strontium-90 and cesium-137 isotopes. Because of their long half-lives and low maximum permissible concentration it can be seen in Table 4-13 that 100 years after the generation of the waste the biological hazard potential fission fuel has dropped by ~3 orders of magnitude but that of fusion has dropped by 3 to 7 orders of magnitude for the case of 316 stainless steel and niobium alloys, respectively. There are essentially no long-term storage problems with controlled thermonuclear reactors constructed of vanadium.

Thus the introduction of fusion into the economy after the year 2000 will tend to reduce both the biological hazard potential in the event of serious fission power plant accidents and will also reduce the long-term storage requirements for radioactive wastes, especially if vanadium, aluminum, and stainless steel are used.

Costs of Fusion Power

Absolute values for the cost of electricity from different types of energy sources are very difficult to obtain because of variations in local economic or climatic conditions (that is, fossil plants at the source of fuel versus fossil plants in the center of a city). However, there are some inherent features that are worth noting about each of these plants.

First, the cost of generating electricity from fossil plants depends quite heavily on the fuel costs which can amount to about 40 per cent of the total costs. The corresponding figure in fission reactors is about 20 per cent, and we have seen that fuel costs for fusion reactors are much less than 1 per cent of the total cost of electricity.

The second point is that because fusion reactors are so capital intensive and have relatively low energy densities and depend on rather expensive reactor materials, one can get an idea (within a factor of about 3) of capital costs of this type of power plant by simply calculating the fabricated cost of the materials in the reactor. Past estimates have shown that the fabricated materials costs for fusion reactors are of the order of $250 per kilowatt electric.[18,29,30]

Some representative values of known

27. *Steam electric plant construction costs and annual production expenses*, US Power Commission Report No. 197, 1972.
28. *Electrical World*, Vol. 180 No. 9, 1973, p. 39.
29. J. Young, University of Wisconsin Report UWFDM-68, Vol. 2, 1974.
30. R. G. Mills, 1974, *Princeton Fusion Reactor Design*.

electrical generating costs are given in Table 4-14 for a coal fired plant and a light water fission reactor and these are compared to estimated fusion reactor costs.[29,30]

Although the make-up of the electrical generating costs varies dramatically from one system to another, the total cost of electricity does not vary significantly with the three forms of energy. One factor which may invalidate the information in Table 4-14 is the cost of environmental protection and long-term radioactive waste disposal. When such costs are properly assessed in relation to fossil, fission, *and* fusion systems, it is felt that the fusion reactor may look better economically.

In summary, it appears that fusion will be able to generate electricity at about the same cost as fission and fossil fuel plants. However, it must be recognized that it is far too early to make any more quantitative statements.

SUMMARY

We will need to develop other sources of energy between now and the year 2000 while fusion technology is being perfected. It appears that 2000 is the earliest date at which we will be able to operate commercial fusion power plants. Given an optimistic penetration of fusion into the electrical generating market, it is possible that 10 per cent of our electrical generating capacity in the year 2010 could be supplied by controlled thermonuclear reactors. This could rise to almost 30 per cent by the year 2020, still below the percentage generated by fission reactors which may account for 40 per cent of the installed capacity.

The fuel requirements for fusion represent its biggest advantage. Both the amount and method of procuring the fuel will represent a negligible impact on the environment. The development of fusion power will considerably relieve the pressure on the environment from the standpoint of strip mining of coal and the depletion of our vital fossil fuels. Because the fuel cost is essentially zero and supplies virtually unlimited, we will have

to concentrate on minimizing the impact on the environment due to structural materials procurement.

Some serious problems for magnetically confined plasma reactors could arise because of the need for refractory metals and relatively expensive alloying elements in iron-based alloys. Deficiencies of niobium, vanadium, chromium, manganese, and nickel in the United States could increase its dependence on foreign suppliers for these metals, thus trading her present problem of fossil-fuel dependence for one of structural-materials dependence in the future.

Fusion reactors could considerably reduce the land requirements for electrical power stations both at the power plant site and because their inherent safety may allow them to be placed closer to, if not inside, large cities. This advantage of fusion will become more important as our land reserves dwindle and a premium is placed on utility land.

The introduction of fusion power, which has no chemical emissions, will greatly reduce the air pollutants from power plants beyond the year 2000. However, when viewed in total, fusion can make only small percentage improvements in air quality because of the large contributions of other energy consuming sectors of the economy.

Fusion reactors can significantly reduce the radioactive emission of power plants to the environment, especially when one views this problem with respect to biological hazard potentials. However, because neutrons are a natural by-product of the deuterium-tritium reaction, there will be large amounts of radioactive isotopes generated during the production of fusion power. These isotopes generate as much activity in curies as do those resulting from fission, but they are chemically more stable and do not have the low maximum permissible concentration levels characteristic of the actinides, strontium-90, iodine-131, or cesium-137. There will be large amounts of tritium in deuterium-tritium fusion reactors and it must be contained with leakage rates not exceeding 10^{-6} per day.

Another advantage of fusion power is that the half-lives of the controlled thermonuclear reactor generated isotopes, with the exception of niobium, are quite a bit shorter than the half-lives of the important fission fuel isotopes. This means that the long-term radioactive storage problems can be significantly reduced by fusion power.

Finally, the cost of generating electricity by fusion does not appear to be significantly higher or lower than that typical of present-day fossil or fission plants. This conclusion could be modified if changes are made in our methods of assessing environmental damage. It would be expected that the corresponding increase in controlled thermonuclear reactor electrical generating costs would be far less than for fossil or fission fueled systems.

If the state of the art advances sufficiently to allow us to tame fusion reactions that emitted no neutrons, it would change many of the conclusions of this article. A similar revision would be in order if we could attain laser-induced reactions or successfully achieve direct conversion of the kinetic energy of the charged particles emitted in fusion reactors to electricity.

All of these are exciting possibilities, but we must be careful that we do not get so fixed on the potential of future developments that we forget to utilize those energy sources at our disposal. Only vigorous effort on our part will ensure that we reach the year 2000 with enough inertia and technical knowledge to take advantage of this potential new source of power.

Solar Energy: Its Time Is Near

WALTER E. MORROW, JR.   1973

Outside of the earth's atmosphere the sun provides energy at the rate of about 1,400 watts per square meter (4,730 BTU per square meter per hour) normal to the sun. By the time sunlight reaches the earth's surface, atmospheric attenuation, clouds, and earth shadowing have taken their toll; the average of solar energy falling on a horizontal surface in southern New England, taken over a long time period, is about 160 watts per square meter. If the solar energy density is to be measured on a platform which is movable so that it may be maintained constantly normal to the sun, this annual average may be improved by a factor of nearly two—to almost 300 watts per square meter in the case of southern New England.

Such energy densities are quite low, and this is essentially why solar energy has to date played a negligibly small role in the United States. However, assuming 30 per cent efficiency, the total present *electric* power demand of the United States could be supplied with solar energy plants having a total area of about 2,000 square kilometers. This is about 0.03 per cent of the United States land area devoted to farming and about 2 per cent of the land area devoted to roads; and it is about equal to the roof area of all the buildings in the United States.   .

But such statements do not comprehend the very considerable hurdles which stand between this apparently bountiful energy supply and its collection, storage, and use.

There has been extensive development of fixed-orientation, low-temperature collectors for building heating and hot water heating, and some systems are now in use. These collectors are usually faced south at inclinations of 45 to 60 degrees above horizontal, with one or two layers of glass or plastic used to reduce convection and radiation losses. Transmission losses through these windows are commonly of the order of 15 to 20 per cent.

Of the solar energy passing through the windows, 80 to 95 per cent can be absorbed with simple black coatings. Thus, between

Walter E. Morrow is associate director of Lincoln Laboratory at Massachusetts Institute of Technology, Cambridge, Massachusetts. During his 24 years at the laboratory, he has been active in research on a wide variety of radio communications techniques, equipment, and systems.

From *Technology Review*, Vol. 76, No. 2, pp. 30–42, December 1973. Reprinted with light editing and by permission of the editor and the *Technology Review*. Copyright 1973 by the Alumni Association of the Massachusetts Institute of Technology. This article is based on a paper prepared for a technical task force of the Federal Power Commission.

70 and 80 per cent of the incident radiation can be collected.

Losses inside the collector are a function of working temperature and occur by reradiation, convection, and conduction. Convection and conduction losses can be made negligible through good design. The reradiation losses are determined by the infrared emissivity of the absorbing surface and the infrared transmission of the windows. Absorbing surfaces with effective infrared emissivities as low as 0.15 are not difficult to achieve. (An emissivity of 0.15 means that the surface will radiate 15 per cent of the infrared energy that would be radiated by a perfect black body of equal area.) The usable heat is usually gathered from the absorbing surface either by water flowing through tubes attached to the surface or by air flowing over the surface itself.

Assuming 75 per cent absorption and an emissivity of 0.15, net collection efficiencies of between 50 and 75 per cent can be proposed, depending on the outlet temperature—the higher the temperature required, the lower the efficiency.

Heat storage for house heating has been accomplished by means of water tanks, bins of rocks, or in hydrated chemicals such as sodium sulfate, the solid-liquid phase change of which adds to the heat energy that can be stored.

Assuming a 40°C temperature change—typical of that required in building and hot-water heating applications—water will store 1.6×10^8 joules per cubic meter; sodium sulfate 3.5×10^8 joules per cubic meter. Analysis of these figures demonstrates that to achieve the large heat storage required for domestic space heating applications—to provide heat through the night and in bad weather—requires substantial volumes of storage. One to a few days' heat storage is typically used in a solar-heated house, with auxiliary heating sources added to provide heat during extended cloudy periods.

Though the temperatures achieved in such low-temperature collector systems as those described above are adequate for heating buildings and water, they are insufficient for the high-efficiency production of electricity or artificial fuels by thermal processes; for these applications high-temperature collectors are required. The achievement of high temperatures depends chiefly on concentrating the solar energy from a relatively large area into a small collector, from which it is carried to storage. Conduction and conversion losses can be made small in high-temperature solar collectors by good design, including, if necessary, the use of vacuum insulation around the heat absorber. Indeed, such solar energy concentrators can be very efficient; used in research, they now yield the highest temperatures available in small furnaces for many applications.

The differences between unfocused and focused collectors are striking: unfocused collectors typically have ratios of concentration to emissivity of 10.0 and are limited to output temperatures of 150°C or less; ratios of 300 and temperatures of 600°C are typical of one-axis-steerable concentrators, and ratios of 10,000 and temperatures of 4,000°C are typical of two-axis-steerable concentrators. These figures suggest that reasonable efficiencies and temperatures can be obtained with a one-axis concentrator, such as shown in Figure 4-9.

Solar energy can also be utilized through its conversion into combustible fuels by photosynthesis in trees, plants, and algae. Conversion efficiencies have been estimated to be in the range of 0.3 to 3 per cent, depending on the vegetation used. It is possible to imagine an energy system built on this conversion: plants grown with sunshine used to fuel furnaces or boilers, for example. Because of the low collection efficiencies, rather large land areas are required to supply significant amounts of energy.

To supply the total current United States energy needs at 3 per cent efficiency would require a land area of about 350,000 square kilometers, about 3 per cent of the total United States land area. Soil depletion and the handling of waste products from the combustion of such fuels are likely to be significant problems for large-scale energy

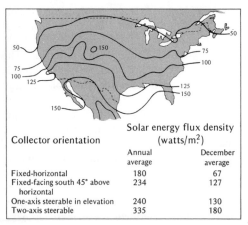

Collector orientation	Solar energy flux density (watts/m²)	
	Annual average	December average
Fixed-horizontal	180	67
Fixed-facing south 45° above horizontal	234	127
One-axis steerable in elevation	240	130
Two-axis steerable	335	180

Figure 4-9. In the northern hemisphere the test of solar energy systems comes in the winter, when energy inputs from the sun reach their minimum. The map shows the average solar energy incidence in the United States in December in watts per square meter, the table shows solar energy densities, also in watts per square meter on collectors with different orientations in an average location in the United States in December. To obtain British thermal units per hour from the figures given, multiply by 3.41.

systems which may be conceived to utilize solar energy in this way. However, an interesting variation of this plan is that of utilizing waste from forestry operations and municipal trash collections as a fuel to produce power or a synthetic fuel such as methanol, or both. For instance, it is estimated that waste from current forestry operations could provide 10 to 20 per cent of United States energy needs projected for 1975.

Photovoltaic conversion of solar energy to electricity using silicon solar cells—direct sunlight striking the cells generates current in a n-p junction in the silicon material—has been widely employed on spacecraft. Typical efficiencies are of the order of 10 per cent, and costs for space-qualified systems can be as much as $1 million per kilowatt. Relatively modest efforts have gone into improving efficiency, and these have yielded silicon and gallium arsenide devices with efficiencies of over 16 per cent. Efficiencies of 20 per cent are believed to be achievable,

compared with a maximum theoretical efficiency for simple photovoltaic converters of about 35 per cent.

High cost is as much a problem as low efficiency: arrays of the silicon cells currently available (with 10 per cent efficiency) cost about $100,000 per kilowatt of peak capacity when engineered for ground installations without solar concentration. Several efforts are under way to reduce the costs of energy from solar cells of this type. Polycrystalline cells would have lower efficiencies than the single-crystal cells now in use in spacecraft, but they should be much less expensive. Use of solar concentrators with special cells designed for high solar intensities would also increase energy output; a system using a one-axis concentrator would require about one-fiftieth of the area of a nonconcentrating system to supply a given amount of energy. If these cells with a concentrator could be provided at the same costs as today's silicon cells, the cost per peak kilowatt might be reduced by fiftyfold—to the order of $2,000.

Given these various alternatives for collecting solar energy at various temperatures and efficiencies, what systems can we envision within the realm of engineering feasibility that will utilize solar energy at capital and operating costs reasonable relative to those of other energy systems? Three types of systems for three different applications have been proposed: (a) domestic space and water heating systems, (b) total energy systems for commercial and industrial buildings, and (c) large-scale solar electric power generation.

SOLAR HOUSES: THE FIRST CHANCE FOR REALITY

A number of experimental solar house-heating systems have been built and there has been substantial production of solar hot-water heaters. (See Figures 4-10 and 4-11.)

A typical house of 1,500 square-foot area requires of the order of 0.7×10^9 joules per day for space heating during December in mid-United States locations (40-degree-day).

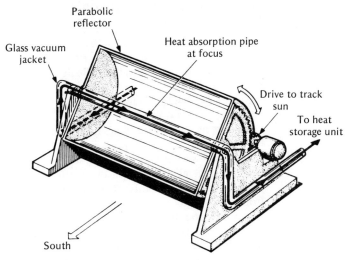

Figure 4-10. When a simple flat-plate fixed collector provides insufficient efficiency and output temperature, a one-axis concentrator is usually proposed for higher efficiency and temperatures at modest cost. This design incorporates a clockwork drive to track the sun and a vacuum-jacketed pipe at the focus of the reflector through which dry nitrogen is circulated to collect heat at 550°C.

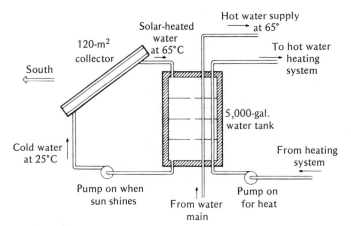

Figure 4-11. The simplest solar energy systems are those designed for domestic space and hot-water heating. In such systems, an inclined, southerly facing flat collector is typically coupled with a heat storage system from which heat can be drawn when required. This diagram shows a combination system that provides domestic hot water at 65°C and hot water, as well, for space heating.

Hot-water heating typically might require another 0.1×10^9 joules per day. In central United States locations a fixed 45° collector can be expected to receive an average of about 127 watts per square meter in December, of which perhaps 60 per cent can be retained; thus, 0.0066×10^9 joules per square meter could be collected in an average 24-hour period. To provide the needed 0.80×10^9 joules per day for space and water heating would require the heat from approximately 120 square meters (1,300 square feet) of collector area. This could be provided by using somewhat more than half the roof area of a single-floor house or slightly more than the roof area of a two-story house.

Four days of heat storage would require 20 cubic meters (5,000 gallons) of water (or 64 cubic meters of rock) heated to 65°C. Smaller storage systems could be used if auxiliary heating were provided for cloudy days.

Current costs for fixed low-temperature collectors range between $20 and $40 per square meter. A 5,000-gallon heat storage system would cost about $1,000. Thus, the total system costs would be between $3,500 and $6,000, not including the usual heat distribution system. If systems of this type were financed as part of house mortgages over a twenty-year period at 7 per cent interest, the increase in annual carrying charges would be between $300 and $550.

How does this compare with current costs for heating by fossil fuel? Such a house requires fuel or heating electricity equivalent to about 10^{11} joules per year (9.5 × 10^7 BTU) assuming a total of 5,000 degree days required. Typical costs in 1972 for 10^{11} joules of heat energy were:

oil at 75 per cent efficiency (at 22¢/gallon)	$280
gas at 75 per cent efficiency (at 23¢/100 cubic foot)	$275
electricity at 100 per cent efficiency (at 1.8¢/kilowatt-hour)	$500

This computation suggests that solar heating systems are competitive with electric heating today in mid-United States locations. Should fuel prices rise significantly faster than the costs of constructing solar heating systems in the future, solar heating could become competitive with gas and oil at some future time.* (See Figure 4-12.)

*Editors' note: The rapid increase in fuel oil prices in 1973–1974 has verified this prediction for oil heat.

TOTAL-ENERGY SYSTEMS FOR THE INDUSTRIAL PARK

Most shopping centers and industrial plants now being built consist of one- or two-story buildings in suburban locations. Such facilities require large amounts of energy for heating, cooling, lighting, and operations. (See Figures 4-13 and 4-14.)

A number of such buildings have recently been constructed with gas- or oil-fueled total-energy systems. In such a system electricity is generated by diesel- or gas-turbine-powered generator units. The waste heat from the engines is used for heating in winter and cooling (by means of absorption air conditioners) in summer. Such systems have the advantage over central power systems of recovering the waste heat from the electric generating process.

A similar arrangement using solar energy can be proposed. Parabolic concentrators could provide 550°C steam for a turbine-alternator plant whose waste heat would be used for heating or cooling, depending on the season.

Approximately 5,000 square meters of one-axis-steerable collectors can be mounted on the roof of a plant occupying 10,000 square meters. About 70 per cent absorption efficiency can be achieved with an outlet temperature of 550°C; thus, the heat col-

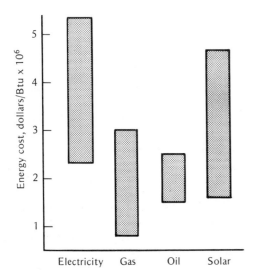

Figure 4-12. Solar systems may now be competitive in total installation and operating cost with gas and oil for domestic space and hot-water heating in the most favorable parts of the United States. As the price of fossil fuel rises faster than the general rate of inflation in the future, solar energy will become increasingly attractive in less favorable areas. Indeed, the author suggests that the balance may have shifted to favor solar energy in much of the United States by 1983.

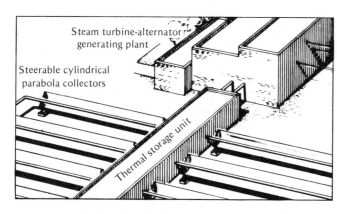

Figure 4-13. The largest solar systems now envisioned are proposed for medium sized base-load electric power installations. This system is for a 1,000-megawatt plant—utilizing large numbers of one-axis-steerable concentrators to collect high-temperature heat to drive conventional steam turbines. An area 16 kilometers square would be occupied by collectors for such a plant.

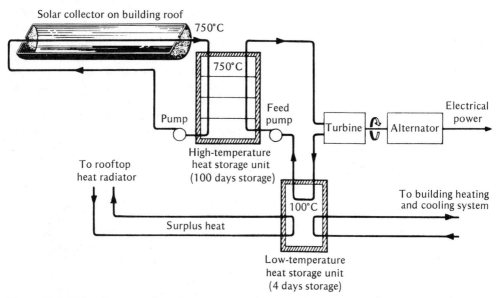

Figure 4-14. This solar-powered total-energy system is projected for use in suburban shopping centers or industrial plants where space for solar collection is available. One-axis-steerable concentrators would be located on the roof; high-pressure dry nitrogen, steam, or sodium chloride would transfer heat at 750°C to a high-temperature storage unit of insulated rocks or molten salt. Heat drawn from the storage unit would drive a turbine to produce electrical power and then, held in a low-temperature storage unit, would be available for space heating or air conditioning.

lected per day will be about 7×10^{10} joules averaged over the year. In order to average out summer to winter solar energy variations, the order of 100 days of heat storage are required. A storage capacity of 7×10^{12} joules can be obtained with 10,000 cubic meters of insulated rock heated to 550°C.

and located along one side of the building. With a thermal electric plant operating at 40 per cent efficiency (which is typical of modern steam generating plants) for perhaps 12 hours out of each 24 that the building is in use, an electric power output of about 670 kilowatts can be delivered. In addition,

an average of 500 kilowatts of low-temperature waste heat would be available over the
full 24-hour period for heating or cooling.

No such system has ever been built on
such a scale, so the costs are very difficult to
predict. Here is a very rough estimate:

collector: 5,000 square meters at $100	$500,000
storage: 6,000 cubic meters at $20	$120,000
1,000-kilowatt steam turbine alternator	$100,000
	$720,000

The retail value of the output energy can
be roughly estimated as:

Electric power:

250 working days
at 12 hours ×
670 kilowatts ≅ 2.00 × 10^6 kilowatt-hours
115 standby days at
24 hours ×
100 kilowatts ≅ 0.27 × 10^6 kilowatt-hours

                    2.27 × 10^6 kilowatt-hours
2.27 × 10^6 at $0.018 per kilowatt-hour ≅
$41,000

Low-temperature heat:

365 days at 24 hr. ×
  500 kilowatts = 4.4 × 10^6 kilowatt-hours
                = 1.5 × 10^{10} BTU
Value of 1.5 × 10^{10} BTU of natural gas
at 23¢/10^5 BTU ≅ $34,000
Total energy value
  per year   ≅ $75,000

The value of this annual energy savings to
a typical suburban building can be estimated
using typical procedures for calculating
industrial plant capitalization. At an 8 per
cent discount rate with inflation of 4 per
cent per year, straight-line depreciation, and
a 48 per cent tax rate, this value of the
energy savings can be calculated to equal
about five times the annual savings or about

$375,000 compared with the $720,000 estimated costs.

Given the assumptions, these calculations
suggest that such total-energy solar plants
are close to being an economic investment
for a typical suburban industrial plant in the
middle latitudes of the United States, although the cost uncertainty is considerably
greater than for the building heating systems
described in the previous section. (See
Figure 4-14.)

BASE-LOAD ELECTRICITY FROM THE SUN

A number of proposals have been made for
the design of large-scale solar-powered electric generation systems. (Such plants could
also generate hydrogen fuel either from electricity through electrolysis or directly from
thermal energy by one of several proposed
processes.) These suggestions generally fall
into one of the following three classes:
ground-based thermal conversion systems
using conventional collectors, ground-based
systems based on photovoltaic arrays, or systems based on photovoltaic arrays mounted
on satellites above the earth.

Proposals for large-scale ground-based
thermal conversion systems suggest plants
similar in design to the total-energy plant
described above. One proposal by Drs. A. B.
and M. P. Meinel of the University of
Arizona includes one-axis-steerable cylindrical parabolas as solar energy concentrators,
with vacuum-insulated heat collection pipes
at the focal points of the collectors. The
energy would be stored as thermal energy in
molten salt or in rock, and conventional
steam turbines and alternators would be
used to produce electricity. Parameters of a
1,000-megawatt continuous-output plant
would be:

area of plant	30 square kilometers
area of collectors	16 square kilometers
outlet temperature of collectors	550°C
collection efficiency	~60 per cent
thermal storage rock	2 × 10^7 cubic meters

thermal plant
  efficiency   ~40 per cent
  over-all efficiency ~25 per cent

The collector efficiency is reduced from the previous case because of the long distance involved in transferring heat from the collector to the steam plant. Assuming that collector costs can be reduced to $60 per square meter and using the same cost assumptions for the other components as for the total-energy plant, a total system cost of about $1.4 billion is calculated, corresponding to $1,400 per kilowatt of capacity. Separate studies of this type of solar plant by Aerospace Corp. have suggested capital costs of $1,000 to $2,000 per kilowatt of capacity for similar designs.

In steady operation over a one-year period, such a 1,000-megawatt plant would produce about 8.8×10^9 kilowatt-hour. At a wholesale value of $0.008 per kilowatt-hour, the year's output would have a value of about $70 million. Using the same capital valuation assumptions as in the case of the total-energy plant, a plant investment of about $350 million could be justified. This is about one-fourth of estimated cost. Obviously, substantial reductions in the solar plant cost, or increases in the value of electricity, would be required before such solar plants would be justified.

An alternative design for a ground-based thermal plant using a two-axis concentrator that consists of a large number of individually movable facets has been proposed by A. F. Hildebrant, G. M. Haas, W. R. Jenkins, and J. P. Colaco of the University of Houston. Each facet would independently track the sun so as to reflect sunlight on a central collector mounted on a tower at one edge of the array. A magnetohydrodynamic thermal-electric system mounted at the collector would be used to produce hydrogen by electrolysis. Tanks of hydrogen at the plant would provide the energy storage. (See Figure 4-15.)

Typical parameters for a plant with

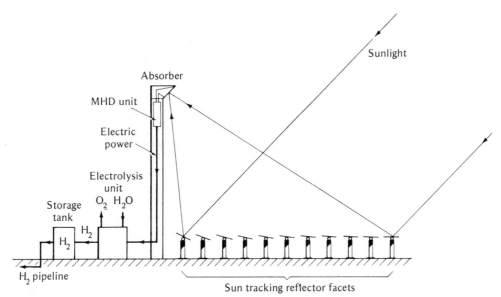

Figure 4-15. If solar heat can be used to drive a magnetohydrodynamic (MHD) power generator, the system might look like this. A two-axis concentrator consisting of a large number of movable reflectors focuses energy at extremely high temperatures to operate a magnetohydrodynamic power unit. Energy from the power unit is used to produce hydrogen by electrolysis, the hydrogen then to be supplied directly to consumers or placed in storage for use when solar energy is not available.

1,000-megawatt continuous output might be as follows:

area of plant	17 square kilometers
area of collector	17 square kilometers
collector efficiency	~60 per cent
conversion efficiency to electricity	~60 per cent
conversion efficiency to hydrogen	~90 per cent
over-all efficiency to hydrogen	~32 per cent

Detailed cost estimates for this class of system are not available, but costs would probably be of the same order as for the previously described ground-based system.

SOLAR CELLS ON THE GROUND AND IN SPACE

Large-scale ground-based plants based on photovoltaic collectors have been proposed, using arrays of silicon photovoltaic cells to energize electrolytic cells producing hydrogen in large quantities. The hydrogen would be piped to consumers either for direct use as a fuel or as an input to a fuel cell to produce electricity. A 1,000-megawatt plant might have the following parameters:

land area	~100 square kilometers
collector area	~ 50 square kilometers
photovoltaic cell efficiency	~12 per cent
electrolyzer efficiency	~95 per cent
net efficiency to produce hydrogen	~11 per cent
fuel cell efficiency	~80 per cent
net efficiency to produce electricity at customer location	~9 per cent

The low net efficiency of such units leads to substantially larger land areas for solar energy collection than those required for thermal-cycle solar systems. A much more difficult problem, however, is the high cost of the photovoltaic cells; at present prices, costs would exceed $100,000 per kilowatt of capacity. Cost reduction by a factor of at least 300 is required to achieve reasonable costs, and many years of intensive development will clearly be required to achieve this goal—if, in fact, it is attainable. An alternate approach to cost reduction would be the use of a one-axis concentrator, yielding reduction by a factor of 50 in the required solar cell area; but cells for such applications would have to be specially designed to accept energies of up to 50 times the sun's intensity.

Solar power plants mounted in satellites in synchronous earth orbit have been proposed by Peter E. Glaser of Arthur D. Little, Inc. Such a system would not be affected by clouds, and the orbit proposed is such that the system would receive solar energy continuously except for two six-week periods a year when earth shadowing would occur for about one hour at what would be midnight in the time zone under the satellite.

The proposal is to use photovoltaic cells to produce electricity as direct current, which would then be converted to radio energy at microwave frequency (3,000 megahertz) for transmission to the earth's surface where conversion to alternating current power would occur. All the components of the system would be massive by any scale with which we are familiar: the satellite-mounted transmitting antenna would be 1.4 square kilometers, and the receiving antenna on earth would cover a square 10 kilometers on each side. The satellite would carry an array of solar cells about 7 kilometers square, giving an output power of 10^7 kilowatts—sufficient, for example, to meet the electrical power needs of New York City. Parameters for such a satellite would be:

size of satellite microwave transmitting antenna	1.4×1.4 kilometers
size of ground microwave receiving antenna	10×10 kilometers
efficiency of solar cells	15 per cent
efficiency of microwave transmission system	70 per cent
projected weight in orbit	2.2×10^6 kilograms

CHANGES Needed

Such a system could be competitive in cost with conventional power plants or even some of the proposed ground-based solar power plants only with very substantial reductions from present satellite launch costs (by a factor of 50 to 1) and solar cell costs (by a factor of 1,000 to 1). Very substantial advances in technology are required to achieve such cost reductions.

ECONOMICS VERSUS THE SUN: WHEN SOLAR ENERGY?

From this range of possibilities for the utilization of solar energy, what applications may we envision for the next few decades? The first point to be made in answering that question is the lesson of basic economics: no application of solar energy will be made unless it is economically advantageous in comparison with available alternatives. The result of this rule applied in the past is that there are now no significant solar energy systems in operation in the United States.

For residential space and water heating applications in central United States solar systems are currently cost competitive with electric energy and within a factor of 1.5 of being cost competitive with gas- and oil-fueled systems. In these same locations total-energy solar systems are probably within a factor of 2 of being cost competitive with fossil-fueled systems, and large-scale ground-based solar electric plants are within a factor of 4 of competing with electric plants fueled with conventional energy sources.

There are good reasons for believing that many of these solar-powered systems may become economically attractive in the years to come. This prediction is based on the probability that the costs of energy derived from conventional sources such as fossil fuels and nuclear fission may rise faster in the future than the costs of building solar-energy systems. Indeed, the latter may be lowered through a vigorous research and development program which places significant emphasis on economical design and production techniques. (See Table 4-15 and Figure 4-16.)

The rationale for the costs of conventional fuels to rise faster than any general rate of inflation is based on two predictions: inexpensive fuel resources will gradually be

Table 4-15. Projected Solar-Energy Application Dates.[a]

	Application Dates Based on Comparative Cost Estimates with Present Technology Only		Application Dates with Intensive Solar Energy Research and Development Program	
	First Use in Favorable Areas	Extensive Use	First Use in Favorable Areas	Extensive Use
Residential space and water heating	1983	1993	1978	1988
Total energy systems in commercial plants	1990	2000	1980	1990
Large-scale power generation	2006	2016	1985	1995

[a]How soon will solar energy become competitive with fossil and nuclear fuels in the United States? The chart shows the author's estimates—the left columns based solely on current solar energy technology in competition with increasingly costly fossil and nuclear fuels, the right columns based on the improved solar technology that might result from intensive research and development beginning immediately.

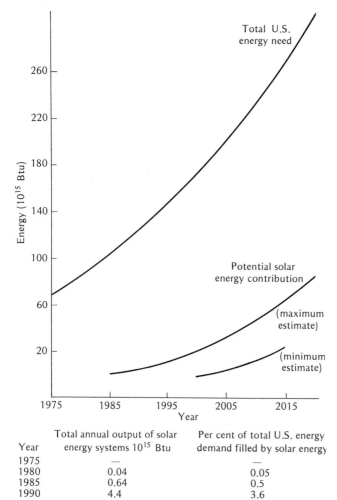

Figure 4-16. A vigorous successful solar-energy research and development program, plus substantial investments in solar-energy systems ($25 billion annually by the year 2000) and the facilities for their manufacture (a total of $11 billion by the year 2000) would result in rapid growth of solar energy utilization by the end of the twentieth century. Indeed, the author proposes that more than 25 per cent of all the energy which the United States requires in 2020 could be drawn directly from the sun.

Year	Total annual output of solar energy systems 10^{15} Btu	Per cent of total U.S. energy demand filled by solar energy
1975	—	—
1980	0.04	0.05
1985	0.64	0.5
1990	4.4	3.6
1995	13	8.7
2000	23	13
2005	33	16
2010	47	20.4
2015	63	23.8
2020	78	26

depleted, and continued increasing demand will force higher prices from all available sources. A Dow Chemical Co. study has projected cost increases in the period from 1974 to 1980 of 8.4 per cent a year for oil and gas, 10 per cent a year for electricity, and 6.7 per cent a year for coal. Current costs of nuclear plants are increasing sharply because of increasingly stringent safety regulations,

and limitations on the supply of uranium can be expected to force fuel price increases. All such projected rates of energy price increase are well in excess of the 4 per cent long-term inflation in materials and services costs which most economists foresee during the current decade. Indeed, only a technical and economic success with the breeder reactor might basically affect these energy cost projections; fusion power systems are not considered to be a factor in the current century.

On the basis of an average annual increase in fuel and electricity costs of about 8 per cent and a general price inflation (including solar energy systems) of 4 per cent, one can estimate the dates when various types of solar energy systems might become less expensive than fossil-fueled alternative energy sources. Such calculations suggest that solar energy, now 1.5 times as costly as fossil fuels for residential space and water heating, may be competitive for this purpose in favorable areas of the United States such as the Southwest—where the climate is moderate and sunshine relatively plentiful—by 1983, and that it may come into extensive use (as a result of a further 50 per cent cost differential) by 1993, when it would be competitive for residential applications in most United States climates. Similar calculations suggest that solar-powered total-energy systems for commercial plants might first appear practical in 1990 and come into more extensive use by the year 2000, and that

comparable dates for large-scale solar-based power plants are 2006 and 2016.

Two factors may affect these estimates. One is any future development of nuclear technology which may act to reduce the cost of nuclear power. The other is the possible effect of an extensive solar energy research and development program. Cost improvements in solar energy systems by a factor of 2 seem relatively easy to achieve given even modestly successful technical innovation, and such improvements might be demonstrated in three to five years in installations of modest size. This could lead to the first extensive use of solar heating systems by 1978 and of total energy systems by 1980. Additional intensive development during the next five years could make large-scale solar plants competitive by 1985. All these estimates are summarized in the accompanying tables. (See Tables 4-16 and 4-17.)

THE GROWTH OF THE SOLAR INDUSTRY

How rapidly might solar energy systems gain

Table 4-16. Projected Annual Savings on Solar-Energy Systems in the United States.[a]

Year	Solar Space and Domestic Hot-Water Systems		Solar Total-Energy Systems		Solar Base-Load Electric Power Plants	
	Number of Dwellings (millions)	Annual Energy Savings (10^{15} BTU)	Floor Area of Buildings (10^6 square meters)	Annual Energy Savings (10^{15} BTU)	Installed Capacity (10^6 kilowatts)	Annual Energy Savings (10^{15} BTU)
1980	0.3	0.04	—	—	—	—
1985	3	0.4	52	0.24	—	—
1990	8.4	1.2	200	0.92	18.7	1.4
1995	14	1.9	400	1.9	69	5.3
2000	20	2.7	610	2.8	137	10.4
2005	27	3.6	850	3.8	208	15.2
2010	34	4.7	1,090	5	284	22
2015	41	5.9	1,350	6.2	363	30
2020	49	7	1,620	7.6	445	37.6

[a]Assuming the most favorable development, annual energy savings of 52.2 × 10^{15} BTU—over 1,160 million tons of oil—may be realized by the use of solar energy by the year 2020. The estimates for dwellings are based on the use of approximately 10^8 BTU annually and a heating plant efficiency of 70 per cent; those for total-energy systems on consumption of 6.67 million kilowatt-hours per year in a plant of 10,000 square meters floor area, with an efficiency of 50 per cent; those for power generation on the construction of solar plants with total capacity of 12.5 × 10^6 killowatt hour each year beginning in 1995, the plants having 40 per cent efficiency.

Table 4-17. Projected Capital Expenditure on Solar-Energy
Systems in the United States.[a]

	Required Annual Capital Expenditures (billions of 1973 dollars)				
Year	House Heating	Total-Energy Systems	Electric Plants	Hydrogen Plants	Total
1978	$0.2	—	—	—	$ 0.2
1980	0.8	—	—	—	0.8
1985	3.2	$0.9	—		4.1
1990	4.2	1.8	$4.4	$4.4	15
1995	4.4	1.9	8.8	8.8	24
2000	4.5	2.0	9.1	9.1	25

[a]If the author's forecasts of technological and economic developments are fulfilled, solar energy will become increasingly competitive with other energy sources during the last 20 years of the
twentieth century, and there will be large capital expenditures in
solar energy systems. Indeed, by the year 2000 such annual expenditures could total at least $25 billion.

acceptance in the United States? The following estimates are made on the basis of two
assumptions: that solar energy is applied
only to new installations (that is, no retro-
fitting is done); and that rates of investment
in new houses, industrial plants, and power
plants continue in the future at approximately present levels.

In the case of residential space and water
heating systems, a total building rate of 2
million units per year is assumed, of which
one-half might be suitable for solar heating.
Installation would begin in 1978; the rate
would rise linearly to 1 million per year in
1988, and thereafter installations would increase in proportion to population growth.
By 1990, with 8.4 million dwellings
equipped with solar systems, annual savings
could be 1.2×10^{15} BTU, equivalent to 30
million tons of oil.

An estimate of the maximum rate of
application of solar total-energy systems to
commercial and industrial buildings can be
made by noting that, on the average,
approximately 2×10^7 square meters of
industrial floor space are constructed every
year. Assuming that half of this construction
is single-story and suitable for solar total-
energy systems and that such systems are
available by 1980, installations associated

with 3.5×10^7 square meters of construction per year might be achieved by 1990. By
then, annual energy savings of 92×10^{13}
BTU, or 23 million tons of oil, could be
achieved.

In 1970 to 1971, electric generating
capacity was being constructed at the rate of
about 25×10^6 kilowatts per year. If it is
assumed that large-scale electric power production by solar energy can begin in 1985
and that by 1995 one-half of all new electric
power installations are solar powered, 69
million kilowatts of base-load electric power
generation will then be from solar energy,
with an annual saving of 131 million tons of
oil or its equivalent.

One may postulate that some of the solar
electric power proposed above would be
redundant with solar home-heating or total-
energy commercial units. But one may also
propose that many other energy requirements—buildings inappropriate for solar
power because of siting or multistory construction, industrial processes, and transportation—could be supplied with energy in the
form of hydrogen or hydrogen-derived fuels
produced by large-scale, high-efficiency
solar-powered plants.

Though there are no proven methods of
directly producing hydrogen at high effi-

ciency (70 per cent or better) from solar energy, a number of multistage thermally driven chemical processes seem to have promise. If one assumes that the costs per unit area of collector will be the same for hydrogen plants as for base-load solar-powered electric plants, that such plants are built at the same rate (rated in terms of solar collection area) as electric plants, and that the efficiencies are of the order of 70 per cent, the energy savings will be about 70 per cent of those achieved by construction of solar-powered base-load electric plants. Construction might be feasible by 1990, and projected energy savings by the year 2000 are some 180 million tons of oil.

The estimates of solar energy utilization developed in the preceding paragraphs can be combined to indicate total possible energy savings for the United States through solar energy development. As Table 4-16 indicates, these savings may begin to be significant during the last decade of this century. The growth in solar energy output projected here is set primarily by the capital investment assumptions given earlier. But other factors may also affect the validity of these estimates. If energy demands can be moderated through conservation programs or under the influence of higher prices, solar sources could meet a larger percentage of total United States energy need than indicated.

If development funding lags or if opportunities for applications are more limited than assumed, a much lower utilization of solar energy would result—the minimum contribution shown in the chart. But it is clear that in either of these situations solar energy has the possibility of contributing very significantly to the country's energy needs. (A third possibility is that a new energy source, such as nuclear fusion, may turn out to be less expensive and more readily available than solar energy systems. In that case, very few if any solar systems will be built.)

One point needs emphasis: no breakthroughs are needed to make solar energy feasible. All that is required is to engineer solar energy systems less costly than current designs by a factor of 1.5 to 4 depending on the type of application— except that considerably greater cost reductions will be necessary to make practical photovoltaic solar plants.

It is fair to say that the engineering problems to be solved in making solar energy practical are considerably simpler than those of the breeder reactor and *far* simpler than those that must be solved in devising a practical fusion reactor. Yet it is also fair to say that a substantial research and development program, together with a vigorous implementation program, will be required to confirm that solar energy may, in fact, provide a significant portion of the nation's future energy needs.

AN AGENDA FOR RESEARCH AND DEVELOPMENT

How much research and development investment can be considered? About 4.5 per cent of sales is typically devoted to research and development in United States industry, and such expenditures are usually required about five years in advance of the beginning of production. Hence one may hypothesize that solar energy research and development expenditures should be computed as 4.5 per cent of the sales anticipated five years hence. Data developed in the following paragraphs suggest sales of $200 million in solar energy equipment in 1978, $15 billion (in 1973 dollars) in 1990, and $25 billion in the year 2000. On this basis one postulates research and development expenditures of $10 million in 1973, $670 million in 1985, and $1.1 billion in 1995. (See Table 4-17.)

The research and development program should include efforts in the nine areas:

(1) Development of low-output-temperature (less than 200°F) solar collectors for space and water heating applications. The emphasis should be on achieving low cost and high collection efficiency in a design compatible with housing construction practices.

(2) Development of inexpensive low-temperature energy storage techniques capable of holding sufficient energy to heat a house for several days, in designs compatible with housing construction practices.

(3) Development of low-cost high-efficiency solar collectors for applications requiring high-output temperatures (500 to 1,000°C) such as electric power plants using thermal cycles and hydrogen fuel production processes. Areas for study include absorbing surfaces with low emissivity, heat mirrors, insulation techniques, inexpensive solar trackers, heat transfer systems, and techniques for cleaning the collectors.

(4) Development of economical high-temperature thermal energy storage systems. Areas of possible effort include molten salts, rocks, and metals as heat storage media; insulation techniques; heat transfer systems; and techniques for combining such devices with solar collectors and thermal-cycle generating plants.

(5) Development of improved photovoltaic solar cells having increased efficiency and lower cost than contemporary designs.

(6) Development of solar concentrators for use with photovoltaic solar cells.

(7) Development of inexpensive electrical energy storage techniques for use with photovoltaic solar cell systems.

(8) Detailed surveys of solar intensity variations in possible areas of application.

(9) Systems studies on various applications of solar energy with emphasis on system costs, financing methods, and the problems of interfacing solar energy with various processes and systems which now consume energy.

In addition to the above, successful research and development on high-efficiency energy conversion techniques for electric power and hydrogen fuel production would contribute significantly to the early realization of solar energy systems.

$36 BILLION FOR SOLAR ENERGY BY THE YEAR 2000

Success in these research and development efforts will make possible the rates of application postulated above. But if these rapid advances are to occur, large capital investments will also be required. To simplify the earlier discussion, the cost of the various types of solar energy systems can be projected as follows:

residential solar heaters	$4,000 each
total-energy plants	$50 per square meter of building supplied, or $100 per square meter of collector area
electric base-load power plants	$700 per kilowatt of electrical output power, or $40 per square meter of collector area
hydrogen production plants	$40 per square meter of collector area

These costs, taken with the application rates developed above, yield the capital expenditure rates shown in Table 4-17—an annual expenditure of $25 billion (1973 dollars) to be invested in solar heating and energy systems by the year 2000.

To these figures must be added the cost of the substantial number of production plants needed to make such large numbers of solar-energy devices and facilities. A crude estimate of this required investment in production facilities can be made by extrapolating from the fact that manufacturing in the United States now requires plant investments equal to about 45 per cent of annual output. The plant is depreciated, on the average, over about 14 years. On this basis an investment of $11 billion (1973 dollars) in solar plant production facilities is indicated by the year 2000.

Combining the cost of research and development, production facilities, and the systems themselves gives a total solar-energy investment of about $300 billion in the next 27 years. As Figure 4-16 shows, investment at that level would mean that 13 per cent of projected United States energy requirements could be filled by solar systems in the year 2000, 26 per cent in 2020.

Though substantial collector areas would be required, the total area involved by the year 2020, about 10,000 square kilometers, would be much less than that used currently for highways. In fact, a substantial fraction of the collector area needed could be accommodated on the roofs of buildings; the rest could be accommodated on land shared with farming or grazing.

The large-scale use of solar energy should have a minimal environmental effect, since such systems operate from an almost inexhaustible energy source external to the earth, produce no pollution products, and can be designed to have minimal effect on the heat balance of the earth.

SUGGESTED READINGS

Brinkworth, B. J., 1972, *Solar Energy for Man*, Compton Press, Salisbury, Wiltshire, England.

Duffie, John A., and William A. Beckman, 1974, *Solar Energy Thermal Processes*, Wiley (Interscience), New York.

Glaser, P. E., 1973, "Space Solar Power," Paper presented at Paris International Congress, July.

Glaser, P. E., 1974, "Power from the Sun," *UNESCO Courier 27*, January, pp. 16–21.

Hay, H. R., 1973, "Energy Technology and Solarchitecture," *Mechanical Engineering*, November, pp. 18–22.

Löf, George O. G., 1973, "Solar Heating and Cooling: Untapped Energy Put to Use," *Civil Engineering*, September, pp. 88–92.

Löf, George O. G., and R. A. Tybout, 1973, "House Heating with Solar Energy," *Solar Energy*, Vol. 14, No. 3, February.

Meinel, Aden B., and Marjorie P. Meinel, 1972, "Physics Looks at Solar Energy," *Physics Today*, February, pp. 44–50.

Metz, William D., 1973, "Ocean Temperature Gradients: Solar Power from the Sea," *Science*, June 22, 1973, pp. 1266–1267.

NSF/NASA Solar Energy Panel, 1972, "An Assessment of Solar Energy as a National Energy Resource," Dept. of Mechanical Engineering, University of Maryland, College Park, Md, 85 p.

Solar Energy, The Journal of Solar Energy Science and Technology, Smithsonian Radiation Biology Laboratories, Rockville, Md.

Tamplin, Arthur R., 1973, "Solar Energy," *Environment*, June, pp. 16–34.

Thekaekara, Matthew P. (ed.), 1974, "The Energy Crisis and Energy from the Sun," Institute of Environmental Sciences, Mt. Prospect, Ill.

Wolf, Martin, 1974, "The Potential Impacts of Solar Energy," *Energy Conversion*, August, pp. 9–20.

Zener, Clarence, 1973, "Solar Sea Power," *Physics Today*, January, pp. 48–53.

Geothermal Energy

L. J. P. MUFFLER and D. E. WHITE   1972

Geothermal energy, in the broadest sense, is the natural heat of the earth. Temperatures in the earth rise with increasing depth. At the base of the continental crust (25 to 50 kilometers), temperatures range from 200°C to 1,000°C; at the center of the earth (6,371 kilometers), perhaps from 3,500°C to 4,500°C. But most of the earth's heat is far too deeply buried to be tapped by man, even under the most optimistic assumptions of technological development. Although drilling has reached 7.5 kilometers and may some day reach 15 to 20 kilometers, the depths from which heat might be extracted economically are unlikely to be greater than 10 kilometers.

White[1] has calculated that the amount of geothermal heat available in this outer 10 kilometers is approximately 3×10^{26} calories, which is more than 2,000 times the heat represented by the total coal resources of the world.[2] Most of this geothermal energy, however, is far too diffuse ever to be recovered economically. The average heat content of each gram of rock in the outer 10 kilometers of the earth is only 0.3 per cent of the heat obtainable by combusting 1 gram of coal and is less than 0.01 per cent of the heat equivalent of fissionable uranium and thorium contained in 1 gram of average granite. Consequently, most of the heat within the earth, even at depths of less than 10 kilometers, cannot be considered a potential energy resource.

Geothermal energy, however, does have potential economic significance where the heat is concentrated into restricted volumes in a manner analogous to the concentration of valuable metals into ore deposits or of oil into commercial petroleum reservoirs. At present, economically significant concentra-

1. White, D. E. "Geothermal Energy." *U.S. Geological Survey Circular* 519. 1965. 17pp.

2. Averitt, Paul. "Coal Resources of the United States, January 1, 1967." *U.S. Geological Survey Bulletin 1275.* 1969. 116pp.

Dr. L. J. P. Muffler is a geologist for the U.S. Geological Survey and coordinator of the Geothermal Research Program at Menlo Park, California.

Dr. Donald E. White is a geologist for the U.S. Geological Survey and managed the U.S. Geological Survey geothermal resource investigations prior to 1972.

From *The Science Teacher,* Vol. 39, No. 3, pp. 40–43, March 1972. Reprinted with authors' revisions and by permissions of the authors and *The Science Teacher.* Copyright 1972 by the National Science Teachers Association.

tions of geothermal energy occur where elevated temperatures (40°C to more than 380°C) are found in permeable rocks at shallow depths (less than 3 kilometers). The thermal energy is stored both in the solid rock and in water and steam filling pores and fractures. This water and steam serve to transfer the heat from the rock to a well and thence to the ground surface. Under present technology, rocks with too few pores, or with pores that are not interconnected, do not comprise an economic geothermal reservoir, however hot the rocks may be.

Water in a geothermal system also serves as the medium by which heat is transferred from a deep igneous source to a geothermal reservoir at depths shallow enough to be tapped by drill holes. Geothermal reservoirs are located in the upflowing parts of major water convection systems (Figure 4-17). Cool rain water percolates underground from areas that may comprise tens to thousands of square kilometers and then circulates downward. At depths of 2 to 6 kilo-

meters, the water is heated by contact with hot rock (in turn, probably heated by molten rock). The water expands upon heating and then moves buoyantly upward in a column of relatively restricted cross-sectional area (1 to 50 square kilometers). If the rocks have many interconnected pores or fractures, the heated water rises rapidly to the surface and is dissipated rather than stored. If, however, the upward movement of heated water is impeded by rocks without interconnected pores or fractures, the geothermal energy may be stored in reservoir rocks below the impeding layers. The driving force of this large circulation system is gravity—effective because of the density difference between cold, downward-moving

3. White, D. E. "Environment of Generation of Some Base-Metal Ore Deposits." *Economic Geology* 63: 301–335; June-July 1968.
4. White, D. E. "Hydrology, Activity, and Heat Flow of the Steamboat Springs Thermal System, Washoe County, Nevada." *U.S. Geological Survey Professional Paper* 458-C, 1968. 109 pp.

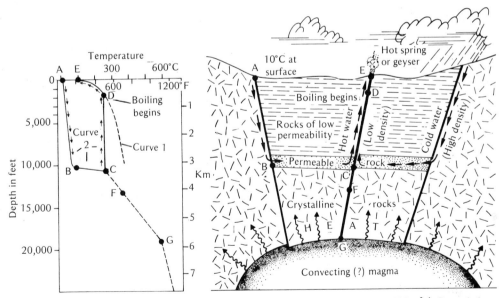

Figure 4-17. Schematic model of a hot-water geothermal system, modified from White.[3,4] Curve 1 shows the boiling point of pure water under pressure exerted by a column of liquid water everywhere at boiling, assuming water level at ground surface. Dissolved salts shift the curve to the right; dissolved gases shift the curve to the left. Curve 2 shows the ground temperature profile of a typical hot-water system.

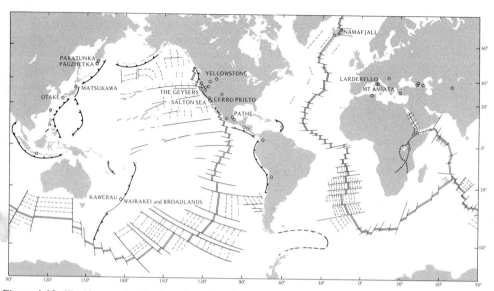

Figure 4-18. World map showing location of major geothermal fields along plate margins. Heavy double lines represent spreading ridges; heavy lines with barbs represent active subduction zones; heavy dotted lines represent rift valleys. Light lines represent transform faults; dashed light lines represent approximate position of magnetic anomalies. (Base map and tectonic features from Coleman, ref. 5, Figure 4.)

recharge water and hot, upward-moving geothermal water.

Many investigators in the past considered the water in geothermal systems to be derived from molten rock at depth. Modern studies of hydrogen and oxygen isotopes in geothermal waters, however, indicate that at least 95 per cent of most geothermal fluids must be derived from surface precipitation and that no more than 5 per cent is volcanic steam.

5. Coleman, R. G. "Plate Tectonic Emplacement of Upper Mantle Peridotites along Continental Edges." *Journal of Geophysical Research* 76: 1212–1222; February 1971.

LOCATION OF GEOTHERMAL SYSTEMS

Geothermal reservoirs are the "hot spots" of larger regions where the flow of heat from depth in the earth is one and one-half to perhaps five times the world-wide average of 1.5×10^{-6} calories per square centimeter per second. Such regions of high heat flow commonly are zones of young volcanism and mountain building and are localized along the margins of major crustal plates (Figure 4-18). These margins are zones where either new material from the mantle is being added to the crust (that is, spreading ridges; see Figure 4-19) or where crustal material is

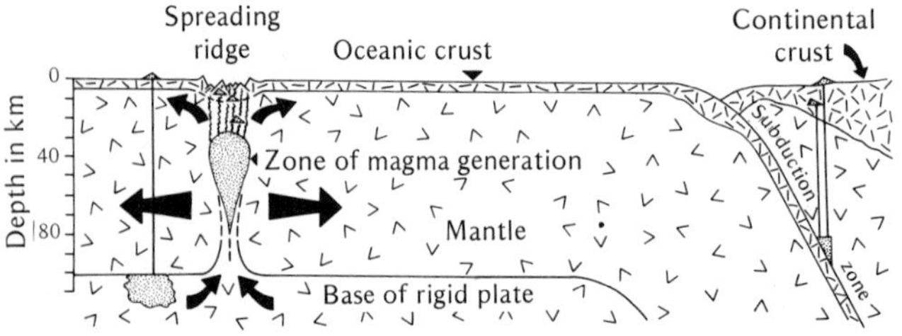

Figure 4-19. Model of development of oceanic crust at spreading ridges and subduction of oceanic crust at consuming plate margins. (Generalized from Coleman, ref. 5, Figure 6.)

being dragged downward and "consumed" in the mantle (subduction zones). In both situations, molten rock is generated and then moves buoyantly upward into the crust. These pods of igneous rock provide the heat that is then transferred by conduction to the convecting systems of meteoric water.

Figure 4-18 shows that the geothermal fields presently being exploited or explored occur in three major geologic environments: (a) along spreading ridges, (b) above subduction zones, and (c) along the belt of mountains extending from Italy through Turkey to the Caucasus. Although this last zone is not a modern subduction zone, it is the zone where the African and European plates are in contact, and it appears to have been a subduction zone in the past. Geothermal fields are absent from the stable, continental shields, which are characterized by lower-than-average heat flow. Although there are no known shallow geothermal reservoirs in

the nonvolcanic continental areas bordering the shields, hot water has been found at depths of 3 to 6 kilometers in the Soviet Union, in Hungary, and on the Gulf Coast of the United States.[6]

USES OF GEOTHERMAL RESOURCES

The primary use of geothermal energy to date is for the generation of electricity. For this purpose, under existing technology, the geothermal reservoir must have a temperature of at least 180°C, and preferably 200°C. Geothermal steam, after separation of any associated water (as much as 90 weight per cent of the total effluent), is expanded into a turbine that drives a con-

6. Jones, P. H. "Geothermal Resources of the Northern Gulf of Mexico Basin." United Nations Symposium on the Development and Utilization of Geothermal Resources, Pisa, Italy, September 1970. Paper I/24.

Table 4-18. World Geothermal Power Production, June 1974.

| Country | Field | Electrical Capacity (megawatts) | |
		Operating	Under Construction
Italy	Larderello	365.1	
	Monte Amiata	25.5	
	Travale	15	
United States	The Geysers	396	106
New Zealand	Wairakei	160	
	Kawerau	10	
Mexico	Cerro Prieto	75	
Japan	Matsukawa	20	
	Otake	13	
	Onuma	10	
	Onikobe		25
Soviet Union	Pauzhetka	5	
	Paratunka[a]	0.7	
Iceland	Namafjall	2.5	
		1097.8	131

Total megawatts: 1228.8

[a]Freon plant.

ventional generator. World electrical capacity from geothermal energy in 1971 was approximately 800 megawatts (Table 4-18), or about 0.08 per cent of the total world electrical capacity from all generating modes. Power from favorable geothermal systems is competitive in cost with either fossil fuel or nuclear power. The production of geothermal power is obviously restricted to areas where geothermal energy is found in sufficient quantity. Unlike coal, oil, gas, or uranium, geothermal steam cannot be transported long distances to a generating plant located near the existing load centers.

Geothermal resources have other uses, but to date they have been minor. Geothermal waters as low as 40°C are used locally for space heating and horticulture. Much of Reykjavik, the capital of Iceland, is heated by geothermal water, as are parts of Rotorua (New Zealand), Boise (Idaho), Klamath Falls (Oregon), and various towns in Hungary and the Soviet Union. Geothermal steam is also used in paper manufacturing at Kawerau, New Zealand, and has potential use for refrigeration. Some geothermal waters contain potentially valuable by-products such as potassium, lithium, calcium, and other metals. Use of geothermal energy to desalt geothermal water itself has been proposed, and the U.S. Bureau of Reclamation and the Office of Saline Water are presently developing a pilot operation for producing fresh water from the geothermal waters of the Imperial Valley, Southern California.

TYPES OF GEOTHERMAL SYSTEMS

There are two major types of geothermal systems: hot-water systems and vapor-dominated ("dry-steam") systems.[7] Among geothermal systems discovered to date, hot-water systems are perhaps twenty times as common as vapor-dominated systems.[8]

Hot-Water Geothermal Systems

Hot-water geothermal systems contain water at temperatures that may be far above surface boiling, owing to the effect of pressure on the boiling point of water (curve 1 of Figure 4-17). A typical hot-water system has temperature-depth relations similar to those of curve 2. Little change in temperature occurs as meteoric water descends from A to B, heat is absorbed from B to C, and from C to D the system contains water at nearly constant temperature (the "base temperature"). From D to E pressure has decreased enough for water to boil, and steam and water coexist. In major zones of upflow, coexisting steam and water extend to the surface and are expressed as boiling hot springs and locally as geysers. Geothermal wells, however, are usually sited in nearby cool, stable ground where near-surface temperatures are controlled by conduction of heat through solid rocks; the temperature-depth curve is therefore initially to the left of curve 1 of Figure 4-17.

Water in most hot-water geothermal systems is a dilute solution (1,000 to 30,000 milligrams per liter), containing mostly sodium, potassium, lithium, chloride, bicarbonate, sulfate, borate, and silica. The silica content and the ratio of potassium to sodium are dependent on temperature in the geothermal reservoir, thus allowing prediction of subsurface temperature from chemical analysis of hot springs.[9,10]

In hot-water geothermal systems, only part of the produced fluid is steam and can be used to generate electricity with present technology. For example, water at 250°C

7. White, D. E., L. J. P. Muffler, and A. H. Truesdell. "Vapor-dominated Hydrothermal Systems Compared with Hot-Water Systems." *Economic Geology* 66: 75–97; January–February 1971.
8. White, D. E. "Geochemistry Applied to the Discovery, Evaluation, and Exploitation of Geothermal Energy Resources." United Nations Symposium on the Development and Utilization of Geothermal Resources, Pisa, Italy, September 1970. Rapporteur's Report, Section V.
9. Ellis, A. J. "Quantitative Interpretation of Chemical Characteristics of Hydrothermal Systems." United Nations Symposium on the Development and Utilization of Geothermal Resources, Pisa, Italy, September 1970. Paper I/11.
10. Fournier, R. O., and J. J. Rowe. "Estimation of Underground Temperatures from the Silica Content of Water from Hot Springs and Wet-Steam Wells. *American Journal of Science 264*: 685–697, 1966.

will produce only about 20 weight per cent of steam when the confining pressure is reduced to 6 kilograms per square centimeter, the approximate well-head pressure commonly used in geothermal installations. The steam and water at this pressure are mechanically separated before the steam is fed to the turbine.

Some attention is currently being directed toward a heat exchange generating system. Heat in the geothermal water is transferred by a heat exchanger to a low-boiling-point fluid, such as freon or isobutane, which is then expanded into a turbine. The geothermal water is not allowed to boil and is reinjected as water into the ground. If this binary fluid generating technology proves economically feasible, it will allow more complete extraction of heat from geothermal fluids and will allow use of hot-water geothermal systems of lower temperature than are presently required for direct geothermal steam generation.

The major known hot-water geothermal fields are Wairakei (160 megawatts) and Broadlands (100 megawatts proposed) in New Zealand, Cerro Prieto (75 megawatts operating; 200 megawatts proposed) in Mexico, the Salton Sea field in California, and the Yellowstone geyser basins in Wyoming. Although the Yellowstone region is the world's most intensive display of hot-spring and geyser phenomena, the area is permanently withdrawn as a national park and will never be exploited for power.

Whereas the salinities of most hot-water fields are 0.1 to 3 per cent, the Salton Sea geothermal reservoir contains a brine with more than 25 per cent by weight of dissolved solids, mainly chloride, sodium, calcium, and potassium. In addition, the brine is rich in a variety of metals.[3] Although temperatures reach 360°C, development of the field has been hindered by problems of corrosion, deposition of silica, and disposal of unwanted effluent. Hot, saline brines also occur in pools along the median trench of the Red Sea where geothermal fluids discharge directly onto the sea floor 2 kilometers below sea level.[3]

Vapor-Dominated Geothermal Systems

Vapor-dominated geothermal systems, in contrast to hot-water systems, produce superheated steam with minor amounts of other gases (CO_2, H_2S), but no water. The total fluid can therefore be piped directly to the turbine. Within the vapor-dominated geothermal reservoir, saturated steam and water coexist, with steam being the phase that controls the pressure. With decrease in pressure upon production, heat contained in the rocks dries the fluids first to saturated and then to superheated steam, with as much as 55°C superheat at a well-head pressure range of 5 to 7 kilograms per square centimeter. Owing to the thermodynamic properties and flow dynamics of steam and water in porous media, vapor-dominated reservoirs are unlikely to exist at pressures much greater than about 34 kilograms per square centimeter and temperatures much above 240°C.[11] Hot brine probably exists below the vapor-dominated reservoirs at depth, but drill holes are not yet deep enough to confirm the presence of such a brine.

Drilling has demonstrated the existence of only three commercial vapor-dominated systems: Larderello, Italy; The Geysers, California; and probably Matsukawa, Japan. Two small fields in the Monte Amiata region of Italy are marginally commercial. Larderello was the first geothermal field to be exploited, starting in 1904, and is still a large producer of geothermal power (380 megawatts). The Geysers at present produces 396 megawatts, but plants under construction will boost capacity to 502 megawatts in 1975, and ultimate potential is in excess of 1000 megawatts.

THE GEOTHERMAL ENERGY RESOURCE

White[1] estimated that the total stored heat of all geothermal reservoirs to a depth of 10

11. James, Russell. "Wairakei and Larderello: Geothermal Power Systems Compared." *New Zealand Journal of Science* 11: 706–719; 1968.

kilometers was 10^{22} calories. This estimate specifically excluded reservoirs of molten rock, abnormally hot rocks of low permeability, and deep sedimentary basins of near "normal" conductive heat flow, such as the Gulf Coast of the United States or Kazahkstan in the Soviet Union. The geothermal resources in these environments are at least ten times greater than the resources of the hydrothermal systems, but they are recoverable only at much more than present costs. Should production of these geothermal resources someday become feasible, the potential geothermal resource in all reservoir types would be at least 10^{23} calories, which is approximately equivalent to the heat represented by the world's potential resources of coal.

For a hot-water geothermal system, approximately 1 per cent of the heat stored in the reservoir can be extracted and converted into electricity under present technology. For the far less abundant vapor-dominated geothermal systems, perhaps 2 to 5 per cent of the heat in the reservoir can be extracted and converted to electricity. Therefore, if the use of geothermal energy continues to be restricted primarily to electrical generation by proven techniques, then the potential geothermal resource to 10 kilometers is only about 10^{20} calories. To a depth of 3 kilometers (the deepest well drilled to date for geothermal power), the resource for electrical generation by proven techniques is even less, approximately 2×10^{19} calories.[1] Use of geothermal resources for other than electrical generation (for example, heating, desalination, horticulture) would greatly increase these geothermal resource estimates, perhaps by ten times, but all these uses involve special geographic and economic conditions that to date have been implemented only on a local scale.

Production of electricity from geothermal energy is presently attractive environmentally because no solid atmospheric pollu-

tants are emitted and no radiation hazard is involved. Geothermal generation is not without environmental effects, however. Effluent from either a hot-water or a vapor-dominated system can pollute streams or ground water. Consequently, federal and state regulations require reinjecting objectionable fluids back into a deep reservoir. Thermal pollution is also a problem, particularly in hot-water systems, but it can be solved in part by reinjection of unwanted water and of residual steam condensate. Noise, objectionable gases, visual impact, and subsidence of the land surface due to fluid withdrawal are other problems that are faced in any geothermal energy development.

Geothermal energy is unlikely to supply more than perhaps 10 per cent of domestic or world electrical power demand. In favorable areas, however, geothermal power may be of major importance, particularly in underdeveloped countries that have few other energy resources.

Although geothermal power was produced in Italy as early as 1904 and in New Zealand by 1955, extensive interest in geothermal resources of the United States has developed only in the past ten years. Large areas in the western United States appear to be favorable for geothermal exploration, but knowledge of the nature and extent of our geothermal resources is inadequate. Further investigations are necessary, not only of the distribution and characteristics of geothermal reservoirs, but also of the various ways in which geothermal energy can be used in the most beneficial and least wasteful manner.

SUGGESTED READINGS

U.S. Dept. of the Interior, 1973, "Final Environmental Impact Statement for the Geothermal Leasing Program," U.S. Dept. of the Interior, Washington, D.C. (Vol. 1–4).

Tidal Power

M. KING HUBBERT   1969

A source of power having a longevity measurable in geologic time is tidal power. Tidal power is similar in all essential respects to hydroelectric power with the following exception. Hydroelectric power is obtained from the energy of unidirectional stream flow, whereas tidal-electric power is obtained from the oscillatory flow of water in the filling and emptying of partially enclosed coastal basins during the semidiurnal rise and fall of the oceanic tides. This energy may be partially converted into tidal-electric power by enclosing such basins with dams to create a difference in water level between the ocean and the basin, and then using the water flow while the basin is filling or emptying to drive hydraulic turbines propelling electric generators.

In order to obtain a quantitative evaluation of the amount of tidal energy potentially obtainable from a given basin, it is useful to determine the maximum amount of energy that can be dissipated into heat during one complete tidal cycle. This is the amount of energy that would be dissipated if the dam gates were closed at low tide when the water in the basin is at its lowest level, and then opened wide allowing the basin to fill at the crest of the tide, and, in a similar manner, by closing the gates when the basin is filled at high tide, and then allowing the basin to empty at low tide.

This maximum possible amount of energy dissipated during one tidal cycle is given by

$$E_{max} = \rho g R^2 S, \qquad (1)$$

where ρ is the density of sea water, g the acceleration of gravity, R the tidal range, and S the surface area of the basin. When all of the quantities to the right in equation (1) are in meter-kilogram-second units, the energy will be in joules.

The maximum possible average power obtainable from such a basin would be obtained if all of the energy E_{max} in equation

Dr. M. King Hubbert is a research geophysicist with the U.S. Geological Survey, Washington, D.C. His scientific work has included: geophysical and geological exploration for oil, gas, and other minerals; structural geology and the physics of earth deformation; physics of underground fluids; and world's mineral and energy resources and the significance of their exploitation to human affairs. He has published 65 technical papers and is author or coauthor of several texts.

(1) were converted into electrical energy. This maximum average power would then be given by

$$\bar{P} = \frac{E_{max}}{T} = \frac{\rho g R^2 S}{T}, \qquad (2)$$

where T is the half period of the synodical lunar day. This is 12 hours and 24.4 minutes, or 4.46×10^4 seconds. When T is in seconds, $\bar{P}$ will be expressed in joules per second, or watts.

The actual energy and power obtainable by means of turbines and electrical generators from such a basin can be only a fraction of the quantities given in equations (1) and (2). In engineering design computations for various tidal-power projects, the amounts of energy and power producible are commonly within the range of 8 to 20 per cent of these maximum amounts, although in one instance, that of La Rance in France, the realizable power approaches 25 per cent.

The source of tidal energy is the combined kinetic and potential energy of the earth-moon-sun system. Hence, as this energy is dissipated on the earth, equivalent changes must occur in the rotational energy of the earth, and in the orbital motions of the moon about the earth and of the earth about the sun. These motional changes, which have been observed astronomically over a period of about three centuries, indicate that the day is lengthening by about 0.001 second per century with a corresponding decrease in the earth's rotational velocity. From such astronomical data, Munk and MacDonald (1960, p. 219) have estimated that the rate of tidal dissipation of energy on the earth is about 3×10^{12} watts.

A considerable fraction of this dissipation occurs in the oceans, especially in the shallow seas, bays, and estuaries, where the tidal ranges and tidal currents, because of inertial effects, become much greater than those in the open oceans. The oceanic tides, as measured on islands in the open oceans, have ranges commonly of less than a meter, whereas those in bays and estuaries have ranges, as shown in Table 4-19, from 1 to more than 10 meters.

A method of estimating the amount of energy dissipated by tides in shallow seas was developed in 1919 by G. I. Taylor and applied to the Irish Sea. The following year, this method was extended by Harold Jeffreys (1920; 1959, pp. 241–245) to most of the shallow seas of the earth for which he estimated a rate of energy dissipation at spring tides of about 22×10^{11} watts, of which 5×10^{11} watts, or two-thirds of the total was accounted for by the Bering Sea alone.

Recently, using oceanographic data subsequently acquired, Munk and MacDonald (1960, pp. 209–221) have re-estimated the energy dissipation in shallow seas. They obtained an average rate of, at most, 10^{12} watts, which is slightly less than the 1.1×10^{12} watts obtained when Jeffrey's rate for spring tides is reduced by a factor of 0.5 to give an average rate. Munk and MacDonald obtained a drastic reduction of Jeffrey's estimate for the Bering Sea from 75×10^{10} (one-half of 15×10^{11}) to only 2.4×10^{10} watts.

The significance of these estimates is that they establish a limit to the maximum amount of power that could possibly be developed from tidal sources. In Table 4-19, which is based on data compiled by Trenholm (1961) and by Bernshtein (1965), a summary is given of the average tidal ranges and basin areas for most of the more promising tidal-energy localities of the world. In addition, the average potential power, and maximum energy dissipation per year, as computed from equations (1) and (2), are given for each locality. The total maximum rate of energy dissipation for these localities amounts to 6.4×10^{10} watts, or 64,000 megawatts. This is about 6 per cent of the Munk and MacDonald estimate of a dissipation rate of 10^{12} watts for all of the shallow seas. If we make a liberal allowance of 20 per cent for the actual average power recoverable at each of these sites, we obtain a result of about 13×10^9 watts, or 13,000 megawatts as the approximate magnitude of the average value of the world's potential tidal-electric power. Comparing this with the

Table 4-19. Tidal Power Sites and Maximum Potential Power.[a]

Location	Average Range R (meters)	R² (square meters)	Basin Area S (square kilometers)	R²S (square meters × square kilometers)	Average Potential Power P (10³ kilowatts)	Potential Annual Energy E (10⁶ kilowatt-hours)
North America						
Bay of Fundy						
Passamaquoddy	5.52	30.5	262	7,990	1,800	15,800
Cobscook	5.5	30.3	106	3,210	722	6,330
Annapolis	6.4	41.0	83	3,440	765	6,710
Minas-Cobequid	10.7	114	777	88,600	19,900	175,000
Amherst Point	10.7	114	10	1,140	256	2,250
Shepody	9.8	96	117	11,200	2,520	22,100
Cumberland	10.1	102	73	7,450	1,680	14,700
Petitcodiac	10.7	114	31	3,530	794	6,960
Memramcook	10.7	114	23	2,620	590	5,170
Subtotal					29,027	255,020
South America						
Argentina						
San José	5.9	34.8	750	26,100	5,870	51,500
Europe						
England						
Severn	9.8	96.0	70	7,460	1,680	14,700
France						
Aber-Benoit	5.2	27.0	2.9	78	18	158
Aber-Wrac'h	5.0	25.0	1.1	28	6	53
Arguenon and Lancieux	8.4	70.6	28.0	1,980	446	3,910
Frênaye	7.4	54.8	12.0	658	148	1,300
La Rance	8.4	70.6	22.0	1,550	349	3,060
Rothéneuf	8.0	64.0	1.1	70	16	140
Mont Saint-Michel	8.4	70.6	610	43,100	9,700	85,100
Somme	6.5	42.3	49	2,070	466	4,090
Subtotal					11,149	97,811
Soviet Union						
Kislaya Inlet	2.37	5.62	2.0	11	2	22
Lumbovskii Bay	4.20	17.6	70	1,230	277	2,430
White Sea	5.65	31.9	2,000	63,800	14,400	126,000
Mezen Estuary	6.60	43.6	140	6,100	1,370	12,000
Subtotal					16,049	140,452
Grand Total					63,775	559,483

[a]Sources: N. W. Trenholm (1961); L. B. Bernshtein, 1965 (1961), Table 5-5, p. 173).

estimate of the world's potential water power of about 2,900,000 megawatts given in Table 4-20, it will be seen that the world's potential tidal power amounts to less than 1 per cent of its potential water power.

Although small tidal mills for the grinding of grain and similar purposes have been used since about the twelfth century, it is only within recent decades that tidal-electric installations have been given serious engineering consideration, and only recently actually brought into operation.

One of the best known of such projects has been that of Passamaquoddy Bay on the United States-Canadian boundary off the Bay of Fundy. This bay has an area of 262 square kilometers and an average tidal range of 5.52 meters, with a maximum potential average power (Table 4-19) of 1,800 megawatts. Plans were drafted for such a project during the early 1930's and construction was actually started before the project was finally killed by lack of Congressional appropriation. In 1948, interest in a Passamaquoddy Tidal Power Project was revived and a new engineering study was authorized by the United States and Canadian governments. This involved the establishment of an International Joint Commission and The International Passamaquoddy Engineering Board to study and draw engineering plans for such a project.

The Engineering Board, in its report of 1959, recommended a two-pool project involving both Passamaquoddy and Cobscook Bays, but with the power obtained solely from Passamaquoddy Bay during its emptying phase. This would have a power plant consisting of 30 unidirectional turbogenerator units of 10,000-kilowatt capacity each, or a total installed capacity of 300,000 kilowatts, with an annual energy production of $1,843 \times 10^6$ kilowatt-hours. Comparing the latter figure with that of $15,800 \times 10^6$ kilowatt-hours given in Table 4-19 as the maximum energy obtainable annually indicates that the proposed system would utilize but 11.8 per cent of the energy potentially available.

After studying this report, the International Joint Commission concluded that the project would be economically infeasible. In response, President John F. Kennedy, by letter of 20 May, 1961, requested the Department of the Interior to restudy the project and propose modifications. This resulted in a recommendation (Udall, 1963) that the power capacity be increased from 300,000 to 1 million kilowatt in order to deliver most of the power during the brief period of peak demand. It also involved a slight reduction from $1,843 \times 10^6$ to $1,318 \times 10^6$ kilowatt-hours in the annual energy production.

Table 4-20. World Water-Power Capacity.[a]

Region	Potential (10^3 megawatts)	Per Cent of Total	Development (10^3 megawatts)	Per Cent Developed
North America	313	11	59	19
South America	577	20	5	
Western Europe	158	6	47	30
Africa	780	27	2	
Middle East	21	1	—	
Southeast Asia	455	16	2	
Far East	42	1	19	
Australasia	45	2	2	
Soviet Union, China and satellites	466	16	16	3
Total	2,857	100	152	

[a]Source: M. King Hubbert, 1962, Table 8, p. 99, computed from data summarized by Francis L. Adams, 1961.

This was recommended to the President for authorization, but as yet no authorization has been obtained.

For the installation of the world's first major tidal-electric plant, that of La Rance estuary which began operation in 1966 (*Engineering,* July 1966, pp. 17–24), honor is due to France. Here, the average tidal range is 8.4 meters, and the power plant is in a dam enclosing an area of 22 square kilometers. The power plant comprises 24 units of 10,000-kilowatt capacity each, and the annual production of energy was estimated to be 544×10^6 kilowatt-hours, which amounts to about 18 per cent of the total energy available (Table 4-19). If the capacity is increased, as planned, to 320,000 kilowatts, this would increase the power utilization to about 24 per cent of that potentially obtainable. This high figure has been made possible by the use of turbines of an advanced design. These are horizontal, axial-flow turbines with adjustable blades permitting operation during both the filling and the emptying of the basin, and also their use as pumps.

The most recent tidal-electric project to go into operation, as reported by *The New York Times* on December 30, 1968, is a small Russian experimental station in the Kislaya Inlet on the Coast of the Barents Sea, 80 kilometers northwest of Murmansk. This consists of a single unit driven by a 400-kilowatt turbine of French manufacture. A second unit is to be installed later, bringing the total power capacity to 800 kilowatts.

According to the same article, a much larger 320,000-kilowatt plant is planned for the Lombovska River (Lumbovskii Bay, Table 4-19) on the northeast coast of the Kola Peninsula, and a 14 million-kilowatt plant for the Mezen Bay on the east side of the mouth of the White Sea. Since the stated capacities of these two plants are both larger than the maximum potential average power obtained from the Bernshtein data in Table 4-19, either the figures are exaggerated, or else it is now planned to enclose larger basins than those given by Bernshtein (1965, Table

5-5, p. 173).

In summary, it may be said that although the world's potential tidal power, if fully developed, would amount only to the order of 1 per cent of its potential water power, and to an even smaller fraction of the world's power needs, it nevertheless is capable in favorable localities of being developed in very large units. It has the additional advantage of producing no noxious wastes, of consuming no exhaustible energy resources, and of producing a minimum disturbance to the ecologic and scenic environment. There are accordingly many social advantages and few disadvantages to the utilization of tidal power wherever tidal and topographic factors combine to make this practicable.

REFERENCES

Bernshtein, L. B., 1965, "Energy for Electric Power Plants," Jerusalem: Israel Program for Scientific Translations.

Engineering, 1966, "Tidal Power Comes to France," *Engineering 202:* 17–24, July.

Jeffreys, H., 1920, "Tidal Friction in Shallow Seas," *Phil. Trans. Roy. Soc., A, 229:* 239–264.

Munk, D. W., and G. J. F. MacDonald, 1960, "The Rotation of the Earth, a Geophysical Discussion." Cambridge Monographs on Mechanics and Applied Mathematics, Cambridge Univ. Press, Cambridge, England.

Taylor, G. I., 1919, "Tidal Friction in the Irish Sea," *Phil. Trans. Roy. Soc., A, 220:* 1–33.

The New York Times, 1968, "Soviet Opens Tidal Power Station," The New York Times Co., December 30, 1968, p. 6.

Trentholm, N. W., 1961, "Canada's Wasting Asset—Tidal Power," *Elect. News Eng. 70*(2): 52–55.

Udall, S. L., 1963, "The International Passamaquoddy Tidal Power Project and Upper St. John River Hydroelectric Power Development," Report to President John F. Kennedy, U.S. Dept. of the Interior, Washington, D.C.

SUGGESTED READINGS

Gray, T. J., and O. K. Gashus (eds.), 1972, "Tidal Power," Proceedings of the International Conference on the Utilization of Tidal Power, 1970, Plenum Press, New York, 630 p.

Wind Power: A Near-Term Partial Solution to the Energy Crisis

WILLIAM E. HERONEMUS   1973

The United States daily sinks deeper into a morass of dependence upon foreign fuel resources, exponentially increasing water and air pollution and addiction to escalating nuclear power costs and safety hazards while ignoring solar energy. Wind power, a solar-energy-driven process, could be developed in the very near future as a partial but significant solution to our energy crisis. The total energy available to this country from the winds via practical wind-power systems complete with necessary storage subsystems could total at least 1 trillion kilowatt-hours per year. The most productive systems would be those installed offshore of New England and the Middle Atlantic coasts in the Westerlies, along the axes of the Great Lakes and through the Great Plains. Benefits starting with near freedom from pollution and ranging all the way through large numbers of factory jobs in areas of high unemployment, together with economical electricity free from future fuel cost and safety system cost escalations, could accrue to a national wind-power program.

THE CONCEPT

Moving particles of air possess momentum and that momentum can be exchanged to do useful shaft work. Winds can turn electrical generators: this has been done for many years in many places at power levels varying from a few watts to over 1 megawatt. The momentum available in the wind is capricious in the short term: the total energy content in a wind at a location measured over the long term is very reproducible. If the electricity produced by the random wind can be used in some process that accepts random energy input, wind power is useful in a most simple system. Most past work in large-scale wind-power planning has been based on the simple process of saving fuel in central generating plants when and as the wind would blow. On that basis the cost of wind power cannot exceed the differential cost of fuel—a very difficult economic situation. If random wind power can be harnessed by some storage scheme, then the total system cost can be compared in terms

Dr. William E. Heronemus is professor of civil engineering at the University of Massachusetts, Amherst, Massachusetts. His research interests are in developing wind energy as a viable means of producing electricity in various parts of the United States.

Presented at the *EASCON* of the Institute of Electrical and Electronics Engineers, Washington, D.C., September 18, 1973. Reprinted by permission of the author.

of the average cost of delivered electricity. If random wind power plus storage can be arranged to meet peak load requirements, then the total system cost can be evaluated in terms of peaking revenue in that system. Many significant wind-power development programs were mounted in different countries in the past: none ever showed a significant economic advantage for wind-power-generated electricity except for very remote regions where oil or coal transportation costs dominated the cost of delivered electricity. The United States is now entering an era where projected fuel costs suggest that wind power may have a significant economic advantage. Scarcely anyone paid any real attention to air and water pollution problems prior to 1968; the United States now, at least, recognizes that any energy conversion system free from air and water pollution effects should be credited with some added value, even through the mechanism for costing externalities has not yet been brought forward. Scarcely anyone has paid any real attention to the moral or ethical consideration involved in this plus the next few generations burning up all the remaining fossil fuels and raising accumulated radioactivity far above the natural background level; many nations are now willing to at least discuss these things, but very little action has yet been taken. A mature wind-power technology, albeit circa 1945 to 1950, exists: assembly-line production of components could start in a relatively short time. Wind power could impact the United States energy market starting in as few as four years if treated as a national priority goal.

THE RESOURCE

Is there enough energy in the wind to make its development worthwhile? One fearless captain of United States Industry recently stated that "wind power could never amount to more than a drop in the bucket." He is wrong—absolutely wrong. Start with the annual average insolation of Earth: income energy averaging 360 quintillion (3.6 X

10^{20}) BTU each year at the surface of the earth, and 5,000 quintillion (50 X 10^{20}) BTU per year at the outer boundary of the atmosphere. This must be placed in perspective by comparing against a projected *total world* energy consumption, in the year 2000, of the order of 10 quintillion BTU per year. A popular approximation of the *average* solar energy reaching the surface of the earth is 1 kilowatt per square meter of surface area. How much of that received energy is converted into kinetic energy in particles of air? The estimates vary between 2 and 20 watts per square meter. Kung[1] computed a mean annual generation rate of 9.5 watts per square meter over North America, and a winter rate of 15.4 watts per square meter. Palmen and Newton[2] state that those values may be high, although it is plausible that North America is an area of considerably stronger kinetic energy production than most of the regions in the northern hemisphere. Other studies show that the distribution of the generation of kinetic energy in the atmosphere varies significantly from the equator toward the poles and, in the northern hemisphere, reaches a maximum flux in the 40 to $60°$ northern latitude belt. Since this article pertains to wind-power systems lying between 42 and $48°$ north latitudes, it is thought that Kung's hemispheric averages applied to that zone might still be conservative. Thus, one might proceed to say that the generation of kinetic energy in that zone represents a conversion of at least 1 to 1.5 per cent of the total incoming solar energy in that zone. The state of Wisconsin has a surface area of about 5.47 X 10^4 square miles (1.42 X 10^{11} square meters). The total kinetic energy available in the atmosphere over Wisconsin might then average 3.53 X 10^6 megawatts. That is a rather large power plant. If 0.1 per cent of its annual yield were extracted, about 30 billion kilowatt-hours

1. E. C. Kung, "Large Scale Balance of Kinetic Energy in the Atmosphere," 1966, *Monthly Weather Review*, Volume 94, pp. 627–640.
2. E. Palmen and C. W. Newton, *Atmospheric Circulation Systems*, 1969, Academic Press, New York and London, L.C. 69–12279.

would flow into the energy market each year. It is suggested that that is more than a drop in the Wisconsin energy budget. If it were possible to extract 1 per cent, the bucket would certainly overflow, even if our wildest projections of electricity demand turn into reality.

Moving on from a physical quantification of the resource to an engineering quantification, one finds the method employed by Golding[3] to be useful. A wind generator of known or estimated net output power versus wind speed is placed in the wind regime at the site of interest, and the annual productivity of that generator is calculated. The results of such calculations for three wind regions follow in a later section.

Several other general comments about the resource are in order. The productivity of a wind is a function of the cube of its speeds, and there is a certain minimum wind speed below which a wind generator will not produce anything. That "cut-in" speed is seldom lower than 6 miles per hour and, unfortunately, may be as high as 15 miles per hour. Hence, much of the wind within the lower 60 feet of the atmosphere is of little use for power conversion. The velocity of the wind increases logarithmically with height above ground. By going up into the air almost anywhere one can find moderate to strong winds. Elevation of 100 to 1,000 feet in the air immediately presents two problems: (a) large numbers of wind generators can become objectionable to the human eye, and (b) the taller the support, the greater is its cost. Here one must face harsh reality: if lofty, visible wind generators are considered to be objectionable in any area, then the entire proposition must be abandoned forthwith. Practical wind-power systems simply cannot assume a low profile. Again, if the high towers are economically unfeasible, they cannot be used.

3. E. W. Golding, *The Generation of Electricity by Wind Power,* 1956, Philosophical Library, Inc., New York.

THE EXTRACTION OR MOMENTUM CONVERSION DEVICE

It has been said that over 3,500 "windmill" patents have been issued. A Chinese vase dated to 4500 B.C. shows a vertical axis windmill not too different from some operating in 1973. Wind momentum conversion devices represent the product of the most practical mechanic and of the most highly respected of the fluid mechanicians produced by society. Many wind generators capable of continuous operation at conversion rates of 70 per cent of the maximum theoretical capacity have existed, and there is good reason to expect 75 to 80 per cent conversion efficiency as new knowledge of high lift, low drag at low Reynolds numbers is created. The last four years have brought a significant improvement in sail-plane theory and practice, for example.

The capability of a wind machine to exchange momentum is a function of its aerodynamic shape (both design and nicety of manufacture) and size; its location in the wind stream; whether its blades are fixed in pitch, controlled in pitch to maintain a synchronous shaft speed, or controlled in pitch to maximize momentum exchange; and whether there is any interference with optimum wake expansion. The best machines are the so-called modern high-speed propeller machines with three or two very-high-aspect-ratio, thin, carefully twisted blades. The less efficient the machine, however, the easier it is to start at a low cut-in speed, and the smaller the machine, the lower the cut-in speed. One can immediately see where cost becomes related to very high efficiency, but machines of very good efficiency can be simple and thus relatively inexpensive and reliable.

The most simple configuration is probably that of a single modern high speed wind generator atop the tallest western cedar pole that can carry it in the expected wind regime, with account taken of storm and ice loadings. The best location for the most simple configuration is in a clear area. If the

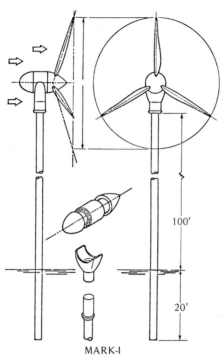

Figure 4-20. MARK-I 20 kw wind generator on single-pole support.

terrain is wooded, the machine will produce well, depending upon how much clearance is provided between tree tops and the swept diameter. From the most simple configuration one can progress down the path of added complexity and cost and added productivity by grouping machines. Where the winds are light to moderate, arrays of large numbers of small machines can produce a significant annual yield, whereas larger machines might not. Larger machines are more economical when the winds are moderate to strong. Concepts for the placement of large numbers of small-to-medium-size wind generators, which share support cost, lead to (a) structural space arrays on top of towers and (b) cable-suspended arrays that would be analogous to hydroelectric dams. The step-by-step transition from a single machine atop a pole to a large number of machines in cable suspension systems can be seen in Figures 4-20, 4-21, and 4-22.

In terms of multiple-land usage, it is contended that wind generators in vast numbers can be located without hindrance to the current use to which land is being placed.

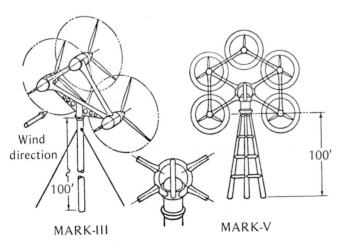

MARK-III   MARK-V

Figure 4-21. MARK-III 60 kw wind generator on stayed 100-ft pole and MARK-V 100 kw wind generator (520 kw machine) on 100-ft tower.

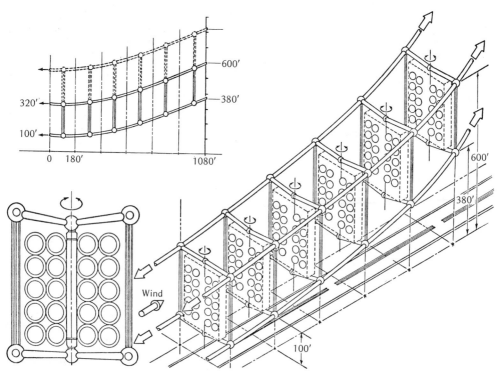

Figure 4-22. Proposed wire rope and kingpost wind generator system. Single bank: 9.6 MW/mile, Double bank: 19.2 MW/mile.

Grazing land, or tilled land, either along fence rows or throughout the fields themselves, could carry large numbers of wind generators. The air space over secondary roads, if oriented essentially normal to the prevailing wind, could carry tens of thousands of miles of cable suspension systems with no adverse effect on the highways below. New tall buildings, if set far enough apart, in some locations could carry enough windpower array to make them self-sufficient for energy. The hundreds of thousands of square miles of forest land, particularly the pulp-wood farms and other commercial forest lands, could be surmounted by wind-power arrays. The vast expanse of east coast continental shelf could support huge arrays of wind-power extraction machines. Again, the contention is that this could be done without detraction from the primary land usage, and could be done so as to avoid aesthetic objection.

THE TOTAL WIND-POWER SYSTEM

The first two elements and probably the most expensive portions of any wind-power system will be the momentum exchange device and the supports. From there nothing more or very much more must be added depending upon what the system's salable product is to be. Some elements that could be built into wind-power systems are given in Table 4-21. Ten salable products are shown in column IX plus four other possible "special" salable products (in which interest will quicken rapidly in the next few years!). To get from column I to column IX many different paths can be followed, depending upon what the salable product is to be. Each additional block entered on Table 4-21 will usually put another efficiency into an already long cascade of efficiencies from resource to product; unnecessary steps must be avoided. The provision of storage can be especially debilitating to over-all efficiency but dependent, of course, upon the amount of storage needed as well as the nature of the storage subsystem. On the other side of the coin, however, lies the possible economic gain from providing storage. Peak power at the bus bar has traditionally been worth three times as much as base load power at that same bus bar: what that ratio will be in the future is a good question. But, one can afford considerable degradation of efficiency for that portion of the product that must go through storage, if one can realize peak power tariffs. Much hard work is required before the economics of very many of the possible paths through Table 4-21 can be set down with certainty. One grand-scale wind-power system has been calculated completely: the proposed offshore wind-power system[4] in which the path through the columns was: I.D to II.G to III.A to IV.G to V.G. to VI.G.8 to VII.G to VII.E producing IX.C. Two by-products IX.F and IX.H exist but were not treated as salable.

SOME SPECIFIC ESTIMATES OF PRODUCTIVITY AND ECONOMICS AT CERTAIN LOCATIONS

Wind-power systems in three different locations have been conceptualized and the available resource data analyzed:

(1) *Region A:* A west-to-east belt of the United States and Canada, starting East of the Rocky Mountains in Montana, proceeding eastward through the Dakotas, Minnesota, Wisconsin, Michigan, lower Ontario and Lake Erie, upper New York and Lake Ontario, on into Vermont, and over Maine out to sea. Within that very large area, specific attention has been given to Wisconsin and the bordering Lake Superior and Lake Michigan.

(2) *Region B:* Upper New York State and the United States portion of Lake Ontario.

4. W. E. Heronemus, "Power from the Offshore Winds," 1972, Proceedings of the 8th Annual Conference of the Marine Technology Society, Sept. 1972, M.T.S., Washington, D.C.

(3) *Region C:* Wisconsin, Upper Michigan, and Northern Minnesota.

Each station has been measured for annual productivity, annual plant factor, and annual storage amount. Annual productivity equals the number of kilowatt-hours of electricity one could expect each year from each 20-kilowatt unit. The annual plant factor is simply the annual productivity divided by 8,760 hours times the generator rating; in other words, the actual productivity divided by the maximum possible productivity if the machine worked at full rated capacity throughout the year. The annual storage amount is the maximum amount of continuous surplus or deficit over or under the line of average productivity per hour. It is the direct measure of the least amount of energy that must be stored if a constant power output is to be achieved from that wind-power system at that site. All of the data are for a height of center line of 100 feet above surface (see Table 4-22).

THREE SPECIFIC WIND-ELECTRICITY SYSTEMS

Wind-power electricity generating systems in three different regions, all using the same storage loop, have been selected to demonstrate the economics of wind generated electricity:

(1) A Prairie States system, feeding both locally and into Chicago, with Sioux City, Iowa, selected as "the generating site": Table 4-22 suggests that Sioux City is representative of a wind field that comes out of the foothills of the Rockies in Montana and Wyoming and down from the prairies of Saskatchewan and blows on out toward the Atlantic.

(2) An Upper Michigan system, feeding into the Green Bay-Fond Du Lac markets: This is a small portion of a very extensive "Wisconsin windpower system."

(3) An upper New York State system, in the Lake Ontario Region, feeding all the way to New York City.

Table 4-21. Elements of Wind-Power Systems.

I Momentum Exchange	II Supports	III Airscrew Driven Device	IV Power Addition or Collection Feature
A. Very large numbers of small (5-kilowatts) wind generators with 10 mile per hour cut-in speed	A. Relatively low single pole	A. Free-wheeling floating voltage direct current generator	A. Single electric generator
	B. Highest practical, single pole	B. Controlled voltage, direct current generator	B. Paralleling direct current generators, large numbers, feeding larger synchronous alternating current generators
B. Large numbers of moderate sized (20–30-kilowatt) wind generators with 10 mile per hour cut-in speed	C. Lattice tower, 150 feet height	C. Network synchronized alternating current generator	C. Paralleling alternating current generators, large numbers, feeding synchronous transmission net
	D. Lattice tower, heights above 150 feet	D. Hydraulic pump–stroked	
		E. Hydraulic pump –not stroked	
C. Fewer numbers of medium sized (100-kilowatt) wind generators with 13 to 15 mile per hour cut-in speed	E. Carrousel bill-board array	F. Air compressor	D. Hydraulic pumps feeding larger hydraulic motor
	F. Tower and cable suspension system	G. Mechanical drive to water pump	E. Air compressors in parallel feeding a common accumulator
	G. Floating supports, relatively low		F. Water pumps feeding a common hydro reservoir
	H. Floating supports, relatively lofty		G. Direct current voltages switched to optimum array of electrolysis cells
D. Smaller numbers of large (2-megawatt) wind generators with 15 mile per hour cut-in speed	I. Floating tower and cable suspension		

V Inherent Storage Feature	VI Storage Feature Variations	VII Energy-Transmission Feature	VIII Storable Reconversion Feature	IX Salable Products
A. Direct current accumulator	E. Compressed Air 1. Small HP gas containers 2. Large HP gas containers 3. LP storage in caverns 4. MP storage in aquifer 5. HP storage in exhausted gas wells 6. Pressure-balanced storage under water	A. Cable B. Cable, collection, transmission and distribution C. Cable, collection, transmission and distribution D. Piping	A. Cold air expansion turbine driving synchronous generator B. Heated air expansion turbine driving synchronous generator C. Hydraulic motor driving synchronous generator D. Hydraulic turbine driving synchronous generator	A. Electric power, fluctuating voltage direct current or fixed voltage into synch. net whenever the wind blows (the fuel saver!) B. Electric power, base load, into synchronous net C. Electric power, base interm. and peak load, into synchronous net
D. Hydraulic accumulator	G. Hydrogen 1. Small HP gas containers 2. Large HP gas containers 3. Tank storage as cryo. 4. Accumulation in metal matrixes as hydride 5. LP storage in caverns 6. MP storage in aquifer 7. HP storage in exhausted gas wells 8. Pressure-balanced storage under water	E. Piping F. Piping and open channel G. Piping	E. Hydrogen-air fuel cell feeding synchronous through inverter F. Hydrogen-oxygen fuel cell feeding synchronous net through inverter G. Hydrogen-air turbine driving synchronous generator H. Hydrogen-air turbine driving synchronous generator I. Hydrogen I.C. engine driving synchronous generator J. Hydrogen-air hot air engine driving synchronous generator	D. Pure Hydrogen gas E. Liquid hydrogen F. Pure oxygen gas G. Liquid oxygen H. Pure by-product water I. Compressed air J. Irrigation water K. Hydrogen and high sulfur coal yields either clean methane or clean liquid coal L. Oxygen and high sulfur coal yields clean high-BTU pipe line gas M. Hydrogen and iron ore yields clean iron production N. Oxygen and iron yield clean steel production
E. Compressed air storage				
F. Pumped hydro storage				
G. Hydrogen gas storage				

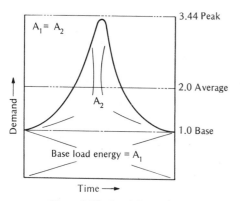

Figure 4-23. Load demand curve.

The load demand curve, (Figure 4-23) is assumed to be the same in all three regions, and is a sharply peaking curve that, according to the 1970 National Survey, will be common by 1990. No seasonal variation in load demand has been taken into account.

The block diagram of the system is shown in Figure 4-24. This system uses gaseous hydrogen as storable, and stores it underwater in pressure balanced storage farms. For the prairie states and the Wisconsin system, the storage is to be in Lake Michigan,

off Kewaunee, Wisconsin, in 850 feet of water. For the upper New York State system, the storage is to be in Lake Ontario in 720 feet of water.

The wind station to be used in the first two cases is the lofty array of Figure 4-22, the king-post and wire-rope suspension system, double banked. The wind station to be used in the upper New York State system is that of a very lofty floating structural grid, in which 200 of the 20-kilowatts wind generators are held (Figure 4-25).

For all three stations, weighted averages for E_1, E_4, E_5, E_7, and $E_{delivered}$ (refer to Figure 4-24 for locations 1, 4, 5, and 7 at which energy per year, E, is calculated) are used.

Those numbers are given in Table 4-23.

For each of these three systems, costs have been calculated using these factors:

(1) A 20-kilowatt wind generator, complete; will cost $110 per kilowatt.

(2) Supports, if on land, of the double-banked king-post and wire-rope type, with electrical collection cables and training drives for each cluster of 20 generators, will cost $40 per kilowatt.

(3) The floating supports, tower, tether, anchor, of the 4-megawatt floating system will cost $200,000.

(4) The over-all efficiency of the storage loop is such that $E_2 = 1.96E_4$. Electrolyzers will cost $25 per kilowatt electric input capacity, and hydrogen-air fuel cells will cost $100 per kilowatt electric output.[5-8] For this to be possible, large but entirely feasible development programs are required. Generators and motors were priced at $20 per kilowatt.

5. Derek P. Gregory, "A New Concept in Energy Transmission," February 3, 1972, *Public Utilities Fortnightly.*
6. D. P. Gregory, D. Y. C. Ng, G. M. Long, "Hydrogen Economy," *The Electrochemistry of Cleaner Environments*, J. O'M. Bockris, Editor, New York, Plenum Press, 1972.
7. W. J. D. Escher, *Helios-Poseidon*, 1972, Escher Technology Associates, St. Johns, Michigan.
8. General Electric Direct Energy Conversion Officer, Lynn, Mass. to W.E.H., Personal Communication, December 1972, Regarding Solid Polymer Electrolyte Electrolyzers.

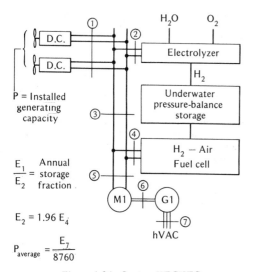

Figure 4-24. System WPGHFC.

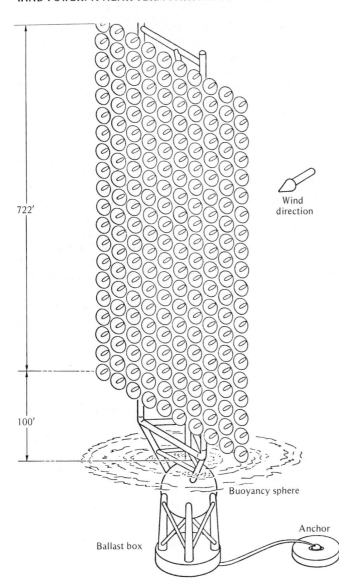

722'

100'

Wind
direction

Buoyancy sphere

Anchor

Ballast box

Figure 4-25. Floating 4 MW wind
station made from low cut-in
wind generators.

Fixed costs of generation have been calculated using a 15.5 per cent fixed charge rate,
which is probably much too high for regions
other than New England, but which has been
used as a standard by the National Science
Foundation Solar Energy Panel. Costs associated with transmission lines and the storage loop, which are strictly parts required

for the generation system, have been
grouped together as "other generation
costs." A cost of 10 mills per kilowatt-hour
were allowed for distribution, General and
Administrative, and profit, in all three cases.

For the three systems, these results were
found for 1973:

A. The Prairie States System, delivering electricity on demand, in the Sioux City, Iowa area: required average revenue = 28 mills per kilowatt-hour.

B. The Prairie States system, delivering electricity on demand, in the Chicago market: required average revenue = 31 mills per kilowatt-hour.

C. The upper Michigan system, delivering electricity on demand, in the Green Bay, Wisconsin market: required average revenue = 28.4 mills per kilowatt-hour.

D. The upper Michigan system, delivering electricity on demand, in the Chicago market: required average revenue = 29 mills per kilowatt-hour.

E. The Lake Ontario system, delivering electricity on demand, in the Oswego, New York market: required average revenue = 31 mills per kilowatt-hour.

F. The Lake Ontario System, delivering electricity on demand in New York City: required average revenue = 32 mills per kilowatt-hour.

Those required average revenues are higher than the 1973 midwestern prices. But midwestern prices are going to rise. It is foolish to think in terms of fuel costing less than 90¢ per million BTU by 1975, and clean fuel from coal, or liquid natural gas will probably reach $1.50 per million BTU by 1980. As for the costs of nuclear generated power, we have witnessed the ease with which that industry has moved from $163 per kilowatt electric at Dresden up to over $450 per kilowatt electric at Yankee, Vermont. And those who really understand the future of that situation (for example, United Engineers and Constructors and EBASCO Services Incorporated) know that George M. Anderson's February 1972 projections of $600-, $700-, and $800-per-kilowatt electric nuclear plants are already on the low side. Also, when the United States public receives a fair deal on the cost of enriched uranium fuel, that will have gone from 2 mills per kilowatt-hour burn-up to closer to 8 mills per kilowatt-hour burnup.

Table 4-22. Characteristics of a Single 32-Foot Diameter X 20-kilowatt Wind Generator at an Axis Height of 100 feet above the Surface.

Station	Annual Productivity (kilowatt-hours per year)	Annual Plant Factor	Annual Storage Amount (kilowatt-hours per year)
A. Prairie States region:			
1. Huron, South Dakota	50,800	0.29	1,928
2. Fargo, North Dakota	53,818	0.31	2,509
3. Casper, Whoming	63,892	0.36	5,543
4. Great Falls, Montana	64,740	0.37	5,297
5. Billings, Montana	46,881	0.27	2,789
6. Wichita, Kansas	77,804	0.44	2,220
7. Topeka, Kansas	45,061	0.26	2,076
8. Tulsa Oklahoma	42,656	0.24	3,189
9. Sioux City, Iowa	69,766	0.40	3,338
B. Upper New York State:			
1. Buffalo, New York	42,189	0.24	3,309
2. Over Lake Ontario	73,861	0.42	10,151
C. Wisconsin, Upper Michigan, Northern Minnesota:			
1. Milwaukee, Wisconsin	42,852	0.24	3,484
2. Green Bay, Wisconsin	57,144	0.33	4,507
3. Duluth-Superior	49,166	0.28	2,331
4. Over Lake Superior	69,031	0.39	8,303

Table 4-23. Efficiencies at Three Selected Locations.

Station	Sioux City, Iowa	Upper Michigan, Point de Tour to Munising	Over Lake Ontario
E_1	91,975 kilowatt-hours per year	94,790 kilowatt-hours per year	21.051×10^6 kilowatt-hours per year
E_4	3,386 kilowatt-hours per year	8,189 kilowatt-hours per year	2.203×10^6 kilowatt-hours per year
E_5	70,128 kilowatt-hours per year	71,390 kilowatt-hours per year	16.412×10^6 kilowatt-hours per year
E_7	63,115 kilowatt-hours per year	64,900 kilowatt-hours per year	14.920×10^6 kilowatt-hours per year
E delivered	60,590 kilowatt-hours per year	63,602 kilowatt-hours per year	14.323×10^6 kilowatt-hours per year
$P_{average}$	8.00 kilowatts	8.15 kilowatts	1874 kilowatts
P_{peak}	13.77 kilowatts	14.02 kilowatts	3222 kilowatts
Installed wind generator capacity	The single 0.20-kilowatt machine	The single 20-kilowatt machine	200 of the 20-kilowatt machines

Attention should be given to large-scale wind power if only from the purest of capitalistic economic motivation.

CONCLUSION

It is proposed that significant quantities of electricity could be generated by wind power over very many portions of the country. In some areas this could be done to a total amount well in excess of predicted local demand and could thus become a very valuable export. The idea of states like Montana, the Dakotas, Wyoming, Kansas, and Oklahoma becoming purveyors of competitive wind-generated electricity to the Chicago energy sink is advanced. Similar generating site-market combinations have been suggested. The resource is huge, and it can be tapped by relatively simple machines in large numbers sharing the existing scene, or by large wind barrages across the prevailing winds. The product is economic by 1973 standards, ranging from 16 to 21 mills per kilowatt-hour at the distribution yard. The latest and largest New England nuclear plant, Vermont Yankee, is charging 22 mills per kilowatt-hour for power delivered to Vermont utilities and more for power delivered to the big market around Boston. The projected costs from future nuclear plants are much higher. There is good reason to feel that every portion of the suggested wind-power systems will be made more efficient with time and much less costly if genuine mass production were undertaken. Every

part of the system mentioned here is amenable to automated production similar to that producing automobile engines for as little as $30 per kilowatt. There is great opportunity here for new business.

And, the windpower resource is renewable, there would be *no* thermal pollution, *no* radioactivity of any kind, *no* air pollution of any kind, *no* fear of catastrophic accident, and *no* vying for the petroleum resources of other nations if it were to be used. It is suggested that United States industry stop its derision of this venerable source of clean energy and start using it, for the benefit of all of us here on earth. Wind-generated electricity could have a significant impact on our energy market in as few as four years if we so desire.

ACKNOWLEDGMENTS

The author is greatly indebted to two of his students, Mr. Theodore Djaferis and Mr. Trent Poole, who did the calculations and the illustrations, respectively, for this paper. Mr. Djaferis joins the author in grateful acknowledgment of the support given to his continued education by Mr. Mark Swann of New Park, Pennsylvania. I also acknowledge with thanks the contributions of Dr. M. M. Miller in advancing the cable-suspension system for large-scale wind-power systems.

9. T. L. Richards and D. W. Phillips, "Synthesized Winds and Wave Heights for Great Lakes," *Climat. Studies No. 17,* Queens' Printer for Canada, Ottawa, 1970.
10. The 1970 National Power Survey, I, Federal Power Commission, U.S. Govt. Printing Office, Washington, D.C., December 1971.

SUGGESTED READINGS

Heronemus, William E., 1972, "The U.S. Energy Crises: Some Proposed Gentle Solutions," *Congressional Record–Senate,* Vol. 118, February 9, 1972, pp. E1043–E1049.
McCaull, Julian, 1973, "Windmills," *Environment 15:* 6–17, January–February.
Wade, Nicholas, 1974, "Windmills: The Resurrection of an Ancient Energy Technology," *Science 184:* 1055–1058, June 7, 1974.

Oil Shale and the Energy Crisis

GERALD U. DINNEEN and GLENN L. COOK   1974

The oil shales of the United States are a potential source of vast quantities of liquid fuels. These sedimentary rocks contain solid organic material that can be decomposed by heat into an oil resembling petroleum, gases, and a carbonaceous residue. Although the shales are widely distributed, only those of the Green River Formation in Colorado, Utah, and Wyoming have received appreciable attention, because these shales are particularly rich in organic material.

Many attempts have been made, since early in this century, to develop commercial utilization of the Green River oil shale. These have typically involved underground mining, with processing in above-ground equipment. This approach is still the most likely to be utilized for developing those portions of the deposit that crop out along cliffs or that are accessible to surface mining. However, an alternative approach—*in situ*

processing—has received increased attention during recent years. This approach may be applicable to deposits of various grades and thicknesses of shale, and it has the additional advantage of avoiding the problem of disposing of large amounts of spent shale.

Although past efforts have not resulted in an oil-shale industry in the United States, the current status of oil-shale technology and economics in the context of the country's need for new energy sources breeds optimism for future development.

OIL SHALE IN THE GREEN RIVER FORMATION

The Green River Formation covers an area of about 17,000 square miles of Colorado, Utah, and Wyoming in four principal basins: The Piceance Creek Basin of Colorado, the Uinta Basin of Utah, and the Washakie and

Gerald U. Dinneen has been research director of the U.S. Bureau of Mines, Laramie Energy Research Center, since 1964. He has concentrated his research on the composition and reactions of shale oil.

Glenn L. Cook, a long-time member of the Laramie Energy Research Center, is supervisor of a research group studying the characterization of oil shale and shale oil. He has worked on identifying the components of oil shales with spectroscopic and nuclear magnetic resonance techniques.

From *Technology Review*, Vol. 76, No. 3, pp. 26–33, January, 1974. Reprinted with light editing and by permission of the authors and *Technology Review*. Copyright 1974 by the Alumni Association of Massachusetts Institute of Technology. This article is adapted from a paper presented at the 1972 annual meeting of the American Society of Mechanical Engineers.

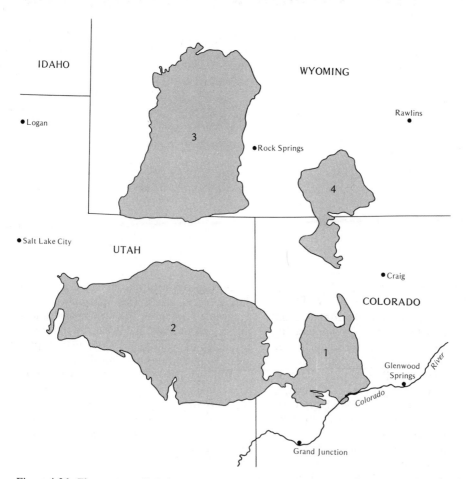

Figure 4-26. The western oil shales, rocks containing up to 10 to 15 per cent organic matter, occur in the Green River Formation in the area where Utah, Wyoming, and Colorado join. The Piceance Creek basin in Colorado (1)—though a small fraction by area of the Green River Formation—contains most of the richest shale and 80 per cent of the total recoverable oil, and it is the area of greatest current interest and potential activity. Other areas which contribute to the total of 600 billion barrels of recoverable oil include the Uinta Basin (2) in Utah (90 billion barrels) and the Green River (3) and Washakie (4) Basins in Wyoming and Colorado with some 30 billion barrels. (Map redrawn from map © 1963 National Geographic Society.)

Green River Basins of Wyoming. (See Figure 4-26.) If only shale at least 10 feet thick and yielding at least 25 gallons of oil per ton is considered, the potential oil yield in place in the three states has been estimated at 600 billion barrels. If leaner shale is included, the oil potential is increased substantially.

The shales of the Green River Formation were laid down during the Eocene geologic period, some 50 million years ago, as sedi-

ments of living organisms and precipitates of minerals at the bottoms of lakes. Deposition continued over a period of several million years. Compaction of the sediments with the elimination of water preserved the organic material, and the high organic content is an indication of the prolific production of living organisms that must have been characteristic of the lakes.

The small area of the Piceance Creek

Basin (about 10 per cent of the area covered by the Green River Formation) contains about 80 per cent of the potentially recoverable oil. Some of the shale crops out in cliffs along the southern edge of the basin, where the Mahogany Zone, the best known interval of the Green River Formation, is about 75 feet thick. In the center of the Piceance Creek Basin continuous sections of oil shale averaging more than 25 gallons of oil per ton are several hundred feet thick but are generally under several hundred feet of overburden. In Utah and Wyoming the sections of rich shale are not as thick, and in Wyoming they are often interspersed with alternating beds of lean shale. Piceance Creek Basin shales were used to develop most of the current technology; they are a resource of tremendous size, and it is probable that they will be the first raw materials utilized in a developing shale-oil industry.

Mineralogically, the oil shales range in composition from rocks composed of illite clay through those containing dawsonite, nahcolite, and potash feldspars as the principal minerals, to those made up of calcite and dolomite. The organic matter is fairly high in hydrogen, so that heat converts about 65 per cent of it to oil, 10 per cent to gas, and 25 per cent to carbonaceous residue. Although shales in the Mahogany Zone have an average oil yield of 25 to 30 gallons per ton the zone is made up of material with yields ranging from about 5 to 75 gallons per ton. This heterogeneity is an important factor that must be considered in developing processes for recovering oil from the Green River Formation.

Oil shale in the Mahogany Zone is a highly consolidated organic-inorganic system with no significant micropore structure, pore volume, permeability, or internal surface. The mineral constituents consist of fine particles with 99 per cent by weight having diameters of less than 44 microns. Only the lower-grade shales typically show any appreciable porosity; and, although there is some variation among the shales of different oil yields, all raw shales are strong rocks.

As would be expected, heating to 950°F

to remove the organic matter produces appreciable porosity, which is roughly proportional to the grade of shale. Further heating to decompose the carbonates produces some additional porosity. After heating, leaner shales retain high compressive strengths, but richer shales lose much of their strength. These characteristics will have to be taken into consideration in recovering the energy values from the shales.

Two general approaches to the recovery of shale oil from the Green River Formation have been proposed: mining, crushing, and above-ground retorting, and *in situ* processing. The first is the traditional approach that has been used in various parts of the world for over 100 years and will probably be used for the first development of the Green River Formation. The second has received serious consideration only in the last few years, but it has potential advantages that make efforts to develop a feasible method worthwhile. Because of the wide variations in conditions under which oil shales are found in the Green River Formation, the two approaches are complementary rather than competitive.

SHALE OIL FROM MINED SHALE

Many attempts have been made during more than the last half-century to mine the Green River oil shale and retort it—treat it by heat—to separate the hydrocarbons from their associated inorganic minerals. Notable among these are the extensive investigations conducted during the past thirty years by the Bureau of Mines under the Synthetic Liquid Fuels Act, by a group of six oil companies utilizing Bureau of Mines facilities, by the Union Oil Co. of California, and by the Colony Development Co. (See Tables 4-24 and 4-25.)

The Bureau of Mines in 1944 opened a demonstration mine near Rifle, Colorado, to tap a 73-foot minable section of the Mahogany Zone, and by 1956 it had been shown with fair assurance that low mining costs and high recovery in a conventional room-and-pillar underground mining operation were possible. The research also indi-

Table 4-24. Shale-Oil Deposits in the Green River Formation.[a]

| | Billions of barrels of oil in place | | | |
	Colorado	Utah	Wyoming	Total
Intervals 10 feet or more thick averaging 25 gallons per ton or more of oil	480	90	30	600
Intervals 10 feet or more thick averaging 10 to 25 gallons per ton of oil	800	230	400	1,430
Total: intervals 10 feet or more thick averaging over 10 gallons per ton	1,280	320	430	2,030

[a]Over 2 trillion barrels of oil are locked in known shale oil deposits in the Green River Formation, but less than one-third of this is in reasonably thick deposits which average more than 25 gallons of oil per ton of shale; only these are generally regarded as potentially exploitable.

Table 4-25. Mahogany Zone Oil-Shale Deposits.[a]

	Weight Per Cent	Per Cent
Mineral matter:		
Content of raw shale		86.2
Estimated mineral constituents:		
Carbonates, principally dolomite	50	
Feldspars	19	
Illite	15	
Quartz	10	
Analcite and others	5	
Pyrite	1	
Organic matter:		
Content of raw shale		13.8
Ultimate organic composition:		
Carbon	80.5	
Hydrogen	10.3	
Nitrogen	2.4	
Sulfur	1.0	
Oxygen	5.8	

[a]The richest oil shales occur in the Mahogany Zone of Colorado (the Piceance Creek Basin near Rifle) and adjacent portions of Utah. Even here the organic matter represents less than 15 per cent of the total shale content; one ton of shale may yield as much as 75 gallons of crude oil, but the average even in this richest shale deposit is more nearly 25 to 30 gallons per ton.

cated that petroleum technology would be adaptable to refining shale oil produced in any of the several retorts studied.

The most promising results were achieved with a gas combustion retort—a vertical, refractory-lined vessel through which crushed shale moves downward by gravity. Recycled gases enter the bottom of the retort and are heated by the hot retorted shale as they pass upward through the vessel. Air is injected into the retort at a point approximately one-third of the way up from the bottom and is mixed with the rising, hot recycled gases. Combustion of the gases and some residual carbonaceous material from the spent shale heats the raw shale immediately above the combustion zone to retorting temperature. Oil vapors and gases are cooled by the incoming shale and leave the top of the retort as a mist. The manner in which retorting, combustion, heat exchange, and product recovery are carried out gives high production and thermal efficiency. The process does not require cooling water, an important feature because of the semiarid regions in which the shale deposits occur. The process appeared to offer the possibility of large-scale operation. (See Figure 4-27.)

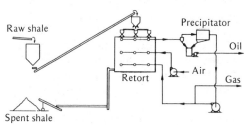

Figure 4-27. Crude oil will be recovered from mined oil shale by some version of a retorting process of which this is a basic flow diagram. Quarried shale, broken into pellets and heated to just under 1,000°F in a retort, gives off gas and vaporized crude oil; from this mixture the crude oil is condensed and the remaining gas is returned to fuel the retort. Spent shale and a carbonaceous residue accumulate at the bottom of the retort and are moved to a disposal site; the latter may be the mine from which the original shale came, except that the volume of the spent shale is somewhat larger than that of the original material, so the mines are inadequate to accommodate all of the solid waste which accumulates. (Drawing: Savage, Holt, and Sims from the *Colorado School of Mines Quarterly,* October, 1968.)

From 1964 to 1968 the Bureau of Mines facilities were leased by the Colorado School of Mines Research Foundation, which operated them under a research contract with six oil companies. The first phase of this research, which lasted approximately two years, was devoted primarily to studying the gas combustion retorting process in two small pilot plants (nominal capacities of 6 and 24 tons per day) that had been constructed by the Bureau. Mining research involved extension of the room-and-pillar method used by the Bureau of Mines with somewhat smaller pillars. Work on retorting in the largest gas-combustion process pilot plant at the facilities showed that oil yields in excess of 85 per cent of the standard assay could be obtained using feed rates of 500 pounds per hour for each square foot of bed area. This was about double the rate previously achieved by the Bureau of Mines, with only a relatively small decrease in oil yield. Although the results indicated a material advance in processing technology, some

of the operating problems associated with the scale-up in the size of the plant were not fully resolved.

Meanwhile, a retort developed by Union Oil Co. of California was tested between 1956 to 1958 on a demonstration scale of about 1,000 tons per day, and at the conclusion of the project it was announced that the process could be commercialized whenever energy demand and economic conditions warranted. This retort, also a vertical, refractory-lined vessel, operates on a downward gas flow principle; the shale is moved upward by a charging mechanism usually referred to as a "rock pump." Heat is supplied by combustion of the carbonaceous residue remaining on the retorted shale and is transferred to the oil shale, as in the gas combustion retort, by direct gas-to-solids exchange. The oil is condensed on the cool, incoming shale and flows over it to an outlet near the bottom of the retort. This process, like that of the Bureau of Mines, does not require cooling water.

A third group—the Colony Development Co., representing Standard Oil Co. of Ohio, Oil Shale Corp., Cleveland Cliffs Iron Co., and more recently Atlantic Richfield Oil Co. as project manager—has for several years been investigating the TOSCO II retorting process, based on a rotary kiln utilizing ceramic pellets heated in external equipment to accomplish retorting. Shale ground into particles not exceeding 0.5 inch is preheated and pneumatically conveyed through a vertical pipe by flue gases from the pellet-heating furnace, entering the rotary retorting kiln with the heated pellets; here it is brought to a retorting temperature of 900°F by conductive and radiant heat exchange with the pellets. Passage of the kiln discharge over a trommel screen permits recovery of the pellets from the shale dust for recycling and reheating. The spent shale is routed to disposal by a screw conveyer. The oil is recovered by condensation from the gases leaving the retort. High shale through-put rates were reported, and considerable research was also done on environmental aspects of oil-shale operations, particularly in regard to handling spent shale deposits. On a commercial scale the system would yield some 50,000 barrels of oil per day.

PROCESSING THE SHALE IN THE MOUNTAIN

Because mining, crushing, and retorting make up about 60 per cent of the cost of producing shale oil by above-ground techniques, recovery of oil from shale by *in situ* processing has received attention during the past 20 years. This approach is attractive because it may be applicable to deposits of various thicknesses, grades of shale, and quantities of overburden. Also, it eliminates the necessity to dispose of large quantities of spent shale.

One concept of *in situ* processing envisions a five-step process: drilling a predetermined pattern of wells into the oil-shale formation, fracturing to increase permeability, igniting the shale at one or more centrally located wells, pumping compressed air down these ignition wells to support combustion of the oil shale and force the hot combustion gases through the fractured rock to convert the solid organic matter to oil, and recovering the oil thus generated through other wells. Research on the *in situ* technique is in the early stages and has so far been devoted to the development of fracturing techniques and methods of underground heating. The fracturing might be achieved by relatively conventional methods, such as hydraulic pressure or chemical explosives, or by unconventional methods, such as a nuclear explosive. Heating may be achieved either by underground combustion or by forcing previously heated gases or liquids through the formation.

One of the earliest investigations of *in situ* processing was by Sinclair Oil and Gas Co., now a part of Atlantic Richfield Co., which conducted a study in 1953 and 1954 at a site near the southern edge of the Piceance Creek Basin. This work confirmed that communication between wells could be established through induced or natural frac-

ture systems, that wells could be ignited successfully, and that combustion could be established and maintained in the shale bed. More recently, over a period of several years in the mid-1960's, Sinclair conducted field research at a site near the center of the Piceance Creek Basin where the shale is much deeper and thicker. The results of this experiment were not promising; the fracturing techniques used apparently did not produce sufficient heat transfer surfaces for successful operation.

Another experiment of particular interest has been conducted by Equity Oil Co., in which hot natural gas was injected into an oil shale formation in order to achieve retorting. This presumably is especially advantageous from the point of view of the type of oil produced.

Two major *in situ* research projects are now in progress, one by the Bureau of Mines and the other by Garrett Research and Development Co. The Bureau's program involves laboratory studies, pilot scale simulation of underground operations, and field experiments. In the field experiments near Rock Springs, Wyoming, several methods of fracturing—hydraulic pressure, chemical explosives, and electricity—are being tested on an oil-shale bed that is 20 to 40 feet thick and under 50 to 400 feet of overburden. Using hydraulic pressure, four horizontal fractures have been produced over a vertical interval of about 35 feet at a depth of approximately 400 feet; tests indicate that these fractures extend at least 200 feet from the injection well. Chemical explosives have also been used—in the liquid form for detonation after being forced into naturally occurring or artificially created fractures, and in pelletized solid form for detonation in well bores. (See Figure 4-28.)

The combination of hydraulic pressure and liquid chemical explosive was used for a test of *in situ* processing of shale in a formation less than 100 feet beneath the surface. Fracturing was completed in a small five-spot pattern, and the shale was then ignited for a combustion test in which about 190 barrels of oil were produced. Preparations

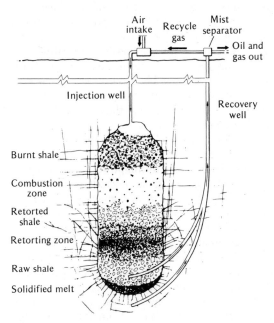

Figure 4-28. This method for *in situ* recovery of crude from oil shale was proposed by D. B. Lombard of the Lawrence Radiation Laboratory and H. C. Carpenter of the U.S. Bureau of Mines in 1968: a nuclear explosive detonated in the oil shale layer creates a "nuclear chimney" with fracturing through the adjacent shale deposits. Air forced into the "chimney" supports combustion, the hot combustion gases move through the fractures to convert the solid organics to liquids and gases, and these products are then drawn off for condensation and—in the case of the gas— to support further combustion. Efficient *in situ* recovery depends on effective fracturing of the shale; the nuclear explosive was unsatisfactory in this (and other) respects, and the method is not presently under active development. But the same general principles of fracturing (by conventional explosives), combustion, and product recovery figure in current *in situ* recovery concepts. (Drawing: Lombard and Carpenter from the *Journal of Petroleum Technology,* June, 1967.)

for a similar, but larger, underground recovery test are in progress.

The Garrett project utilizes a combined mining and explosive fracturing technique to prepare the oil shale for *in situ* processing. In this approach shale is mined from the lower part of a room to provide porosity when the shale above this mined portion is collapsed into it. The broken shale in the room is then retorted from the top down. The present field experiment on the southwestern edge of the Piceance Creek Basin was started about a year ago. The company has indicated that preliminary results are encouraging. As the project progresses, it should answer many of the questions concerning the feasibility of processing large quantities of broken shale *in situ.*

HOW TO REFINE CRUDE SHALE OIL

The fuels and chemicals normally produced from petroleum can also be obtained from shale oil. However, an adaptation of petroleum technology based on the properties peculiar to shale oil is required. Shale oil produced by retorting oil shale from the Green River Formation is usually a dark, viscous material with a relatively low sulfur content but a high pour point (high viscosity) and high nitrogen content; indeed, the pour point is so high that viscosity reduction is necessary before the oil is amenable to pipeline transportation. The high nitrogen content complicates the refining process. At present, it appears that special treatment of shale oil will be required to lower the nitrogen content before it is acceptable for such refining methods as catalytic cracking. The Bureau of Mines has recently demonstrated that a premium refinery feedstock can be made from *in situ* shale oil by a processing sequence suggested by the National Petroleum Council.

With the present emphasis on *in situ* processing, there is considerable interest in how the composition of oils produced in this way may compare with those from above-ground retorting. Recent research suggests that the specific gravity, pour point, viscosity, and contents of heavy gas oil and residuum will be lower in oil produced *in situ;* the content of naphtha, light distillate, and light gas oil will apparently increase under these conditions. The nitrogen and sulfur contents of the oils do not seem to vary systematically with the retorting variables. On the basis of these results, it seems that shale oil produced *in situ* may be more valuable than that produced in above-ground retorts.

THE ENVIRONMENTAL IMPACT OF A SHALE OIL INDUSTRY

Can the production of shale oil be done without significant environmental degradation? The environmental effects directly associated with oil-shale processing are expected to be from retorted or burned shales, contaminated waters, and gases. The present emphasis on *in situ* processing is partly because this technique would obviate the necessity of disposing of large quantities of retorted or burned shale. However, before *in situ* processing is finally proved feasible, the effect of leaching of the in-place retorted shale on ground waters will have to be determined. A start toward investigating this problem has been made by drilling a series of wells at the Bureau's experimental site near Rock Springs to compare hydrologic conditions in the area before and after an *in situ* processing experiment.

As indicated earlier, some portions of the Green River deposit appear to be most amenable to mining and above-ground

processing. For this reason, and because of the unresolved technical problems confronting *in situ* processing, both industry and the Bureau of Mines are both investigating problems associated with disposal of retorted or burned shales. The Colony group has done substantial research on the vegetation of retorted shale and has shown that this can be accomplished.

Retorting oil shale produces water from the combustion of fuel used to heat the shale and from decomposition of the organic matter in the shale. Because these retorting waters have been in contact with shale oil, they contain substantial amounts of organic materials in addition to the usual inorganic compounds. Experiments to treat water from gas combustion retorting and from the *in situ* retorting experiment near Rock Springs indicate that nearly complete removal of the inorganic ions can be achieved from water from the gas combustion process, but slightly less complete removal was obtained from water from the *in situ* process.

Gases generated by the combustion inherent in oil-shale processing are expected to be amenable to gas treating methods developed for other industries.

THE ECONOMICS OF SHALE OIL

Predictions of the probable cost of producing shale oil must be based on engineering estimates; data on plants of the size required to produce commercial quantities of oil are not available. Recent estimates—including those by the Bureau of Mines, the National Petroleum Council, and the Oil Shale Corp.—vary widely. Without extensive discussion of the assumptions involved in the different estimates, comparison of them is impossible. However, all estimates agree that to construct a plant to produce about 100,000 barrels per day of shale oil (about 0.6 per cent of our present petroleum demand) will involve a capital investment of several hundred million dollars, and the oil will have to sell in the range of $4 per barrel to yield a reasonable return on investment.

CAN WE BEGIN A HIGH-RISK NEW INDUSTRY?

Green River oil shale represents an immense potential source of energy, and there is now mounting evidence that the political and economic climate may be favorable for a modest beginning of a new industry—the winning of the energy values from the oil shales of the Green River Formation of Colorado, Utah, and Wyoming. However, the development is contingent on the willingness of industry to commit substantial sums of money to what must still be termed a high-risk venture; among the costs will be the necessary techniques of environmental control to prevent the degradation of air, water, and land in an area that is still largely in the state in which it was created by nature.

SUGGESTED READINGS

Burwell, E. L., T. E. Sterner, and H. C. Carpenter, "Shale Oil Recovery by *In Situ* Retorting, a Pilot Study," *J. Petrol. Technol. 22,* December 1970, p. 1520–1524.

Duncan, Donald C., and Vernon E. Swanson, "Organic-Rich Shale of the United States and World Land Areas," U.S. Geological Survey Circ. 523, 1965, 30 p.

East, J. H., Jr., and E. D. Gardner, "Oil Shale Mining, Rifle, Colorado, 1944–56." *U.S. Bureau of Mines Bulletin* 611, 1964.

Matzick, A., R. O. Dannenberg, J. R. Ruark, J. E. Phillips, J. D. Lankford, and Boyd Guthrie, "Development of the Bureau of Mines Gas-Combustion Oil-Shale Retorting Process." *U.S. Bureau of Mines Bulletin* 635, 1966.

U.S. Dept. of the Interior, "Final Environmental Impact Statement for the Prototype Oil Shale Leasing Program," U.S. Dept. of the Interior, Washington, D.C., 1973 (Vol. 1–6).

Van West, Frank P., "Green River Oil Shale," in *Geologic Atlas of the Rocky Mountain Region, U.S.A.* Denver: Rocky Mountain Association of Geologists, 1972.

Coping with the Oil Sands

A. R. ALLEN   1974

The reserves of readily available liquid hydrocarbons have been predicted with reasonable accuracy over the last five years, the production capability has been analyzed, and the consumption rate has followed the expected trend. Everyone has been nervous about the vulnerability of our economy to the international behavior of the Organization of Petroleum Exporting Countries from which we have been purchasing ever-increasing amounts of hydrocarbons. These concerns have now been amply justified.

Though this awareness has existed in the minds of the producers and the enlightened economists, the oil companies have been bombarded with irrational accusations of price fixing, gouging the public, excess profits, and lethargy in developing new sources.

Many, many millions of dollars are being spent each year in looking for oil and gas. Although the success rate has been very low, oil is being found in the less accessible regions of our continent and further and further offshore. Some oil companies have also devoted a great deal of money to examining the economic potential for hydrocarbon production out of oils sands, oil shales and coal.

One such company, Sun Oil Company, after some years of research and pilot-plant work embarked upon a commercial scale of operations to produce 45,000 barrels of synthetic crude oil per day from Athabasca Tar Sands. This operation lies near the center of the Athabasca Tar Sand region* at latitude 57° north and longitude 111° west in the province of Alberta. (See Figure 4-29.)

*Editors' note: The Athabasca Tar Sands cover an area of about 30,000 square miles and contain an estimated reserve of 300 billion barrels of recoverable crude oil—an amount closely approximating the world's total known reserve of conventional oil. Commercial recovery of oil from the oil sands are currently being accomplished by open-pit mining the sands and transporting the sand to a separation plant. However, this process will only recover 35 to 36 billion barrels of this estimated reserve. It is anticipated that recovery of bitumens from the sands in place (in situ mining) will become technically and economically feasible in the future.

A. R. Allen is Director of Operation for Great Canadian Oil Sands Limited, Fort McMurray, Alberta, Canada.

Reprinted with light editing and by permission of the author. This article is based on an address delivered at the American Chemical Society, 167th National Conference, Los Angeles, California, April, 1974.

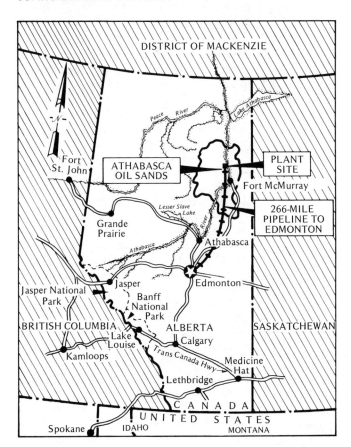

Figure 4-29. Athabasca tar sand location, Alberta, Canada.

This article deals mainly with the solids handling and some related problems encountered in producing oil from the tar sand. A quick look at the dimensions involved will help one to have a more complete grasp of the whole picture. The ultimate reserves of bitumen in the Athabasca Tar Sands have been estimated at about 700 billion barrels. The sands are discontinuous, erratic both in thickness and depth below the land surface, and highly variable in terms of oil saturation and particle size. The most important factors affecting economic development of the tar sands are the depth below surface, thickness of the ore body, and bitumen saturation. Figures 4-30 and 4-31 show roughly how these factors apply to the Athabasca region.

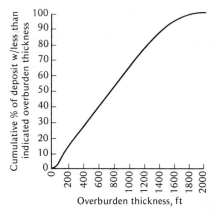

Figure 4-30. Athabasca tar sands overburden thickness.

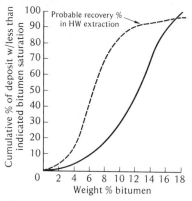

Figure 4-31. Athabasca tar sands bitumen saturation and probable recovery rate.

Recovery may be affected by two methods, by mining the sands and transporting the sand to a separation plant, or by using ways to recover bitumen from the sands in place.

The recoveries which might be anticipated before processing the sand, are estimated to conform roughly to Table 4-26. The restraints applied to these guesses are that only one open pit mining operation has been attempted on a large scale, and it has encountered many problems; underground mining may be completely uneconomic at any scale due to high gas concentrations and mobility of tar sand and aquifers above and below the tar sands beds; *in situ* processes have been under investigation for many years and produce high-cost oil, and early assessments may have been overly-optimistic.

The figures mentioned are generalized.

Table 4-26. Ore Recovery Estimates before Sand Processing

Method	Depth to Top of Ore	Recovery Per Cent of Ore
Open-pit mining	0–150	85–90
Underground mining	>300	50–60
In situ extraction	>600	20–30

However, when an operation is to be undertaken, the data required for the decision to invest over a billion dollars a plant must be specific. The changes in depth, thickness and grade of tar sand can be very abrupt and unless carefully evaluated in detail before beginning an operation, could result in financial disaster. All of the money has to be spent before a drop of oil is produced.

With this in mind, look again at Figure 4-30 and realize that at the present price structure for oil it is unlikely that more than 10 per cent of the bitumen in this vast reserve will be recovered by open pit mining methods. This would represent about 43 billion barrels of synthetic crude oil and would take about 100 years to produce from 12 plants each handling up to 400,000 tons a day of solids.

Let us now examine the Great Canadian Oil Sands operation. The lease of 4,000 acres of this area contained about 700 million barrels of recoverable bitumen. The operation now produces just over 50,000 barrels of synthetic crude oil per day. This entails production from one open-pit mine of around 140,000 tons of tar sand and 130,000 tons of overburden per stream day of these activities. The temperature varies between 90°F down to −50°F, and to add a local touch of geography much of the area is covered by muskeg swamp, a semifloating mass of decaying vegetation.

This is not a conventional strip mining operation. The restraints imposed by environmental considerations, administrative regulation, confining boundaries, weather, and the nature of the materials to be handled create severe limitations.

All overburden and tailings must be confined within the operators lease and must not be placed on recoverable tar sand within the lease. The overburden and tailings occupy a considerably larger volume after the oil has been removed than they did before being disturbed; also, space is required in the pit for excavators, conveyors, and other equipment. Accordingly, much long-term planning is required to permit progressive dumping of overburden and tailings without hindering the subsequent mining operations.

The initial tailings pond was constructed on the river flat with a starter dyke of overburden and subsequent sand dyke construction to allow the mine to progress unhindered. While this is going on, the overburden is being used to prepare a second tailings pond inside the mined-out portion of the pit by the time the initial tailings pond is filled.

In order to accommodate the volume increase and allow space for conveyors the terminal or new land surface must be at least 70 feet above the original land surface.

OVERBURDEN REMOVAL

The sequence of operations prior to the actual mining of the tar sand involves an unusually long time span. The first step is to clear off the scraggly growth of tamarack and black spruce trees that grow in the muskeg. This and the preparation for muskeg removal can only be done efficiently when the muskeg is frozen deeply enough to support equipment. Early operations tended to concentrate more on the retrieval of sunken equipment than removal of material. So, being thus restricted to the coldest part of the winter keeps people pretty well on their toes. The hours of daylight are short, the time period is limited, visibility is restricted by fogs, and exposure conditions for men are at their worst. Morale and efficiency require a lot of attention under these conditions.

In order to remove the muskeg, drainage networks are blasted to run off as much water as possible during the summer periods. Ideally, two years drainage are needed prior to muskeg excavation—even then, muskeg material can only be handled efficiently in winter. Different methods have been tried—finally settling down to two-pass ripping for fragmentation and frost penetration followed by the combined efforts of fifteen yard front-end loaders and 150-ton trucks. By working a long muskeg face it is possible to maintain a trafficable frozen bottom on which the vehicles operate. Icing up and freezing of the damp fibrous material in the

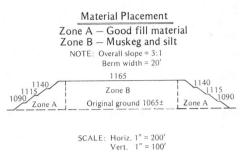

Figure 4-32. Typical section of muskeg dump.

loader buckets is a nuisance affecting efficiency but more tolerable than the previous difficulties with different equipment arrays.

The frozen muskeg cannot be used in constructing the tailings dykes and so has to be piled up in special dumps. When the muskeg piles thaw, a mobile, liquid, stringy mess results. Earth dams are constructed to keep the material under control. Muskeg dumps, as shown in Figure 4-32, are up to 100 feet high. The bulk of the overburden, removed after the muskeg operation, is used to build the earth dams inside the floor of the pit after the tar sand has been mined. The material consists of various kinds of sand, clay, boulders and low-grade tar sand. The distribution of these members is determined and the excavation sequence must be planned to conform to the civil engineering requirements of the massive 300-foot-high dams. Figure 4-33 shows a typical cross section of the overburden dams.

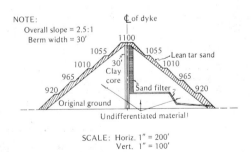

Figure 4-33. Typical section of overburden dyke.

The amount of overburden that must be removed varies greatly with location. The minimum objective is to have one year of "prepared" tar sand available for mining at the end of the stripping season; with additional room for maneuvering the stripping fleet. The overburden dams can only be built using unfrozen material; hence, this job also has a time restraint of about 8 months in the year.

Figure 3-34 shows the actual amount of cover to be removed and the "average line" to meet the long-term requirements of a stable work force and fleet size.

In the overburden removal, again several methods have been tried, many more have been studied. Some of the overburden members are very wet fine silts. The base of overburden is tar sand. The major problems associated with the job are the scarcity of

road building material, and very poor soil-bearing properties. Trafficability both on the burrow floors and the dyke being constructed is poor. Long hauls, big boulders, ground moisture, and compaction difficulties make this an expensive operation.

The present method of removing overburden uses 7×15 cubic yard front-end loaders, 21×150-ton rear dump trucks, graders, bulldozers, and compactors. For five months out of the eight, the operation is supplemented by the loading capability of a Bucyrus-Erie bucket wheel—in all $10 million worth of mobile equipment. Round trips are about 3 miles: 14 million cubic yards, or 24 million tons should be moved each year, from here on; in 8 months of elapsed time per year. Rain and silts are our worst enemies.

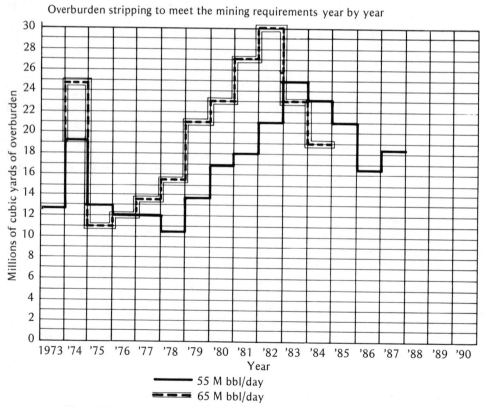

Figure 4-34. Overburden stripping to meet the mining requirements year by year.

MINING OPERATION

Nothing quite like this material has ever been mined before in such quantities, and this operation will probably be the smallest of its kind—the next two operations will be at least twice the size of the Great Canadian Oil Sands operation.

The average thickness of the tar sand in the lease now being mined is about 140 feet. It varies from about 90 to 220 feet. The mine is designed to extract the tar sand in two parallel benches up to 75 feet in height with a trailing bench of varying depth. The floor of the mine is limestone, but it is often not possible to run the main excavating machinery on the limestone base due to the pronounced rolling shape of the floor and the travel-grade limitations of the machinery.

The bucket-wheel excavators deliver the tar sand via traveling conveyors to a shiftable face conveyor. These conveyors discharge to collector belts feeding the extraction facility. The excavators are of German manufac-ture and have a theoretical digging capacity of about 5,000 tons per hour each. Service factors are below 80 per cent, and they are in a chain of apparatus, each one having an individual service factor. The two systems are barely capable of supplying the plant. Feed capability is supplemented by a smaller wheel and trucks, loaders and trucks, or in special cases, scrapers.

The logistics of large bins or surge piles with this kind of material are a special kind of nightmare—and so not in the design. This means that with the 30- to 40-minute surge coupling, which is designed through the extraction plant feed bin, all phases of the bitumen production step are tightly inter-dependent. Operators and supervisors throughout must be in close communication and constantly alert to the operating conditions and delays which might cause a set of problems upstream or downstream. The domino effect rules with an iron fist.

The tar sand is difficult stuff to deal with; its properties are not constant and these properties vary with particle size, oil satura-

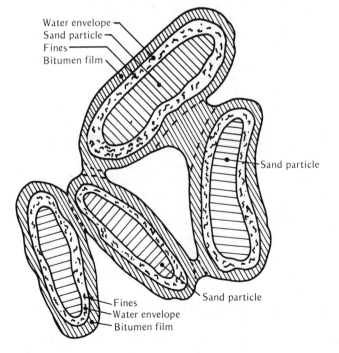

Water envelope
Sand particle
Fines
Bitumen film

Sand particle

Fines
Water envelope
Bitumen film

Figure 4-35. Typical arrangement of tar sand particles.

tion, moisture content, temperature, and, as it seems sometimes, even with the time of day. Figure 4-35 shows the typical arrangement of the components of tar sand.

The problems of temperature variation are of immense importance in the mining operations. The bitumen viscosity varies from 35,000 centistokes at 90°F to 10^6 centistokes at 40°F. At 0°F we have a solid—but not quite. The sand particles are quartz with a mohs hardness of about 5.5 to 6.0. When these particles are bound together with the bitumen matrix at low temperatures, extremely high pressures are required to penetrate the formation. The teeth of the digging apparatus take the brunt of the punishment. On undisturbed tar sand in wintertime a 100-pound alloy tooth can be worn out in less than 8 hours—each of the bucket wheels has 120 of these cutting teeth. This was indeed part of the initial experience—teeth were being airlifted to the operation from anywhere in the world they could be obtained. It was obvious something had to be done about this, and it has now been done. Another early problem was related to the large dimension of the jointing planes—blocking out of huge frozen chunks of tar sand, which cascaded from the crest—occasionally wiping out the digging heads on the machines; and yet another—that minute water envelope around the grains of sand. After penetrating the 8 feet of frozen material the tar sand in place averages 40°F; at this temperature the shear strength is much lower—and the water is not ice—but on exposure to the metal surface of buckets and transfer chutes or the cold surface of conveyor belts it promptly freezes, then stays there. As the newly planted frozen surfaces continue to cool, more sand builds up until openings of 20 square feet or more, designed to handle the passage of 10,000 tons an hour become completely closed off.

The resulting mass of frozen bitumen, water, and sand is far worse than concrete to remove. Many frigid hours have been spent trying to break, burn, drill, or scrape this out of chutes. Obviously, something had to be done about this too—and has been.

Then comes the summertime and occa-sional thunderstorms. Tar sand in summer can only be described as smelly and sticky; it hangs on with appalling tenacity to everything it touches. The material on the floors of the working benches no longer has the confining stress of the load it once bore, it is black, and the ground surface of exposed tar sand has a temperature higher than the air temperature. Some gas escapes from the tar sand. Within minutes of a freshly cut surface being exposed, the surface becomes noticeably softer and in the rich sands the surface will bleed bitumen and crumble to a sticky mess. The trafficability of the softened skin is very poor and the resistance to forward motion extremely high. Repeated passes of equipment cause the thickness of the softened layer to increase and the muckiness of the surface to intensify. Equipment is rapidly bogged down. This thixotropic characteristic of the tar sand disappears with the cool weather—but a sponginess remains until the frost penetrates deep into the floor. Throughout this period of each year there are some problems with equipment. In addition to the continuous cleaning load on mobile equipment (a 4,000-pound pickup truck will weigh in empty at 6,000 pounds), there are occasional incidents due to differential settlement under the tracks of the 1,800-ton bucket wheels. One further problem associated with getting tar sand to the extraction plant in summertime is the conveying systems. The bitumen sticks to the conveyor belts, and partly transfers itself to the idlers and drive pulleys carrying with it some sand. This causes tremendous vibration, differential slip on drives, rapid wear of shells, failure of bearings, ripping of belts, and summer madness in the maintenance staff. Again the cleaning problems have enormous dimensions. We have managed to solve some of these problems to a degree: the cost is unbelievably high and we still have a long way to go.

THE EXTRACTION OF BITUMEN FROM TAR SAND

Bitumen extraction is accomplished by a very simple process which is sometimes diffi-

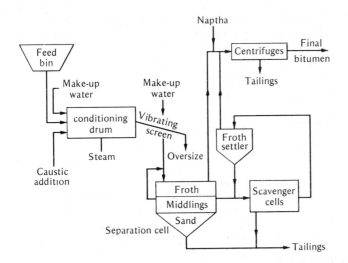

Figure 4-36. Schematic flow diagram of the primary extraction process.

cult to control. Figure 4-36 is a simplified flow sheet. Tar sand is removed from the 5,000-ton surge bin, which it does not want to leave, by eight pan feeders, which deliver in pairs to four parallel belt conveyors. The plant has four parallel processing lines each one capable of handling about 2,300 tons of tar sand per hour. The fundamental steps in processing are feed conditioning, separation of the bitumen, waste disposal, and cleaning bitumen concentrate.

Conditioning is done in a very short period of time by mixing feed with water and caustic soda at 180°F. The vessel used is a slowly rotating drum into which live steam is sparged below the slurry surface. In this process the tar sand disintegrates liberating bitumen from sand and clay particles. The sand is water-wetted in its natural state and stays water wet, allowing fairly clean separation of bitumen in the next process. The clays in the feed report largely as layers interbedded with the tar sand. These clay layers do not ablate or disintegrate as completely as the tar sand does, but remain intact as lumps or balls of clay. In addition to clays, large pieces of rock are encountered in the feed that also remain as competent chunks. The slurry leaving the conditioning drum is passed over a vibrating screen. The unablated clays and the rocks are eliminated by this screening process. All screen-undersize material drops into a sump pump

and is delivered by 18-inch pumps into separation cells.

In the separation process the sand particles settle quickly to the bottom of the cell, which has fast thickener type rakes for conveying the sand to spigots controlled by three large air-operated valves. The bitumen floats to the surface of the cell, and in between the bitumen and the sand layer is a layer wherein separation is taking place. This layer consists predominantly of clays. In order for the process to be effective, the temperature of the cell must be controlled. This control is achieved in the conditioning process: the density caused by the presence of clay suspended in water must also be controlled so that there is an effective differential between the density of warm bitumen and the density of the clay slurry, and last, the viscosity of the clay suspension in the middle of the cell is important. Control of viscosity is achieved by managing the pH with caustic soda. This control is also effected in the conditioning cell.

The bitumen flows from the top of the cell, whence it is further heated and pumped into the upgrading plant for removing the last water and mineral particles from the bitumen froth. The layer in the center of the cell is tapped from the side of the cell and this side stream containing clay, silty particles, and minute droplets of bitumen that cannot make it through this medium is then

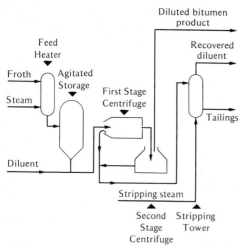

Figure 4-37. Schematic flow diagram of the final extraction process.

delivered to banks of standard sub-'A' flotation cells where ordinary air flotation is used to scavenge the minute portions of bitumen. The concentrate from the scavenger cells is delivered along with the primary bitumen froth from the separation cell into the centrifuge plant.

The bitumen from primary separation is delivered first into a holding tank where it is deaerated. The deaerated heated slurry is then mixed with a diluent and fed to primary centrifuges (scroll type). In Bird centrifuges the coarse mineral load is removed from the diluted bitumen along with some of the water. The concentrate from the Bird centrifuging operation is then passed into several banks of Westfalia separators. These are high speed centrifuges which take out most of the remaining water and the clay particles. The Westfalia concentrate, now containing about 0.5 per cent of mineral particles and 2 or 3 per cent water, is pumped into diluted bitumen storage tanks, whence it is fed to the refining process.

Figure 4-37 shows the schematic flow in the centrifuge plant. The tailings from the two-stage centrifugation of bitumen are delivered into the tailings pond separately from the main tailings stream. The entire

centrifuge plant is a separate building compartmented for protection against fire and operated with very strict rules because of the dangerous atmospheres that could accidentally be generated by a leak of the light hydrocarbons.

TAILINGS DISPOSAL

The tailings stream, about 24,000 gallons per minute, is delivered to the pond by pumps in a separate building from the main extraction plant. These are multistage installations of centrifugal pumps. In the tailings area the coarse fraction of the tailings sand is used to build the dyke while the slimes portion flows into the center of the pond. There are four lines across the top of the tailings dyke. Construction of the dyke by decantation methods and compacting can be carried out simultaneously in three locations. Normally, only three banks of pumps are needed to handle the flow. Because of the final purpose of the tailings dyke, the quality of construction is monitored in considerable detail and supervised in general by a consultant in soils mechanics. The top layer of the liquid portion of the pond is recycled to the extraction plant for use in the processing. There is a slight imbalance in water management due to the fact that the clays will not settle, and also that there is a small amount of bitumen floating on the pond. Certain

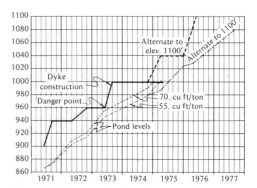

Figure 4-38. Relationship between the rate of construction of the dyke and the rate of rise of the water level contained by the dyke.

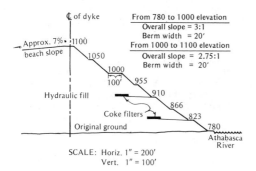

Figure 4-39. Typical section of sand dyke.

controls must be imposed upon the quality of the water recycled because prior to its use it must pass through several tubular heat exchangers. Figure 4-38 shows the relationship between the rate of construction of the dyke and the rate of rise of the water level contained by the dyke. Figure 4-39 shows a typical section through the dyke.

LAND RECLAMATION

This topic, if dealt with in proportion to the number of words uttered over the past two years would fill a library and take seven hours to discuss. This article can only touch on the highlights at the Great Canadian Oil Sands operation.

The climate for the growth of government departments and controlling agencies is far more favorable than the climate we have for the rapid reestablishment of fields and forests. Research studies are in progress to define the chemical, physical, mineralogical, and microbiological properties of tailings deposits. Concurrent, but dragging somewhat, are field studies to determine how tailings can be treated to improve their quality by addition of topsoil, leaching experiments, and the use of synthetic fibers and muskeg and manure. Field revegetation experiments and "high-speed" laboratory growth experiments are in progress. The quality of supernatant waters is being scrutinized to determine its effect upon migratory birds, fish populations, and thermal regimes with adjustment possibilities for the development of aquatic life systems. Then more general over-all conditions must be brought into the picture so that some direction may be given to the operators.

Great Canadian Oil Sands began its program of rehabilitation three years ago, sloping the dyke walls and establishing grasses and shrubs, and is now into the experimental stages of tree growth. In this work several thousand trees of different kinds, in different areas, and with different early life support, are under observation. The cooperation and assistance we have received from research councils, government departments and universities has been excellent. It is not easy to set the pace for a regional program which will later affect all producers in this area; but with the number of people living in this region twenty years from now, their recreational needs, and the desired quality of their surroundings, one cannot leave these matters to settlement by chance or public pressure. A well-designed and carefully planned future must be mapped out now.

SUMMARY

In summary, Great Canadian Oil Sands Ltd. is the only "synthetic oil" production installation of its type in the world. It is the largest single investment ever made by Sun Oil Company. Production of oil from mined tar sand is inherently a high cost method of producing oil, both capital and labor intensive. Tar sands are an extremely low-valued, nonhomogeneous ore requiring very careful investigation. The cost of overburden removal is high relative to the value of the tar sands beneath.

These characteristics make profitability highly dependent upon careful evaluation, and the soundness of a number of both long term and short term mining decisions. The extraction process is new and continuously improving.

There is a great deal of interest in the detail of these works and the techniques of evaluation, the decision processes involved to produce proper decisions with interacting features of pit limits, lease lives, selection of ore, and methods analyses.

It is hoped that this article has led to a more complete understanding of why you are going to have to pay more for your future feedstocks.

SUGGESTED READINGS

Allen, A. R., 1973, "The Mining and Extraction of Bitumen from the Athabasca Tar Sands,"
Great Canadian Oil Sands, Ltd., Fort Mc-Murray, Alberta, 24 p.

Carrigy, M. A. (ed.), 1974, "Guide to the Athabasca Oil Sands Area," Alberta Research, Information Series 65, Edmonton, Alberta, 213 p, (collection of 8 articles).

Humphreys, R. D., 1973, "Great Canadian Oil Sands, Ltd., Tar Sands Pioneer," Petroleum Society of CIM, Edmonton, Canada.

Kaminsky, V. P., 1973, "Selection of a Mining Scheme for a Tar Sands Extraction Plant," Canadian Society of Petroleum Geologists, "Oil Sands Symposium," 6 p.

Coal Conversion Technology

HARRY PERRY   1974

Even before the Arab oil embargo of fall, 1973, it was obvious that indigenous oil and gas resources could not satisfy the burgeoning United States demand for fuel much longer. Petroleum had become a dominant force in the economy, but it was far too vulnerable.

Increases in the absolute amounts of petroleum consumed per year were even more indicative of the supply problems that were certain to occur at some future time. In 1956, the peak year of exploration, the United States consumed 2.93 billion barrels of oil and 10.1 trillion cubic feet of gas. By 1973, this had increased to 6.3 billion barrels of oil and 22.8 trillion cubic feet of gas.

At the same time, the contribution of oil and gas to the total energy supply increased from 67 to 75 per cent. Nearly all of this growth was at the expense of coal. Because it is more difficult to handle than the other fuels, leaves a residue that must be disposed of, and creates dust and dirt during its use, coal was displaced in the residential and commercial markets first by oil and then by gas.

Traditional coal markets were lost one by one. Following World War II, the coal industry lost a large market when the railroads converted their inefficient, coal-burning steam locomotives to more efficient diesel engines. The remaining coal markets were for the production of coke as a fuel, for the rapidly growing electric utility market, and for large industrial users such as cement mills. By 1973, the electric utility industry consumed 69 per cent of all the coal used in the United States.

Prices of bituminous coal remained relatively static between 1948 and 1969, ranging only from $5.08 per ton to $4.39 per ton, the lower price generally prevailing at the end of the period. In the face of inflation during this period, real prices for coal de-

Harry Perry has been a consultant working part time for the National Economic Research Associates, Inc. and for Resources for the Future, 1755 Massachusetts Ave., N.W., Washington, D.C., since 1973. He is former director of coal research for the U.S. Bureau of Mines (1940–1967), research advisor to the assistant secretary of the interior for mineral resources (1967–1970) and senior specialist for energy in the environmental policy division of the Congressional Research Service of the Library of Congress (1970–1973). His experience has been in mining, combustion, carbonization and gasification of coal, the abatement of air pollution, and the large-scale production of helium.

From *Chemical Engineering*, Vol. 81, No. 15, pp. 88–102, July 22, 1974. Reprinted by permission of the author and *Chemical Engineering*. Copyright © 1974 by McGraw-Hill, Inc., New York.

clined significantly. Starting in 1973, there were a number of adverse economic developments: inflation picked up; the field price allowed for natural gas escalated sharply; steep price rises in imported oil were imposed by the Organization of Petroleum Exporting Countries (OPEC) nations; and productivity declined in the underground coal mines. The result: eastern United States bituminous coal prices also increased sharply, from an average mine price of $4.99 per ton in 1969 to $7.66 per ton in 1972, and an estimated $8.50 per ton in 1973. During the energy shortage in the spring of 1974, spot prices were as high as $25 to $35 per ton.

FOUR CONVERSION TECHNIQUES

Such price volatility indicates a great turnabout for coal. But this time, interest is centering on converting the mineral into liquid and gas products to substitute for oil and gas.

Four distinct conversion routes are possible. These include: pyrolysis, solvation, hydrogenation, and production of synthesis gas. The amount and type of products made by each of these methods depend upon the coal properties and process conditions.

In pyrolysis, the coal is heated in some manner to break it down into solids, liquids, and gases. Higher heating rates yield greater amounts of liquids and gases.

In solvation, the coal is dissolved and, with the addition of small amounts of hydrogen, can be filtered and converted into an essentially ash- and sulfur-free solid or liquid, depending upon the degree of hydrogenation.

In hydrogenation reactions, coal and hydrogen are reacted together directly. If this is done in the presence of a catalyst at 850°F and at elevated pressures, a liquid product can be made. If a catalyst is not present, the coal can react directly with hydrogen at even higher temperatures (1,500–1,800°F) and pressures to form methane.

In the production of synthesis gas (car-

bon monoxide and hydrogen), coal is usually reacted with an oxidizing agent and steam. However, the heat for the steam-carbon reaction can be supplied in ways other than by using an oxidizing agent (see discussion on coal gasification). The synthesis gas produced can then be used to make a high-BTU gas by reacting the purified synthesis gas over a nickel catalyst. The purified gas can also be used as the raw material for production of alcohols, ammonia, synthetic gasoline, and a variety of other petrochemicals. The product obtained depends on the raw materials, processing conditions, and the catalyst used.

GASIFICATION—BACKGROUND

Gas was first made from coal by heating the coal in the absence of air. The gas was distributed in many cities for lighting streets, homes, and buildings. Late in the nineteenth century, the major market for gas shifted to cooking as the lighting market was replaced by electricity.

As the gas market grew, it became necessary to find ways to supplement the low yield from coal distillation. Over 70 per cent of the coal remains as a solid when it is heated, and this portion had to be sold if the price were to be competitive. When slot-type ovens for making coke for the steel industry became widely used, coke-oven gas—the composition of which is similar to that of distillation gases—was used as a supplement.

Other methods that used all of the coal rather than just the distillation products were devised to further supplement coal-gas supplies. The most widely used process made a gas known as "water gas" or "blue gas." The process was cyclic. First, a bed of anthracite or coke was heated by burning a part of the carbon in the bed with air. The evolving hot gases heated the rest of the carbon. The basic reaction was

$$C + O_2 + 4N_2 \rightarrow CO_2 + 4N_2$$

As the carbon in the upper part of the bed became heated, the reverse reaction started to occur:

$$CO_2 + C + 4N_2 \rightarrow 2CO + 4N_2$$

When the CO value in the existing gas became too high to be acceptable, the air blast was stopped and the hot bed of carbon reacted with steam according to the reaction

$$C + H_2O \rightarrow CO + H_2$$

The heat for this endothermic reaction was supplied by the hot carbon. When the carbon bed cooled to the point at which rate of steam and carbon reaction were too low, the steam flow was stopped and the "blow" cycle repeated, using air.

Gas produced during the steam period ("run gas") had a heating value of 300 BTU per cubic foot and had to be enriched to make it compatible with the other gases being distributed by the gas utilities. The enrichment was also carried out in a cyclic process. Gases from the last part of the "blow" cycle, which contained CO, were burned to heat a vessel filled with refractory bricks. When the refractory was sufficiently hot, the gas was shut off and an oil was introduced onto the hot bricks, where it was "cracked" into lower-molecular-weight hydrocarbons, which were then mixed with the "run" gas to adjust the heating value to the desired level.

The "water gas" process had many limitations. Cyclic processes are always expensive and are troublesome to maintain and operate. The process required coke or anthracite, both noncoking sources of carbon. Over-all efficiency of the process was low, the oil used for enrichment greatly increased cost, and the operation was inevitably dirty.

To avoid some of these problems, the "producer gas" process was used at industrial plants needing a clean source of fuel but for which the cost of transport and distribution of the gas was not important (the gas was consumed at the point of manufacture). In this process, the bed of carbon was reacted continuously with a mixture of air and steam. The simultaneous reactions were

$$C + O_2 + 4N_2 \rightarrow CO_2 + 4N_2$$
$$C + CO_2 \rightarrow 2CO$$
$$C + H_2O \rightarrow CO + H_2$$

The product was a mixture of CO, H_2, CO_2, N_2, and whatever distillation gases remained in the final gas product. Heating value was 135 to 150 BTU per cubic feet. Despite its low heating value, producer gas was much less costly per unit of heat than was water gas.

In the United States, manufactured gas played a major role in supply to markets far from natural-gas fields but near coal fields. The residential gas market in the eastern United States was supplied mainly with manufactured gas as late as 1932. With the introduction of long-distance natural-gas pipelines during World War II, however, natural gas rapidly became more widely distributed, and by the end of the war it was supplying nearly 90 per cent of the heating value of gas being sold. The last year for which data were reported for manufactured gas was 1968, and by then the amount was too inconsequential to matter.

Although interest in new coal-gasification technology disappeared in the United States as a result of the shift to natural gas, interest continued high in petroleum-short Europe, where coal remained the chief source of energy supplies. Mixtures of carbon monoxide and hydrogen (synthesis gas) also became increasingly important as the basic raw material for ammonia and a whole range of organic chemicals needed to supply the exploding markets for plastics.

In the past several years, a relatively large number of United States firms have announced plans to construct commercial coal-gasification plants, using proven foreign technology, but none are yet under construction.

MEDIUM- AND HIGH-BTU GAS

Most of the interest in coal gasification has centered on medium- or high-BTU gas that could be piped economically over long dis-

Table 4-27. Coal Gasification Processes for Production of Medium- and High-BTU Gas.

Process	Reactor Bed Type	Gasifying Medium	Nature of Residue	Pressure, (atmospheres)	Temperature (°F)	Capacity (tons per day)
Commercial						
Lurgi gasifier	Fixed	Oxygen-steam (air being tested)	Dry ash	30–35	500–800 (top), <2,000 (bottom)	1,000
Koppers-Totzek	Entrained	Oxygen or limited oxygen-steam (air being tested)	Dry ash	1	1,750–2,350	850
Winkler	Fluidized	Oxygen-steam and air-steam	Dry ash	1	1,500–1,800	100
Small Demonstration						
Hygas[a] (Institute of Gas Technology)	Fluid	Hydrogen in hydro-gasifier	Dry char	75–100	1,200–1,500 first stage; 1,700–1,800 second stage	80
Carbon dioxide acceptor	Fluid	Air in regenerator, steam in the gasifier	Dry ash	10–20	1,575	30
Pilot Plants						
Synthane	Fluid	Oxygen-steam	Dry char	40–70	1,100–1,450 first stage; 1,750–1,850 second stage	75
Bigas	Entrained	Oxygen-steam	Slag	70	2,700 first stage; 1,700 second stage	120
Cogas	Fluid	Steam in gasifier	Dry ash	1–3	1,600–1,700	50
Carbide-Battelle	Fluid	Steam in gasifier	Dry ash	6	1,600–1,800	25
Institute of Gas Technology-electrothermal	Fluid	Steam in gasifier	Dry char	75–100	1,900	25
Institute of Gas Technology-steam oxygen	Fluid	Steam-oxygen	Dry ash	100	1,500	25
Institute of Gas Technology-steam iron	Fluid	Steam in oxidizer	Dry ash	100	1,500	25
Hydrane	Fluid	Hydrogen in hydro-gasifier	Dry char	70	1,600–1,700	0.25

[a]The Hygas process is a hydrogasification route. Three alternative methods to produce the hydrogen required have been proposed. These are: electrothermal, steam-oxygen and steam-iron processes; they are described under "Pilot Plants."

tances. High-BTU gases might even substitute for natural gas.

The state of these processes and some of their major characteristics are shown in Table 4-27. Most use oxygen instead of air, and all except one (the electrothermal gasifier) use some form of heat carrier to supply heat for the endothermic steam-carbon reaction.

Three types of gas-solid contact have been used: fixed, fluid, and entrained beds. Fixed-bed processes require a sized coal and have been operated smoothly only on noncoking or weakly coking coals. These techniques tend to preserve the methane and other distillation products of the fresh coal, and thus have a decided advantage over fluid and entrained processes for making high-BTU gas—unless carried out in two stages, in which case the distillation products are consumed by the gasifying media. On the other hand, entrained processes can use all sizes and any kind of coal, and will produce fewer pollutants. The fluid-bed processes have operating characteristics that are intermediate to the fixed and entrained beds. In fixed-bed processes, the gases leave at low temperatures in the range of 500 to 800°F; in fluid processes, they leave at intermediate temperatures (1,500 to 1,800°F); while in entrained routes (unless staged), the gases exit at over 2,000°F.

HIGH-BTU GAS

Lurgi Process

This is a fixed-bed process. As shown in Figure 4-40, sized noncoking coal is fed by lock hoppers into a pressure gasifier having a diameter of up to 12 feet. Steam and oxygen are introduced below the grate at the bottom of the gasifier in amounts that will cool the grate and prevent clinkering of the ash. The grate is rotated and the ash is collected in a lock hopper, from which it is removed periodically. The coal is spread evenly over the entire bed by a distributor located near the top of the gasifier. Raw gases leave the top at about 850°F and are scrubbed and cooled before further treatment.

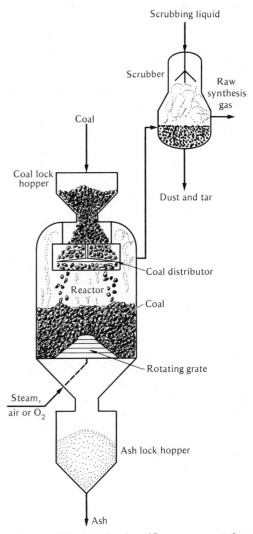

Figure 4-40. The Lurgi gasifier uses a rotating grate underneath the coal bed for feeding oxygen and steam.

One of the advantages of this process is that the countercurrent flow of reactants in a fixed-bed reactor allows the efficient use of the heat released during the oxidation of the coal near the base of the gasifier. The hot gases transfer a large part of this heat to the incoming coal as they pass through the coal in the upper levels of the gasifier. This results in a smaller oxygen consumption per unit of gas produced, and—since the gases distilled from the coal leave the gasifier with the raw product gas—in a higher-BTU-content raw gas. This makes it a more desirable feedstock for further upgrading to high-BTU gas, but less desirable if synthesis is wanted.

Another advantage of the Lurgi method is that it operates under pressure (unlike the other two commercially available processes). This leads to significant over-all process economies, because in most cases the gas will be processed further at elevated pressure. Over-all thermal efficiencies of the Lurgi method are said to be in the order of about 70 per cent.

The disadvantage of the Lurgi route is that it requires a sized coal. Fines produced in mining must either find suitable outlets in other parts of the gasification plant or they must be briquetted to produce a suitable sized feed, which adds to coal costs. The Lurgi gasifier, as it is presently designed and used, can only handle noncoking (or very weakly coking) coals. This would eliminate its use with virtually all of the United States coals east of the Mississippi River. Finally, the fixed-bed gasifiers are basically low-through-put devices that require a large number of gasifiers occupying a large area. This, in turn, requires more complex and costlier piping than if fewer gasifiers were needed. One way to reduce the number and reduce piping would be to increase the diameter of

the individual gasifiers, but this would require extensive development work.

Koppers-Totzek Process

As shown in Figure 4-41, the Koppers-Totzek process reacts the coal, steam, and oxygen in an entrained state at atmospheric pressure. Most of the ash leaves the gasifier with the raw product gas, but a portion leaves as a slag and is collected in a receiver at the bottom of the gasifier. Commerical gasifiers use either two or four opposing burners, and the four burners can handle up to 850 tons per day.

Because of the entrained mode of operation, the raw gases leave the gasifier at very high temperatures (up to 3,300°F) so that the consumption of oxygen per unit of product gas is significantly higher than for fixed-bed operations. Another disadvantage when compared to the Lurgi unit is the atmospheric operation of the Koppers-Totzek process. There appears to be no reason why the Koppers-Totzek unit should not be capable of pressure operation, although this would require developing a low-cost, reliable method to feed fine coal into a pressure vessel, and modification in the design of the slag-removal system.

The Koppers-Totzek gasifier, however, can use all of the coal, including the fines, and can gasify any type or rank of coal.

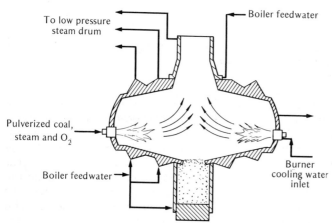

To low pressure steam drum

Boiler feedwater

Pulverized coal, steam and O₂

Boiler feedwater

Burner cooling water inlet

Figure 4-41. The Koppers-Totzek reactor contains an entrained bed of reactants. Gas flows out top; ash to bottom.

Moreover, it has fewer environmental problems because the tars, phenols, light oils, etc. collected during operation of the Lurgi gasifier are never produced at the higher temperature of the process. Over-all thermal efficiencies are said to be about 77 per cent.

Winkler Process

This process is an atmospheric fluid-bed route in which the gasifying media are oxygen and steam. The fluid bed operates at 1,500 to 1,850°F, and most of the ash is carried over with the product gas. The unreacted carbon that is also carried out of the bed is reacted with additional steam and oxygen in the disengaging space above the fluid bed. To prevent slagging of the ash, the gases are cooled by a radiant boiler section in the upper portion of the gasifier. The process was constructed at 16 plants in a number of countries, using a total of 36 generators. These plants are still operating, with the largest having a capacity of 1.1 million standard cubic feet per hour. The last installation was in 1960.

Like the Koppers-Totzek process, the Winkler route can handle the entire size range of coal, but without pretreatment it would be difficult, if not impossible, to operate with a strongly coking coal. Oxygen consumption is intermediate between that of the fixed-bed Lurgi and the entrained Koppers-Totzek process. Though the Winkler does not produce the tars, phenols, and light oils that the Lurgi does, this process like Koppers-Totzek, has been operated commercially only at atmospheric pressure. Studies of estimated results under conditions of 1.5 atmospheres pressure have been made. Over-all thermal efficiencies are said to be about 75 per cent.

Hygas Process

The Hygas process is one of the two large pilot plants that have been constructed under contract with the Office of Coal Research and which are being tested at the present time. Hydrogen (produced in another process step) is reacted directly with coal at 1,000 to 1,500 pounds per square inch to make a high-BTU gas. The hydrogasifier is operated in a countercurrent fashion with coal being fed into the top and hydrogen at the bottom. It consists of two fluid beds in series; the upper bed is operated at about 1,200°F and the lower bed at 1,700°F. Operating in this fashion optimizes the process with respect to reaction rate and the amount of methane at equilibrium in the product gas. The easily gasified part of the coal reacts in the upper zone, the more refractory portion in the lower zone, and the most refractory part of the coal is removed and used to prepare the hydrogen for the hydrogasification reaction.

Three methods to produce the hydrogen required for Hygas have been proposed by the Institute of Gas Technology: electrothermal, Institute of Gas Technology steam-oxygen, and Institute of Gas Technology steam-iron. The characteristics of these three processes are shown in Table 4-27. The Hygas pilot plant is still in the early stages of operation.

Carbon Dioxide-Acceptor Process

This is the second of the two fully constructed, Office of Coal Research-supported, large pilot plants for producing high-BTU gas. The flow sheet is shown in Figure 4-42. Coal is fed into the top of the gasifier, and after being devolatilized, the char is reacted with steam in a fluid-bed gasifier operating at 150 to 300 pounds per square inch, into which hot dolomite also has been introduced. The dolomite provides the reaction heat in two ways—by the sensible heat that it brings into the gasifier (obtained during calcination) and by the heat released when the calcium oxide reacts with part of the CO_2 made in the gasifier by the char-steam reaction.

The product gas leaves at the top of the gasifier, and a stream of spent dolomite and unreacted char is removed from the bottom. These solids are introduced into a second vessel, where the unreacted carbon is burned

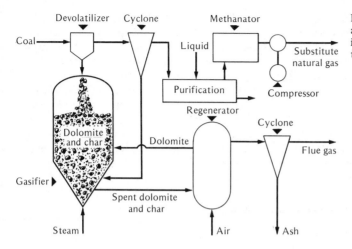

Figure 4-42. The carbon dioxide acceptor process operates from air instead of pure oxygen because of the dolomite regeneration loop.

with air and the heat produced calcines and regenerates the dolomite.

The major advantage of the process is that it avoids the use of oxygen—a relatively expensive raw material. The major disadvantage is that it can probably only be utilized successfully with reactive coals such as lignite, and may not be suitable for higher rank coals.

The pilot plant is still in the early stages of operation.

Synthane Process

This process, as with the balance of high-BTU gas processes that will be described, has only been operated in a small pilot plant, although a 70-ton-per-day plant is now under construction at Bruceton, Pennsylvania for completion by late 1974.

The Synthane process, shown in Figure 4-43, is a two-stage pressurized gasifier developed by the Bureau of Mines, in which the coking properties of the coal are destroyed with oxygen and steam, either in a free-fall stage or in a fluid bed. The coal then enters a carbonization zone and is finally gasified in a lower zone using steam and oxygen. Char and ash are removed from the bottom of the gasifier and raw product gas (containing the effluent gas from the pretreater) leaves at the top. The upper part of the fluid bed is

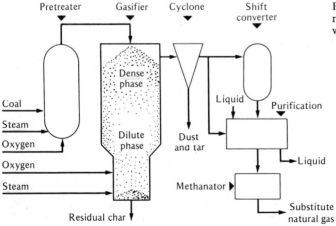

Figure 4-43. The synthane route requires a pretreated feed to prevent slagging in the reactor.

operated at about 1,100°F and the lower part at 1,750 to 1,850°F.

Bigas Process

This route is a two-stage entrained, super-pressure method developed by Bituminous Coal Research, Inc. The basic design data were obtained in a 100-pounds-per-hour unit, and a 120-tons-per-day plant is now under construction at Homer City, Pennsylvania. It is expected to start operation in late 1974 or early 1975.

Coal is introduced into the upper section of the gasifier, where it is heated by the hot gases produced in the bottom of the unit. The coal is carbonized, and the distillation gases leave the gasifier with the gases produced in the lower section of the vessel. Part of the char leaves with the raw gas and is recirculated to the lower section, where it reacts in an entrained state with steam and oxygen to produce synthesis gas. The bottom stage is operated at 2,800°F and the upper stage at 1,400 to 1,700°F.

Cogas Process

The Cogas process is a privately funded research and development effort, supported by a consortium of six companies. No detailed flow sheets or cost estimates have been released, but it has been reported that two different types of pilot plants are being investigated, neither one of which uses oxygen for the gasification step. One pilot plant is at Princeton, New Jersey, and the other is in England. Gasification occurs in a fluid bed, and ash is removed either dry or as a slag. In April 1974, the first successful run was reported on the British pilot plant. Construction of the Princeton plant has been about completed.

Union Carbide-Battelle Process

This is a two-stage fluid bed system. In one stage, part of the carbon is burned with air in a fluid-bed combustor, during which the coal ash agglomerates to form a heat barrier.

These hot pellets are circulated to a second vessel into which coal and steam are introduced to provide the heat for the gasification reaction between steam and carbon. Part of the heat carrier is separated from the char, returned to the first vessel and reheated. The balance of the heat carrier is removed as ash from the gasifier. The process is operated at 100 pounds per square inch. As with the Cogas process, the Union Carbide-Battelle process does not require oxygen. Construction of a 25-ton-per-day development unit was started in late 1973.

Hydrane Process

This is the Bureau of Mines version of the hydrogasification step incorporated in the Hygas pilot plant. It is a two-stage counter-current process. Coal is introduced at the top of the first stage, where it is devolatilized in a dilute phase to destroy its coking properties using hot gas produced in the hydrogasification section at the bottom of the vessel. Here the char produced in the upper section is reacted at 1,650°F with hydrogen (produced in a separate vessel), using the residual unreacted char remaining from the hydrogasification operation. The method, which is expected to be used to make hydrogen, is similar to that for the Synthane process and is a pressurized fluid-bed gasifier using steam and oxygen as the reactants.

Bench-scale tests have been completed and currently a small 10-pounds-per-hour pilot plant is being operated.

Several other processes have been investigated. These include the Molten Salt process of the Kellogg Co., the Atgas reactor of the Applied Technology Corp., and the nuclear coal gasification process of Stone and Webster-Gulf General Atomics. In addition, a limited number of commercial plants using other types of gasifiers were constructed abroad for a period following World War II, and large-scale pilot tests were run on several different types of gasifiers in the United States in the late 1940's and early 1950's.

The molten salt process was studied on a

small scale some years ago but was discontinued because of problems arising from the corrosive nature of the salt. Currently a corrosion resistant design is being tested for its long-term effectiveness. The early version of the process used two vessels. In the first, steam and coal were reacted in a molten bath of sodium carbonate to make synthesis gas. The unburned carbon and sulfur in the coal were discharged with the circulating molten carbonate solution. In the second vessel, the residual carbon in the salt was burned with air to reheat the sodium carbonate, which was recirculated.

In a more recent development, M. W. Kellogg Co. has reported investigation of a single-vessel molten carbonate gasification system using two different methods of operation. In one, the molten carbonate moves between two sides of a divided vessel. On one side, steam and carbon are reacted to make synthesis gas while on the other, the oxidation reaction—formerly carried out in the second vessel—occurs. In the other variation of the single-vessel process, steam and oxygen are introduced into the molten bath of carbonate, where both the steam-carbon reaction and the carbon-oxygen reaction occur simultaneously.

The Atgas process, also in the laboratory stage, uses a molten iron bath to carry out the reaction in a single vessel. The process operates at low pressures, and either air or oxygen can be used as the oxidizing medium. Limestone is injected into the bath to control the sulfur compounds that would otherwise appear in the product gas. When air is used, the product gas is a sulfur-free, low-BTU gas (195 BTU per cubic foot); with oxygen, the product is a medium- or high-BTU gas, depending upon how much subsequent upgrading is used.

As a result of a favorable study for the State of Oklahoma on the possibility of adapting the nuclear high-temperature gas-cooled reactor to coal gasification, Gulf General Atomics and Stone and Webster are cooperating on a joint program to develop this process. The over-all concept involves making both liquid products and pipeline gas. The hydrogen for the process is made by reforming a part of the product methane with steam. The heat required for the steam-methane reforming reaction is supplied by a hot stream of helium from the gas-cooled reactor.

A number of private firms have announced that they have pilot-plant studies under way. General Electric is investigating a fixed-bed atmospheric gasifier using moderately coking coals. The unit will eventually be used to make synthesis gas, but is now running with air (see section on low-BTU gas). Exxon has announced construction of a $40-million pilot plant for production of 20 billion BTU per day of medium-BTU gas (about 400 BTU per cubic foot) from 500 tons of coal. The gas will be piped to an Exxon refinery and used as fuel there. The plant should be completed in late 1976.

A consortium of American companies has been formed to test a slagging, fixed-bed gasifier at the Westfield gasification plant of the Scottish Gas Board. A bottom section will be added to one of the existing Lurgi gasifiers so that it can be operated in a slagging mode. This work will be an extension of a series of experiments conducted several years ago by the British Gas Council at Solihull.

METHANATION STUDIES

Although at least three commercial processes exist for the production of synthesis gas or medium-BTU gas from coal, the methanation step required to make a high-BTU gas has never been operated on a commercial scale. In making a gas of pipeline quality, it will be necessary to use a methanator to upgrade the gases that are produced in any of the commercially proved gasifiers.

A wholly owned subsidiary of the Continental Oil Co. has joined with 15 other industrial firms to demonstrate the purification and methanation steps on a large scale. The tests are being conducted at the Westfield plant of the Scottish Gas Board. The output of a single Lurgi generator, approximately 10 million cubic feet per day of medium-BTU gas, is purified and then methanated in a fixed-bed reactor to produce 2.6 million cubic feet per day of high-BTU gas.

The tests were started in 1973 and are continuing. No detailed reports have been issued, but published data suggest that, although the investigation is not complete, the program has achieved a degree of success. High yields of methane have been produced at 85 to 90 per cent of designed capacity.

CCI has been operating a small methanation pilot plant for several years with private sponsorship. This plant is able to produce about 150,000 cubic feet per day of pipeline-quality gas. It consists of several fixed-bed reactors. Temperature control for the very exothermic process is achieved by recycling part of the hot gases. Much of the effort has been devoted to developing improved catalysts that can function over a wide range of operating conditions and that can recover from short periods of unbalanced operation.

Other projects include a fixed bed being studied at the Institute of Gas Technology Hygas plant, a liquid-phase methanator that will be built on a small, skid-mounted scale by Chem Systems (under contract from Office of Coal Research), and a fluid-bed unit constructed by Bituminous Coal Research. No results have yet been reported for the fluid-bed method.

The Bureau of Mines has a project going to study the use of Raney nickel catalysts that are flame sprayed on tubes and plates. Two types are being tested on a small scale: in one, the catalyst is sprayed on steel plate over which the synthesis gas is passed. Temperature control is obtained by recycling cold product gas. In the other reactor, the catalyst is sprayed on the inside of the tubes, which are surrounded by a boiling liquid that removes the heat of the methanation reaction.

LOW-BTU GAS

The process steps involved in the manufacture of low-BTU gas from coal are similar to those for high-BTU gas through the gasification step, except that in low-BTU gas, air is substituted for oxygen. The product gas is diluted with the nitrogen in the air, producing a gas with a heating value in the range of 135 to 200 BTU per cubic foot, depending on how much of the gases distilled from the raw coal are retained in the product.

The manufacture of low-BTU gas is less complicated and thus less expensive than making high-BTU gas. The shift conversion is not needed, and a simpler gas cleanup system can be used because there is no need to clean the gas for the protection of the methanation catalyst. The methanator is also eliminated, thus increasing the overall thermal efficiency significantly.

If the major societal objective was to make a clean fuel from coal for heating purposes, methods to produce low-BTU gas should have been investigated first. However, the opposite actually occurred. High-BTU gas investigations were conducted many years in advance of low-BTU gas studies. The reasons are understandable: The gas transmission and distribution companies—with large capital investments of their facilities—were concerned, at an early date, that supplies of indigenous gas would be insufficient to keep their pipelines operating at high capacities. They needed a substitute natural gas, not just a clean gas from coal, since the cost of transporting and distributing low-BTU and even medium-BTU gas to most of their customers would have been prohibitive. Moreover, a return to a significantly lower-BTU gas would require replacement of millions of gas burners. These companies were willing to support a high-BTU gas at an early date.

Unfortunately, there has been little research on processes to make clean, low-BTU gas from coal using air as an oxidant. Of the processes in Table 4-27, only those of Lurgi and Winkler have actually operated in a full-scale gasifier using air. The Lurgi is just completing the experimental operations stage. The stirred fixed-bed and the General Electric processes listed in Table 4-28 have also been operated on a reasonable scale using air

Table 4-28. Coal Gasification Processes for Production of Low-BTU Gas.

Process	Reactor Bed Type	Nature of Residue	Pressure (atmosphere)	Temperature (°F)	Capacity, (tons per day)
Commercial					
Winkler	Entrained	Dry ash	1	Approx. 1,500	2,000
Demonstration					
Lurgi	Fixed	Dry ash	20	1,000	2,000
Pilot Plants					
Stirred fixed producer (U.S. Bureau of Mines)	Fixed	Dry ash	20	1,000	20
General Electric fixed bed	Fixed	Dry ash	8	1,000	0.25
Combustion Engineering-Consolidated Edison gasifier	Entrained	Slag	1	>2,100	180
Westinghouse Electric Corp. gasifier	Multiple fluid beds	Dry ash	10–16	1,300–1,700 and 2,000	15
Pittsburg-Midway gasifier	Entrained (two-stage)	Slag	4–35	>2,100	1,200
Institute of Gas Technology U-gas	Fluid bed	Dry ash	20	1,900	30–50

to produce a low-BTU gas. In addition to these four, the Atgas process has been tested on a small scale using air to produce a 190-BTU-per-cubic-foot gas.

The balance of the processes shown in Table 4-28 are now under study, but there are no operating pilot plants. The Office of Coal Research has awarded contracts for preliminary design of the low-BTU gasifiers to Combustion Engineering, Westinghouse, and Pittsburg & Midway Coal companies. The Institute of Gas Technology is actively seeking support for its "U-gas" project.

The large difference between the state of development of high-BTU and low-BTU gasification processes is shown by comparing Tables 4-27 and 4-28. A large number of processes have been extensively tested for high-BTU gas while most of the new low-BTU gas processes have yet to be operated—construction on some has not even been started. The fiscal-year-1975 federal coal research and development budget has attempted to correct this imbalance by increasing the low-BTU gas budget to $49 million, compared to only $31.8 million for the high-BTU gas budget. Five atmospheric-pressure and five high-pressure, low-BTU projects are to be supported.

Air-Blown Winkler Generator

A number of the commercially installed Winkler generators were operated using air instead of oxygen to supply the heat. No special problems with air operation have been reported. Costs per million BTU are estimated to be about 20 per cent lower for gasification of United States lignite with air rather than with oxygen.

Air-Blown Lurgi Gasifier

An air-blown Lurgi gasifier has been operated intermittently for several years at a German power plant designed to use the make-gas in a combined gas-steam-turbine plant. The coal is gasified with air at 20-atmospheres pressure. The exit gases leave the gasifier at about 1,000°F and are scrubbed to remove tar and dust. Hydrogen sulfide would be removed at this stage, if necessary. The gases are expanded in a turbine and then burned in a pressurized boiler. The exhaust gases from the boiler are again expanded in a gas turbine and then used to heat the feedwater to the steam turbine.

After modification to correct some design deficiencies, testing was resumed in late 1973 and is continuing.

Commonwealth Edison Co., in cooperation with the Electric Power Research Institute, is planning to operate an air-blown Lurgi gasifier. It would process 60 tons per hour of coal to supply a boiler fuel for a 70-megawatt generating unit. The tars made during gasification would be used to form briquettes of the fine coal produced in mining that is unsuitable as a Lurgi feed. Original plans called for construction of the Lurgi to begin in late 1974, and operation to begin in 1976.

Stirred Fixed-Bed Producer

The fixed-bed gasifier of the Bureau of Mines, shown in Figure 4-44, has an inner diameter of 3.5 feet and is 24 feet long. The gasifier is equipped with a stirrer to break up any coke that is formed in the upper section. The stirrer both rotates on the shaft and moves vertically in the gasifier.

Unlike any other previously tested fixed-bed gasifier, it was found possible to gasify highly-coking coals using this stirrer. Additional research is to be directed toward higher pressure operation and to gas cleanup at high temperatures and pressures.

General Electric Fixed-Bed Producer

General Electric has studied the gasification of strongly coking coal in a small-diameter fixed-bed gasifier operated at atmospheric pressure. The coal is fed by extrusion into the gasifier and is shaped into uniform cylinders. A larger gasifier, 2 feet in diameter, is being designed to operate at 20 atmospheres and gasify 12 tons per day.

The unique feature of the process is the

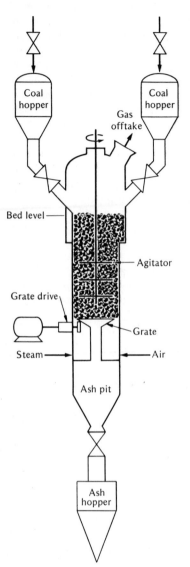

Figure 4-44. Another Bureau of Mines development, the stirred fixed-bed producer, can gasify highly coking coals.

extrusion feeding system. It would simplify feeding of fine coal into a pressure vessel by eliminating lock hoppers.

The following four processes are still in the design and construction stage and no operating data are available.

Combustion Engineering

This is an entrained process supported by the Office of Coal Research, which will be operated in the range of 1 to 10 atmospheres. It will produce a clean, low-BTU gas suitable for steam/gas combined-cycle electric plants.

The experimental pilot plant is designed to handle 120 tons per day—the equivalent of the heat requirements of a 10-megawatt power plant. The plant is thought to be large enough so that, if successful, design of a full-scale plant would follow.

Westinghouse Electric Corp.

This is also supported by the Office of Coal Research and is being directed by a five-member government/industry team. The process selected for study consists of two fluid-bed gasifiers in series. In the first fluid bed, operating at 1,300 to 1,700°F, the dried coal is devolatilized, desulfurized, and partially hydrogasified with gas produced in the second fluid bed. The fresh coal is prevented from coking by contacting it with very large streams of hot recycled char and lime absorbent from the second vessel. Fresh limestone is added to the vessel and a stream of spent limestone removed. The char carried out with the spent limestone is stripped from it and returned to the system. The product gas leaving the first vessel is separated from carryover char and consists of a clean, low-BTU gas suitable for gas turbine use. The overflow char from the first fluid bed is then reacted in a second vessel with air and steam. Gasification of the char occurs in the upper section of the fluid bed in the second vessel at 1,800 to 2,000°F. The heat for the gasification reaction is generated by the combustion (at 2,100°F) of part of the char in the lower section of the gasifier. At this temperature, the ash agglomerates and settles to the bottom of the combustor where it can be removed.

Pittsburg and Midway Coal Mining

This program is cosponsored by the Office of Coal Research and four industrial companies, and is designed to develop an entrained type, two-stage, slagging pressure gasifier. The process is fashioned after the two-stage Bigas process. The fresh coal is devolatilized and partially gasified in the upper stage of the gasifier as it comes into contact with the hot gases produced from the combustion of char with air in the lower section. Char carried over with the product gas is separated and reinjected into the bottom part of the gasifier. The exit temperatures are sufficiently high so that they contain no tars or oils.

IGT-U Gas Process

This is a fluid-bed pressurized gasifier developed by the Institute of Gas Technology to produce a clean gas with a heating value of about 140 BTU per cubic foot. The process is designed to be able to use coking and high-sulfur coals. If it is necessary to destroy coking properties, the coal would be pretreated with air at $800°F$ in a separate vessel, and the heat released during pretreatment would be recovered as steam. The pretreated coals and the gas produced flow into the gasifier vessel. The char reacts with steam and air at temperatures at which the ash agglomerates, then settles to the bottom and can be removed from the char-ash fluid bed (similar to the Westinghouse ash-removal method). The gas leaving the fluid bed has a residence time above the bed of 10 to 15 seconds at 1,500 to $1,900°F$, in order to crack the tars and oils that may have been formed. Eliminating the tars and oils will reduce heat exchanger fouling and simplify gas cleanup. The gas will then be cleaned using a new process that is said to be capable of removing sulfur and particulates at high temperatures and pressures. The Institute of Gas Technology is working with Ralph Parsons Co. to prepare a cost estimate for a demonstration plant large enough to permit direct scaleup to a commercial size.

UNDERGROUND GASIFICATION OF COAL

The idea of gasifying coal without mining can be traced back as far as 1868. However, it was not until after World War II that world-wide interest developed in the technique, and experimental work was conducted until about 1960 in at least eight countries. The main advantage of underground gasification is that the heat in the coal could be extracted with little or no underground manpower required and with limited mining costs. Moreover, it might be possible to extract energy from coal seams that are either too thin or of too poor a quality to be mined by conventional methods.

The only commercial applications of underground gasification have been in the Soviet Union, where production in 1956 was reported to be 116 billion cubic feet of fuel having an average heating value of 85 BTU per cubic foot. This represents less than 0.05 per cent of United States demand for gas in 1972.

Because of the enormous potential for increasing safety and reducing extraction costs, underground gasification is again receiving attention in the United States. In the spring of 1973, the Bureau of Mines initiated a small-scale underground gasification experiment in a subbituminous coal seam in Wyoming. Before the experiment was initiated, the direction of the maximum natural permeability of the bed was determined. It was found that this permeability coincided with the natural fractures of the seam. A number of holes were drilled from the surface to the coal seam and the natural permeability of the bed was increased by various fracturing methods. The system was ignited using forward burning, and some combustible gas has been produced. The experiment is still underway, and continuous operation for one year was achieved in March 1974. Coal consumption has been 15 tons per day, and the product gas reached an output of 3 million cubic feet per day at a heating value of 140 to 170 BTU per cubic foot.

COAL LIQUEFACTION—BACKGROUND

In the past, the degree of interest in liquefaction technology has largely depended on estimations of the adequacy of proved oil reserves, both domestic and worldwide.

Extensive liquefaction research was conducted in Germany in the 1920's and 1930's. In the absence of indigenous liquid fuel supplies, Germany had to convert its large coal resources to remain a major power.

In the United States, the Bureau of Mines conducted small-scale feasibility studies of German technology in the 1930's, but the effort was largely moot because of the huge east Texas oil discoveries in 1930. During World War II, interest revived in the United States, and in 1944 Congress passed the Synthetic Liquid Fuels Act that provided $60 million in funds until the program's expiration in 1955. Then, just as in 1930, another big oil discovery killed interest in coal liquefaction—this time the discovery was in the Middle East.

Until recently, liquefaction research has concentrated on producing a substitute refinery feedstock. Now the emphasis is on getting the maximum amount of low-cost, clean fuel suitable for direct use in boilers.

COAL TO REFINERY FEEDSTOCKS

The Bergius Process

The Bergius process is one of the two processes used industrially during World War II. This route produced about 85 per cent of all synthetic liquid fuels made in Germany. It is a hydrogasification reaction. Coal is first ground to a fine size and mixed with a hydrocarbon liquid, which has been produced by the process itself, and with a catalyst. This mixture is reacted with hydrogen (produced by the gasification of coal) at 10,000 pounds per square inch at 850°F. The products from the first reactor are separated into light, middle, and bottom fractions. The middle fraction is further treated over a catalyst in a vapor phase and under relatively mild conditions to produce petroleum-like products. The bottom fraction is filtered to remove solids (unreacted coal, catalyst, ash), and the remaining liquid is

used to mix with the fresh coal being processed in the first reactor.

The extreme operating conditions in the primary reactor (resulting from the poor quality of the catalysts available) and the large volume of hydrogen consumed, which was expensive to produce from coal using existing technology, made the liquid products very expensive. Overall process efficiency was about 55 per cent. No commercial Bergius plants are currently operating.

The Fischer-Tropsch Process

The Fischer-Tropsch process was the other one used in World War II by the Germans to produce industrial quantities of synthetic hydrocarbon liquids. It is a synthesis gas process. The major difference in the products resulting from the Bergius and Fischer-Tropsch processes was that the coal hydrogenation oil was aromatic whereas the Fischer-Tropsch was paraffinic (with cobalt) or highly olefinic (with iron catalysts).

Coal is first gasified completely to synthesis gas. The ratio of the hydrogen to carbon monoxide can be adjusted to produce different final products for use with different types of catalysts. The purified gas is passed over a catalyst at temperatures of 570 to 640°F at pressures of about 450 pounds per square inch. A mixture of straight-chain paraffinic and olefinic products are made that can be further refined to give synthetic liquid fuels.

Whereas the Fischer-Tropsch process as used in Germany avoids the extreme processing conditions of the Bergius method, the complete gasification of coal and its reconstitution to liquid products gives an overall conversion efficiency of only 38 per cent. One small industrial Fischer-Tropsch plant has been in operation in South Africa since about 1955. It uses fixed-bed Lurgi gasifiers to make the synthesis gas and two different methods of contacting the purified gas with catalyst: a modified fixed-bed German process and an entrained circulating-bed catalyst. The major engineering problem is the removal of the large volumes of heat that are released when the gas is converted to a liquid by the catalyst.

The plant produces a wide range of hydrocarbons from methane to light oils. In recent years the methane and other lighter weight hydrocarbons have been used increasingly as a pipeline gas substitute.

Numerous alterations have been made in the plant since it was first designed, and plant management now believes that the use of the historical data to calculate costs is misleading. With larger plants and improved design, costs of liquid and gaseous fuels from a Fischer-Tropsch process are said to be similar to other conversion processes.

REFINERY FEEDSTOCKS— DEVELOPMENTAL

Several projects are underway to test new technology, or at least untried technology. Ambitious feasibility studies are in progress to determine the economics of methanol production from gasified coal. The methanol would be used as fuel in boilers, and possibly as a gasoline extender in automobiles. A typical process scheme would couple a gasification route (yielding synthesis gas) with existing methanol production technology.

Processes under investigation to produce nonalcoholic hydrocarbon fuels include the COED process (FMC Corp.), Project Gasoline (developed by Consolidation Coal Co.), and the H-Coal process (developed by Hydrocarbon Research, Inc.). In addition, there is the Synthoil process of the Bureau of Mines, which is still in the early experimental stage.

The COED Process

Though the liquid produced by this process can be used as a refinery feedstock, the process is basically a pyrolysis method aimed at removing valuable liquids from the coal before it is burned. The amount of liquid and char produced will vary with both the coal used and processing conditions, but in all cases the char is the major product. As a result, COED must be viewed as a way to supplement supplies of liquids from coal and not as a primary coal liquefaction process, unless methods to liquefy the char are found.

The process consists of a series of fluidized, low-temperature, carbonization reactors. The number of reactors and the processing conditions necessary to optimize liquid yields will vary for different coals. Typically, four stages operating at 600, 850, 1,000, and 1,600°F are used. Heat for the process is generated by burning char in the fourth stage and by using the hot gases and hot char to supply the heat for the other stages. Some medium-BTU gas is produced during the carbonization, but the two major products are a hot char and liquids. The oil is filtered to remove solids and hydrotreated at pressure to make a synthetic crude oil. Hydrotreating tests have been conducted with favorable results.

A number of coals have been tested in a small unit with a capacity of 100 pounds per hour. A few coals have been tested in a larger pilot plant with a capacity of 1.5 tons per hour, and there are plans to test additional coals in this pilot plant. A western Kentucky coal has been successfully carbonized, and even more strongly coking coals are to be tested in fiscal years 1975 and 1976.

The usefulness of the COED process as a source of liquids from coal will depend on the successful development of a method for using the char in an environmentally acceptable way. If the over-all economics justify it, there should be no major problems in engineering and constructing a large commercial plant.

Project Gasoline

Following successful laboratory tests, this process was to have been tested on a 1-ton-per-hour scale, but the plant constructed at Cresap, West Virginia, was plagued with a large number of engineering difficulties, only some of which were directly related to the hydrogenation process itself. Moreover, operations were handicapped by labor unrest so that only a limited amount of experimental data was obtained from the pilot plant before work was terminated.

In the process, which uses both solvation and catalytic hydrogenation, the coal is first partially converted to liquid by a hydrogen-transfer recycle solvent. Then solids are separated, and the liquid product is treated with hydrogen in a fluid-bed catalyst reactor. The solvent (which is recycled) must then be separated from the product, which is a satisfactory refinery feedstock. The solids separation has been one of the most persistent engineering problems encountered. The hot filters originally used have never operated satisfactorily. When cyclones were used for solid-hot liquid separation they gave a product somewhat higher in ash and sulfur than desired.

The maximum amount of liquids is produced at an extraction temperature of 810 to 815°F, but to reduce the sulfur content to the levels desired, operations must be conducted under conditions that yield more gas and distillates while consuming more hydrogen. The residue from the extraction step can be utilized to generate the hydrogen and to rehydrogenate the solvent.

Standard Oil of Ohio, which still retains some interest in the background patents of Project Gasoline, has proposed that a 900-ton-per-day pilot plant be erected, supported by 15 or more energy companies. The plant is expected to produce 500 tons per day of clean solid fuel and 400 barrels per day of low-sulfur distillates. Total capital costs of the project are estimated at $70 million.

H-Coal Process

This process has been under study by Hydrocarbon Research, Inc. and represents a modification of their H-Oil process of catalytic hydrogenation—which is being used commercially—so as to be able to handle coal.

Coal is mixed with a recycle oil and fed into a hydrogenator containing a granular catalyst of cobalt molybdate, which is maintained in an ebullient state by the flow of the liquid-solid mixture and hydrogen. The

product is first treated in a hot atmospheric flashdrum and the bottoms stream is further treated to produce a bottoms slurry product, part of which is used for the recycle stream. A number of coals have been tested in an 8-inch-diameter reactor handling 200 pounds per hour of coal. The catalyst can be added or removed continuously as required. A high yield of low-sulfur liquids is produced that are suitable for a refinery feedstock. Tests have also been made to produce a low-sulfur boiler fuel rather than a refinery feedstock. Because of its high cost, the catalyst that is attrited must almost all be recovered from the unconverted coal and coal ash.

As with Project Gasoline, one of the major problems has been solids separation. Several methods have been investigated, including the use of a hydroclone, centrifuge, magnetic separation and filtering. Hydroclones removed only two-thirds of the solids, centrifuges were too expensive, and the results of magnetic separation were not encouraging. Filteration rates were not as cost effective as treating the solid-liquid mixture by vacuum distillation, followed by coking.

Construction of a larger plant to hydrogenerate 300 to 700 tons per day of coal has been proposed to the Office of Coal Research for financing. Three years will be required for design and construction, and a two-year operating program is believed to be required.

Synthoil Process

This particular process has been under development by the Bureau of Mines for the past three years. The key feature of this one-step hydrodesulfurization process is the use of rapid, turbulent flow of hydrogen to propel a coal slurry through an immobilized bed of cobalt-molybdate catalyst pellets. The reactor is operated at 2,000 to 4,000 pounds per square inch and 800°F, and with an excess of hydrogen (which is recycled). The slurry vehicle for conveying the coal is a recycled portion of its own product oil. The combined effect of the hydrogen, turbulence, and catalyst is to liquefy and desulfurize the coal at high yields and high throughput. Sulfur is removed as hydrogen sulfide, which is converted into elemental sulfur for sale or storage.

Both a low-sulfur liquid suitable for boiler fuel use and a refinery grade feedstock have been produced from five different high-sulfur, high-ash coals. The nature of the product, whether it is to be a boiler fuel oil or refinery feedstock, depends upon the pressure and residence time to which the coal is exposed.

Operations with 100-pound-per-day and 0.5 ton-per-day development units have proven long-term operability. The smaller reactor was 5/16 inches in diameter and 68 feet long; the larger is 1 inch in diameter and 14 feet long, both containing a fixed bed of 1/8-inch catalyst pellets. Coal reaction residence time is only 2 minutes.

The process can be operated to produce a gasoline at much less severe operating conditions (4,000 versus 10,000 pounds per square inch) than the Bergius process because of improved catalysts (cobalt, molybdenum on alumina) that have been developed, and because of the rapid reaction rates obtained through the use of a highly turbulent regime. Thus, it should be much less costly than feedstock made by the Bergius process.

An 8-ton-per-day pilot plant is currently being designed.

Other Research Programs

Although no details have been revealed, it has been reported that at least three oil companies now have coal liquefaction processes currently under study, and other companies were known to have had programs in the past. Gulf Oil Co. is said to be investigating on a relatively small scale at least one, and possibly two, different coal hydrogenation technologies. Exxon has said that it has at least one process under study, in which a donor solvent is used to transfer hydrogen to the coal in one vessel, followed by catalytic hydrogenation in a second vessel. The pilot plant will be operated until 1973 and, if

successful, a 300-ton-per-day demonstration plant will be constructed.

The Oil Shale Corp., which has developed a retort for producing oil from oil shale, has also tested coal in that unit. The process involves low-temperature pyrolysis of coal using an inert heat carrier. Tests have been conducted on subbituminous coals, and liquid product yields of 13 to 22 gallons per ton have been obtained. Gas yields varied between 1,250 and 1,650 standard cubic feet per ton. The higher liquid and gas yields were at a retorting temperature of 970°F and the lower yields were at 800°F. The chars produced have the same bulk density as the raw coal, but because their heating value is higher (12,000–13,000 BTU per pound for chars compared to 7,200–9,500 BTU for the coals), they can be transported at a much lower cost per million BTU. The chars have been evaluated by several boiler manufacturers who have indicated the material can be used satisfactorily as a fuel.

Tests will also be conducted for the Office of Coal Research on a zinc chloride catalyst designed to produce a 90-octane (lead-free) gasoline from coal. Earlier tests on a small scale have been successful, with the only major problem being recovery of the catalyst. In addition, a new type of reactor—a centrifugal one—will be tested. It should provide a more rapid and direct contact of the gas and solids and promote a more rapid reaction rate. In this way, a very short contact time can be used, and this prevents the repolymerization of the product and favors the production of a high yield of lighter products.

RESEARCH AND DEVELOPMENT: LOW-ASH, LOW-SULFUR FUELS

A certain parallel exists between research and development paths followed by gasification and liquefaction investigators: in gasification, the premium, high-BTU gas was sought first; in liquefaction, the refinery feedstock has been the first choice. A more widely useful liquid product would be a low-ash, low-sulfur one that could be used directly as fuel.

While all of the liquefaction processes already discussed may be able—with modifications—to produce a low-sulfur boiler fuel from coal at lower cost than a refinery feedstock, only the Synthoil and the H-Coal processes have reported the results of such experiments. In addition, extensive experimental work has been carried out on the Pamco process, which makes a low-sulfur boiler fuel that solidifies at ambient temperatures, although it can be modified to also make a low-sulfur liquid product.

Pamco Process

As originally conceived, fine coal is dissolved in a recycled stream, hydrogenated with small quantities of hydrogen at modest pressures (1,000 pounds per square inch) and filtered to remove the solids. These process steps eliminate part of the sulfur and add some hydrogen to the coal. The product is a relatively low-sulfur, low-ash material, solid at room temperature but liquid at 350°F. It could be burned in a boiler, either as a solid or as a liquid, if it is first heated. With more stringent processing conditions, or with a second step involving catalytic hydrogenation of the product, a satisfactory refinery feedstock can be produced.

After a number of years of small pilot operations, a 75-ton-per-day coal pilot plant was completed in late 1973, and testing on this scale is now being initiated. Another smaller pilot plant using essentially the same process is being tested at Wilsonville, Alabama by the Edison Electric Institute and Southern Electric Generating Co.

Synthoil to Low-Sulfur Boiler Fuel

This process, described earlier for producing a liquid fuel from coal in a fixed-bed turbulent reactor, has also been operated so as to make a low-sulfur, low-ash boiler fuel. Coals with a sulfur content of as high as 5.5 per cent have been treated at pressures of 2,000 pounds per square inch and temperatures of

840°F. Oils with sulfur contents in the range of 0.2 to 0.4 per cent have been produced. Production of 3 barrels per ton of low-sulfur oil has been achieved using only 3,000 cubic feet of hydrogen per barrel of product. Overall energy conversion efficiencies of 75 to 78 per cent have been obtained.

H-Coal Process

This route can also produce a low-sulfur boiler fuel rather than a refinery feedstock by changing operating conditions. These modifications result in a major increase in reactor space velocity and a greatly reduced hydrogen consumption per ton of coal processed—both of which result in lower product costs. Low-sulfur boiler fuel (0.5 per cent sulfur) has been produced from an Illinois No. 6 coal containing 5.0 per cent sulfur, using only two-thirds the amount of hydrogen required when making a synthetic crude oil from the same coal.

COSTS

In giving any estimate of the costs of synthetics from coal it must be recognized that they are just that—estimates. No commercial plant of any kind is currently in operation in the United States, so that none of the published cost data is based on actual plant operation.

There are a number of other caveats that need to be stressed in making meaningful comparisons among cost data. *First,* in a period of rapidly escalating capital costs, the date of the estimate and the method used (if any) to inflate costs during construction must be known. *Second,* the source of the estimate must be identified, since the promoters of a new process tend to be optimistic and those with competitive processes pessimistic about both capital and operating costs. *Third,* in comparing different estimates, the cost of the coal should be the same, or the effect of any differences in coal costs on product price should be known. *Fourth,* the degree to which the estimate has included the costs of meeting environmental standards must be described. *Fifth,* even the best and most careful estimates tend to increase as one approaches the actual construction of the facility. *Sixth,* the early estimates of product costs are now known to have been grossly understated, and there is still no way to be certain that present estimates, although much higher, may not still be too low. *Finally,* even the most careful estimates by firms using commercially available technology have been sharply escalated. For example, the cost of a Lurgi plant was estimated at under $300 million in the spring of 1971. The latest estimate for the same size plant filed with the Federal Power Commission by El Paso Natural Gas in the fall of 1973 was just under $500 million.

The available information on various synthetic processes with costs computed to 1973 prices is presented in Table 4-29. Published capital investments were converted to 1973 dollars. No changes were made in the contingencies or offsites claimed on the original paper, but all coal costs were standardized at 32¢ per million BTU for bituminous coal ($8 per ton) and 25¢ per million BTU for western coal ($3 per ton).

The hazards noted earlier of using published data are illustrated in Table 4-29, in which four different estimates for a Lurgi plant using the same available commercial technology give capital costs ranging from $214 to $427 million, and all of them are lower than the latest estimate of nearly $500 million.

Table 4-30 presents in condensed form the data from Table 4-29, adjusting all the plants to the same size—250×10^9 BTU per day. For high-BTU gas this is equivalent to 250 million cubic feet per day, for hydrocarbon liquids it is equivalent to 40,000 barrels per day, and for methanol it is equivalent to 12,500 tons per day. The scaling factor used to calculate investment costs for the common-sized plant was the 0.9 power of capacity. The investment was also standardized to include on sites, auxiliaries and off sites. The figures also include 15 per cent for contingency, 5 per cent for startup, 7.5 per cent for working capital and 15 per cent

interest on construction loans. A plant stream factor of 90 per cent, labor costs of $5.50 per hour, 2.5 per cent of plant investment for taxes and insurance, and 4.5 per cent for plant maintenance were all assumed.

Cost of high-BTU gas made from bituminous coal in a Lurgi generator is estimated at $1.35 to $1.50 per million BTU, and at $1.00 to $1.27 per million BTU using western coals. Low-BTU gas ranged from $1.10 to $1.25 per million BTU using bituminous coals and from $0.90 to $1.05 per million BTU using western coals.

Estimates of liquids made from coal range from $1.12 to $1.66 per million BTU ($6.40–$9.60 per barrel). The lower values are for a low-sulfur boiler fuel and the higher values are for a refinery feedstock. Methanol made from a bituminous coal was estimated at $1.58 to $1.74 per million BTU or the equivalent of $9.15 to $10.10 per barrel of oil.

All of these values are probably lower than what will actually result when plants are constructed and operated. This will be the result partly of continuing inflation in construction costs and partly from the other reasons already given. For example, if the most recent Lurgi capital cost estimates are used, the costs of high-BTU gas shown in Table 4-30 are about 25¢ too low. On the other hand, some of the newer gasification and liquefaction processes now in the pilot plant stage, if successful, might result in lower product costs.

These tables, even with all the uncertainties about the accuracy of the calculated values, demonstrate that if we decide to rely on synthetics from coal, the price of energy will increase sharply. The well-head price of oil used to be about $3.25 per barrel before the recent Organization of Petroleum Exporting Countries price rises. The estimates for making it from coal would increase this price two or three times. To the consumers of gasoline this would mean that, if the increased cost of syncrude were simply allowed to be passed along, an increase in price from the former retail price of 35¢ per gallon (when crude oil was $3.25 per barrel)

Table 4-29. Studies of Plant Costs for Various Substitute Fuels Processes (1973 prices).[a]

Study	Output Fuel (billion BTU/day)	Daily Plant Output	Total Capital Investment ($/million)	Annual Operating Costs ($ million)	Capital Charge 15 per cent per year (¢/million BTU)	Operating Cost (¢/million BTU)	Coal Cost (¢/million BTU)	Product Cost (¢/million BTU)
Coal Gasification								
FPC[1][b] Lurgi-bituminous	237.2	250 million cubic feet pipeline gas	347	23.1	66.5	29.5	54.1	150.1
FPC New-bituminous	240.0	250 million cubic feet pipeline gas	296	16.2	56.1	20.5	53.1	129.7
FPC Lurgi-Western	238.8	250 million cubic feet pipeline gas	313	24.6	59.6	31.2	26.1	116.9
FPC New-Western	240.0	250 million cubic feet pipeline gas	261	16.1	49.4	20.3	26.5	96.2
NPC[2][c] Lurgi-bituminous	243.0	270 million cubic feet pipeline gas	285	20.8	53.3	25.9	47.1	126.3
NPC Lurgi-Western	243.0	270 million cubic feet pipeline gas	241	18.6	45.1	23.2	23.6	91.9
NAE[3,4][d] Lurgi	235.0	240 million cubic feet pipeline gas	318		61.5			
Fluor[5] Lurgi-Western	252.1	257 million cubic feet pipeline gas	427		77.0	24.5	28.5	130.0
Coal Liquefaction								
NPC[2] H-Coal-bituminous	240.0	30,000 barrels syncrude 60-billion BTU fuel-gas	260	26.8 excluding coal	49.2	33.9	43.2	126.3
NPC PAMCO-bituminous	180.0	30,000 barrels "de-ashed product"	187	13.4 excluding coal	47.2	22.6	42.7	112.5
Amoco, COED-bituminous	405.0	29,175 barrels syncrude 250 million cubic feet pipeline gas	500	53.8 excluding coal	56.0	40.0	56.0	152.0
Foster-Wheeler,[6] CONSOL A-Bit	358.8	284.3 billion BTU liquid 74.5 billion BTU gas	309	36.0 excluding coal	39.1	30.4	45.1	114.6
Foster-Wheeler, CONSOL B-Bit	421.0	282.1 billion BTU liquid 138.9 billion BTU gas	405	43.3 excluding coal	43.7	31.2	47.8	122.7
Low-BTU Gas from Bituminous Coal								
NAE[4] low-to-intermediate-BTU gas	235.0		165–210		31.9–40.6		40.0–45.7	110–125
FPC[1] low-BTU gas	235.0		189–191		36.6–36.9		20.0–22.9	
Methanol from Coal								
ORNL[7][e]-bituminous	391	20,000 tons methanol	416	58.4 excluding coal	48.4	45.3	47.8	141.5

[a]Source: "Project Independence: An Economic Evaluation," MIT Energy Laboratory Policy Study Group, Mar. 15, 1974. Printed in Technology Review, May 1974. Reprinted by permission.
[b]FPC—Federal Power Commission.
[c]NPC—National Petroleum Council.
[d]NAE—National Academy of Engineering.
[e]ORNL—Oak Ridge National Laboratory.

References for Table 4-29:

1. "Final Report–The Supply-Technical Advisory Task Force–Synthetic Gas-Coal" National Gas Survey, Federal Power Commission, Apr. 1973.
2. "U.S. Energy Outlook–Coal Availability" Report of the Fuel Task Group on Coal Availability, National Petroleum Council Committee, 1973.
3. "Evaluation of Coal Gasification Technology, Part I, Pipeline Quality Gas," National Academy of Engineering, 1972.
4. "Evaluation of Coal Gasification Technology, Part II, Low- and Intermediate-BTU Fuel Gases," National Academy of Engineering, 1973.
5. Moe, J. M., "SNG from Coal via the Lurgi Gasification Process," Symposium on "Clean Fuels Firm Loaf," Institute of Gas Technology, Chicago, Ill., Sept. 10–14, 1973.
6. Foster-Wheeler Corporation, "Engineering Evaluation and Review of Consol Synthetic Fuel Process," R&D Report No. 70, Office Coal Research, Feb. 1972.
7. Methanol from Coal for the Automotive Market, ORNL Feb. 1974.
8. Wen, C. Y., "Optimization of Coal Gasification Processes," R&D Report No. 66, Office of Coal Research, 1972.
9. Siegel, H. M., and T. Kalina, "Technology and Cost of Coal Gasification," Mech. Engr., May 1973, pp. 23–28.
10. Shearer, H. A., "The COED Process Plus Char Gasification," Chem. Engr. Prog. 69, No. 3, 43 (1973).
11. Michel, J. W., "Hydrogen and Synthetic Fuels for the Future," 166th Nat. Mec. Amer. Chem. Soc., Div of Fuel Chem Preprints, 18 No. 3, 1, Aug. 1973.

Table 4-30. Plant Costs for 250 Billion BTU per Day of Product.[a]

Process Type	Input Fuel (tons per day)	Thermal Efficiency (per cent)	Total[b] Capital (1973) ($ million)	Annual Operating Cost ($ million)	Costs (¢/million BTU) (330 days per year onstream)			
					Capital (at 15 per cent per year)	Operating Costs	Coal Cost	Total Cost
Gasification (Lurgi, gasifier)								
Bituminous coal[c]	14,700–17,900	56–68	334–390	21.4–22.2	60.7–70.9	25.9–26.9	47.1–54.1	135–150
Western coal[d]	19,600–23,800	56–68	290–390	19.5–23.0	52.7–70.9	23.6–28.5	23.6–28.5	100–127
Liquefaction	13,200–17,500	60–75	233–373	22–35	43.4–67.8	26.7–35.0	42.7–56.0	112–166
Low BTU Gas								
Bituminous[c]	12,500–14,300	70–80	195–208	–	35.5–37.8	–	40–45.7	110–125
Western[d]	16,700–19,000	70–80	195–208	–	35.5–37.8	–	20–22.7	90–105
Methanol								
Bituminous[c]	14,900	60–67	279–364	44	50.7–66.2	–	53.4	158–174

[a]Source: "Project Independence: An Economic Evaluation," MIT Energy Laboratory Policy Study Group, Mar. 1974.
[b]Includes on sites, off sites, auxiliaries, 5 per cent startups, 15 per cent interest during construction, 7.5 per cent working capital.
[c]32¢/million BTU, 25 million BTU per tons, $8 per ton.
[d]16¢/million BTU, 18.75 million BTU per ton, $3 per ton.

to 47 to 59¢ per gallon would result.

High-BTU gas prices for a residential consumer in Washington, D.C., would increase from about $1.70 per million BTU to $2.95 per million BTU (on a pass-along basis) if all of the gas were produced from synthetics. For gasoline, the increase in retail price to the consumer would be 35 to 70 per cent (assuming gasoline taxes did not change), and for the residential gas consumer in Washington, D.C., an increase of 70 per cent.

Since the higher priced gas and liquids would be "rolled in" slowly with existing lower cost supplies as synthetic plants were constructed, prices would increase slowly, and it would be a number of years before the maximum values indicated above would be reached.

REFERENCES

A.E.C., Sr. Management Comm., "Synthetic Fuels From Coal," Feb. 1974.

Institute for Gas Technology, "Clean Fuels From Coal," Symposium, Sept. 10–14, 1973.

Perry, H., "The Gasification of Coal," *Scientific American 230,* No. 3, March, 1974, pp. 19–25.

Perry, H., "Coal Liquefaction," *Physics and the Energy Problem–1974,* American Institute of Physics Conf. Proceedings, No. 19, Feb. 1974, pp. 43–56.

"Project Independence: An Economic Evaluation," M.I.T. Energy Laboratory Policy Study Group, Mar. 15, 1973.

"Synthetic Fuels," Cameron Engr., Mar. 1974.

U.S. Dept. of the Interior, Office of Coal Res., Energy Res. Program, Feb. 1974.

SUGGESTED READINGS

Siegel, H. M., and T. Kalina, 1973, "Technology and Cost of Coalgasification," *Mechanical Engineering 95,* May, pp. 23–28.

Squires, Arthur M., 1974, "Clean Fuels from Coal Gasification," *Science 184,* pp. 340–346.

The Black Box

Fuel Cells

TERRI AARONSON   1971

A clean, safe means of generating electricity may result from the development of the fuel cell, a kind of "black box" that has virtually no moving parts, makes no noise, is free of vibration, and generates only innocuous products and electricity. The fuel cell is actually an electrochemical device; that is, it converts the chemical energy of various fuels directly into electrical energy. Ordinarily, heat produced by burning fuel or by nuclear reaction is used to produce steam to turn large generating units that produce electricity. This process of using chemical energy to produce heat energy to produce mechanical energy to produce electrical energy is inherently inefficient. The newest conventional generating devices operate at about 40 per cent efficiency.[1] Fuel cells theoretically approach 100 per cent efficiency in converting chemical energy to electrical energy.[2] Because of their greater efficiency, fuel cells require less fuel to generate an equal amount of electricity than do conventional or nuclear power plants. Another advantage of fuel cells is that they remain efficient over a wide range of power levels and can be built in large or small units. Fuel cells are being developed to relieve peak power loads for conventional power stations, for vehicular use, and for potential use as large (1,000-megawatt) central power stations. (A megawatt is the equivalent of one million watts, or 1,000 kilowatts of electrical energy.) A major project (discussed below) is now under way to test the feasibility of using small fuel-cell units to provide the entire energy needs of an individual home, apartment complex, or industrial area.

Electric power generation contributes significantly to the degradation of air quality throughout the nation. In 1966 it was estimated[3] that electric power plants contributed a total of 20 million tons of sulfur oxides, nitrogen oxides, particulate matter,

1. Angrist, Stanley W., *Direct Energy Conversion,* second edition, Allyn and Bacon, Inc., Boston, 1971, p. 20.
2. Bockris, J. O'M., and S. Srinivasan, *Fuel Cells: Their Electrochemistry,* McGraw-Hill Book Company, St. Louis, 1969, p. 2.

3. *Sources of Air Pollution and Their Control,* Public Health Service Publication No. 1548, Washington, D.C., 1966.

Terri Aaronson is former news editor of *Environment* magazine.

From *Environment,* Vol. 13, No. 10, pp. 10–18, December 1971. Reprinted by permission of the author and the Committee for Environmental Information. Copyright © 1971 by the Committee for Environmental Information.

carbon monoxide, and hydrocarbons to the atmosphere. Nuclear power plants emit low levels of radiation to the atmosphere, generate larger amounts of radioactive wastes, and contribute to thermal pollution by the need for vast quantities of cooling water that are discharged to the environment at high temperatures. The use of fuel cells for power generation would avoid producing the above-mentioned pollutants. And, by using fuel cells in automobiles the major source of air pollution in urban areas—the internal combustion engine—would be eliminated.

The fuel cell was discovered by Sir William Grove in 1839,[4] but the devices have become practical only within the past decade. Grove's fuel cell consisted of separate tubes of hydrogen and oxygen in a dilute solution of sulfuric acid. Grove placed strips of platinum foil into the tubes of gas and recorded that "a shock was given which could be felt by five persons joining hands, and which when taken by a single person was painful."[5] After Grove's initial experiments, though, little attention was paid to fuel cells because of the costliness at that time of producing hydrogen and oxygen, in addition to the expense of the platinum. Fifty years after Grove's first fuel cell, which was intended solely for limited scientific purposes, Ludwig Mond and Carl Langer recognized the fuel cell's potentiality as a new means of generating electricity. Although Mond and Langer developed a device superior to Grove's first fuel cell, it was overshadowed by the steam-powered dynamo. Work on electrochemical cells once again proceeded slowly, with researchers looking for fuels other than hydrogen.

Modern fuel-cell technology was pioneered by Francis T. Bacon in the 1930's. In 1959, Bacon announced that he and J. C. Frost of Cambridge University had developed the first practical fuel cell. It could be used to power a fork-lift truck, a circular saw, and a welding machine. Just two months after Bacon's announcement, H. K. Ihrig of Allis-Chalmers Mfg. Company demonstrated a fuel-cell tractor that he had developed.

The biggest boost to fuel-cell development, though, came when it was realized that, because of its high power output and low weight, the fuel cell was well suited to space applications. Government-sponsored research on fuel cells soared, reaching a high of $15.9 million in 1963.[6] However, government interest has waned, and it seems that fuel-cell research has been shelved. In 1969 only $2.2 million in government funds was spent on fuel-cell research, climbing to slightly more than $3 million for 1970.[7] This seeming lack of interest on the part of the government is somewhat ironic in light of professed concern with pollution and the reference, made by William Ruckelshaus, administrator of the federal Environmental Protection Agency, to the "enormous unrealized potential" of fuel cells.[8]

VERSATILITY

A single fuel cell technically is only a single set of two electrodes, joined by an electrolyte (which conducts current between the two electrodes) and an external circuit. Each set of electrodes is composed of a fuel (such as hydrogen or natural gas) electrode and an oxygen (or air) electrode; thus, a hydrogen/oxygen fuel cell is one that has a hydrogen electrode and an oxygen electrode (see later discussion of fuel-cell components). Only a small amount of current can be drawn from a single fuel cell. A primary attraction of fuel cells is that they can be joined together in series to attain the amount of power

4. Liebhafsky, H. A., and E. J. Cairns, *Fuel Cells and Fuel Batteries, A Guide to Their Research and Development,* John Wiley & Sons, Inc., New York, 1968, p. 18.
5. Grove, W. R., quoted in Liebhafsky and Cairns, *op. cit.,* p. 19.

6. Kordesch, K. V., "Outlook for Alkaline Fuel Cell Batteries," Battelle Institute Seminar, "From Electrocatalysis to Fuel Cells," Seattle, Dec. 9–11, 1970.
7. Kordesch, K. V., personal communication, Oct. 29, 1971.
8. Ruckelshaus, W. D., "Energy and the Environment," address to the World Energy Conference, Washington, D.C., Sept. 24, 1971.

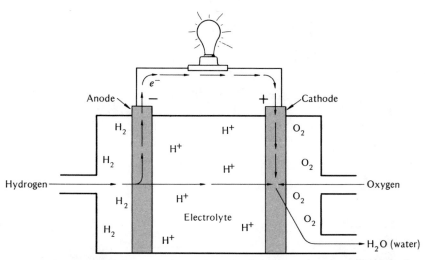

Figure 4-45. Electricity from a fuel cell. In a simple hydrogen/oxygen fuel cell, hydrogen is fed to the anode where it forms hydrogen ions (H^+), giving up electrons (e^-) that travel along an external circuit to the cathode. The energy produced by the fuel cell is tapped by placing an electric load, such as an electric light, in the circuit. At the cathode, oxygen combines with electrons and hydrogen ions (that were formed at the anode) to produce water. The water is then drawn off from the cell. The reactions taking place in a hydrogen/oxygen fuel cell are:

At the anode: $H_2 \rightarrow 2H^+ + 2e^-$
At the cathode: $O_2 + 4H + 4e^- \rightarrow 2H_2O$
Overall: $2H_2 + O_2 \rightarrow 2H_2O$

desired for a particular task. There is no loss of efficiency in joining a great many fuel cells together, nor is there a significant reduction in efficiency by joining only a few cells together. Since fuel cells can be made very thin, they can be stacked, occupying little space. When connected together, they technically are called a fuel-cell battery, or sometimes a fuel-cell power plant. (Fuel cells described herein are actually fuel-cell batteries.) (See Figure 4-45.)

Because the power output of fuel cells can be adjusted by adding or subtracting units, fuel cells are extremely versatile. They have been envisioned for every size power plant imaginable. Domestic units may someday provide all the electricity an individual home might need. Or, a moderate-sized unit might provide power to a single neighborhood. Or, a slightly larger unit might power an entire community in a remote area. Or, if it is more convenient, very large fuel-cell power plants could be assembled to provide large blocks of power to whole areas, much as the present conventional power plants do. (See Figure 4-46.)

Fuel cells can provide electricity for a host of uses other than power plants. The Army, for instance, is interested in fuel-cell applications for small generating units to be used in the field. Fuel cells are attractive tactically because they make no noise and because they can be operated at low temperatures that are undetectable by infrared sensors. Fuel cells have proved to be efficient in space applications, and someday biochemical fuel cells may offer a simple solution to the problem of human waste disposal from spacecraft, by transforming the wastes into electrical energy. Another intriguing possibility for biochemical fuel cells is a pacemaker for persons with deficient hearts; such a pacemaker would run off of substances found within the body. It has been

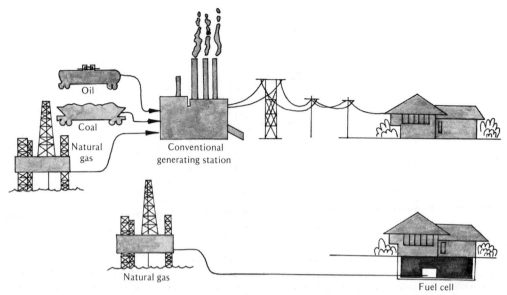

Figure 4-46. Fuel cell power. In contrast to conventional generating stations, which must use the chemical energy of fuel to produce heat energy to produce mechanical energy to produce electrical energy that must then be transmitted to users, fuel cells can directly convert the chemical energy of fuels to electrical energy. This direct conversion can take place in individual homes, such as is shown in the illustration. There, natural gas is delivered via pipeline and is first transformed to hydrogen, which is then fed to the fuel cell to produce electricity. By avoiding extra steps in the conversion process and by eliminating the need for long-distance transmission of electricity, fuel cells are much more efficient than conventional generating plants.

suggested that blood can provide the fuel and necessary oxidant, in addition to acting as an electrolyte, for a totally implanted pacemaker.[9]

Fuel cells have already been installed in a variety of vehicles, from fork-lift trucks to tractors, and may someday be used to power nonpolluting automobiles, buses, trains, or even submarines. In the mid-1960's the General Motors Corporation designed and built a van that was powered by hydrogen/oxygen fuel cells. The fuel cell powered van compared favorably with a conventional van in terms of road tests,[10] but the "electrovan" was not without disadvantages. Most notably,

9. Angrist, op. cit., p. 402.
10. Marks, C., E. A. Rishavy, and F. A. Wyczalek, "Electrovan—A Fuel Cell Powered Vehicle," Society of Automotive Engineers, Automotive Engineering Congress and Exposition, Detroit, Jan. 9–13, 1967.

the electrovan was heavier than the conventional one, and the fuel-cell materials were expensive. These disadvantages may be overcome by future development, and General Motors is continuing to work with fuel cells. A more difficult problem, though, is the hazard of storing high-energy fuels under pressure. Safety requirements similar to those stipulated for commercial gasoline trucks might be necessary for a hydrogen/oxygen fuel-cell vehicle.

A more recent fuel-cell-powered vehicle that has bypassed some of the difficulties of the electrovan has been designed and built by K. V. Kordesch of the Consumer Products Division of the Union Carbide Corporation. This vehicle is actually powered by a hybrid system composed of a fuel cell and a conventional battery. Currently the car contains a hydrogen/air fuel-cell power plant and seven lead-acid batteries for additional short bursts of power.[11] The potential for overcoming the explosion hazard of fuel rests in the possibility of operating the car with an ammonia/air fuel-cell power plant, again supplemented by batteries.[12] Liquid ammonia does not need to be stored under high pressure, and thus is less subject to explosion.

The fuel cell/battery hybrid vehicle is an Austin sedan that was originally converted from an internal combustion engine to an all lead-acid battery power source. With the present hybrid system, the car presents a possible alternative to polluting automobiles that are now driven in urban areas. The hybrid car can run for 200 miles without refueling, and has a top speed of 50 miles per hour, thus making it a feasible urban vehicle, but not acceptable for long-distance driving. Dr. Kordesch, who often drives the car for personal use, has logged more than 2,500 miles on it since its conversion to fuel cells in 1969. The features of the car[13] suggest that such a power-plant design is quite practical, although widespread acceptance is unlikely in the near future considering that the proven internal combustion engine will probably be cleaned up sufficiently to meet projected air pollution abatement standards. However, as standards become more stringent, and more controls are placed on the combustion engine, the cost and efficiency of a fuel-cell-powered vehicle may appear more attractive.

The fuel cell/battery car takes 1 to 2 minutes initial start-up time for the fuel cell to attain full power. While the fuel cell is "warming up," the batteries can be used to drive the car. The batteries can be recharged by the fuel cell at a stop—a traffic light, for instance. Recharging the hydrogen tanks carried atop the car is little problem, with an average time of 3 minutes to refill the tanks. And car maintenance should be minimal. The fuel cells themselves present little problem, and because they are completely shut down during long off periods (the car is designed so that the electrolyte drains from the fuel cells during lengthy shutdown) and little power is lost during idling, the fuel cells should remain capable of producing sufficient power for quite some time. If fuel cells for automobiles were mass produced, the cost of the cells might be acceptable, and the cost of running the car would be small, an estimated 0.5¢ per mile.

TWO POWER-PRODUCTION PROJECTS

The electric utility industry seems to be watching the development of fuel cells, but is not involved in any large-scale projects to develop fuel-cell technology. However, development of fuel cells for commercial power production is not at a standstill. Two projects are testing the feasibility of fuel cells for generation of electric power on a mass scale. The first project is sponsored by the Office of Coal Research, an agency of the federal Department of the Interior. The second, and larger, project is sponsored by a coalition of gas utilities and the Institute for Gas Technology.

The Office of Coal Research has spent approximately $3.75 million during the past seven years on developing a fuel cell that will be run on coal. Although there was no active work on a coal-reacting fuel cell in the past year, it is likely that Westinghouse Electric Corporation (prime contractor of the Office of Coal Research) will soon begin construction of a very small-scale (100-kilowatt) power plant that will test the practicality of a complete coal-reacting power plant. (The kilowatt is a unit of power equivalent to 1,000 watts or about 1.34 horsepower.) The coal-reacting power plant is actually an "indirect" fuel cell. The coal is first gasified (producing hydrogen) in a high-temperature gas generator; the hydrogen is then reacted

11. Kordesch, K. V., "Hydrogen-Air/Lead Battery Hybrid System for Vehicle Propulsion," *Journal of the Electrochemical Society, 118*(5): 812–817, 1971.
12. Kordesch, K. V., personal communication, Oct. 12, 1971.
13. Kordesch, K. V., "City Car With H_2-Air Fuel Cell/Lead Battery (One Year Operating Experiences)," Society of Automotive Engineers, Inc., 1971 Intersociety Energy Conversion Engineering Conference Proceedings, p. 38.

with air to produce water and electricity. Sulfur is scrubbed from the gas before it enters the fuel cell, and nitrogen oxides are not produced. By-products of this type of fuel cell are carbon dioxide and water. The fuel cell itself utilizes a solid electrolyte, and because it is operated at high temperatures, it is not expected to need expensive "noble" metal catalysts. (Noble metals are those such as silver, platinum, and gold that are good conductors of electricity.) The Office of Coal Research has estimated that a complete fuel-cell power system utilizing coal will be 60 per cent efficient, and that nearly 1.5 times more electrical energy can be obtained from each ton of coal used in a fuel cell than in a conventional plant.[14] Thus, fuel cells would contribute greatly to conserving our coal resources. Other advantages of the coal-reacting fuel cell are that it does not require cooling water (high temperatures generated in the fuel cell can be used effectively in the coal-gas generator), and thus sites for fuel cells should not be dependent upon the availability of large bodies of water. Fuel cells have no moving parts, and thus should prove reliable in operation. And, they have a high power density per cubic foot; it is estimated that a sufficient number of fuel cells to generate one megawatt of power would fit into an area 7 feet in diameter and 6 feet high.[14]

If the Office of Coal Research is successful in demonstrating that coal-reacting fuel cells are a practical means of producing energy, it is possible that such cells will be built. Economics will dictate that these power stations be large and centrally located and that the electricity produced be transmitted over the present grid of high-tension wires. Or, it is possible that by the time central fuel-cell power plants are built, new, more efficient transmission lines, as are now envisioned, will have come into existence. However, no matter how the energy is transmitted, some energy will be lost in the process, thus cutting down on the efficiency of the total

system of power generation and delivery.

The most obvious way of avoiding loss of energy by transmission is to produce power near the consumer. That is, in fact, the goal of the project that is being pursued by the Team to Advance Research for Gas Energy Transformation, Inc., known as TARGET, a coalition of 32 member companies, most of which are gas utilities. Together with the Pratt & Whitney Aircraft Division, TARGET has developed a fuel-cell power system which is designed to be used for localized power production.[15] Now in its fourth year, the TARGET program is in a stage of on-site testing. Plans call for fuel-cell power plants to be installed in 37 locations throughout the country. The fuel cells will provide power for homes, apartments, stores, restaurants, and office and industrial buildings.

Each fuel-cell system has the capability of generating 12.5 kilowatts of power; that should be more than sufficient to meet the peak demands of an individual dwelling unit. For the industrial installations, it may be necessary to arrange numerous fuel-cell systems in series to provide adequate power. There will be no loss of power or efficiency with multiple cells.

TARGET's fuel-cell system runs on natural gas. It could, of course, be run on a synthetic gas. The system is again an indirect fuel cell, using a steam reformer to decompose the natural gas (methane) to hydrogen and carbon dioxide. The gases are then fed into the fuel cell itself, which acts as a hydrogen/air cell. Water formed as a product of the fuel-cell reaction is circulated back to the steam reformer. Electricity produced by the system is direct current power, as is the power produced by all fuel cells. The TARGET fuel cell is attached to an inverter that transforms the direct current into alternating current. Losses of efficiency occur in both the reformer and in the inverter; the remaining over-all efficiency of the TARGET system is about 45 per cent—equal to or better

14. Office of Coal Research, *Annual Report 1971*, p. 48.

15. "Target for Tomorrow," Connecticut Natural Gas Corporation, Hartford, Conn.

than all other generating systems (not including losses during transmission).

The TARGET fuel-cell system, like other fuel cells, is very clean, producing carbon dioxide, water, and some heat as products. It is quite compact, about the size of a furnace, and commercial models are expected to be reduced to about the size of a gas air conditioner. Because there are very few moving parts (pumps and fans) in the fuel cell, it should produce practically no noise, and should be quite reliable. The efficiency of the TARGET fuel cell is expected to permit the system to provide a given amount of power to a site using one-third less fuel than any other power system. The cost of the system should be reasonable because, although the initial installation of a fuel cell may be relatively expensive, the savings on fuel and the low cost of gas transmission (as compared to electricity transmission) should more than compensate installation costs. Transmission and distribution of energy in the form of gas in most parts of the country cost only one-fifth as much as the transmission of an equal amount of energy in the form of electricity. However, the economics of the entire TARGET system have not yet been worked out in detail; projected costs will not be estimated until completion of the installation program, probably in late 1972. It has been suggested, though, that "successful large-scale use of fuel cells will depend more on resolving economic problems than technical ones."[16] Technical problems seem to have been minimal in the first installation of the TARGET program. A fuel cell was installed in a model home in Connecticut in July 1971 and ran through September 1971. Full evaluation of the first practical test is not complete, however.

COMPONENTS

The basic fuel cell consists of two electrodes—the anode, or fuel, electrode; and the cathode, or oxygen, electrode (air may be substituted for oxygen). Both the fuel and

16. *Ibid.*, p. 14.

oxygen are continuously fed to the electrodes, which need some sort of catalyst for accelerating the chemical reactions. Electrode catalysts can be made of many materials; however, most fuel cells work best with noble metals, usually platinum. Unfortunately, platinum and the other noble metals are relatively scarce resources, making them quite expensive. Thus, the need for noble metals in fuel cells is one of the most difficult problems to overcome before fuel cells can be practical for widespread use. The one other main component of a fuel cell is the electrolyte, which conducts current between the two electrodes. Electrolytes may also be many substances, either acid or alkaline, liquid or solid.

The modern fuel cell is similar to Grove's original cell. Fuel is fed to the anode (which, in the case of fuel cells, is negative, unlike the ordinary electrolysis battery) where the fuel is oxidized, giving up electrons, thereby forming ions of the fuel. (An electron is one of the particles of an atom and has a charge of negative electricity. An ion is an atom, group of atoms, or molecule that has acquired a net electric charge by gaining electrons in, or losing electrons from, an initially neutral configuration.) The electrons travel along an external metallic path, or circuit, to the cathode. The energy of the fuel cell is tapped by placing some sort of electric load in the circuit. At the cathode either oxygen or air normally is fed to the fuel cell, although other reducing agents may be used. The oxygen combines both with ions of the fuel that have formed at the anode and with electrons to form a new compound that can be drawn off from the cell. For an illustration of a simple hydrogen/oxygen fuel cell, see Figure 4-45.

Many fuels other than hydrogen can be used in simple fuel cells. Some combinations other than hydrogen/oxygen that are being considered for fuel cells are: hydrogen/air; carbon or carbon monoxide/air; natural gas (methane)/air; ammonia/air; hydrazine/oxygen or air; sodium amalgam/oxygen; zinc/oxygen; and aluminum/oxygen. Not all of these combinations have yet been proved to be practical. Further considerations are the choice of materials for the electrode catalysts and the electrolyte. None of the combinations mentioned above is expected to produce uncontrollable pollutants of any sort. For instance, when air is used in place of oxygen, the nitrogen in air does not form harmful nitrogen oxides, but rather is emitted simply as nitrogen gas.

The most glamorous use of fuel cells to date has been in space applications. In the Gemini and Apollo spacecraft, hydrogen/oxygen fuel cells were used. Quite naturally, then, this type of fuel cell is currently the most technically advanced. However, it is unlikely that the fuel-cell design used for either the Gemini or Apollo programs will ever be used on a mass scale. The Apollo fuel cell has a relatively short lifetime, and the Gemini fuel cell needs expensive platinum catalysts for the electrodes. Recent developments, though, have reduced to a very low level the amount of noble metal catalysts necessary for the operation of a fuel cell, and it has now been shown that nonnoble metal catalysts, such as nickel compounds, borides, carbides, and organic catalysts, can work efficiently in some fuel cells.[12]

KINDS OF FUEL CELLS

Rather than using pure hydrogen (which has limited availability) in a fuel cell, it may be advantageous to use some hydrocarbon compound, such as natural gas, which, for example, is used in the TARGET program. Natural gas is already widely available through a nationwide network. Although resources of natural gas are declining, it is relatively inexpensive, and a synthetic natural gas can be produced from a variety of materials, including coal and sewage. There are two kinds of fuel cells that can use natural gas as a fuel—the direct cell and the indirect cell.

The direct fuel cell is one in which the hydrocarbon fuel is fed to the anode, in either a gaseous or liquid state, and then the fuel cell operates as described above, except that carbon dioxide is formed as a reaction product and must be drawn from the cell so that it does not clog the workings of the electrodes. Direct hydrocarbon fuel cells are still in the early developmental stages[17] but have already been shown to be more efficient than conventional engines using the same fuels.[18] However, until electrode catalysts other than the noble metals can be found that will operate in direct hydrocarbon fuel cells, the direct cells will probably be too expensive for common acceptance.

Indirect fuel cells involve a process by which the hydrocarbon undergoes a steam-reforming process outside of the cell to obtain pure hydrogen which is then fed into the fuel cell. (More energy is produced from the fuel cell than is required to decompose the hydrocarbon.) Once the hydrogen is produced, the fuel cell functions as any ordinary hydrogen/air cell. The main advantage of the indirect fuel cell is that it does not require costly platinum electrocatalysts, but can operate with less expensive materials. The primary disadvantage of the indirect cell is that the reforming process requires energy, which cuts down on the efficiency of the cell.

Two fuel cells that also seem very promising use a nitrogen-hydrogen compound, either hydrazine or ammonia, as fuel. Hydrazine is an attractive fuel because it is highly reactive and it does not require expensive electrocatalysts. Unfortunately, hydrazine itself is very expensive. There are indications, however, that large-scale production methods of fabricating hydrazine may bring down the cost. Even if that is not accomplished, the high energy output of hydrazine is likely to allow the fuel to be used in specialized fuel-cell applications.[19] Unlike hydrazine, ammonia is quite inexpensive. The performance of ammonia in fuel cells is, however, markedly inferior to hydro-

17. For a review of current work on both direct and indirect fuel cells, using a variety of electrode catalysts and electrolytes, see Bockris, *op. cit.*, Chapter 10.
18. Bockris, *op. cit.*, p. 21.
19. Liebhafsky and Cairns, *op. cit.*, p. 414.

gen/air cells.[20] Therefore, it seems logical that if ammonia is to be used as a fuel, it should be in an indirect fuel cell, which would first reform the ammonia to hydrogen and nitrogen gas.

Some fuel cells have been designed so that the products of the cell can be decomposed into the initial fuel and oxidant; for example, water may be dissociated into hydrogen and oxygen. These fuel cells are called *regenerative*. In order to regenerate the fuel, energy must be added to the system. This can be accomplished by thermal, electrical, chemical, radioactive, or photochemical methods. Most research on regenerative fuel cells has been done on electrically regenerative cells using lithium/chlorine or hydrogen/bromine as fuel combinations. The substances are fairly expensive and thus are practical only if they are reused. Electrically regenerative fuel cells are actually a kind of rechargeable battery, in that they must be fed electricity to continue working; as such, if they were to come into large-scale use (they have been suggested for use in automobiles),[21] they would necessitate greatly enlarged capacities for producing electric power. Unless that power were produced by fuel cells or some other clean method, electrically regenerative fuel cells could intensify, rather than alleviate, pollution by central power stations.

Regenerative fuel cells other than those powered by electricity have not yet been carefully studied, but other means of regeneration present some interesting possibilities. Radiochemical regenerative fuel cells would work by radiolysis, that is, the product of the fuel cell would be dissociated by irradiation. J. O'M. Bockris,[22] of the University of Pennsylvania, has suggested that fuel cells could use radioactive nuclear wastes as a means of regeneration. Because the regenerative fuel cell is a kind of closed system, little or no radiation should escape. However, a radiochemical regenerative de-

vice has yet to be proven powerful enough to be practical, much less safe.

The only other type of regeneration that seems plausible currently is photochemical regeneration. Such a system would utilize solar energy to transform the fuel-cell product to its reactants. Photochemical regeneration is most likely to find uses in space applications, where exposure to solar energy is assured.

One of the most fascinating fuel-cell concepts is that of a biochemical cell. In the biochemical fuel cell the reactions at one or both of the electrodes are promoted or catalyzed by biological processes.[23] Although the biochemical fuel cell sounds outlandish at first, it can be an efficient, reliable process. It is the process, in fact, that powers the human body. In the body, enzymes catalyze reactions wherein food (fuel) is oxidized in cells to produce energy, some of which is electrical.[24] Biochemical fuel cells may be either of the direct or indirect types. The indirect biochemical fuel cell operates on the waste products of a living organism. Fuel (such as formic acid) is fed to bacteria (such as *Escherichia coli*) that then produce a waste product (such as hydrogen) which can be used directly as the anode fuel in a fuel cell.[25] Conversely, algae can produce oxygen for the cathode. Or, an indirect fuel cell can be one in which an enzyme acts on a substance that is then fed into the cell.

Direct biochemical fuel cells are ones in which all the reactions take place within the fuel cell. For example, a direct biochemical fuel cell might use glucose as the fuel electrode, with the enzyme glucose oxidase as the electrode catalyst. Or, a direct biochemical fuel cell might actually have living organisms within the fuel cell to provide the enzymes required by the biochemical-electrochemical process. One of the most interesting biochemical fuel cells is one designed

20. Bockris, *op. cit.*, p. 591.
21. Bockris, *op. cit.*, p. 603.
22. Bockris, *op. cit.*, p. 608.

23. Liebhafsky and Cairns, *op. cit.*, p. 401.
24. Mitchell, Will, Jr., *Fuel Cells*, Academic Press, New York, 1963, p. 1.
25. Williams, K. R., ed., *An Introduction to Fuel Cells*, Elsevier Publishing Company, New York, 1966, p. 254.

to run on urea (in urine) and air, catalyzed by the enzyme urease. It is possible that such fuel cells, or others designed to run on organic wastes, might be useful in remote areas of underdeveloped nations. However, it has yet to be proven that any feasible biochemical fuel cell can produce a significant amount of power.

ECONOMICS

Evaluating the cost of fuel cells for various applications is largely a matter of conjecture at this stage of development. However, it is possible to project expected costs of fuels and to foresee reductions in the costs of fuel-cell manufacturing as the technology is advanced. The most comprehensive review of fuel-cell economics to date suggests that fuel cells are most likely to find application in the large-scale production of power.[26] Fuel cells may also be used economically in small, specialized vehicles such as fork-lift and delivery trucks and in other specialized areas such as railway engines and naval propulsion. General automobile applications of fuel cells are not at present or in the near future economically attractive. In comparison to conventional engines, fuel cells in the power range considered for automobiles are most economical when used intensively. Thus, it would be more economical to use fuel cells for taxicabs than for general-use automobiles; however, even taxi applications do not yet compare favorably with the internal combustion engine. According to the review mentioned above, for fuel cells to become appealing for automotive use, the cost of fuel would have to be drastically lowered or the cost of manufacturing the fuel cells would have to be reduced beyond present estimates. Dr. Kordesch, however, expects that if fuel cells were mass produced for automobile use, they would become economically competitive with the internal combustion engine.[12] Catch 22, though, is

that Kordesch does not envision fuel cells being mass produced until their cost is reduced. The one aspect that may confuse all these projections is that future air-quality standards may require pollution controls on the internal combustion engine, which may lower the efficiency and increase the cost sufficiently to make fuel cells competitive, or even preferable, economically. Reduction in the supply of fossil fuels may further enhance the economics of fuel-cell use.

Large fuel-cell power plants, on the other hand, will probably be economically attractive, even if future air pollution regulations are not considered. Depending on the size and type of the fuel-power system, fuel-cell electric power may cost as much as 0.627¢ per kilowatt-hour less than power from future conventional power plants.[27] (The kilowatt-hour, a common unit of electric power consumption, is a unit of energy equal to that expended in 1 hour at a steady rate of 1 kilowatt.) Another analysis of fuel-cell power costs, considered only for very large central generating stations, estimates that, depending on the cost of initial investment for fuel cells, fuel-cell-generated power may cost as much as 0.71 ¢ per kilowatt-hour less than other means of producing large blocks of electricity (see Table 4-31), and that even with high estimates of initial investment, fuel-cell-generated power will be less costly than light- or boiling-water reactors, fast-breeder reactors, high-temperature gas turbines, magnetohydrodynamics, or fossil-fuel power plants.[28] Estimates made for the Office of Coal Research indicate that a coal-fired fuel-cell plant of large size (1 million kilowatts—or 1,000 megawatts) will provide energy for only 0.299¢ per kilowatt-hour[29] —a cost far below the one indicated in Table 4-31. This discrepancy

26. Verstraete, J., et al., "Fuel Cell Economics and Commercial Applications," *Handbook of Fuel Cell Technology*, C. Berger, ed., Prentice-Hall, Englewood, New Jersey, 1968, p. 496 ff.

27. Berger, *op. cit.*, p. 592.
28. Baron, S., "Options in Power Generation and Transmission," presented at Brookhaven National Laboratory Conference, "Energy, Environment and Planning, The Long Island Sound Region," Oct. 5—7, 1971.
29. News release, Department of the Interior, Office of Coal Research, Washington, D.C., Dec. 12, 1966.

Table 4-31. Comparison of Fuel Cell Costs with Other Means of Electric Power Production (in cents per kilowatt-hour).[a]

Types of Costs	Fuel Cell Plants		Fossil Fuel Plants		Water		Fast Breeder	High-Temperature Gas Turbine	Magnetohydrodynamic Power Plant
	1980[b]	1990[c]	1980	1990	1980	1990	1990	1990	1990
Annual capital	0.67¢	0.34¢	0.54¢	0.67¢	0.84¢	1.05¢	1.05¢	0.62¢	0.62¢
Fuel	0.36¢	0.36¢	0.57¢	0.71¢	0.23¢	0.29¢	0.10¢	0.48¢	0.48¢
Maintenance and operation; insurance	0.04¢	0.04¢	0.03¢	0.04¢	0.06¢	0.07¢	0.07¢	0.04¢	0.04¢
Thermal effects	0.02¢	0.02¢	0.03¢	0.04¢	0.04¢	0.06¢	0.04¢	0.03¢	0.03¢
Total	1.09¢	0.76¢	1.17¢	1.46¢	1.17¢	1.47¢	1.26¢	1.17¢	1.17¢

[a]Source: S. Baron, "Options in Power Generation and Transmission" presented at Brookhaven National Laboratory Conference, "Energy, Environment and Planning, The Long Island Sound Region," Oct. 5–7, 1971.
[b]These figures consider that the initial cost equals that of fossil fuel plants.
[c]These figures consider that the high efficiency and ease of manufacturing of fuel-cell plants brings their initial cost to one-half that of fossil fuel plants.

emphasizes the hazards in projecting cost for new technologies when there are so many variables.

Costs of fuel cells will undoubtedly decrease, though, as the technology advances. New, lower-cost electrodes will be developed, and costs for electrolytes, particularly solid ones, will decrease as the science is perfected. There are certain inherent advantages, too, that will help fuel cells become economically competitive with other means of power production. The efficiency of fuel cells permits them to use less fuel to produce a given amount of power than do other methods. As fuel prices increase with time, as they are projected to do, the fuel cell's efficiency will be emphasized, and expected increases in the practical efficiency of fuel cells will enhance this aspect. But the most significant advantage of fuel cells is their inherent cleanliness. The products of fuel cells are usually water, carbon dioxide, and nitrogen. Of these products, the only one that might have a detrimental effect on the environment is carbon dioxide, which has been suggested by some scientists to cause an atmospheric "greenhouse" effect, which may produce weather changes. However, the amount of carbon dioxide released from fuels, be they used in conventional power generators or in fuel cells, is constant. There-

fore, fuel cells would release no more carbon dioxide into the atmosphere than do conventional generators; in fact, fuel cells would probably release less because they need less fuel to produce an equal amount of energy.

Fuel cells seem to hold the answer to the question of how to produce energy economically, at least for some major uses. Although the technology of fuel cells has improved vastly in the last decade, more developmental work must be done before these devices are commercially practical. It is somewhat ironic that, in these times of increasing environmental concern, little governmental funding for fuel-cell development is forthcoming. The black box that is so close to being practical is not quite off the shelf.

SUGGESTED READINGS

Cairns, Elton J., "Physics Opportunities in Electrochemical Energy Conversion," *Physics and the Energy Problem—1974,* Amer. Inst. of Physics Conf. Proceedings, No. 19, Feb. 1974.

Lueckel, W. J., L. G. Eklund, and S. H. Law, "Fuel Cells for Dispersed Power Generation," IEEE Trans. on Power Apparatus and Systems, Vol. 92, No. 1, 1973, pp. 230–236.

Maugh, Thomas H., "Fuel Cells: Dispersed Generation of Electricity," *Science 178:* 1273–1274, December 22, 1972.

Magnetohydrodynamic Central Power:
A Status Report

J. B. DICKS   1972

On June 4, 1971, President Nixon released a message to the Congress concerning the energy crisis. The main thought of the message was to ask for more money for the nuclear breeder reactor. It is obvious from the timetable given, which sets a goal of 1980 for demonstration of the breeder reactor, that such devices will not be available in time to alleviate the impending uranium shortage discussed in this paper. The president, at the same time, argued the necessity of a more balanced research and development attack on the energy problem and requested an increase in funds for the coal-gasification problem. Although this is a step in the right direction, it does not go far enough in anticipating the role of fossil fuel during the next fifty years in the United States.

Some important factors were neglected, particularly the promise of magnetohydrodynamic central power, both technologically and economically. The fiscal year 1972 proposed budget goes somewhat further in recommending increased expenditures for coal gasification and includes $3 million to begin a magnetohydrodynamic central-power program. This $3 million amount is significant but inadequate when compared to the national programs conducted in other countries.

AN OLD PRINCIPLE

Magnetohydrodynamic power generation is achieved when an easily ionized metal, such as potassium or cesium, is introduced into high-temperature combustion gas, which is expanded to high velocity through a nozzle and then directed into a magnetic field with properly arranged electrodes and external circuit.

In this situation, a moving conductor cuts magnetic-field lines, and a useful electromotive force is generated. Although this kind of electrical configuration was described by Faraday over 100 years ago and was one of the first generator configurations invented, the problems asssociated with high

Dr. John B. Dicks is director of the Energy Conversion Division at the University of Tennessee Space Institute, Tullahoma, Tennessee.

From *Mechanical Engineering*, Vol. 94, No. 5, pp. 14–20, May 1972. Reprinted with light editing by permission of the author and the American Society of Mechanical Engineers. Copyright 1972 by the American Society of Mechanical Engineers. This article is based on a paper contributed by the American Society of Mechanical Engineers Energetics Division.

temperature have prevented its application to combustion-gas plasmas until recently. Through the use of current high-temperature technology and some ten years of research and development in magnetohydrodynamics, the state of the art has reached the point such that 10 more years of work can produce large power plants in the 2,000-megawatt range for practical use. The impetus for developing such plants lies in their high thermal efficiency, between 50 and 60 per cent as compared with 40 per cent for conventional fossil fuel and 32 per cent for nuclear power plants. This makes magnetohydrodynamic-type steam plants attractive from the standpoint of economics, thermal pollution, and air pollution.

Within the past two years a whole new technical situation has arisen within the context of a changing social climate, so that the current status of magnetohydrodynamics is quite different from that set forth in the August, 1969, issue of *Mechanical Engineering* (see ref. 5). No longer is the future of magnetohydrodynamic technology or any other technology a simple estimate of technical feasibility and economic benefit. The public acceptance of power plants, the future power-demand curves, the cost of power-plant construction, and the effect of all these factors on power sources other than magnetohydrodynamics must be considered in order to adequately describe the status of the technology. The posture of the federal government and its organization with respect to central power will profoundly affect the future of any technology and thus needs to be examined as well.

It is now, therefore, a good time to review the status of magnetohydrodynamics central power. A good place to start is at the international meeting concerning magnetohydrodynamic power generation held in Munich, Germany, in April, 1971. Of particular interest was the announcement of an operating Soviet magnetohydrodynamic experimental facility, U-25. Extensive Soviet experiments on long-duration preheaters, magnetohydrodynamic channels, and other components have been performed. Smaller, but significant, experiments on central power components have also been constructed in Japan.

FUTURE OF MAGNETOHYDRODYNAMICS

The prime question should be: Is the expenditure of some $282 million necessary to acquire magnetohydrodynamic power-generation technology a reasonable technical risk in which the people of the United States can expect a large return in the future? If this question can be answered in the affirmative, then the discussion will turn to the acceptability of magnetohydrodynamic power generation from the standpoint of safety to the public, pollution of the environment, and other peripheral economic effects to be reasonably expected. Figure 4-47 shows a version of the traditional power-demand curve for the United States until the year 2000. It is possible to avoid answering questions concerning the competition between magnetohydrodynamic fossil-fuel plants and a system of nuclear power plants by merely calling attention to the fact

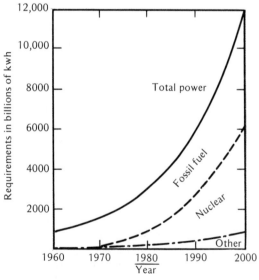

Figure 4-47. Projected power requirements.

that nuclear plants by their very nature must be base-load plants and that the rest of the power needs might be satisfied largely by magnetohydrodynamic power plants. Thus, some 30 per cent of the power plants might be magnetohydrodynamic plants, the rest being nuclear. Such a power system is not necessarily the optimum one for the country, however. Leaving aside for a moment the question of the acceptance of the conventional nuclear plant and the breeders by the public, it is worthwhile to take a look at the economics of the nuclear system as compared to a system where fossil-fuel magnetohydrodynamic plants take over a large portion of the power production.

The lower curve in Figure 4-48 shows the projected cost of uranium in 1969 dollars if nuclear reactors are put into service as estimated from the curve in Figure 4-47. This is not a realistic cost curve because it assumes that uranium is bought at the time it is consumed, which is not the usual practice. If we assume that the utilities will follow the usual custom of obtaining contracts for nuclear fuel for all (30 years) or a large part of the lifetime of the reactors, our uranium reserve would be committed to fueling reactors as they are built. The effect is shown in Figure 4-48 for 10, 20, and 30 years of uranium supply committed to the reactor when it is built. One sees that the reactors will be priced out of competition after 1985, because the 1969 price of $6.50 per pound will have increased by a factor of three to four for new reactors. The standard answer from the nuclear establishment to all who point out this obvious future uranium shortage is that additional exploration will turn up the required uranium supply. However, anyone familiar with the current oil and gas situation will have grave reservations concerning the assumption that mineral resources can always be found when needed.

Another answer—this one from the Atomic Energy Commission—is that breeder reactors, when installed, will alleviate the uranium supply shortage. But even optimistic estimates of a fuel-doubling time of 10 years in the breeder leads to a prediction that it would require 30 years to fully install a breeder system that would supply most of the nation's power needs. One must add to this 30 years the fact that it will probably take at least 10 years to site and build the first generation of breeders and that the breeder, of course, is not developed as yet and may require 10- to 15-years development time. We finally come up with the fact that it will be 50 years from now before the breeder can fully supply the uranium required for a completely nuclear central power system in the United States. It is obvious that breeder reactors will not be on the line in appreciable numbers before 1995, and that long before this time the cost of uranium will have risen by a factor of three or four.

REDUCED POWER BILL

The yearly savings in the nation's power bill, if magnetohydrodynamic fossil-fuel plants

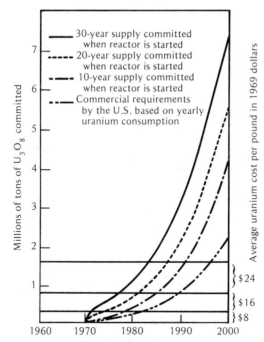

Figure 4-48. Cumulative projected uranium requirements.

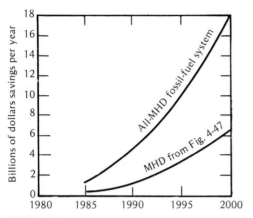

Figure 4-49. Contrast between savings brought about by MHD from the fossil-fuel system predicted in Figure 4-47 with an all-fossil-fuel system for plants constructed after 1985.

were installed beginning in 1985 instead of ordinary fossil-fuel plants, are shown in Figure 4-49. The upper curve represents the savings to be realized if fossil fuel takes over completely from nuclear fuel in 1985, and the lower curve indicates the savings if the split between nuclear and fossil-fuel power generation is as shown from the curves in Figure 4-47. If magnetohydrodynamic central power plants of 55 per cent efficiency are developed, one would expect the savings in the power bill to lie somewhere between these two curves. The competition might very well be effective in lowering the cost of nuclear power as well. It is assumed in making these cost estimates that sulfur dioxide is virtually eliminated from the magnetohydrodynamic exhaust, regardless of the type of coal burned, because of the seed-recovery process.

UNITED STATES EFFORT

Since our 1969 status report, no new magnetohydrodynamic facilities of significance have been reported in the United States. Old open-cycle facilities at Avco, Stanford, and the University of Tennessee have undergone modifications, and results of basic research have been reported. Technical

progress of note in the United States includes the achievement of new high thermal efficiency and power density by Avco[1] for the special case of a very-low-density combustion gas. These parameters are significantly higher than those achieved by any investigator in open-cycle magnetohydrodynamic generators to date. The studies show that high power density and thermal efficiency can be obtained without aerodynamic choking at low density. This work is of more interest to special-purpose generators than it is for central power, where the density must be higher and the velocity much lower than with the conditions of the Avco experiments.

At Stanford University, basic experimental work indicates that the conducting-side-wall generators produce slightly more power than insulated-side-wall generators running at the same gas conditions.[2] This result is also predicted by theoretical work at the University of Tennessee Space Institute.[3] At the University of Tennessee Space Institute, the first generator experiments in which coal and char were directly fired were reported, but the experimental duration of 12 seconds is only sufficient to show that no difficulties occur from ash buildup for this short time.[4] Contracts were granted by the Office of Coal Research to Avco, Massachu-

1. Sonju, O. K., and Teno, J., "An Experimental and Theoretical Investigation of a High Interaction Combustion Driven MHD Generator," *Proceedings of the Fifth International Conference on Magnetohydrodynamic Electrical Power Generation, Vol. 1,* Munich, Germany, 1971, p. 199.
2. Eustis, R. H., and Kessler, R., "Measurement of Current Distributions and the Effect of Electrode Configuration on MHD Generator Performance," *Proceedings of the Fifth International Conference on Magnetohydrodynamic Electrical Power Generation, Vol. 1,* Munich Germany, 1971, p. 281.
3. Koester, J. K. et al., "The Influence of Electrode Drop on Current Distribution in Diagonal Conducting Wall Generators," *Proceedings of the Fifth International Conference on Magnetohydrodynamic Electrical Power Generation, Vol. 1,* Munich, Germany, 1971, p. 265.
4. Wu, Y. C. L. et al., "Factors Affecting the Performance of Diagonal Conducting Wall Open Cycle MHD Generators," *Proceedings of the Fifth International Conference on Magnetohydrodynamic Electrical Power Generation, Vol. 1,* Munich, Germany, 1971, p. 213.

setts Institute of Technology, and the University of Tennessee Space Institute for work on central-power magnetohydrodynamics. At Avco and the University of Tennessee Space Institute long-duration facilities began operation in the spring of 1972. These facilities should give much-needed data on long-duration operation, but are relatively small devices because of the lack of funding. Government funding in this area is approximately $2 million in fiscal year 1971 with an expectation of $3 million in fiscal year 1972. In addition, the work at Avco and the University of Tennessee Space Institute is also being supported by contributions from the utilities.

INTERNATIONAL STATUS

By far the most spectacular results were announced by the delegation of the Soviet Union when it was stated that an announcement had been made in Moscow at the twenty-fourth Party Conference in March, 1971, that a new kind of power plant was in operation on the Moscow power network. This plant is the U-25, the prospective design of which was described in *Mechanical Engineering,* August, 1969, issue.[5,6,7,8] Conjecture in the United States was that this plant would begin operation somewhere around November, 1971; so it appears to be ahead of our original estimates. The plant is complete, except for the steam turbine of the

bottoming unit which would be of no importance in the experimental plant. A new set of specifications for this plant was presented as shown in Table 4-32. The author and several other people from the United States had an opportunity to inspect this plant in conjunction with the Joint International Atomic Energy Agency/ENEA International Magnetohydrodynamic Liaison Group meeting in Moscow in December of 1971.

The plant's exterior air preheaters consist presently of aluminum oxide, and are heated by natural gas and then used to heat the incoming air. Such heaters will be periodically cycled to provide a continuous flow of air at 1,200°C. Such preheat is necessary in the magnetohydrodynamic cycle in order to make the combustion products conducting. In the U-25 additional temperature is gained through the addition of a small amount of pure oxygen preheated at 1,200° C to the air. The preheaters have been in operation for some time, though it is not completely clear for how long they have been operated. Others at the High Temperature Institute have been cycled for 8,000 hours. The combustion chamber is drastically smaller than the combustion chambers used with conventional power plants of the same size, because of the high temperature and pressure. Their seed-recovery process is quite successful, as the Moscow group claims 99.9 per cent seed recovery. Other technical triumphs in this program include successful operation of boiler tubes for long periods of time in a potassium-seed combustion gas.

Soviet Effort: $200-million Bread Board

It is interesting to speculate on the rationale behind this approach by the High Temperature Institute to develop magnetohydrodynamics central-power technology. The approach is all the more interesting since no large-scale development in nonnuclear power plants has been undertaken before. In general, rather than taking a revolutionary approach, power technology has crept slowly year by year up to higher powers (13 megawatts) at slightly increasing efficiency.

5. Dicks, J. B. et al., "MHD Power Generation: Current Status," *Mechanical Engineering,* Vol. 91, No. 8, Aug. 1969, pp. 18-25.

6. Kirillin, V. A. et al., "Investigations at U-02 MHD Plant—Some Results," *Proceedings of the Fifth International Conference on Magnetohydrodynamic Electrical Power Generation, Vol. 1,* Munich, Germany, 1971, p. 353.

7. Gnesin, G. G. et al., "Research on Materials for Manufacturing of Open-Cycle-MGDT Electrodes," *Proceedings of the Fifth International Conference on Magnetohydrodynamic Electrical Power Generation, Vol. 1,* Munich, Germany, 1971, p. 393.

8. Zhimerin, D. G. et al., "Investigation of Cooled Channel on Enin-2 Installation," *Proceedings of the Fifth International Conference on Magnetohydrodynamic Electrical Power Generation, Vol. 1,* Munich, Germany, 1971, p. 249.

Table 4-32. Principal Characteristics of the U-25 Installation.

Compressor outlet temperature	147°C
Air-heater outlet temperature	1200°C
Combustion temperature	2800°C
General outlet pressure	1.07 atmospheres
Field strength	2 tesla
Magnetohydrodynamic electric power	25 megawatts
Turbine power	86,000 kilowatts
Net plant output	81,000 kilowatts
General outlet temperature	2300°C
Stack temperature	150–180°C
Mass flow	50 kilograms per second
Duct inlet cross section	0.38 × 0.766 meters
Duct outlet cross section	0.38 × 0.783 meters
Duct velocity	850 meters per second
Duct length	5 meters
Compressor power	5.6 megawatts
Auxiliaries power	5000 kilowatts
Oxygen plant power	16,000 kilowatts
Magnet power	2.8 megawatts
Plant efficiency	30 per cent
Thermic power	250 megawatts
Central power steam	50 megawatts
Total power	75 megawatts
Combustible	natural gas
Combustant	enriched oxygen
Temperature of preheated enriched oxygen	1200°C
Seed	potassium
Combustion gas supply	50 kilograms per second
Entrance pressure	275 atmospheres
Entrance velocity	850 meters per second
Steam pressure	100 atmospheres
Temperature of the superheater	540°C
Feedwater temperature	150°C

In Professor Scheindlin's method a gigantic experimental bread board has been constructed. The power-plant components are widely separated and housed in a large building devised so that experimental changes can be made with ease. Because of the problem of radioactivity, it is not possible to develop nuclear power along these lines, but magnetohydrodynamics suffers from no such limitations and the bread-board approach will give the Soviet Union an optimum experimental program. For example, the question most frequently asked is, What is the optimum channel design for the magnetohydrodynamic generator, and what is its capability of endurance? The U-25 is so designed that a number of trial channels can be placed within its magnet and tried in succession. We have seen pictures of such channel construction and believe that a number already exist constructed with cold walls, hot walls, and intermediate temperatures. The only design that we have examined in detail is a water-cooled channel designed for Faraday operation containing many water-cooled copper hemispherical electrodes.

One photograph that we have seen of these devices was the corner of such a channel shown in a motion picture. It appeared to be a steep-diagonal-wall design with relatively large insulator spacing. We expect that in addition to the diagonal-wall electrical design, Faraday and Hall channels will be tried as well, so that in the near future the High Temperature Institute will have information on which channel works best. Not only is the magnetohydrodynamic channel removable in this setup, but other components are as well. We expect that the conventional magnet will be replaced by a superconducting magnet at some time. We have been told that the seed-removal and exhaust-cleanup device has been used at some other location. We were also informed that the performance of the preheaters was not satisfactory, and some improvements will be made in these devices.

We have been told that there are 1,000 people at work on this magnetohydrodynamics project alone, and we believe that the project itself is skillfully and intelligently organized so that the Soviet Union will acquire the necessary technology for central power in a short period of time at an optimum cost. Questions of endurance and electrical efficiency will be solved in good time, and the High Temperature Institute should be congratulated on its ability to put such a plant in operation so soon. In the United States, because of cost limitation, we are at least five years away from a plant of this type. The hardware not including design cost is valued at $50 million. The very large auxiliary oxygen plant would add a substantial amount to this.

West German Magnetohydrodynamics Program

The West Germany open-cycle magnetohydrodynamics program is divided into two parts. One group is at the Max Planck Institut für Plasmaphysik in Garching near Munich, with its work centered around generators designed for operation of times less than 1 hour, and is cooperating with the MAN Corp. of Munich.[9] Magnetohydrodynamic generators have very quick starting characteristics, and without special effort

can be brought to full power in less than 1 second. Some of the smaller utility companies in West Germany are very interested in generators having this characteristic, as they are now having to buy peak-load power from larger utilities for very short times at very high rates. It is also thought by the group in Garching that such generators will have utility in fusion power plants, especially in the development program for such devices.

This second effort is being conducted on magnetohydrodynamic generators with operating times greater than 1 hour.[10] This effort is being carried on by cooperation between the Institut für Technische Physik in Julich and the Forschungsinstitut des Steinkohlenbergbauvereins, Bergbau-Forschung GmbH at Essen. Experiments up to 20 hours have been run with a small magnetohydrodynamic channel in Julich. In Essen, work is underway on a unique process for inexpensive enrichment of air with oxygen. This air-enrichment process could have a profound effect on magnetohydrodynamic generator systems if it develops as currently projected from initial work.

In Garching, the first very-high-magnetic-field magnetohydrodynamic generator ever built is being tested with magnetic fields up to 50 kilogauss. This work is yielding very important data in a magnetic field range that cannot be reached by other investigators. The magnetohydrodynamic channel is of diagonal design at a 45-degree angle similar to those that have been investigated previously.[4,11,12] As it is projected that both

9. Bunde, R. et al., "Theoretical, Experimental and Technical Investigations for the Development of a Pulsed Combustion MHD Generator," *Proceedings of the Fifth International Conference on Magnetohydrodynamic Electrical Power Generation, Vol. 1*, Munich, Germany, 1971, p. 229.

10. Bohn, Th. et al., "Theoretical and Experimental Studies for the Development of a 30 MWth Open Cycle MHD Generator," *Proceedings of the Fifth International Conference on Magnetohydrodynamic Electrical Power Generation, Vol. 1*, Munich, Germany, 1971, p. 243.

11. Dicks, J. B. et al., "MHD Generator in Two-Terminal Operation." *AIAA Journal*, Vol. 6, 1968, pp. 1651–1657; also, *Proceedings of the IEEE*, Vol. 56, 1968, pp. 1555–1562.

12. Dicks, J. B. et al., "Experimental Study of Diagonal Conducting Wall Generators Using Solid Propellants," *AIAA Journal*, Vol. 6, 1968, pp. 1647–1651; also *Proceedings of the IEEE*, Vol. 56, 1968, pp. 1574–1578.

peaking plants and central power plants will operate with superconducting magnets in the range of 50 kilogauss, the results of the Garching experiments are of great interest in the field. The expenditure on development work in West Germany is of the order of $2 to $3 million per year.

The Japanese Effort

A very extensive effort directed toward open-cycle magnetohydrodynamic [13], [14], [15] power generation with experiments on all phases of magnetohydrodynamic central power plants is being conducted in Japan. The generator channels tested have been largely the segmented Faraday design with some work performed on the Hall generator configuration as well. A number of preheater designs have been run with special attention paid to the seed buildup that occurs under some working-gas conditions in these devices. Methods of improving channel duration and preheater operation are being investigated presently. The approach in Japan to the preheater problem differs from that in the Soviet Union in that tubular preheaters are being investigated, rather than the regenerator type. Some progress has been made in all of these areas. Figure 4-50 shows the estimated power-demand curve for Japan as given in "The Present Status of Magnetohydrodynamic Power Generation in Japan" by the Magnetohydrodynamic Power Generation Study Team and T. Sekiguchi at the University of Tokyo. This shows a relatively modest power requirement by the

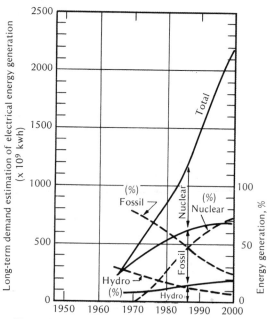

Figure 4-50. Projected Japanese power demand.

13. Fushimi, K. et al., "Experiments on MHD Generation with Hot-Air Combustion ETL Mark II," *Proceedings of the Fifth International Conference on Magnetohydrodynamic Electrical Power Generation, Vol. 1,* Munich, Germany, 1971, p. 187.
14. Fushimi, K. et al., "Development of a Long Duration MHD Channel," *Proceedings of the Fifth International Conference on Magnetohydrodynamic Electrical Power Generation, Vol. 1,* Munich, Germany, 1971, p. 371.
15. Mawatari, K. et al., "Experiments on Tubular Heat Exchangers for an MHD Power Plant," *Proceedings of the Fifth International Conference on Magnetohydrodynamic Electrical Power Generation, Vol. 1,* Munich, Germany, 1971, p. 503.

year 2000, approximately equal to that expected by the United States by the year 1980. It is very plain that the Japanese expect to get uranium from the United States, as does much of the western world. Agreement to furnish the western world with uranium for their reactors is contained in the June 4, 1971 message to Congress by President Nixon when he said, "Plant expansions are required so that we can meet the growing demands for nuclear fuel in the late 1970's—both in the United States and in other nations for which this country is now the principal supplier."

The president is speaking here of the uranium-enrichment plants; however, we must also supply the uranium to be enriched. The Japanese currently view magnetohydrodynamics as chiefly a peak-load technology. This stems from the Japanese economy in which the main source of fossil fuel is expensive oil from Indonesia. Thus, as in many countries other than the United States, the burning of uranium rather than fossil fuel presents an economic advantage. In the United States, the reverse is true.

Other Countries

Numerous other open-cycle magnetohydrodynamics efforts exist in other countries, such as the large one in Poland and the beginning of a significant national program in Canada. In the British Isles and in France the magnetohydrodynamics effort has been reduced partially because of the economy and partially because there fossil-fuel power, in general, does not have the advantage of very low fuel costs that it does in the United States.

The Status of Magnetohydrodynamics in the United States

Within recent years in the United States there has been literally no central-power magnetohydrodynamics program other than the small efforts that could be maintained in industries and the universities using their own funds to work on central power on the side. The vast majority of the work has been in basic research on basic phenomena and development work for the Defense Department. During 1971, funds became available to start a minimal amount of central-power magnetohydrodynamic work. This is largely being funded by the Office of Coal Research in cooperation with power companies. The largest such effort is under a contract let to Avco and a group of utility companies to work on clean-fuel peaking plants with a small amount of coal-burning included. This contract is of the order of magnitude of $2.6 million to be spent over three years. Additional amounts would come from Avco and the associated utilities. The next largest contract is with the University of Tennessee Space Institute, with $324,000 to be spent over one year on power generation with coal and char fuels. This work includes a small investigation of chemical regeneration. Of the total funds, $261,000 is being furnished by the Office of Coal Research, $50,000 by the Tennessee Valley

Authority, and $30,000 by the university. It is expected that a contract for approximately $100,000 per year will be let to the Massachusetts Institute of Technology to perform some basic research studies and to advise the Office of Coal Research. In addition to this, STD Corp. of Los Angeles may receive approximately $90,000 to direct and operate a master computer program designed for magnetohydrodynamic power-system analysis. At Stanford University there will be a research program funded by the Electric Research Council and the Bureau of Mines.

Avco, Stanford, and the University of Tennessee Space Institute all have a long history of continuous research and development on open-cycle magnetohydrodynamic power generation and have additional magnetohydrodynamic open-cycle work funded from other sources. The total central-power program in the United States is inadequate to make appreciable progress in this area, but there is the anticipation that additional money will be available in the fiscal 1972 appropriation by Congress and from the Electric Research Council to expand this program. As a matter of fact, all of the efforts enumerated here are preliminary to a national program to be agreed upon by the Office of Coal Research and the Electric Research Council. The participants in the initial program have plans for such expansion when the resources are made available.

It is very unlikely that a decision to develop power generation at a minimum cost will ever come about in the United States. It is thus impossible for us to follow the breadboard plan of experimentation that is being pursued at the High Temperature Institute in Moscow. In general, programs in the United States, which do not have heavy support within the government, have to start with low-level funding on relatively inefficient feasibility demonstrations. As the need for the end product of development nears and becomes more evident, the pace is stepped up and extra money must be appropriated to make up for lost time. A better development plan is shown in Table 4-33, where feasibility of components is demonstrated at the

Table 4-33. Magnetohydrodynamics Development Cost Plan.

Fiscal Year	1971	1972	1973	1974	1975	1976	1977	1978	1979	1980	1981	1982
Research and development	0.6	4.0	4.0	5.0	6.0	6.0	6.0	6.0	6.0	7.0	7.0	7.0
20-megawatt plant			4.1	11.9	10.0	2.0	1.0	1.0				
100-megawatt plant					3.4	11.4	28.8	25.0	4.0	3.1		
1,000-megawatt plant							2.0	26.2	60.3	101	72.0	2.9
Yearly total	0.6	4.0	8.1	16.9	19.4	19.4	37.8	58.2	70.3	111	79.0	9.9
Gross cost of program												434.7
Less sale of power												20.2
Less residual worth of plant at one-half of construction cost												132.0
Net program cost												282.5

more realistic 20-megawatt size for magneto-hydrodynamic power. In the next stage, overlapping somewhat, is a pilot plant to be designed to obtain efficiency data that can be extrapolated to full-size construction. Finally, a 1,000-megawatt plant is included at twice the estimated cost of such plants when they are produced in numbers. The necessity of a development plant of 1,000-megawatt size has been amply demonstrated by breakdowns and other problems of recently developed large plants in both the nuclear and fossil-fuel programs. Detailed estimates of the eventual cost of magneto-hydrodynamic generator plants are 30 per cent less than the comparable cost of a fossil-fuel plant. In comparing the cost of development outlined here with other power-development programs, one should make certain that the other program contains development plants of the 1,000-megawatt size.

If instead of a program such as the one advocated here, the country embarks on several years of basic work performed with apparatus of very small size, some of the important cost savings from magnetohydro-dynamic development outlined in Table 4-33 and the accompanying text will be lost because of the resulting delay.

As of February, 1972, the line item in the administration's budget for magnetohydro-dynamic development on central power for fiscal 1973 is $2.6 million for work outside of the government and $400,000 for work in the Bureau of Mines. There is, therefore, some chance that a government appropriation of $2.6 million will be available in fiscal 1973 to match with money obtained from the utilities.*

Important for the future of energy development in the United States is the energy study being conducted by the Senate Committee on Interior and Insular Affairs. There have been many energy studies instituted during the past year, but none has been satisfactorily constituted, authoritative, definitive, and suitable from the standpoint of providing an information basis suitable for new legislation. An increasing indication that the development being presently conducted is unbalanced is evident from the fact that tremendous amounts of development are going on in the nuclear field and almost none in the fossil-fuel field. A department of national resources covering all forms of energy would be the best solution to the balance problem, but many difficulties now lie between the conception of such a department and its realization. The increased interest in energy on the part of Congress is a bright spot for the future, and we expect that the central-power situation, including magnetohydrodynamics, may be profoundly affected by legislation introduced by Senator Metcalf in the Senate and various members of the House of Representatives which is aimed at acquiring a national central-power distribution system.

The future of central power is cloudy, with the uranium supply and price difficult to forecast, the breeder reactor uncertain in its development time and acceptance by the public, the conventional fossil-fuel plant now asymptotically approaching its highest efficiency, and the cost of power-plant construction steeply rising along with the price of fossil fuel. All of these conditions make the future of central power in the United States uncertain, and predictions exceedingly difficult. It does seem clear, however, that magnetohydrodynamic fossil-fuel power generation, if acquired, would do several important things. It would provide economic competition for the nuclear system, give a possible alternative for relatively pollution-free power production if the breeder reactor fails to gain public acceptance, and extend the lifetime of our coal reserves.

*Editors' note: Magnetohydrodynamics research is part of an $18 million line item on "Advanced Cycles, Fuel Cells and Other Concepts" in the fiscal year 1975 budget.

The Hydrogen Economy

DEREK P. GREGORY   1972

The gas industry today is taking very significant steps to supplement the nation's supply of natural gas. In the long term, however, importation of liquid natural gas and gasification of coal and oil cannot continue to satisfy our energy demands indefinitely because the resource bases from which they are obtained are themselves limited. Even the nation's huge coal reserves are subject to economic limitation because the deeper we bite into these reserves, the more expensive the extraction process will be. In the long run, we must look toward a new form of energy to meet our needs.

Nuclear energy is already here, and its use will expand rapidly. As we head toward the nuclear age, let us take a look at the role the gas industry can play in the long-term future. Many energy forecasts do not extend beyond the year 2000.

Nevertheless, by early in the next century, our energy supply and demand patterns will undergo some revolutionary changes. Before we dismiss the year 2000 as unworthy of serious concern, remember that today's high school students will only be halfway through their working lives at the turn of the century!

TIME FOR THE "HYDROGEN ECONOMY"

Because most of the present research on nuclear energy is directed toward converting this energy to electricity, there is good reason to believe that we will continue to move toward an "electric economy." Even without this nuclear energy stimulus, this trend already had been taking place for some years. Figure 4-51 shows how the total United States electricity generation each year is growing far faster than the total over-all use of United States energy. It is worth noting that our other "unlimited" energy sources—solar, geothermal, wind, and tide—are likely to be exploited, if at all, by conversion into electricity.

The huge anticipated growth of nuclear power stations will only take place by the construction of large plants, 1,000 mega-

Derek P. Gregory is director of energy systems research at the Institute of Gas Technology, Chicago, Illinois; he previously spent four years as research manager of Energy Conversion Ltd. in England, and several years in other positions in the United States and England on fuel cell research.

From *American Gas Association Monthly*, June 1972. Reprinted by permission of the author and the American Gas Association.

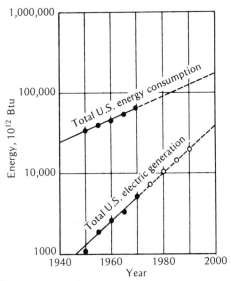

Figure 4-51. Electrical energy and over-all energy use.

watts and larger, at fairly remote locations where cooling water is available and where concern over safety is minimized. Now we cannot have a lot more power stations without a corresponding increase in transmission lines. These lines already create a problem in obtaining the extensive right-of-way required. The environmentalists raise objections on aesthetic grounds when transmission lines clutter the landscape.

However, there is another way of moving energy around. We believe that it is possible to convert electrical energy into hydrogen, which is a gas; to move this gas in underground pipelines that are out of sight; and to deliver it to customers wherever it is needed. This possible use of hydrogen as a universal fuel provides the concept that we call the "hydrogen economy." Just as we are moving to an electric economy, we could at the same time move toward a "hydrogen economy."

THE ULTIMATE "ECOLOGY FUEL"

Hydrogen is what we might term the "ecology fuel." It is the cleanest burning fuel we know. The top of Figure 4-52 shows the cycle of present fossil fuels. We dig up fossil fuels from the ground and use them as a means of transporting and storing energy. Once they arrive at their destination, we then burn the fuels to produce the energy in the form we want it and deposit the products of combustion into the environment. These products, which are, of course, pollutants, are scavenged from the environment by natural processes—for example, by growing vegetation—and in principle can be recycled by a long, slow process to recreate fossil fuels. The only problem here is that this recreation of fossil fuels takes many millions of years, which is why we are running out of fossil fuels and why we are polluting our environment with their waste products.

The bottom part of Figure 4-52 shows an

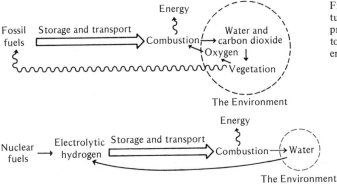

Figure 4-52. "No deposit-No return." The upper part shows the present energy cycle and the bottom part shows the hydrogen energy cycle.

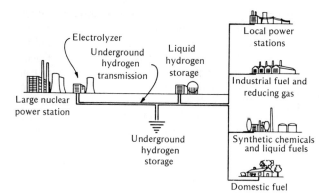

Figure 4-53. The complete hydrogen economy delivery system.

alternative system. Hydrogen would be produced, for example, by electrolysis of water, at a nuclear power station. This hydrogen would be our new energy storage and transmission medium. We would deliver the hydrogen to the point of use and burn it to produce energy in the desired form. The only combustion product is water. We can dump as much water as we like into the atmosphere without upsetting the natural balance because the earth's crust has plenty of water already. The net result is that we have an energy cycle that can go on forever without polluting the atmosphere. Thus, we have the making of a perpetual gas industry.

USES OF THE "ECOLOGY FUEL"

Figure 4-53 shows how we might consider using this hydrogen. Hydrogen made at a nuclear power station could be transmitted in underground pipelines. We could provide storage capacity by underground storage fields analogous to gas storage fields or by liquid hydrogen storage analogous to liquid natural gas storage tanks. We could deliver the hydrogen to the same customers that use natural gas today—for example, as a domestic fuel to do all the heating and cooking jobs around the home. It could also be used (a) to run small local power stations to generate electricity right where it is needed, and (b) as an industrial fuel and as a reducing gas in the metallurgical industry. Hydrogen is also valuable as a synthetic fuel for

making chemicals and liquid fuels such as methanol. We will need liquid fuels of some sort for our automobiles and aircraft. We might even consider hydrogen itself as a transportation fuel; studies are going on right now to run automobiles, airplanes, trains, and buses on hydrogen.

WHERE WILL WE BUILD THEM?

Let us step back and take a look at the siting problems of the nuclear power stations and indeed of the large fossil fuel power stations. One of the more important aspects of deciding where we put a power station is providing it with the necessary cooling water because a power station is only 30 to 40 per cent efficient and most of the power produced from it is in the form of low-grade heat, which has to be dissipated either into the sea, a river or a lake, or into the air. As we require more and more power stations, pressure will increase to site them along our coast lines or even offshore, as is seriously being proposed by Westinghouse right now.

If, as it appears, we will have to site our power stations remotely from the cities in order to find acceptable locations for them, the existing electric transmission network will have to be expanded. At present, the gas industry already has a gas transmission network with a far greater energy-carrying capacity than the electric system. Thus, the gas industry has a very important trump card to play because it can move energy in its under-

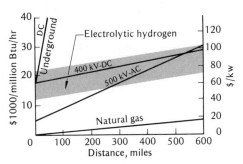

Figure 4-54. Cost of energy transmission facilities.

current transmission, the cost of underground electric lines is incredibly expensive, while the cost of natural gas lines of a similar capacity is relatively cheap. We believe we could install an electrolyzer for about the same cost as a transformer station and that cost of the hydrogen line would be a little greater than the cost of the gas line.

The costs of the two fall into the shaded area of Figure 4-54 and would be competitive with overhead power lines over distances of 400 to 600 miles. The cost of a hydrogen line compared to an underground electricity line would be competitive within about 50 miles.

Figure 4-55 compares the operating costs of hydrogen transmission and electricity transmission. The electrical figures are published Federal Power Commission data on the costs of moving energy, once it has been generated, at different voltages by different types of overhead lines.

Again, the underground transmission cost can be seen to be very high indeed. We have superimposed on this diagram the Atomic Energy Commission's estimate of the operating cost of an electrolyzer, which would turn the electricity into hydrogen at the generating station, and our estimate of the cost of hydrogen transmission pipelines. Thus, Figure 4-55 again shows that after a few hundred miles we could deliver energy as hydrogen more cheaply than we can deliver electricity.

ground gas pipelines far more cheaply than the electrical industry can move energy in its overhead electrical lines.

COMPARING TRANSMISSION COSTS

Let us take a closer look at the relative costs of gas, electricity, and hydrogen transmission. First, the installation costs of electricity lines are much higher than those of natural gas pipelines. Figure 4-54 shows the actual capital installation costs of all high-voltage alternating-current and direct-current electrical lines and transformers compared with the costs of installing all 36-inch-diameter trunk gas lines and compressors during 1969. The intercepts on the zero-miles axis are the transformer costs. We can see that, compared to overhead alternating-

WILL HYDROGEN BE CHEAP ENERGY?

Figure 4-56 shows the over-all picture of the prices of delivered energy based on relative selling prices. We have taken the average selling prices of electricity and natural gas to all customers during 1969 and broken these prices down into the shares taken by the production, transmission and local distribution sectors of the business. The first two columns clearly show, first, that it is much cheaper today to buy energy in the form of natural gas than it is to buy electrical energy and, second, that the cost of transmission and distribution of electricity is considerably

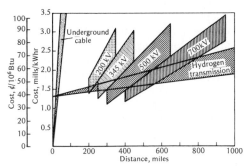

Figure 4-55. The relative costs of transmitting hydrogen and electricity.

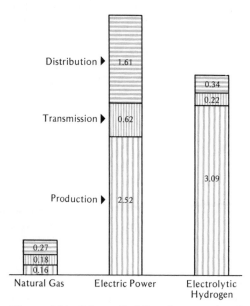

Figure 4-56. Prices of delivered energy based on relative selling prices.

ADDED ENVIRONMENTAL VALUE

Hydrogen has another advantage that cannot be given a dollar value. This is the environmental advantage of not only its use as a clean fuel but also of its help in winning the environmental battle of transmission.

Let us now consider some present-day facts. The photograph at the beginning of Part 5, was taken in an Illinois corn field. Beneath the ground is a 36-inch-diameter natural gas line. The only indications that it is there at all are the markers where the pipeline crosses a road. In electrical energy terms, a 36-inch natural gas line can carry 11,000 megawatts of electricity. Behind the marker there are four overhead power lines on four sets of towers. The total capacity of all four lines is less than one-third of the energy capacity of the one gas line. The large tower in the center can carry about 1,000 megawatts, or one-eleventh of the power carried by the underground gas line.

There is no reason why hydrogen cannot be carried in underground gas lines. At present, the United States has industrial pipelines carrying hydrogen which are up to 25 miles long and West Germany has one about 125 miles long. The energy-carrying capacity of the same size line at the same pressure is about the same for hydrogen as it is for natural gas.

UNDERGROUND STORAGE AND LIQUEFACTION

Hydrogen can be stored, too. There seems to be no reason why hydrogen cannot be stored in depleted gas fields in the same way that we store natural gas today. Although hydrogen is a low-density gas and is therefore more prone to leakage, a depleted gas field is inherently leak tight. Helium is currently being stored in a depleted gas field, without apparent leakage.

Another way of storing hydrogen is by liquefaction. Liquid hydrogen is, of course, the rocket fuel used to send men to the moon. As a result, the aerospace industry has developed a great deal of technology in the handling of liquid hydrogen. A spherical

higher than the cost of transmission and distribution of natural gas.

The third column is an estimate of the typical selling price hydrogen would have today if it were to be made on a very large scale by electrolysis, using today's production cost of electricity and assuming cost figures for large-scale electrolysis plants estimated by the Atomic Energy Commission. We have then estimated the cost of transmitting and distributing hydrogen on the basis of the known costs of doing the same thing for natural gas.

This estimate shows that hydrogen energy could be sold more cheaply today than could electrical energy. But clearly there will be no market for this hydrogen as long as natural gas is as cheap as it is. We might expect natural gas prices to increase at a more rapid rate than nuclear electricity prices because the supply of natural gas is finite. We hope that, with the help of the breeder reactor, we will have an almost unlimited supply of nuclear electricity. One day, then, hydrogen will be economically attractive as a fuel.

65-foot-diameter liquid-hydrogen tank at Cape Kennedy has a capacity of 900,000 gallons of liquid. In energy terms, this is equivalent to 38 billion BTU or 11 million kilowatt-hours of energy. Although this is only about one-twentieth of the energy equivalent of a typical peak-shaving liquid natural gas plant, it still represents a large amount of energy storage.

By way of comparison, the only way we have of storing large quantities of electrical energy is in the pumped hydroelectric storage system in which a lake is filled up during periods of low demand by running the hydroelectric station in reverse. In periods of high demand, it is run as a conventional hydroelectric generator. Not many locations have the terrain suitable for such a system. The world's largest pumped storage system is located at Ludington, Michigan. With an artificial lake about 2.5 miles long, 1.5 mile across, and 6 miles in perimeter, it drives an 1,800-megawatt hydroelectric station and is capable of operating for an eight-hour period. This system, which, of course, occupies many acres of land and costs over $300 million, will hold 15 million kilowatt-hours of electricity. By comparison, the 65-foot-diameter hydrogen storage tank at Cape Kennedy holds the equivalent of 11 million kilowatt-hours.

HYDROGEN PRODUCTION TECHNIQUES

How will the hydrogen for such energy production schemes be made? Water can be split by adding energy to convert it to hydrogen and oxygen. Essentially, this can be done in three ways. The first is electrolytic, where we pass a direct electric current through a conducting water solution, liberating hydrogen and oxygen separately at the electrodes. Second, if we heat water to a high enough temperature, it will spontaneously split into hydrogen and oxygen. But such temperatures are inaccessible to use from nuclear reactors, and the subsequent physical separation of hydrogen and oxygen would be very difficult.

A third method, which we call thermochemical splitting, would be to react the water in a series of chemical reactions in which the intermediate chemicals are always recycled and the over-all reaction produces hydrogen and oxygen in two separate reaction steps. A few such processes under study appear, on paper at least, to be capable of operating at temperatures well within the reach of nuclear reactors or solar furnaces.

The electrolytic production method is already in use and has been in use for many years in areas where electrical energy is relatively cheap. Large-scale electrolysis plants are in use in Canada, India, Egypt, and Norway. The efficiency of these plants is only about 60 per cent, but it appears that, with proper research and development, these efficiencies could be increased.

The quantity of hydrogen that we are talking about for widespread hydrogen use is, of course, immense. For example, if we are trying to produce the same amount of energy as is contained in the 21-trillion cubic feet of natural gas produced in 1971, we would require an electric generating capacity of about a million megawatts. The total United States generating capacity at the moment is only about 350,000 megawatts. During the past five years, it has only grown by 98,000 megawatts.

The formidable task of increasing the electrical generating capacity by a factor of 4, for example, is rather frightening, but we must remember that revolutionary changes of this magnitude will have to take place, in any case, to meet our burgeoning energy needs.

END USES OF HYDROGEN AS A FUEL

Let us now take a look at the end use of hydrogen as a fuel. Its first most significant difference from natural gas is that its heating value is only 325 BTU per cubic foot, about one-third of that of natural gas. This means that we will have to handle three times the volume of hydrogen to produce the same amount of heat. However, since the density and viscosity of hydrogen are so much lower, we can get about the same amount of energy through a pipe with the same pressure drop whether we use natural gas or hydrogen.

The next most important difference between the two fuels is the flame velocity of hydrogen, which is very much greater than that of natural gas. This means that burners designed to operate on natural gas will not operate with hydrogen and will have to be modified. The gas industry went through this modification once when it turned from manufactured gas to natural gas, and it will have to modify gas equipment again if we convert to hydrogen.

On the positive side, when the hydrogen is burned, it cannot possibly form carbon monoxide or carbon dioxide because it contains no carbon. The only reaction product possible is water vapor. Nitrogen oxides could be formed because the air going into the flame will be heated up, but even this can be suppressed if we can use low-temperature catalytic combustion. The development of catalytic burners for natural gas has been difficult, but it is a much easier problem for hydrogen.

Since the only reaction product of hydrogen combustion is water vapor, we should be able to operate many of our appliances without a flue. We might be able to get away from the concept of central heating where furnaces are restricted to one location because of the presence of only one chimney. A revolutionary change in heating and cooking could be brought about by catalytic combustion in unflued appliances. A foretaste of these new appliances was seen in the "Home for Tomorrow" exhibited by the American Gas Association and Institute of Gas Technology in 1969.

THE SAFETY FACTOR AND THE FUTURE

Perhaps the most important consideration of all is that of safety. Hydrogen is considered by many as being too dangerous for use by the man in the street. It only takes one disaster with hydrogen for people to remember how dangerous it can be. In 1936, the airship *Hindenburg* burned at Lakehurst, New Jersey, as a result of accidental ignition of vented hydrogen from the airship, which had just flown through a thunderstorm.

We have not been allowed to forget this disaster. The universal fear of hydrogen has been called the "*Hindenberg* syndrome." Indeed, hydrogen is a hazardous material and must be handled with all due precautions, but, as a result of its extensive use in industry and more recently in the aerospace area, we know how to handle it. More important, the gas industry operated for many years on manufactured gas or town gas, which contained as much as 50 per cent hydrogen. This was just as dangerous as the pure hydrogen that we are now considering.

The fear of hydrogen has been inflated unreasonably by such spectacles as the *Hindenburg* disaster. What does not get recorded, of course, is that the *Hindenburg* was designed to operate with helium, not hydrogen, and that, of 97 people on board, 62 escaped, probably because the hydrogen fire, which only lasted just over a minute, produced very little radiation energy, most of it going straight upward. The gas industry *will* handle hydrogen safely. We will make sure that hydrogen is safe to use before we would consider putting it into the hands of the public.

At the Institute of Gas Technology, we have an active program studying the problems that have to be faced to bring the "hydrogen economy" and a perpetual gas industry into being. This program is supported by the American Gas Association. In looking at long-range future projects like this, it is appropriate to dwell on some comments that were made by Alvin Tofler in his book, *Future Shock*. He says, "In dealing with the future, it is more important to be imaginative and insightful than to be 100 per cent right." He cites the example of early map makers, who were inaccurate and who produced maps which by today's standards are very poor. But without their maps, the New World would not have been discovered when it was. Tofler defines "future shock" as the "premature arrival of the future." If we are indeed on a collision course with tomorrow, the consideration of the "hydrogen economy" is at least one way to prepare for it.

SUGGESTED READINGS

Gregory, Derek, P., 1973, "The Hydrogen Economy," *Scientific American 228,* January.

Gregory, Derek, P., D. Y. C. Ng, and G. M. Long, 1972, "The Hydrogen Economy," in J. O' M. Bockris (ed.), "The Electrochemistry of Cleaner Environments," New York, Plenum Press.

Jones, Lawrence W., 1971, "Liquid Hydrogen as a Fuel for the Future," *Science 174:* 367–370.

Maugh II, Thomas H., 1972, "Hydrogen: Synthetic Fuel of the Future," *Science 178:* 849–852.

Stewart, W. F., and F. J. Edeskuty, 1974, "Alternate Fuels for Transportation, Part 2: Hydrogen for the Automobile," *Mech. Eng.,* June pp. 22–28.

Power from Trash

WILLIAM C. KASPER   1974

Ever larger amounts of urban refuse, instead of becoming an increasing waste disposal problem, could provide the United States with a low-cost and low-polluting fuel, particularly to run electric utilities. Methods both technically feasible and economically sound have been developed to use the potential heat in our solid waste. This article presents an overview of various processes that have been developed and emphasizes those projects that would most probably be operated by or in conjunction with an energy-supplying utility.

There are two basic methods to dispose of the solid waste collected in our urban areas. The simplest and most widely used method is to bury the solid waste in what is today euphemistically referred to as a sanitary land fill.

The other method involves the burning of refuse in an incinerator; recently emphasis has centered on developing incinerators which not only greatly reduce the physical volume of the solid waste, but also convert the heat produced by the process into an energy source such as steam. Other methods which recognize the heat potential of refuse include processes which convert solid waste into a liquid or gaseous fuel and processes which use refuse as a supplementary fuel in large utility boilers.

Another step may be added to some of these disposal methods to recover the non-combustible metals and glass in municipal refuse for future use. This recovery may take place before or after combustion; generally, recovery before the refuse has been burned is likely to result in materials with a higher value than those recovered from the residue of combusted refuse. It is even possible to recover and recycle most, if not all, of the material found in typical municipal refuse without any of it being used as a fuel.

The most prevalent material found in municipal refuse, accounting for more than one-half of the weight of a typical ton of solid waste, is paper in various forms, such as newspaper, cardboard, magazines, and brown paper. An analysis to determine the composition of typical municipal refuse made by the federal Environmental Protection Agency found that paper products ac-

William C. Kasper is principal economist, Office of Economic Research, New York State Public Service Commission.

From *Environment*, Vol. 16, No. 2, pp. 34–38, March 1974. Reprinted with light editing by permission of the author and the Scientists' Institute for Public Information. Copyright © 1974 by the Scientists' Institute for Public Information.

counted for 53 per cent of the weight; food wastes such as meat scraps and vegetable food waste accounted for 8 per cent; glass comprised another 8 per cent; ferrous and nonferrous metals made up 7 per cent; and the remaining 24 per cent consisted of grass clippings, rags, leather goods, and various other items.

Stated another way, the Environmental Protection Agency study found that 52.7 per cent of the solid waste consisted of volatile matter; 7.3 per cent was fixed carbon, 20 per cent was ash and metals; and the final 20 per cent was moisture. The combustible portion of the waste (volatile matter and fixed carbon), which made up 60 per cent of the total weight, had a heat content of 8,766 BTU per pound. This would mean that the overall heat content of a typical pound of refuse, including the moisture, ash, and metal fractions as well as the combustible material would be 5,260 BTU per pound or slightly over 10.5 million BTU per ton. This heat content is most nearly comparable to that of low-rank lignites which have a heat content of slightly less than 12 million BTU per ton. The refuse, however, has a lower moisture level (20 per cent) compared with the average of 50 per cent found in a low-rank lignite.

A desirable quality of solid waste is its low sulfur content, usually less than 0.12 per cent by weight. A heat content of 5,260 BTU per pound and a sulfur level of 0.12 per cent translates into 0.23 pound of sulfur per million BTU. This means that refuse used as a fuel in new or modified steam plants would meet the federal emission standards for these plants. These standards essentially call for the use of coal with a sulfur content of 0.6 pound per million BTU or the use of oil with a sulfur content of 0.4 pound per million BTU. On an equivalent heat basis, the burning of refuse would be equivalent to burning bituminous coal with a sulfur content of 0.3 per cent.

BURNING VERSUS RECYCLING

These "qualities" of refuse as determined by the Environmental Protection Agency study may be considered typical but not universal since the components in refuse vary from area to area and even from season to season within the same area. When refuse is used as a fuel, paper products, which make up 53 per cent of the weight, provide 71 per cent of the heat. Thus, burning solid waste physically destroys this paper and converts it to relatively useless ash; this ignores the possibility that paper, like the metals and glass found in refuse, might be recycled.

One point that deserves consideration is that the paper industry purchases almost 14,000 BTU of various forms of energy to produce a pound of paper of the type typically found in municipal refuse. However, this purchased energy amounts to only 64 per cent of the total energy used by the paper industry. This purchased energy plus energy produced from the industry's own process wastes and electric generating facilities result in a total energy input of 22,000 BTU per pound of paper. When the 7,500 BTU released by the burning of a pound of paper is balanced against the 14,000 BTU of energy purchased to produce it, an energy deficit of 6,500 BTU results. This means that when paper is burned, the energy released amounts to 54 per cent of that purchased for its production. When this same 7,500 BTU is balanced against the total energy input of 22,000 BTU per pound, the energy deficit amounts to 14,000 BTU per pound. Thus, when the paper found in municipal solid waste is burned, we get back only 34 per cent of the total energy used to produce it.

Considering that the paper industry in this country uses about 8 per cent of all the energy consumed by the industrial sector, and since the burning refuse for the most part means burning paper, we should determine if recycling the paper in refuse represents a better alternative. The recycling of paper is not an academic possibility, it is a large-scale reality, since at the present time wastepaper accounts for slightly over 20 per cent of the fiber consumed by the American paper industry. However, even this level represents a proportional decline from earlier years when wastepaper used in paper production reached a high of 36 per cent in the early 1940's.

In light of the current energy shortage, and the current paper shortage, it should be determined whether it would be advantageous to increase the amount of wastepaper that is recycled. A basic question is whether the energy required to increase the amount of paper being recycled would be less than that required to produce an equivalent amount of paper products from virgin materials. The environmental effects of both processes also need to be compared. Even if the weight of evidence indicates that the amount of paper recycled should be increased substantially, it is unlikely that this could be accomplished in the near future. The possibility that all the wastepaper produced in this country could be recycled also appears highly improbable. In light of the above, the development of processes to utilize the presently wasted potential heat from the solid waste produced in this country should be encouraged.

THE POTENTIAL

The urban population in the United States annually produces slightly over 136 million tons of refuse. The combustible portion of this refuse has a heat content of just over 1,433 trillion BTU. If used as a fuel for electric generation, this quantity of heat could produce 136 billion kilowatt-hours, which is about 10.9 per cent of the electric energy produced by conventional steam generating equipment in 1970. The heat content that could be supplied by this refuse is equivalent to that released by the burning of 54.7 million tons of low-sulfur coal, or 228 million barrels of low-sulfur residual fuel oil.

Total United States refuse production (excluding industrial and agricultural wastes) is probably about 185 million tons per year. However, the refuse produced by the 25 per cent of the population living in rural areas could probably not be collected economically enough to warrant its inclusion in estimates of the amount of refuse available for possible use as a fuel source.

The New York State 1970 urban population of 15.6 million would produce some 14.2 million tons of refuse per year.* This refuse has a heat potential of 150 trillion BTU, which could produce some 13 billion kilowatt-hours of electrical energy per year. This amount of electricity is 20 per cent of the electrical energy produced by conventional steam generating equipment in New York State in 1970. The heat available from New York State's refuse is equivalent to the amount that would be produced by burning 5.7 million tons of low-sulfur coal or 23.8 million barrels of low-sulfur residual fuel oil.

---

*The 1970 census indicated that the urban population of the United States and New York State was as follows:

United States   149,325,000
New York State   15,602,000

If we assume an average level of production for solid waste of 5 pounds per capita per day, the total amount of refuse produced by the above urban populations would be:
United States:

$$\frac{149,325,000 \times 5 \times 365}{2,000} = 136,259 \text{ million tons per year}$$

New York State:

$$\frac{15,602,000 \times 5 \times 365}{2,000} = 14,247 \text{ million tons per year}$$

Figures from the Environmental Protection Agency indicate that a typical pound of municipal solid waste has a heat content of 5,260 BTU. Using this figure, we can determine the heat content of the refuse annually produced by the urban populace of the United States and New York State:
United States $136.259 \times 10^6 \times 5.260 \times 10^3 \times 2 \times 10^3 = 1,433.4 \times 10^{12}$ BTU. New York State $14.237 \times 10^6 \times 5.260 \times 10^3 \times 2 \times 10^3 = 149.8 \times 10^{12}$ BTU.
Using the U.S. 1970 heat rate of 10,508 BTU per kilowatt-hour, we find that the 1,433.4 trillion BTU could produce 136.4 billion kwh and that this would amount to 10.9 percent of the 1,254.4 billion kwh produced by conventional steam generating equipment in the United States during 1970. The 1970 heat rate for New York State was 11,376 BTU per kilowatt-hour. This means that the 149.8 trillion BTU in New York State's urban refuse could have produced 13.2 billion kwh, which is 20.0 per cent of that produced by conventional steam generating equipment in New York State during 1970. The equivalent amounts of coal and residual fuel oil were determined by using 26.2 million BTU per ton of coal and 6,287 million BTU per barrel of residual fuel oil.

METHODS

The difference between using refuse as a fuel source and incineration of refuse is that in using refuse as a fuel, priority is given to the production of heat with the mass and volume reduction being secondary. The use of refuse as a fuel falls into two categories. In the first, direct burning of refuse results in the release of heat for the production of steam. This usually means that the refuse is used as a supplementary fuel source and provides only a portion of the total heat input. In the second, refuse is converted to a fuel, and this fuel is then combusted to produce heat.

In Europe, the use of solid waste as a fuel is fairly common. Germany, with more installations than any other European country, has more than twenty refuse-burning installations equipped with steam heat recovery systems. Presently, these plants utilize the refuse produced by some eleven million people. It is anticipated that by 1975 the number of people served will reach fourteen to fifteen million, or approximately 25 per cent of Germany's population at that time. In Milan, Italy, a second trash-burning power plant with an electrical output of 11.4 megawatts is expected to be operational late in 1974. This plant will handle some 660 tons of refuse per day in addition to that being combusted in the 7.3-megawatt plant which has been operating since 1968. Milan expects to have three more refuse-fired power plants of about 15 megawatts each operating by 1980.

In these European installations, refuse is generally fired in a separate boiler using a fossil fuel as a supplementary fuel source. These boilers supply only a portion of the steam requirements of the power plants they serve. It should also be noted that in Europe the disposal of refuse and the furnishing of electricity are usually the responsibility of the same governmental entity. This combination of responsibilities may have been a leading factor in determining that the burning of refuse by power plants represented a viable solution to two problems.

The most nearly comparable approach to the European systems is the project currently nearing completion in Nashville, Tennessee. Beginning in 1974, the Nashville Thermal Transfer Corporation, a nonprofit public corporation, will be burning solid waste to produce steam for heating and for driving the turbines that power large water chillers. The steam and chilled water produced will be piped to some 27 buildings in downtown Nashville. Initially, a daily output of some 720 tons of refuse produced in homes and commercial establishments in the mid-city area of Nashville will be used as a fuel. It is expected that the refuse processed will within a few years include the entire 1,300 tons per day produced in metropolitan Nashville.

The refuse will be delivered to the processing plant without charge to the processor. However, the city will pay the Nashville Thermal Transfer Corporation $150,000 per year, or about 60 cents per ton of refuse. Combustion of the refuse will result in a 95 per cent reduction in volume, thus considerably reducing the land-fill requirements for the area served. This is expected to save the city some $3 per ton of refuse collected or a total of $1.25 million annually. Also, the plant's central location is expected to substantially reduce hauling time and result in a savings of $72,000 per year in transportation costs.

Twelve state buildings, including the State Capitol, will be served as well as four buildings of the metropolitan government, eleven private structures and a church. Estimates indicate users will pay some 25 per cent less for purchased heating and cooling compared with operation of their own equipment. The state will save an estimated $164,000 each year for space conditioning its buildings in Nashville.

Probably the most publicized project in which refuse has been used as a fuel is that being carried out by the consulting engineering firm of Horner and Shifrin using the refuse of the city of St. Louis, Missouri at the Meramec plant of the Union Electric Company (see "Fuel from City Trash,"

Environment, September 1973). The solid waste processed consists of materials discarded by households and excludes bulky material, such as appliances or furniture, and commercial and industrial waste. The project was designed to allow 10 to 20 per cent of the pulverized coal burned at the Meramec plant to be replaced by refuse. The household refuse received at the processing station is milled to reduce particle size to a maximum of 1.5 inches. An electromagnetic separator then removes magnetic material from the milled refuse. From the processing plant, the milled refuse is trucked some 18 miles to the power plant. The refuse is fired in two boilers, each with a nominal rating of 125 megawatts and combustion capability of 56.5 tons of Illinois bituminous coal per hour at rated load.

Plans call for the addition of an air classifier at the processing plant to remove nonburnable and heavy particles such as nonmagnetic metals, glass, dense plastics, and heavy rubber material. It is anticipated that after the air classifier is installed the total weight of materials removed from the refuse will amount to about 20 per cent by weight of the refuse processed. The installation of the air classifier may make it possible to increase the percentage of processed solid waste as a proportion of the total fuel mix.

FUEL FROM REFUSE

Another approach employed to utilize the potential heat content of refuse involves the conversion of the solid waste to a fuel which can then be utilized by an electric utility or some other user. Section 208 of the National Resource Recovery Act provided grants to Baltimore and San Diego for the construction of large-scale demonstration projects which will convert refuse to fuel.

In Baltimore, a plant which will process 1,000 tons per day of solid waste is being built under a contract with Monsanto Enviro-Chem Systems, Incorporated. This plant will utilize Monsanto's "Langard" system which employs a pyrolysis process (chemical decomposition by heat) to convert most of the organic matter in the solid waste into a low-BTU gas. This gas will then be burned on site to produce steam which will be sold to Baltimore Gas and Electric for use in its steam distribution system.

The Baltimore plant will process only the solid waste produced by residential and commercial users, not by industries. The first step after the refuse is brought to the plant is to remove large bulky products such as appliances, auto engine blocks, and so on. The refuse is then shredded and fed into a storage bin from which it is in turn fed into a kiln in which the organic components of the refuse decompose into various gaseous products, carbon, and ash. The hot residue is then separated into carbon char; the magnetic materials are removed; the balance, composed mainly of nonferrous metals and glass, is referred to as glassy aggregate. The very fine carbon char that is produced could be converted to fuel briquettes; however, its usefulness would probably be limited owing to its 50 per cent ash content. Present plans are to mix the char with sewage sludge and use this mixture as a fertilizer. It is hoped that the ferrous metals received can be sold for scrap and that the glassy aggregate can be mixed with asphalt and used for street paving.

One limitation of this process is the generally low quality of the gas which is produced. The gas has a heat content of only 100 BTU per cubic foot. In addition, the chemical composition of the gas precludes its use as a gas turbine fuel. The production of a rather "dirty" fuel with a low heat content necessitates its use at a point close to the source of production. It would not be economical to transport such a fuel over long distances. There is, of course, the possibility that this gas could be upgraded, but at the present time Monsanto officials state they have no plans to do so.

The San Diego facility will use a pyrolysis process developed by the Garrett Research and Development Company, a subsidiary of Occidental Petroleum. The fuel produced by this process is a heavy oil-like liquid somewhat similar to No. 6 fuel oil. A gallon of

Garboil, which is the trade name for this liquid fuel, weighs about one-third more (32.6 per cent) than a gallon of No. 6 oil, but has only 77 per cent of the heat content. Thus a pound of Garboil has 58 per cent of the heat content of a pound of No. 6 oil.

It should be noted that while this "oil" can be used as a substitute for No. 6 fuel oil by electric utilities or for other applications where a heavy industrial heating oil is burned, it has different chemical properties than a petroleum-derived fuel and cannot be refined to produce gasoline, lubricating oils, and so on.

The Garrett process is similar to the Monsanto process in that both involve pyrolysis. However, the Garrett process requires a greater degree of refuse sorting since the system requires an input that is 95 per cent organic. Also, the refuse must be shredded to a rather fine one-eighth inch by one-eighth inch maximum size. In the Garrett process, most of the inorganics are removed before pyrolysis, whereas in the Monsanto process the inorganics are removed after pyrolysis. The glass recovered from the raw refuse using the Garrett process is claimed to have a purity of better than 99.7 per cent, which would make it readily usable for the production of new glass containers. The recovery of magnetic materials is readily accomplished, and since most of the non-magnetic metals and other inorganic materials are also removed during processing, it is hoped that eventually metals such as aluminum, copper, and brass will also be recovered. The final product produced by the pyrolytic process is char, but here again its usefulness is somewhat limited by its ash content, which is at least 40 per cent.

The Horner and Shifrin, Monsanto, and Garrett processes reviewed in this report have specific application to electric utility use. Pilot plants based on the Monsanto and Garrett processes have been built and operated. Information obtained from these pilot plants is now being used to develop design and operating criteria for the large scale demonstration plants being built. The Horner and Shifrin firm has been evaluating the results of the experience gained from the demonstration project currently underway in St. Louis and has extrapolated the experience gained to date to a possible application at the facilities of the Consolidated Edison Company in New York. Expectations are that in the near future an agreement will be worked out whereby Consolidated Edison will be burning some of the refuse produced in New York City.

OTHER PROCESSES

Another process which should be mentioned is that developed by the United States Bureau of Mines' College Park Metallurgy Research Center in Maryland. The research center is operating a rather sophisticated five-ton-per-hour pilot plant which separates raw refuse into a number of components. These clean components consist of chopped paper, chopped plastics, shredded light iron, heavy iron, heavy nonferrous metal, chopped aluminum, organic wastes, and green, colorless, or amber glass. The Black-Clawson Company has also developed a material recovery process, the end products of which include paper pulp, magnetic metals, glass, and aluminum. The advantage of these systems is that with the degree of separation achieved, the possibility of recycling and reusing the various components found in solid waste, including paper, is greatly enhanced.

The Bureau of Mines' Research Center has also developed a process to recover the iron, nonferrous metals, and glass from incinerator residue. A plant designed to process 250 tons of incinerator residue per day is presently being built in Lowell, Massachusetts, with funds provided by the federal government through the Office of Solid Waste Management. Obviously, processing incinerator residue means that only that fraction which contains recoverable products (listed previously in connection with other systems) is processed. This means that less than 20 per cent of the material found in refuse is used.

EVALUATING THE PROJECTS

This article has outlined various processes that have been developed to provide a new approach to the problem of disposing of the constantly increasing quantities of solid waste being produced by our society. There are other processes undergoing development that, at some time in the future, may prove to be of greater significance than those that have been reviewed. At the present time, however, none has reached the level of development of those covered in this report.

One step in the evaluation of the processes discussed in this report is the comparison of the costs associated with each. Except for the Horner and Shifrin process, the cost estimates which are available are based on extrapolation of experience gained from somewhat limited pilot plant operation. The cost estimates are, of course, just that and only after full-scale operation will we have reliable information which will allow reasonable cost comparisons to be made. (See Table 4-34.)

The main emphasis of this paper has been

Table 4-34. Estimated Costs per Ton of Solid Waste For Refuse Processing Systems (in dollars).

Process	Total Cost per ton[a]	Revenue per Ton From Recovered Products[b]	Net Cost per Ton
Monsanto (pyrolysis)	11.00	4.67	6.33
Garret (pyrolysis)	7.35	6.10	1.25
Bureau of Mines (refuse separation)	4.53	3.44	1.09
Horner and Shifrin (estimates for burning trash as fuel for electricity generation, New York City)	6.93	0	6.93

[a]These cost estimates include both operating and capital recovery costs. The capital costs were amortized over a 10-year period using an interest rate of 5 per cent per year. All processes were assumed to be government owned and thus no local taxes are included in the operating costs. The Horner and Shifrin costs are based on an operating rate of 1,170 tons per day. If the full capability of the system (1,750 tons per day) were used to compute the costs, the cost per ton would be $4.62.
[b]Revenue per ton includes revenue received from the sale of recovered materials including the value of the fuels produced. The Horner and Shifrin process is expected to provide at no cost to Consolidated Edison a fuel with a value of $3.70 per ton of solid waste processed. However, offsetting this are the operating and capital costs incurred by Consolidated Edison which would be associated with the burning of processed solid waste. The Bureau of Mines' analysis used a value of $1.23 per ton for the shredded paper and other combustibles produced, which translated into a value of $0.65 per ton of solid waste. If the shredded paper produced could by recycled, it would have a value of at least $5.00 to $10.00 per ton. Assuming a paper content of 55 per cent per ton of solid and a value of $5.00 per ton for wastepaper would mean a value of $2.65 for the paper found in a ton of refuse. This would increase the value of the revenue per ton for the Bureau of Mines' process to $5.44 and result in a net cost of −$0.91 per ton of refuse.

to describe and compare those processes which would most likely be operated by or in conjunction with an energy supplying utility. Of the three projects which have this type of utility as an active participant, namely those referred to as the Horner and Shifrin, the Monsanto, and the Garrett process, it is the author's opinion that the Garrett process could prove to be the most efficacious because of the final product produced. Garboil, the oil-like liquid fuel, which is the main output of this process, can be produced, stored, and transported independently of the needs of the final user. Because of this, the Garrett process does not require the close interdependence between the producer and the user that is necessitated by the Monsanto and Horner and Shifrin processes. Also, the use of Garboil would probably not require major changes in utility type boilers presently using residual fuel oil.

Any final conclusion as to which process will be the most viable solution to the twin problems of disposing of increasing amounts of refuse and of providing a low-cost, low-polluting fuel must await the construction and operation of large-scale demonstration facilities. After these facilities become operational, we should be able to determine if any one of them will prove to be significantly more economical and reliable than the others.

If the cost estimates indicated in Table 4-34 prove to be reasonably accurate, the use of solid waste as a fuel might become questionable since the Bureau of Mines' process for the recovery of the materials found in solid waste would seemingly render the other processes economically unjustifiable. However, due to the time lag between demonstrated technical feasibility and large-scale implementation, it does not seem likely that in the near future any significant quantity of the nation's refuse will be recycled by using the Bureau of Mines' process. It is also possible that all the paper in the nation's refuse could not be recycled. This will require the use of other methods if we are going to utilize the presently wasted assets present in our garbage.

Implementation of methods which use the potential heat in solid waste will undoubtedly require financing at some governmental level. An official of the Environmental Protection Agency stated that a "good deal of the solutions should be addressed at the state and local levels." This point of view was reflected by the reduction of the Environmental Protection Agency request for solid waste control programs from $30 million in the 1973 fiscal year to $5.8 million in the 1974 fiscal year. The author is more inclined to agree with the position recently adopted by the National League of Cities that "solid waste management transcends local jurisdictions and funding capabilities and that satisfactory progress toward adequate improvements will require continued federal financial support."

SUGGESTED READINGS

Bohn, Hinrich, L., 1971, "A Clean New Gas," *Environment 13:* 4–9.

Gibney, Lena, C., 1974, "Liabilities into Assets," *Environmental Sci. and Tech. 8:* 210–211, March.

Lowe, Robert A., 1973, "Energy Recovery from Waste–Solid Waste as a Supplementary Fuel in Power Plant Boilers," U.S. Environmental Protection Agency, 24 p.

Maugh II, Thomas H., 1972, "Fuel from Wastes: A Minor Energy Source," *Science 178:* 599–602.

Wilcox, Denny, 1973, "Fuel from City Trash," *Environment 15:* 36–42, September.

Wisely, F. E., 1972, "The Rationale of Solid Waste Disposal by the Recovery of Energy," Solid Waste Disposal Seminar, Union Electric Co., St. Louis, Missouri, October 26–27, 1972.

5

ENERGY POLICY RECOMMENDATIONS AND FUTURE DIRECTIONS

Location of underground natural gas lines. Such pipelines will be able to carry as much energy, much more economically, than the unsightly, easily damaged overhead lines. (Photo courtesy of D. P. Gregory, Institute of Gas Technology.)

INTRODUCTION

We have surveyed the general problems of energy management, the dilemmas posed by available fossil and nuclear fuels, and the nature and promises of various energy alternatives. With this background on the boundary conditions within which we must operate, let us examine some of the recommendations being put forward to resolve our energy dilemmas and consider how these will shape the world of the future.

The theme running through several of the analyses presented here is that energy flow in a society is a very complex, multiply connected subject requiring a systems-analysis approach rather than piecemeal, "Band-Aid" solutions. In fact, the tendency, particularly on the part of governments, to turn to *ad hoc* solutions rather than global, comprehensive policies of energy management is the source of many of the dislocations we have experienced recently.

A prime example of this *ad hoc* approach to problem solution is that of the automobile industry. We have, instead of an energy-saving, comprehensive coordinated plan incorporating mass transit, a stop gap—the catalytic muffler—a "Band-Aid" for the internal-combustion engine. This problem and the direction that development of automobile power may take in the future is examined by Janice Crossland in *Cars, Fuel, and Pollution*. Since the private automobile is the single largest end user of energy (over twice as large as the primary metal industry, which is the second largest end user), such developments will have a significant impact on the energy situation.

The problems of the present internal combustion engine are described and alternative engine designs are discussed. As is so often the case, environmentally acceptable designs usually result in much higher fuel consumption, though such accessories as air conditioners also reduce fuel economy. Certain designs, namely the diesel and external-combustion engines, do show promise for meeting pollution standards while maintaining reasonable efficiencies.

In *Efficiency of Energy Use in the United States,* Eric Hirst and John Moyers examine three specific areas (transportation, space heating, and air conditioning) and demonstrate how relatively minor changes in our life styles would result in significant savings in the energy budget. The calculations presented indicate that a not unreasonable shift from car and plane transporation to bus and train transportation would result in a saving of at least 4 per cent of the total energy expenditure. Optimum insulation of all private homes would result in roughly the same energy savings. If all room air conditioners ran at the presently available efficiency of 10 BTU per watt-hour rather than the present average 6 BTU per watt-hour, we would save 1,500 acres per year from the strip mining of coal. Suggestions are presented on how we might encourage such savings.

G. A. Lincoln looks at many of these same issues in a somewhat broader social perspective in *Energy Conservation.* The question of how to attain desired social and economic objectives while using less energy is addressed directly. The trend toward increased speed and convenience in transportation, for instance, is shown to result directly in increased energy consumption. How do we develop social incentives to reverse such trends? What are the ethical implications of such attempts at social engineering? If we do not develop such social incentives, will the inevitable law of supply and demand do it for us? These and several other intriguing social issues related to conservation are tackled in this article.

The historical background of many of our present energy policies is put into better perspective by John Steinhart in his article, *The Impact of Technical Advice on the Choice for Nuclear Power.* The development of the federal scientific advisory apparatus with its preponderance of nuclear physicists and engineers is traced from World War II and show to have had a significant effect on the development of energy policy. Though no scapegoats are blamed for past energy decisions, one cannot help but wonder how much different the history of energy development might have been had the advice come primarily from biologists, chemists, or geologists.

From his first-hand experience in directing Wisconsin's Office of Emergency Assistance, Stanley York presents a penetrating analysis, in *Role of the Government in Resolving Energy Problems.* He sees continuing dislocations caused by supply and distributions problems and highlights the need for a broad and coordinated set of national energy policies. The crucial role of uniform information collection and public understanding of such information is emphasized. A "Division of Thinking and Planning" is advocated to deal with policy issues and people problems associated with energy dislocations. Many such problems are indicated and positive recommendations made for dealing with them.

Glenn T. Seaborg presents a very hopeful perspective on what the world of 1994 may look like in his article *The Recycle Society of Tomorrow.* He foresees a highly disciplined future society in which cooperation, conservation, and ingenuity are the dominant values. The "use and discard" ethic will have given away to the "conserve and recycle" ethic, with virgin raw materials used as make-up material to replace losses in the recycle process. Energy development will have progressed in all of the areas previously discussed, with communications relieving the burdens on energy resources that transportation had imposed. Certainly, this steady-state world, even with its tightly controlled social and physical environment, is vastly preferable to the grim social collapse predicted by less optimistic world models. The greatest challenge to our society in facing the limitations of a finite world is to resolve the profound ethical dilemmas involved in such social decisions with compassion and wisdom.

Cars, Fuel, and Pollution

JANICE CROSSLAND   1974

Mass production of a nonpolluting and energy-efficient car is unlikely in the near future without major new incentives or requirements for the automobile industry. Federal laws have increased the pressure to produce clean cars, but so far practical results have been few. Not only do cars produce toxic exhaust gases, but the gas-guzzling engines and increased weight of cars marketed by Detroit in recent years have greatly increased the fuel consumed by automobiles. The major barrier to improvement has been the commitment of United States automobile makers to what is basically an unchanging technology, the conventional internal-combustion engine (see "Looking Forward," *Environment*, May 1973). Itself resistant to change, the automobile industry has been notorious for carefully acquiring patent rights to innovations that might improve emissions controls or performance and that might permit competitive challenge to Detroit's chosen technology.

Clearing the air of our cities and saving energy may require a determined effort to improve public, as opposed to private, transportation to meet the needs of the public as a whole. As to cleaning up the automobile itself, two pollution-control strategies are now being tested. Using the present-day engine as a start, the first approach has been to add devices onto the tail pipe in an attempt to treat pollutants after they have been formed. The alternative approach is to develop an inherently less-polluting engine. Though the automobile manufacturers have decided on the first approach, the latter means of achieving low emissions has not been overlooked. Advances in the development of a number of alternative engine designs were discussed at a recent conference, the First Symposium on Low Pollution Power Systems Development, held in Ann Arbor, Michigan, in October 1973. The United States Environmental Protection Agency was host to official government and industry representatives from England, France, Sweden, the Netherlands, Germany, and Italy. Voicing a common concern for cleaner air and disillusionment with modifications which can be made on today's internal-combustion engine, participants centered

Janice Crossland is a research associate with *Environment*, St. Louis, Missouri.

From *Environment*, Vol 16, No. 2, pp. 15–20, 25–27, March 1974. Reprinted by permission of the author and the Scientists' Institute for Public Information. Copyright © 1974 by Scientists' Institute for Public Information.

on alternative automobile engines presently under development by manufacturers, in some cases under the sponsorship of governmental agencies. No radically new designs were proposed, but progress was reported in a number of areas.

The impetus for development of clean engines in the United States comes from 1970 amendments to the Clean Air Act. In 1970, the Environmental Protection Agency set up an Advanced Automotive Power Systems Program (now the Alternative Automotive Power Systems Program) to demonstrate, by 1975, a practical, efficient alternative to the spark-ignited, reciprocating internal-combustion engine. The major aim of the program is reduced exhaust emissions, and the Environmental Protection Agency has set as its goal test engines which produce only one-half of the pollutant levels allowed for 1976. The need for performance and fuel economy at least comparable to present day engines is also being taken into consideration.

Talks at the October conference described progress in this effort, which is aimed at the study of only a few engine designs. At the outset of the program, in 1970, two hybrid engines, the heat engine/battery and the heat engine/fly wheel, as well as battery-powered electric systems, were eliminated from consideration. The fly wheel car, for example, would have lower emissions and better fuel economy than today's car because of the ability of the fly wheel to store energy, making a lower energy output necessary by the engine. (See *Environment*, "The Windup Car," June 1970, and "A Lift for the Auto," December 1972.) According to the Environmental Protection Agency, however, hybrid systems would need exhaust after-treatment devices, would be complex, and would be costly to produce. Battery-powered autos were eliminated because, according to the Environmental Protection Agency, the battery would need more than the five-year development time set as the goal of the Alternative Automotive Power Systems Program.

Selected for development in the United

States program were versions of the Rankine engine (an external-combustion engine) and the gas turbine; the stratified charge engine, previously under study in a Department of Defense and Army program; and the diesel engine. Since the program began in 1970, good fuel economy for all alternative engines has become a high-priority goal.

Other systems, developed abroad, such as Honda's version of the stratified charge engine and the Stirling external-combustion engine, can be added to the list of possibilities. Each of these possibilities was discussed at the October symposium, which was the first to be held under an agreement signed by several members of the NATO Committee on the Challenges of Modern Society to exchange information on the technical aspects of development of a low-polluting engine. The following summary of clean engine technology draws on information from this symposium and a number of other sources.

INTERNAL COMBUSTION

Any effort to develop a less-polluting conventional gasoline engine means finding a way to reduce the toxic substances produced in the combustion process. When fuel is burned completely, only harmless carbon dioxide and water are products of the reaction, but this requires a sufficient supply of air. Better performance is achieved at an air-to-fuel ratio of approximately 15 to 1, which is not high enough to burn all the fuel. Carbon monoxide and incompletely burned fuel (hydrocarbons) result.

Nitrogen oxides, the third major auto pollutant, form when oxygen and nitrogen in the air combine under conditions of high temperature and pressure. The more oxygen present (the leaner air-to-fuel ratio), the greater the amount of nitrogen oxides produced. Lean mixtures therefore favor thorough combustion but are responsible for the production of greater amounts of nitrogen oxides. (Higher temperatures also favor increased nitrogen oxide production.) This complicates emissions control technology.

The internal-combustion engine is an inherently dirty machine, and its pollution-producing capacity, as well as its capacity to consume fuel, has increased over the years. The automobile was first powered by low-speed, low-compression engines. Engine size increased gradually, and between 1954 and 1959 average horsepower doubled. Increases in engine size led to increases in the quantity of exhausts. Compression-ratio increases meant higher temperatures and increased production of nitrogen oxides. And, to obtain more power, a richer air-to-fuel ratio (11 or 12 to 1 instead of 15 to 1) was needed, resulting in increased quantities of unburned hydrocarbons and carbon monoxide. (See "The Tailpipe Problem," *Environment,* June 1970.)

In the late 1960's, when it became clear that engines would have to be altered to reduce pollution, Detroit's initial effort was to make a number of relatively minor modifications. One of these was simply to reverse the trend of equipping everyone's automobile with a racing car engine. Thus, since 1970 decreased compression ratios and leaner air-to-fuel mixtures have become standard. Other modifications include the addition of positive crankcase ventilation valves to reduce hydrocarbon emissions by routing oil vapors back into the engine rather than allowing them to escape to the atmosphere from the crankcase. These have been standard on most cars since 1968. More recently, many 1973 cars have been equipped with an exhaust gas recirculation device. As the name implies, a part of the exhaust gas is recycled back into the engine, reducing the combustion temperature and as a result reducing formation of nitrogen oxides. Taken together, these and a number of other modifications have reduced emissions sufficiently to meet 1973 standards.

Meeting the standards set for 1975 and 1976 in the clean air law of 1970 has been another matter. To achieve these lower emissions levels, Detroit turned to a variety of tack-on devices. Major emphasis was placed on two types of catalytic converters: a nitrogen oxide reduction converter and an oxida-tion converter to help eliminate carbon monoxide and hydrocarbons. The unreliability of this approach was an important factor in the government's decision to suspend the goals set in the 1970 law for at least one year.

Both types of converters help to change nitrogen oxides, carbon monoxide, and hydrocarbons into harmless substances. The nitrogen oxide converter is attached to the tail pipe first. This device consists of an inert (for example, ceramic) material coated with an active substance, such as platinum. Next an oxidation catalyst, which oxidizes hydrocarbons and carbon monoxide to carbon dioxide and water, may be added to the tail pipe. It is similar to the nitrogen oxide converter in that it also consists of an inert base coated with a thin layer of active substance, such as platinum. More work has been done so far on the oxidation catalyst as this device is needed to meet the original 1975 emissions standards. The nitrogen oxide catalyst would be needed to meet 1976 standards. (Both standards have now been suspended.)

Will these devices do the job? An extensive report, issued by the National Academy of Sciences Committee on Motor Vehicle Emissions in February 1973, outlined numerous problems. This committee was formed as required by the 1970 Clean Air Amendments to carry out studies, under contract with the Environmental Protection Agency, on the technical feasibility of meeting auto emissions standards. A major portion of the report was devoted to various pollution control devices which were at the time considered by the automobile manufacturers.

The committee reported that add-on devices depend on optimum conditions for good emissions control. According to the report, catalysts have worked in the chemical and petroleum industries because they are operating under "steady-state" conditions. An automobile catalyst, however, is subject to a variety of conditions, such as constantly changing temperature, gas composition, and rate of gas flow. For instance, catalytic converters for hydrocarbon and carbon monoxide reduction may not be effective when the engine is cold; on the other hand, overheating the engine (for example, when going uphill) contributes to the deterioration of the catalyst surface.

The latter factor, deterioration, is a special problem. Tests on cars with dual catalyst systems (which reduce nitrogen oxides by one catalyst and hydrocarbons and carbon monoxide by another) show that while emissions levels may be within 1976 limits initially, in most cases these limits are exceeded before the car has accumulated 10,000 miles.[1] Environmental Protection Agency regulations state that emissions control systems must be effective for 50,000 miles, or for five years.

In addition to mechanical difficulties, catalytic converters are themselves potential sources of pollution. One experimenter has found that metal-containing particles are formed when automobile exhaust gases are passed over catalysts that contain nickel, chromium, and copper. These particles are small enough to be taken into the lungs and are therefore a potential health hazard.[2] The Environmental Protection Agency workers have also learned that a sulfuric acid mist is produced by some catalysts intended for installation on many 1975 cars. The level to which pedestrians would be exposed along the roadside has been measured by different investigators and has varied from 5 to 150 micrograms per cubic meter. It is thought that exposure to 8 to 10 micrograms per cubic meter for 24 hours could adversely effect asthmatics and the elderly.[3] The source of the sulfuric acid is sulfur which occurs naturally in oil.

Further criticism of the catalytic converters stems from indications that these devices will cause a decline, or penalty, in

fuel economy as determined by increased fuel consumption. There have been various estimates as to the extent of the reduction in fuel economy due to pollution control devices which have been placed on cars so far. According to one estimate, based on Environmental Protection Agency test figures, the average reduction in miles per gallon due to emissions control devices in 1972 cars, on the basis of the average fuel consumption for cars in the pre-control 1957 to 1967 period, is 5.3 per cent; for 1973 cars, this fuel penalty is 6.6 per cent. Since a number of factors can contribute to loss of fuel economy it is difficult to pin down precisely the contribution of emissions control devices. Vehicle weight, however, is the major factor in determining fuel economy.

The controversy surrounding the Clean Air Act has thus centered on the effectiveness of catalytic converters. According to 1970 amendments to the Clean Air Act (Table 5-1), hydrocarbons and carbon monoxide levels must be reduced 90 per cent from their 1970 levels by 1975; nitrogen oxides must be reduced 90 per cent from their 1971 levels by 1976. Some leeway was written into the amendments that allows this timetable to be set back one year if Environmental Protection Agency officials feel that there is sufficient reason to do so. That is, the company applying for a suspension of the standards must prove that although a "good faith effort" to meet the standards has been made, effective technology is not available. The act also stipulates that the decision of the Environmental Protection Agency must be made on the basis of studies carried out by the National Academy of Sciences.

Under these conditions, both the 1975 and 1976 deadlines were extended for one year, and interim standards were set. (See Table 5-1.) Thus, original 1975 standards will not have to be met until 1976; original 1976 standards will not have to be met until 1977. The 1975 interim standards make oxidation catalysts mandatory for cars sold in California but not in the rest of the nation. As of March 1974, a further extension of

1. *Report of the Committee on Motor Vehicle Emissions*, National Academy of Sciences, Feb. 12, 1973, Table 3-4, pp. 39–41.
2. Balgord, William D., "Fine Particles Produced From Automotive Emissions—Control Catalysts," *Science, 180:* 1168–1169, June 15, 1973.
3. Shapley, Deborah, "Auto Pollution: EPA Worrying That the Catalyst May Backfire," *Science, 182:* 368–371, Oct. 26, 1973.

Table 5-1. Automobile Exhaust Emissions Standards (grams per mile).

	1972	1973	1974	1975 Original	1975 Interim	1976 Original	1976 Interim	1977
Hydro-carbons								
federal	3.4	3.4	3.4	0.41	1.5	0.41	0.41	0.41
California	3.4	3.4	3.4	1.0	0.9	0.41		
Carbon monoxide								
federal	39.0	39.0	39.0	3.4	15.0	3.4	3.4	3.4
California	39.0	39.0	39.0	24.0	9.0	3.4		
Nitrogen oxides								
federal	–	3.0	3.0	3.0	3.1	0.4	2.0	0.4
California	3.2	3.2	2.0	1.5	2.0	0.4		

auto emission standards has been proposed. The Senate has passed amendments to the Clean Air Act that would delay for one more year the requirements for meeting 1975 standards. If this action becomes law, original 1975 standards will not have to be met until 1977.[4]

Having thus committed ourselves to cleaning up the internal-combustion engine rather than to developing new technologies, we have been told by the automobile manufacturers that there has not been enough time to work out all of the technical problems presented by pollution control devices. However, several government reports, which anticipated the 1970 Clean Air Amendments, question industry's commitment to clean autos. A National Academy of Sciences report in 1966, for instance, stated that: " . . . attempts have been made for ten years or more to persuade the auto industry to devise exhaust control devices. Results have been slow in coming."

And there was pressure at the time for developing a nonpolluting, alternative engine. Said a 1965 report by the President's Science Advisory Committee[5]: "The pollution from internal-combustion engines is so serious, and is growing so fast, that an alter-

native nonpolluting means of powering automobiles, buses and trucks is likely to become a national necessity."

Nearly ten years later we still have not cleaned up the air or produced a viable alternative to today's car. The only important engine change planned in recent years, development of the Wankel rotary internal-combustion engine, actually represents a step backwards. Despite all of this, however, some progress has been made in research on alternative engine designs. These engines are discussed below and can be divided roughly into two categories: internal-combustion engines, which include the rotary, stratified charge, diesel, and gas turbine engines; external-combustion engines, which include Rankine and Stirling cycle engines. Reports on each of these engines were given at the October 1973 Low Pollution Power Systems Development symposium from which much of the following information was gathered.[6]

ROTARY POWER

General Motors not only expects to market a rotary engine soon, but predicts that it will

4. "Senate Breeches Auto Emissions Standards," National News Report, Capital Summary, Vol. 5, no. 49, Dec. 21, 1973.
5. Macinko, John, "The Tailpipe Problem," Environment, 12(5): 7, June 1970.

6. Some of the background information on alternate engines is also taken from Ayres, Robert U., and Richard P. McKenna, Alternatives to the Internal Combustion Engine: Impacts on Environmental Quality, Johns Hopkins University Press, Baltimore, Maryland, 1972; and Report of the Committee on Motor Vehicle Emissions, loc. cit.

make up a substantial part of the car market by 1980. Because the engine has a rotating, rather than reciprocating piston arrangement, it is a more well-balanced, smoother running engine and should be more durable in the long run. It is also smaller and weighs less than a reciprocating engine with the same power output. General Motors originally planned to introduce the Wankel on some cars in 1974, but its poor fuel economy has led to postponement of these plans.

Though the rotary engine is expected to provide good performance, and may be a boon to race car enthusiasts, it has little advantage for pollution control or for saving fuel. Compared to the conventional internal-combustion engine, the Wankel rotary engine produces two to five times the level of hydrocarbons, one to three times the level of carbon monoxide, and about 30 per cent less gas mileage. Nitrogen oxides, on the other hand, are 25 to 75 per cent lower, in part because of a more rapid cooling of gases as they expand through the piston chamber. Data show that 1975 emissions standards can be met only with the aid of pollution control devices. It is thought that it might be possible to adapt the stratified charge concept to the rotary engine to reduce emissions, but this idea has not yet been tested. A low-emission rotary engine, thus, has all of the problems inherent in the control strategy now being employed for conventional internal-combusion engines and would create an even greater demand for fuel.

STRATIFIED CHARGE

Perhaps one of the most promising of the various alternatives, particularly for the immediate future, is a modification of the internal-combustion engine, which effectively solves the problem of finding the air-to-fuel ratio that will reduce hydrocarbons and carbon monoxides and not, at the same time, increase the nitrogen oxides. In ordinary engines, the air-to-fuel ratio remains the same throughout the entire combustion cycle while in the stratified engine the air-to-fuel ratio is varied during different parts

of the cycle. In one type of stratified charge engine, for example, the area around the spark plug has a rich air-to-fuel mixture, whereas a lean mixture elsewhere in the combustion chamber enables any leftover unburned hydrocarbons or carbon-monoxide to be burned. The concept of the stratified charge has been known since the 1920's.

What makes the stratified charge engine a promising solution is not only its potential for low emissions, but its good fuel economy and performance. And, because the stratified charge engine is a modification of the basic internal-combustion engine, its production would not require special materials or extensive retooling on Detroit's part. The 1973 National Academy of Sciences study, cited earlier, compared in some detail the advantages and disadvantages of stratified charge engines and the conventional internal-combustion engine with catalytic converters and concluded that:

> The system most likely to be available in 1976 in the greatest numbers—the dual-catalyst system—is the most disadvantageous with respect to first cost, fuel economy, maintainability, and durability. On the other hand, the most promising system—the carburated stratified-charge engine—which may not be available in very large numbers in 1976, is superior in all these categories.

The version which has received the most publicity in recent months, and which is the most highly developed, is the compound vortex controlled–combustion engine produced by Honda of Japan. Preliminary tests, in 1971, indicated that the Honda engine might be able to meet original 1975 clean air standards. These early predictions have been borne out in subsequent tests on the Honda Civic (a small Japanese economy car), the Chevrolet Vega, and the Chevrolet Impala. The Civic was shown to meet 1975 standards in December 1972, and in October 1973, the Environmental Protection Agency reported that a 1973 Chevrolet Impala, modified with a compound vortex controlled-combustion engine, met the 1976 interim standards for nitrogen oxides and original 1976 standards for hydrocarbons and carbon monoxide. The tests on the Impala were carried out on a car which had accumulated 3,000 miles; further tests will be needed to assure that emissions remain low. However, in tests on the Honda Civic, which have run up to 50,000 miles, little deterioration in emissions control was noted.

Unlike the Honda engine, which requires no after-treatment devices, a stratified-charge engine developed in an Army and Defense Department program can meet air pollution standards only with the help of an exhaust gas recirculation device and oxidation catalysts. The original purpose of the Army program, however, was to increase the fuel economy of light Army vehicles (such as jeeps). Before the addition of emissions-control devices, they had succeeded in showing a 30 per cent fuel economy gain over the conventional internal-combustion engine; with addition of controls, which reduce nitrogen oxides to 0.33 gram per mile, fuel economy is comparable to conventional engines without pollution control devices. When the Army program began in 1963, 40 different systems were reviewed as possible candidates. Versions of the stratified-charge developed by Texaco and by Ford have received the most attention by the Army and the Environmental Protection Agency. By 1978, the Army hopes to produce 10,000 stratified-charge vehicles: none is planned by Detroit.

RANKINE ENGINE

The steam engine, one type of Rankine engine, was a major competitor of the internal-combustion engine during the early years of the automobile industry. It is the only engine which thus far meets original 1976 emissions standards. In a Rankine engine, a fluid is alternately boiled and condensed in a closed cycle; power is provided to turbine blades or pistons when the fluid is in the vapor stage. Heat to boil the fluid is provided by an external flame. While water and steam provided the power in the early forms of the Rankine engine, other fluids may be used. The Environmental Protection Agency's Alternative Automotive Power Systems program presently has under development four versions of the basic Rankine principle: a steam piston engine, being developed by Scientific Energy Systems Corporation of Massachusetts; a steam turbine, under development by Lear Motors of Nevada; an organic fluid piston engine, developed by Thermo Electron Corporation of Massachusetts; and an organic fluid turbine developed by Aerojet Liquid Rocket Company of California.

The reason for interest in the steam engine is reflected in recent Environmental Protection Agency data on emissions levels for all four Environmental Protection Agency development versions. Table 5-2 shows

Table 5-2. Preliminary Emissions Data for Rankine Cycle Engines.

Pollutant	1976 Federal Standards	Aerojet	Lear	Scientific Energy Systems	Thermo-Electron Corporation
Nitrogen oxides	0.40	0.149	0.38	0.18	0.275
Hydrocarbons	0.41	0.053	0.21	0.18	0.17
Carbon monoxide	3.40	0.317	0.70	0.43	0.21

that the emissions levels of Rankine systems are well below the original 1976 federal standards. (It should be noted that these figures are from laboratory tests on engines, not vehicle road tests.) A particular advantage of the Rankine cycle is that combustion is continuous and takes place at relatively low temperatures and pressures, effectively reducing the quantity of nitrogen oxides produced.

Considerable testing has so far been carried out on various components of the four engines, but testing in an automobile has not yet been done. The Environmental Protection Agency planned, in late 1973, to select one of the four it considers most promising for tests in a vehicle, and a working model is expected to be ready by late 1975. This timetable, of course, means that any serious thought of mass production of a Rankine engine, based on research by the Environmental Protection Agency, is many years in the future.

One of the drawbacks that has always been cited for the Rankine engine is its poor fuel economy. According to the Environmental Protection Agency, however, this is no longer a valid criticism. Their engines have approximately the same fuel economy as the conventional internal-combustion engine at low speed, but do not do as well at higher speeds (above 45 miles per hour). The Environmental Protection Agency predicts, however, that with planned modifications, the Rankine engine will show better fuel economy than the internal-combustion engine. Since 1976 emissions standards have already been met with the Rankine engine, a large portion of the remainder of the Environmental Protection Agency program will be devoted to this task.

The Environmental Protection Agency also is concerned about the potential toxicity of organic liquids which are used in two of their test designs. A number of possible fluids have been rejected because of their high flammability or toxicity. A fluorinated alcohol (trifluoroethanol) now used in one of its engines, is being tested by the Environmental Protection Agency for possible chronic toxicity effects should leakage of the fluid occur.

Another serious effort to produce a working model of a steam-powered car within the next few years is being undertaken by two California firms under the sponsorship of the California State Legislature, which previously sponsored the California Steam Bus Project. The purpose of the steam bus project (financed by the U.S. Department of Transportation) was to test the feasibility and public acceptance of a new type of propulsion system. The project succeeded in identifying many of the technical problems in steam engine development and helped to educate the public on the need for alternative engines. The steam buses met 1975 California heavy-duty vehicle standards but had poor fuel economy. The purpose of the program, however, was not to produce a production prototype, and authors of the project's final report concluded that approximately ten more years of work were necessary before the steam buses tested could be mass produced.[7]

A report on a new undertaking, the California Clean Car project, was given at the October symposium. The purpose of the project, begun in November 1972, is to build and test cars equipped with a steam piston engine and a steam turbine engine. Unlike the Environmental Protection Agency program, in which engines are being built for relatively large cars, the California program will produce engines for subcompact cars which will be suitable for urban driving. Thus, the maximum speed for one model will be 75 miles per hour on flat terrain, and 50 miles per hour on a 5 per cent grade. Cars, outfitted with the steam engine, were scheduled for testing in spring 1974.

GAS TURBINES

A feasible gas turbine engine for cars has not yet been developed, although the idea of

7. Lane, James A., Kerry Napuk, and Roy A. Renner, *California Steam Bus Project: Final Report*, The Assembly Office of Research, California Legislature, Sacramento, Jan. 1973.

putting such an engine into an auto is not new. Chrysler several years ago built a few turbine cars which were given much publicity and were entered in auto races, but they were never put on sale. Most research by the automobile companies has been directed towards putting gas turbines into buses, trucks, and construction vehicles.

The gas turbine, which is essentially an internal-combustion engine in which combustion gases drive turbine blades rather than pistons, has a number of advantages. The gas turbine is mechanically simpler than a piston internal-combustion engine and therefore is expected to need less maintenance; it is also smoother running and thus is expected to have a longer life. Additionally, the gas turbine can be run on lower-grade fuels. From the standpoint of emissions controls, the gas turbine has the advantage of producing lower levels of hydrocarbons and carbon monoxide. This is due to the fact that large amounts of air are injected into the compressor (lean air-to-fuel mixture); combustion is continuous and takes place at high temperatures. (In the piston internal-combustion engine, combustion is intermittent, and some unburned products are left on the walls of the cylinder.)

The gas turbine is now under study at the Environmental Protection Agency because of its potential as a low polluter. Several problems remain to be solved, however, before a low-polluting turbine car could be marketed. Because of the high temperature of combustion, the turbine provides no advantage over the piston internal-combustion engine for nitrogen oxide reduction, and at their present level of development, the turbines cannot meet 1976 standards for nitrogen oxides. (The turbine under study by the Environmental Protection Agency is one built by Chrysler, which began working on development of a gas turbine car in 1953.) In addition to reducing nitrogen oxides, fuel economy must also be improved. Turbine cars are uneconomical when idling or running at low speeds, because the engine itself turns at high speeds over the entire operating range of the car.

Representatives of Volkswagen of Germany reported at the Low Pollution Power Systems Development meeting the progress their company has made in development of a gas turbine. They have produced an engine which fits into the Volkswagen minibus, but, as in the case of the Chrysler engine, high nitrogen oxide emissions and high fuel consumption are major problems which remain to be solved.

DIESEL

While the diesel engine is used to power trucks and buses in the United States, in Europe it has also been adapted to smaller vehicles such as delivery vans, taxicabs, and some passenger cars. In London, for instance, taxicabs are diesel-powered. Diesel engines have good fuel economy (up to 50 per cent greater economy than comparable gasoline engines while idling) but have not gained widespread popularity for passenger cars, even in Europe, because they are noisy and have less power than gasoline vehicles of equal weight. They also cost more, although according to Diarmuid Downs, managing director of Ricardo and Company Engineers Limited, England: "The price differentials between gasoline and diesel cars in Europe bear little relation to the cost of production but are based on estimates of what the current market will bear."

According to one Environmental Protection Agency estimate, it is even expected that the diesel could save the American consumer money in the long run, more than $90 billion over a ten-year period, when compared to the present cost of 1973 cars.[8]

In the past, interest in diesels stemmed from their good fuel economy. However, diesels are also gaining favor as a low-polluting alternative. The diesel engine is a piston-driven, internal-combustion engine. It differs from ordinary auto engines in that

8. "Transportation Controls," (Background Material on Public Hearings for Transporation Control Strategies) *Environmental Facts,* U.S. EPA, Washington, D.C., July 3, 1973, p. 23.

ignition results from the heat produced by compression of air, rather than from a spark. A British study has shown that diesel engines emit about 8 per cent of the carbon monoxide produced by gasoline-powered vehicles, 12 per cent of the unburned hydrocarbons, and 46 per cent of the nitrogen oxides.

Environmental Protection Agency officials are now evaluating the diesel for low emissions. They have found, for instance, that the Mercedes 220D, with some modifications of the fuel injection system, produces the following emissions: hydrocarbons, 0.28 gram per mile; carbon monoxide, 1.08 grams per mile; and nitrogen oxide, 1.48 grams per mile. (Original 1976 standards are 0.41 gram per mile for hydrocarbons; 3.4 grams per mile for carbon monoxide; and 0.4 gram per mile for nitrogen oxides.) The problem remaining, as far as emissions controls are concerned, is to meet 1976 standards for nitrogen oxides. At its present stage of development, the diesel combustion process could be modified to reduce nitrogen oxides to an estimated 0.7 gram per mile. Further reductions could be made with the use of an exhaust gas recirculation device. These estimates, however, are for a small (65-horsepower, 2,800-pound) European car. Adapting the diesel engine to a larger, more powerful, American car will present more problems since larger diesel engines produce higher emissions levels.

The Environmental Protection Agency has contracted with both American and British research firms to study these and other problems. For example, diesels produce smoke and unpleasant odors and are heavy, although the weight of the diesel could be reduced by building part of the engine of aluminum. Another criticism leveled against the diesel has been the production of 3,4-benzopyrene, a suspected carcinogen. According to George Donald, superintending engineer with the Department of Environment in London, tests show that under proper operating conditions this compound is not found in the exhaust gases.

Despite some of its disadvantages, the diesel is now considered a feasible low-polluting and fuel-saving alternative to the spark-ignited internal-combustion engine. According to one study, improved diesel engines could be put into limited production in the United States in 1980.

STIRLING ENGINE

In 1816, Robert Stirling received a patent for an engine that today shows some promise as an alternative to the internal-combustion engine. A number of Stirling engines were built in the nineteenth century, but they never became highly developed because the materials needed to build an efficient engine were not available at the time. Major developments on this external-combustion engine have since been carried out in Holland, Sweden, and Germany. Ford Motor Company in the United States has also shown some interest and is working with the G. V. Phillips Company of Holland in a program in which a Ford car will be equipped with a Stirling engine by 1975. A Stirling engine car could probably be produced in the United States by 1982.

The Stirling engine works essentially as follows: heat is transferred from a combustion chamber, through the wall of a closed piston chamber to the gas inside, preferably a light gas such as helium or hydrogen; heating and cooling the gas causes it to expand and contract; a special piston system is driven when the gas is expanded. Because the combustion process takes place entirely outside of the piston chamber, the Stirling engine is an external-combustion engine.

Combustion in the Stirling engine is continuous and highly efficient, resulting in low levels of carbon monoxide and hydrocarbons. The United Stirling Company of Sweden reports that the emissions from one of their engines meets 1976 standards for unburned hydrocarbons (0.1 gram per mile) and carbon monoxide (0.9 gram per mile). High temperatures, however, make it more difficult to reduce the quantity of nitrogen oxides, and their level was measured at 0.6 gram per mile (exceeding by 0.2 gram per mile the 1976 standards). Tests have shown that lower nitrogen oxide levels can be achieved by using exhaust gas recirculation.

Another advantage of the Stirling engine is that because combustion is external, a variety of fuels could be used. Being efficient, this type of engine has good fuel economy. According to a representative of G. V. Philips Company of Holland, a company which has been working on the Stirling engine since 1938, the Stirling engine has slightly better gas mileage than the conventional internal-combustion engine. In one test, an American car with emissions control devices obtained 10.7 miles per gallon; a comparable car, with a Stirling engine, had 14.7 miles per gallon.

Despite many advantages, there are problems which will have to be worked out before such an engine could be mass produced. For instance, it is not yet known whether or not emissions levels will remain low after 50,000 miles of driving. Also, whereas conventional materials may be used for certain parts of the engine, special heat-resistant materials are needed to withstand the high temperatures developed. These materials are at the present time costly to produce. These and other technical problems mean that there are several years of work ahead before such a car can be mass produced.

ALTERNATE FUELS

Changing engine design is only one way to reduce emissions levels. Brief mention should also be made of the advantages of use of a variety of alternate fuels. Methane (natural gas), for example, has been used to power automobiles and produces much cleaner exhaust gases than gasoline. One California company has reported that when its automobiles run on methane, carbon monoxide emissions are down 93 per cent, hydrocarbons down 45 per cent, and nitrogen oxides down 87 per cent.[9] Using methane to power automobiles could help to solve a number of problems, in addition to helping reduce air pollutants produced by

cars. Although natural gas is presently in short supply, methane can be produced by the anaerobic digestion of sewage and other organic wastes. Propane has been considered as an alternative and is already used in over one-half million cars in the United States. Reports indicate that lower emissions levels can also be obtained with propane.

Methyl alcohol is used as racing car fuel and produces reduced amounts of nitrogen oxides because of lower-combustion temperatures. Since it does not misfire so readily at lean air-to-fuel mixtures, hydrocarbons and carbon monoxide are also reduced. A recent report suggests that using methyl alcohol as an additive to gasoline would reduce emissions from present automobiles. In one series of investigations it was found that carbon monoxide could be decreased from 14 to 72 per cent and fuel economy increased by 5 to 13 per cent by the addition of 5 to 30 per cent methanol to gasoline used in an unmodified internal-combustion engine.[10]

Methane, propane, and alcohol, being hydrocarbon fuels like gasoline, produce the same types of pollutants. Carbon monoxide and unburned hydrocarbons, however, could be eliminated altogether by using hydrogen gas as a fuel. Nitrogen oxides would still be produced because of the reaction between atmospheric nitrogen and oxygen within the engine. Hydrogen gas would provide a large energy output (61,500 BTU per pound versus 21,500 BTU per pound for propane). According to the National Academy of Sciences report, hydrogen has been used successfully as an alternative fuel in conventional internal-combustion engines. Hydrogen gas, however, would be difficult to store and ship. It is highly explosive and can cause severe damage to the pipes used for its distribution.

A report on an NASA-sponsored project was given at the Low Pollution Power Systems Development symposium which claimed that a hydrogen injection system for

9. Ayres and McKenna, *op. cit.*, p. 135.

10. Reed, T. B., and R. M. Lerner, "Methanol: A Versatile Fuel for Immediate Use," *Science, 182:* 1299–1304, Dec. 28, 1973.

internal-combustion engines has the potential for meeting original 1976 auto emissions standards while at the same time increasing fuel economy. In their system, small amounts of hydrogen which are generated in the vehicle, are mixed with gasoline. In tests on model engines, carbon monoxide and nitrogen oxide levels are within 1976 limits. Further work will be needed to reduce hydrocarbon levels, reduce the size of the hydrogen generator, and determine how the engine will work under actual driving conditions since it has not yet been tested in a vehicle. The advantages of such a system are that conventional engines can be used and since only small amounts of hydrogen are needed, there is no problem in storing this gas. The results of this project are still a few years in the future.

FUTURE PROSPECTS

Given the number of alternatives to the conventional internal-combustion engine, then, what are the prospects for a clean car? Much depends on achieving some of the improvements mentioned, such as increasing fuel economy in some cases and developing reasonably priced components in others. Further evaluation of individual components for reliability and durability, particularly for some of the newer designs, is also essential. Predictions are that it would be feasible to produce a stratified-charge engine in the near future, even for 1976-year models, and at least limited production of some of the more advanced engines could begin in the late 1970's or early 1980s.

Technical feasibility, however, is only one facet of the problem of introducing new technology into society. Since three major automobile manufacturers have a virtual monopoly over the production of cars in the United States, marketing a clean car can only come about through the efforts and cooperation of this industry. Having created a market for large, high-powered cars, and having spent most of its development money on refining the internal-combustion engine, much of the industry's criticism of alterna-

tive engine designs is based on the fact that they will not be equivalent to the internal-combustion engine in power or driveability. Hence, criticism of the diesel engine because of its low power-to-weight ratio. Part of the development effort in diesels will be spent adapting this engine to the American market; that is, attempting to increase horsepower without increasing emissions.

This is not to say that the automobile manufacturers have not carried out research efforts of their own on alternatives. They have. The decision to mass produce the Wankel engine, however, is illustrative of the type of incentives needed by the industry before it will actually market something new. Said David Cole, a University of Michigan professor of engineering and son of Edward Cole, president of General Motors, in a *Fortune* magazine interview on the development of the Wankel engine: "There is a point in savings where you can afford to obsolete tooling. You wouldn't do it for two cents a unit, but for $200 you could write it off or push it into the river."[11]

The creation of a new market and projected savings in labor costs are the motivating factors for marketing the rotary engine at a time when there should be a premium on developing an auto with inherently low emissions levels and high fuel economy.

With initiatives to produce a clean car on the part of industry lacking, it has fallen to government to attempt to secure industry cooperation. There have been major setbacks, however, in the attempts of the Environmental Protection Agency to meet goals set by the 1970 Clean Air Amendments. It has been necessary for Environmental Protection Agency officials not only to determine appropriate emissions standards and set up emissions testing procedures, but also to oversee research on various technological alternatives. Stemming from this situation has been, in some cases, a close working relationship between the Environmental Protection Agency and the industry it intends

11. Burck, Charles G., "A Car that May Reshape the Industry's Future," *Fortune,* July 1972.

to regulate. The Environmental Protection Agency receives much of its advice from the oil and auto industries, which have a vested interest in the outcome of its research projects and regulatory decisions.

In this regard, one report recently described the relationship between the Environmental Protection Agency and the Coordination Research Council-Air Pollution Research Advisory Committee.[12] The Air Pollution Research Advisory Committee is one of a number of committees formed by the Coordinating Research Council, a trade organization which coordinates research between the petroleum and transportation industries. The Coordinating Research Council sponsors automotive research which is in many cases pertinent to Environmental Protection Agency decision-making. At the same time, Environmental Protection Agency representatives sit on the Air Pollution Research Advisory Committee committee which decides who will receive its research funds. Most of these funds have come from the American Petroleum Institute and the Motor Vehicles Manufacturers Association.

The Environmental Protection Agency itself, in its Alternative Automotive Power Systems program, has contracted with several firms which have a vested interest in the outcome of that research. Chrysler, for instance, is under contract to help Environmental Protection Agency with a large portion of the effort to develop and evaluate the gas turbine. The Environmental Protection Agency has also recently begun studies on the feasibility of using alternative fuels in autos. One study will be carried out for the Environmental Protection Agency by Esso Research and Engineering Company, a subsidiary of Exxon.

Another reason cited for the inability to market a clean car is that the federal government simply has not as yet committed itself to this goal in terms of money spent. A bill introduced to the Senate, in 1973, would authorize funding in grants and loans to develop an alternative engine over a three-year period.[13] This program would receive considerably more funding than the Environmental Protection Agency's Alternative Automotive Power Systems program. The program would be carried out by the Low Emission Vehicle Certification Board, a group set up under the Clean Air Act and made up of representatives from the Environmental Protection Agency, the Department of Transportation, Council on Environmental Quality, and the General Services Administration. This type of program, of course, once again shifts the burden of cleaning the auto to the tax payer, rather than those industries responsible for air pollution in the first place.

Unfortunately, we have exhausted the possibilities for devising more of the relatively simple engine modifications which have been responsible for the decreases in emissions achieved to date in the United States. Further reductions in carbon monoxide and hydrocarbon emissions, and reversing the increase in nitrogen oxide production, will involve a much more serious effort than has yet been made. This may require reevaluation of the transportation system itself, with widespread changes in our overall strategy. A greater emphasis on mass transportation or other forms of public transit may result. A complete reordering of the nation's transport system could produce energy savings, reduced traffic congestion, and less noise, as well as cleaner air.

12. Shapley, Deborah, "Auto Pollution: Research Group Charged with Conflict of Interest," *Science, 181:* 732–735, Aug. 24, 1973. Letters, *Science, 182:* 774–775, Nov. 23, 1973.

13. S. 1055, Automotive Research and Development Act, Introduced by Senators Tunney and Magnuson, Feb. 1973. The bill, approved by the Senate Commerce Committee, has been added to an energy conservation bill sponsored by Senator Henry Jackson.

Efficiency of Energy Use in the United States

ERIC HIRST AND JOHN C. MOYERS   1973

Conflicts between the demand for energy and environmental quality goals can be resolved in several ways. The two most important are: (a) development and use of pollution control technologies and of improved energy-conversion technologies and (b) the improvement in efficiency of energy use. Increased efficiency of energy use would help to slow energy growth rates, thereby relieving pressure on scarce energy resources and reducing environmental problems associated with energy production, conversion, and use.

Between 1950 and 1970, United States consumption of energy resources (coal, oil, natural gas, falling water, and uranium) doubled,[1] with an average annual growth rate of 3.5 per cent—more than twice the population growth rate.

Energy resources are used for many purposes in the United States[2] (Table 5-3). In 1970, transportation of people and freight consumed 25 per cent of total energy, primarily as petroleum. Space heating of homes and commercial establishments was the second largest end use, consuming an additional 18 per cent. Industrial uses of energy (process steam, direct heat, electric drive, fuels used as raw materials,* and electrolytic processes) accounted for 42 per cent. The remaining 15 per cent was used by the commercial and residential sectors for water heating, air conditioning, refrigeration, cooking, lighting, operation of small appliances, and other miscellaneous purposes.

During the 1960's, the percentage of energy consumed for electric drive, raw materials, air conditioning, refrigeration, and electrolytic processes increased relative to the total. Air conditioning showed the largest relative growth, increasing its share of total energy use by 81 per cent, whereas the other

1. Bureau of Mines, *U.S. Energy Use at New High in 1971* (News Release, 31 March 1972).
2. Stanford Research Institute, *Patterns of Energy Consumption in the United States* (Menlo Park, Calif., November 1971).

*In this article all fuels used as raw materials are charged to the industrial sector, although fuels are also used as feedstocks by the commercial and transportation sectors.

Eric Hirst and John C. Moyers are research staff members in the Oak Ridge National Laboratory-National Science Foundation environmental program, Oak Ridge National Laboratory, Oak Ridge, Tennessee.

From *Science*, Vol. 179, No. 4079, pp. 1299–1304, March 23, 1973. Reprinted by permission of the authors and *Science*. Copyright 1973 by the American Association for the Advancement of Science.

Table 5-3. End-use of Energy in the United States.

Item	1960[a] (per cent)	1970[b] (per cent)
Transportation	25.2	24.7
Space heating	18.5	17.7
Process steam	17.8	16.4
Direct heat	12.9	11.0
Electric drive	7.4	8.1
Raw materials	5.2	5.6
Water heating	4.0	4.0
Air conditioning	1.6	2.9
Refrigeration	2.1	2.3
Cooking	1.5	1.2
Electrolytic processes	1.1	1.2
Other[c]	2.7	4.9

[a]Data for 1960 obtained from Stanford Research Institute.
[b]Estimates for 1970 obtained by extrapolating changes in energy-use patterns from Stanford Research Institute data.
[c]Includes clothes drying, small appliances, lighting, and other miscellaneous energy uses.

uses noted increased their shares of the total by less than 10 per cent in this period.

The growth in energy consumption by air conditioners, refrigerators, electric drive, and electrolytic processes—coupled with the substitution of electricity for direct fossil fuel combustion for some space and water heating, cooking, and industrial heat—accounts for the rapid growth in consumption of electricity. Between 1960 and 1970, whereas consumption of primary energy[1] grew by 51 per cent, the use of electricity[3] grew by 104 per cent. The increasing use of electricity relative to the primary fuels is an important factor accounting for energy growth rates because of the inherently low efficiency of electricity generation, transmission, and distribution which averaged 30 per cent during this decade. In 1970, electrical generation[1,3] accounted for 24 per cent of energy resource consumption as compared to 19 per cent in 1960.

3. Edison Electric Institute, *Statistical Yearbook of the Electric Utility Industry for 1970* (Edison Electric Institute, New York, 1971).

Industry, the largest energy user, includes manufacturing, mining, and agriculture, forestry, and fisheries. Six manufacturers—of primary metals; of chemicals; of petroleum and coal; of stone, clay, and glass; of paper; and of food—account for half of industrial energy consumption,[4] equivalent to 20 per cent of the total energy budget.

Energy consumption is determined by at least three factors: population, affluence, and efficiency of use. In this article we describe three areas in which energy-efficiency improvements (the third factor) might be particularly important: (a) transporation of people and freight, (b) space heating, and (c) space cooling (air conditioning).

Energy efficiency varies considerably among the different passenger and freight transport modes. Shifts from energy-intensive modes (airplanes, trucks, automobiles) to energy-efficient modes (boats, pipelines, trains, buses) could significantly reduce energy consumption. Increasing the amount of building insulation could reduce both space-heating and air-conditioning energy consumption in homes and save money for the homeowner. Energy consumption for air conditioning could be greatly reduced through the use of units that are more energy efficient.

TRANSPORTATION

Transportation of people and goods consumed 16,500 trillion British thermal units* in 1970 (25 per cent of total energy consumption).[1] Energy requirements for transportation increased by 89 per cent between 1950 and 1970, an average annual growth rate of 3.2 per cent.

Increases in transportation energy con-

4. U.S. Bureau of the Census, *1967 Census of Manufactures, Fuels and Electric Energy Consumed* (MC67 (S)-4, Government Printing Office, Washington, D.C., 1971).
*Conversion factors are: from British thermal units to joules (1,055), from miles to meters (1,609), from inches to meters (0.0254), from acres to square meters (4,047), and from tons to kilograms (907).

Table 5-4. Energy and Price Data for Intercity Freight Transport.

Mode	Energy (BTU per ton-mile)	Price (cents per ton-mile)
Pipeline	450	0.27
Railroad	670	1.4
Waterway	680	0.30
Truck	2,800	7.5
Airplane	42,000	21.9

sumption[5] are due to: (a) growth in traffic levels, (b) shifts toward the use of less energy-efficient transport modes, and (c) declines in energy efficiency for individual modes. Energy intensiveness, the inverse of energy efficiency, is expressed here as British thermal units per ton mile for freight and as British thermal units per passenger-mile for passenger traffic.

Table 5-4 shows approximate values[6] for energy consumption and average revenue in 1970 for intercity freight modes; the large range in energy efficiency among modes is noteworthy. Pipelines and waterways (barges and boats) are very efficient; however, they are limited in the kinds of materials they can transport and in the flexibility of their pick-up and delivery points. Railroads are slightly less efficient than pipelines. Trucks, which are faster and more flexible than the preceding three modes, are, with respect to energy, only one-fourth as efficient as railroads. Airplanes, the fastest mode, are only one-sixtieth as efficient as trains.

The variation in freight prices shown in Table 5-4 closely parallels the variation in

energy intensiveness. The increased prices of the less efficient modes reflect their greater speed, flexibility, and reliability.

Table 5-5 gives approximate 1970 energy and price data for various passenger modes.[6] For intercity passenger traffic, trains and buses are the most efficient modes. Cars are less than one-half as efficient as buses, and airplanes are only one-fifth as efficient as buses.

For urban passenger traffic, mass-transit systems (of which about 60 per cent are bus systems) are more than twice as energy efficient as automobiles. Walking and bicycling are an order of magnitude more efficient than autos, on the basis of energy consumption to produce food. Urban values of efficiency for cars and buses are much lower than intercity values because of poorer vehicle performance (fewer miles per gallon) and poorer utilization (fewer passengers per vehicle).

Passenger transport prices are also shown in Table 5-5. The correlation between energy intensiveness and price, though positive, is not as strong as for freight transport. Again, the differences in price reflect the increased values of the more energy-intensive modes.

5. E. Hirst, Energy Consumption for Transportation in the U.S. (Oak Ridge National Laboratory Report ORNL-NSF-EP-15, Oak Ridge, Tenn., 1972); R. A. Rice, "System energy as a factor in considering future transportation," presented at American Society of Mechanical Engineers annual meeting. December 1970.
6. Energy efficiency and unit revenue values for 1970 are computed in E. Hirst, Energy Intensiveness of Passenger and Freight Transport Modes: 1950–1970 (Oak Ridge National Laboratory Report ORNL-NSF-EP-44, Oak Ridge, Tenn., 1973).

Table 5-5. Energy and Price Data for Passenger Transport.

Mode	Energy (BTU per passenger-mile)	Price (cents per passenger-mile)
Intercity[a]		
Bus	1,600	3.6
Railroad	2,900	4.0
Automobile	3,400	4.0
Airplane	8,400	6.0
Urban[b]		
Mass transit	3,800	8.3
Automobile	8,100	9.6

[a]Load factors (percentage of transport capacity utilized) for intercity travel are about: bus, 45 per cent; railroad, 35 per cent; automobile, 48 per cent; and airplane, 50 per cent.
[b]Load factors for urban travel are about: mass transit, 20 per cent; and automobile, 28 per cent.

Table 5-6. Actual and Hypothetical Energy Consumption Patterns for Transportation in 1970.

	Total Traffic	*Percentage of Total Traffic*						Total Energy (10^{12} BTU)	Total Cost (10^9 dollars)
		Air	Truck	Rail	Waterway and Pipeline	Auto	Bus[a]		
		Intercity freight traffic							
Actual	2,210[b]	0.2	19	35	46			2,400	45
Hypothetical	2,210	0.1	9	44	46			1,900	33
		Intercity passenger traffic							
Actual	1,120[c]	10		1		87	2	4,300	47
Hypothetical	1,120	5		12		58	25	3,500	45
		Urban passenger traffic							
Actual	710[c]					97	3	5,700	68
Hypothetical	710					49	51	4,200	63
		Totals							
Actual								12,400	160
Hypothetical								9,600	141

[a]Intercity bus or urban mass transit.
[b]Billion ton-miles.
[c]Billion passenger-miles.

The transportation scenario for 1970 shown in Table 5-6 gives energy savings that may be possible through increased use of more efficient modes. The first calculation uses the actual 1970 transportation patterns. The scenario—entirely speculative—indicates the potential energy savings that could have occurred through shifts to more efficient transport modes. In this hypothetical scenario, half the freight traffic carried by truck and by airplane is assumed to have been carried by rail; half the intercity passenger traffic carried by airplane and one-third the traffic carried by car are assumed to have been carried by bus and train; and half the urban automobile traffic is assumed to have been carried by bus. The load factors (percentage of transport capacity utilized) and prices are assumed to be the same for both calculations. The scenario ignores several factors that might inhibit shifts to energy-efficient transport modes, such as existing land-

use patterns, capital costs, changes in energy efficiency within a given mode, substitutability among modes, new technologies, transportation ownership patterns, and other institutional arrangements.

The hypothetical scenario requires only 78 per cent as much energy to move the same traffic as does the actual calculation. This savings of 2,800 trillion BTU is equal to 4 per cent of the total 1970 energy budget. The scenario also results in a total transportation cost that is $19 billion less than the actual 1970 cost (a 12 per cent reduction). The dollar savings (which includes the energy saved) must be balanced against any losses in speed, comfort, and flexibility resulting from a shift to energy-efficient modes.

To some extent, the current mix of transport modes is optimal, chosen in response to a variety of factors. However, noninternalized social costs, such as noise and air pollution and various government activities (regulation, subsidization, research), may tend to distort the mix, and therefore present modal patterns may not be socially optimal.

Present trends in modal mix are determined by personal preference, private economics, convenience, speed, reliability, and government policy. Emerging factors such as fuel scarcities, rising energy prices, dependence on petroleum imports, urban land-use problems, and environmental quality considerations may provide incentives to shift transportation patterns toward greater energy efficiency.

SPACE HEATING

The largest single energy-consuming function in the home is space heating. In an average all-electric home in a moderate climate, space heating uses over half the energy delivered to the home; in gas- or oil-heated homes, the fraction is probably larger because the importance of thermal insulation has not been stressed where these fuels are used.

The nearest approach to a national standard for thermal insulation in residential construction is "Minimum Property Standards for One and Two Living Units," issued by the Federal Housing Administration. In June 1971, the Federal Housing Administration revised the minimum property standards to require more insulation, with the

stated objectives of reducing air pollution and fuel consumption.

A recent study[7] estimated the value of different amounts of thermal insulation in terms both of dollar savings to the homeowner and of reduction in energy consumption. Hypothetical model homes (1,800 square feet) were placed in three climatic regions, each representing one-third of the United States population. The three regions were represented by Atlanta, New York, and Minneapolis.

As an example of the findings of the study, Table 5-7 presents the results applicable to a New York residence, including the insulation requirements of the unrevised and the revised minimum property standards, the insulation that yields the maximum economic benefit to the homeowner, and the monetary and energy savings that result in each case. The net monetary savings are given after recovery of the cost of the insulation installation, and would be realized each year of the lifetime of the home. A mortgage interest rate of 7 per cent was assumed.

The revised minimum property standards provide appreciable savings in energy consumption and in the cost of heating a residence, although more insulation is needed to

7. J. C. Moyers, *The Value of Thermal Insulation in Residential Construction: Economics and the Conservation of Energy* (Oak Ridge National Laboratory Report ORNL-NSF-EP-9, Oak Ridge, Tenn., December 1971).

Table 5-7. Comparison of Insulation Requirements and Monetary and Energy Savings for a New York Residence.

Insulation Specification	Unrevised MPS[a]		Revised MPS[a]		Economic Optimum	
	Gas	Electric	Gas	Electric	Gas	Electric
Wall insulation thickness (inches)	0	1.875	1.875	1.875	3.500	3.500
Ceiling insulation thickness (inches)	1.875	1.875	3.500	3.500	3.500	6
Floor insulation	No	No	Yes	Yes	Yes	Yes
Storm windows	No	No	No	No	Yes	Yes
Monetary savings (dollars per year)	0	0	28	75	32	155
Reduction of energy consumption (per cent)	0	0	29	19	49	47

[a]Minimum property standards (MPS) for one and two living units.

minimize the long-term cost to the home-owner. A further increase in insulation requirements would increase both dollar and energy savings.

The total energy consumption of the United States[1] in 1970 was 67,000 trillion BTU, and about 11 per cent was devoted to residential space heating and 7 per cent to commercial space heating. Table 5-7 shows reductions in energy required for space heating of 49 per cent for gas-heated homes and 47 per cent for electric-heated homes in the New York area by going from the minimum property standards required insulation in 1970 to the economically optimum amount of insulation. The nation-wide average reductions are 43 per cent for gas-heated homes and 41 per cent for electric-heated homes. An average savings of 42 per cent, applied to the space heating energy requirements for all residential units (single family and apartment, gas and electric), would have amounted to 3,100 trillion BTU in 1970 (4.6 per cent of total energy consumption). The energy savings are somewhat under-stated—as insulation is added, the heat from lights, stoves, refrigerators, and other appliances becomes a significant part of the total heat required. The use of additional insulation also reduces the energy consumption for air conditioning as discussed later.

Electrical resistance heating is more wasteful of primary energy than is direct combustion heating. The average efficiency for electric power plants[1] in the United States is about 33 per cent, and the efficiency[3] of transmitting and distributing the power to the customer is about 91 per cent. The end-use efficiency of electrical resistance heating is 100 per cent; therefore, the over-all efficiency is approximately 30 per cent. Thus, for every unit of heat delivered in the home, 3.3 units of heat must be extracted from the fuel at the power plant. Conversely, the end-use efficiency of gas- or oil-burning home heating systems is about 60 per cent (claimed values range from 40 to 80 per cent), meaning that 1.7 units of heat must be extracted from the fuel for each unit delivered to the living area of the home.

Therefore, the electrically heated home requires about twice as much fuel per unit of heat as the gas- or oil-heated home, assuming equivalent insulation.

The debate about whether gas, oil, or electric-resistance space heating is better from a conservation point of view may soon be moot because of the shortage of natural gas and petroleum. The use of electricity generated by nuclear plants for this purpose can be argued to be a more prudent use of resources than is the combustion of natural gas or oil for its energy content. Heating by coal-generated electricity may also be preferable to heating by gas or oil in that a plentiful resource is used and dwindling resources are conserved.

The use of electrical heat pumps could equalize the positions of electric-, oil-, and gas-heating systems from a fuel conservation standpoint. The heat pump delivers about 2 units of heat energy for each unit of electric energy that it consumes. Therefore, only 1.7 units of fuel energy would be required at the power plant for each unit of delivered heat, essentially the same as that required for fueling a home furnace.

Heat pumps are not initially expensive when installed in conjunction with central air conditioning; the basic equipment and air handling systems are the same for both heating and cooling. A major impediment to their widespread use has been high maintenance cost associated with equipment failure. Several manufacturers of heat pumps have carried out extensive programs to improve component reliability that, if successful, should improve acceptance by home owners.

SPACE COOLING

In all-electric homes, air conditioning ranks third as a major energy-consuming function, behind space heating and water heating. Air conditioning is particularly important because it contributes to or is the cause of the annual peak load that occurs in the summertime for many utility systems.

In addition to reducing the energy re-

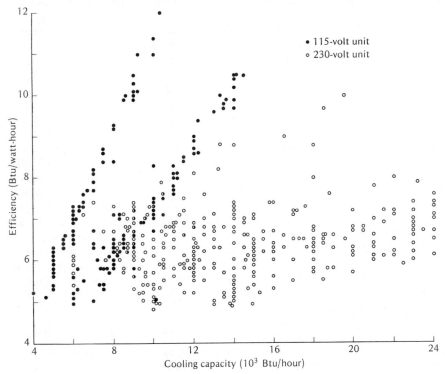

Figure 5-1. Efficiency of room air conditioners as a function of unit size.

quired for space heating, the ample use of thermal insulation reduces the energy required for air conditioning. In the New York case, use of the economically optimum amount of insulation results in a reduction of the electricity consumed for air conditioning of 26 per cent for the gas home or 18 per cent for the electric home, compared to the 1970 minimum property standards-compliance homes.

The popularity of room air conditioners is evidenced by an exponential sales growth with a doubling time of 5 years over the past decade; almost 6 million were sold in 1970. The strong growth in sales is expected to continue since industry statistics show a market saturation of only about 40 per cent.

There are about 1,400 models of room air conditioners available on the market today,

sold under 52 different brand names.[8] A characteristic of the machines that varies widely, but is not normally advertised, is the efficiency with which energy is converted to cooling. Efficiency ranges from 4.7 to 12.2 BTU per watt-hour. Thus, the least efficient machine consumes 2.6 times as much electricity per unit of cooling as the most efficient one. Figure 5-1 shows the efficiencies of all units having ratings up to 24,000 BTU per hour (see ref. 8).

From an economic point of view, the purchaser should select the particular model of air conditioner that provides the needed cooling capacity and the lowest total cost

8. Association of Home Appliance Manufacturers, *1971 Directory of Certified Room Air Conditioners,* 15 June 1971.

(capital, maintenance, operation) over the lifetime of the unit. Because of the large number of models available and the general ignorance of the fact that such a range of efficiencies exists, the most economical choice is not likely to be made. An industry-sponsored certification program requires that the cooling rating and wattage input be listed on the name plate of each unit, providing the basic information required for determining efficiency. However, the name plate is often hard to locate and does not state the efficiency explicitly.

The magnitude of possible savings that would result from buying a more efficient unit is illustrated by the following case. Of the 90 models with a capacity of 10,000 BTU per hour, the lowest efficiency model draws 2,100 watts and the highest efficiency model draws 880 watts. In Washington, D.C., the average room air conditioner operates about 800 hours per year. The low-efficiency unit would use 976 kilowatt-hours more electricity each year than the high-efficiency unit. At 1.8 cents per kilowatt-hour, the operating cost would increase by $17.57 per year. The air conditioner could be expected to have a life of 10 years. If the purchaser operates on a credit card economy, with an 18 per cent interest rate, he would be economically justified in paying up to $79 more for the high-efficiency unit. If his interest rate were 6 per cent, an additional purchase price of $130 would be justified.

In the above example, the two units were assumed to operate the same number of hours per year. However, many of the low-priced, low-efficiency units are not equipped with thermostats. As a result, they may operate almost continuously, with a lower-than-desired room temperature. This compounds the inefficiency and, in addition, shortens the lifetime of the units.

In addition to the probable economic advantage to the consumer, an improvement in the average efficiency of room air conditioners would result in appreciable reductions in the nation's energy consumption and required generating capacity. If the size distri-bution of all existing room units is that for the 1970 sales, the average efficiency[8] is 6 BTU per watt-hour, and the average annual operating time is 886 hours per year, then the nation's room air conditioners consumed 39.4 billion kilowatt-hours during 1970. On the same basis, the connected load was 44,500 megawatts, and the annual equivalent coal consumption was 18.9 million tons. If the assumed efficiency is changed to 10 BTU per watt-hour, the annual power consumption would have been 23.6 billion kilowatt-hours, a reduction of 15.8 billion kilowatt-hours. The connected load would have decreased to 26,700 megawatts, a reduction of 17,800 megawatts. The annual coal consumption for room air conditioners would have been 11.3 million tons, a reduction of 7.6 million tons, or at a typical strip mine yield of 5,000 tons per acre, a reduction in stripped area of 1,500 acres in 1970.

OTHER POTENTIAL ENERGY SAVINGS

Energy-efficiency improvements can be effected for other end uses of energy besides the three considered here. Improved appliance design could increase the energy efficiency of hot-water heaters, stoves, and refrigerators. The use of solar energy for residential space and water heating is technologically feasible and might some day be economically feasible. Alternatively, waste heat from air conditioners could be used for water heating. Improved design or elimination of gas pilot lights and elimination of gas yard lights would also provide energy savings.[9] Increased energy efficiency within homes would tend to reduce summer air-conditioning loads.

In the commercial sector, energy savings in space heating and cooling such as those described earlier are possible. In addition, the use of total energy systems (on-site generation of electricity and the use of waste heat for space and water heating and absorp-

9. Hittman Associates, *Residential Energy Consumption—Phase I Report*, No. HUD-HAI-1, (Columbia, Md., March 1972).

tion air conditioning) would increase the overall energy efficiency of commercial operations.

Commercial lighting accounts for about 10 per cent of total electricity consumption.[10] Some architects claim that currently recommended lighting levels can be reduced without danger to eyesight or worker performance.[11] Such reduction would save energy directly and by reducing air-conditioning loads. Alternatively, waste heat from lighting can be circulated in winter for space heating and shunted outdoors in summer to reduce air-conditioning loads.

Changes in building design practices might effect energy savings.[11] Such changes could include use of less glass and of windows that open for circulation of outside air.

Waste heat and low temperature steam from electric power plants may be useful for certain industries and for space heating in urban districts.[12] This thermal energy (about 8 per cent of energy consumption in 1970)[13] could be used for industrial process steam, space heating, water heating, and air conditioning in a carefully planned urban complex.

The manufacture of a few basic materials accounts for a large fraction of industrial energy consumption. Increased recycle of energy-intensive materials such as aluminum, steel, and paper would save energy. Savings could also come from lower production of certain materials. For example, the production of packaging materials (paper, metal, glass, plastic, wood) requires about 4 per cent of the total energy budget. In general, it may be possible to design products and choose materials to decrease the use of packaging and to reduce energy costs per unit of production.

IMPLEMENTATION

Changes in *energy prices,* both levels and rate structures, would influence decisions concerning capital versus life costs, and this would affect the use of energy-conserving technologies. *Public education* to increase awareness of energy problems might heighten consumer sensitivity toward personal energy consumption. Various local, state, and federal *government policies* exist that, directly and indirectly, influence the efficiency of energy use. These three routes are not independent; in particular, government policies could affect prices or public education (or both) on energy use.

One major factor that promotes energy consumption is the low price of energy. A typical family in the United States spends about 5 per cent of its annual budget on electricity, gas, and gasoline. The cost of fuels and electricity to manufacturers is about 1.5 per cent of the value of their total shipments. Because the price of energy is low relative to other costs, efficient use of energy has not been of great importance in the economy. Not only are fuel prices low, but historically they have declined relative to other prices.

The downward trend in the relative price of energy has begun to reverse because of the growing scarcity of fuels, increasing costs of both money and energy-conversion facilities (power plants, petroleum refineries), and the need to internalize social costs of energy production and use. The impact of rising energy prices on demand is difficult to assess. According to one source[14] : "In the absence of any information, we assume a long-run price elasticity of demand of −0.5

10. C. M. Crysler, General Electric Company, private communication.
11. R. G. Stein, "Architecture and energy," presented at the annual meeting of the American Association for the Advancement of Science, Philadelphia, 29 December 1971.
12. A. J. Miller et al., *Use of Steam-Electric Power Plants to Provide Thermal Energy to Urban Areas* (Oak Ridge National Laboratory Report ORNL-HUD-14, Oak Ridge, Tenn., 1971).
13. R. M. Jimeson and G. G. Adkins, "Factors in waste heat disposal associated with power generation," presented at the American Institute of Chemical Engineers national meeting, Houston, Texas, March 1971).

14. National Science Foundation RANN Program, *Summary Report of the Cornell Workshop on Energy and the Environment, 22 to 24 February 1972,* Senate Committee on Interior and Insular Affairs, No. 92-23, May 1972 (Government Printing Office, Washington, D.C., 1972).

(meaning that in the long run a doubling of energy prices will reduce demand by a factor of the square root of 2, namely to about 70 per cent of what it would have been otherwise)."

The factors cited above (fuel scarcity, rising costs, environmental constraints) are likely to influence energy price structures as well as levels. If these factors tend to increase energy prices uniformly (per BTU delivered), then energy price structures will become flatter; that is, the percentage difference in price between the first and last unit purchased by a customer will be less than that under existing rate structures. The impact of such rate structure changes on the demand for energy is unknown, and research is needed.

Increases in the price of energy should decrease the quantity demanded and this is likely to encourage more efficient use of energy. For example, if the price of gasoline rises, there will probably be a shift to the use of smaller cars and perhaps to the use of public transportation systems.

Public education programs may slow energy demand. As Americans understand better the environmental problems associated with energy production and use, they may voluntarily decrease their personal energy-consumption growth rates. Experiences in New York City and in Sweden with energy-conservation advertising programs showed that the public is willing and able to conserve energy, at least during short-term emergencies.

Consumers can be educated about the energy consumption of various appliances. The energy-efficiency data for air conditioners presented here are probably not familiar to most prospective buyers of air conditioners. If consumers understood energy and dollar costs of low-efficiency units, perhaps they would opt for more expensive, high-efficiency units to save money over the lifetime of the unit and also to reduce environmental impacts. Recently, at least two air-conditioner manufacturers began marketing campaigns that stress energy efficiency. Some electric utilities have also begun to urge their customers to use electricity conservatively and efficiently.

Public education can be achieved through government publications or government regulation, for example, by requiring labels on appliances which state the energy efficiency and provide estimates of operating costs. Advertisements for energy-consuming equipment might be required to state the energy efficiency.

Federal policies, reflected in research expenditures, construction of facilities, taxes and subsidies, influence energy consumption. For example, the federal government spends several billion dollars annually on highway, airway, and airport construction, but nothing is spent for railway and railroad construction. Until recently, federal transportation research and development funds were allocated almost exclusively to air and highway travel. Passage of the Urban Mass Transportation Act, establishment of the National Railroad Passenger Corporation (AMTRAK), plus increases in research funds for rail and mass transport may increase the use of these energy-efficient travel modes.

Similarly, through agencies such as the Tennessee Valley Authority, the federal government subsidizes the cost of electricity. The reduced price for public power customers increases electricity consumption over what it would otherwise be.

Governments also influence energy consumption directly and indirectly through allowances for depletion of resources, purchase specifications (to require recycled paper, for example), management of public energy holdings, regulation of gas and electric utility rate levels and structures, restrictions on energy promotion, and establishment of minimum energy performance standards for appliances and housing.

The federal government spends about $0.5 billion a year on research and development for civilian energy, of which the vast majority is devoted to energy supply technologies[14]:

Until recently only severely limited funds were available for developing a detailed understanding of the ways in which the nation uses energy.... The recently instituted Research Applied to National Needs (RANN) Directorate of the National Science Foundation ... has been supporting research directed toward developing a detailed understanding of the way in which the country utilizes energy.... This program also seeks to examine the options for meeting the needs of society at reduced energy and environmental costs.

Perhaps new research on energy use will reveal additional ways to reduce energy growth rates.

SUMMARY

We described three uses of energy for which greater efficiency is feasible: transportation, space heating, and air conditioning. Shifts to less energy-intensive transportation modes could substantially reduce energy consumption; the magnitude of such savings would, of course, depend on the extent of such shifts and possible load factor changes. The hypothetical transportation scenario described here results in a 22 per cent savings in energy for transportation in 1970, a savings of 2,800 trillion BTU.

To the homeowner, increasing the amount of building insulation and, in some cases, adding storm windows would reduce energy consumption and provide monetary savings. If all homes in 1970 had the "economic optimum" amount of insulation, energy consumption for residential heating would have been 42 per cent less than if the homes were insulated to meet the pre-1971 Federal Housing Administration standards, a savings of 3,100 trillion BTU.

Increased utilization of energy-efficient air conditioners and of building insulation would provide significant energy savings and help to reduce peak power demands during the summer. A 67 per cent increase in energy efficiency for room air conditioners would have saved 15.8 billion kilowatt-hours in 1970.

In conclusion, it is possible—from an engineering point of view—to effect considerable energy savings in the United States. Increases in the efficiency of energy use would provide desired end results with smaller energy inputs. Such measures will not reduce the *level* of energy consumption, but they could slow energy growth *rates*.

SUGGESTED READINGS

Hirst, Eric, "Pollution-Control Energy Costs," *Mech. Engineering 96:* 28–35, September 1974.

Energy Conservation

G. A. LINCOLN 1973

As the energy crisis looms ever larger, energy conservation is beginning to receive increasing attention.[1,2] Energy conservation can make a substantial contribution in ameliorating or postponing the potential energy shortages faced by the United States over the next several decades. To realize this contribution, however, will require not only the political will to implement the necessary conservation measures but also the imagination and intellectual resources of the scientific community to develop new technologies to increase the efficiency of energy use.

This article is directed to provoking thought on how to attain economic, social, and other objectives while using less energy resources. Its purpose is not so much to answer questions of energy conservation as to raise them. The discussion is provocative in places, deliberately so. It is not intended to suggest any policy commitments on the part of the author or any of those whose advice and suggestions have contributed to the discussion. Rather, the objective is to enlist the interest of thinking people, and particularly the scientific community, in the energy conservation effort.

The history of civilization is, to a large extent, the story of man's progress in harnessing energy. Discovery of the controlled use of fire was certainly a major milestone in man's emerging domination of other forms of life. Development of the sail to utilize the energy of wind to propel watercraft opened up the rest of the world to curious and acquisitive societies around the Mediterranean basin. Windmills and watermills represented early attempts to harness energy sources for direct work. The industrial revolution, one of the great landmarks of our present culture, consisted essentially of the large-scale replacement of muscle power by controlled mechanical energy derived, in turn, from thermal energy.

A less noted, but equally significant, impact of the industrial revolution was the

1. P. H. Abelson, *Science* 178, 355 (1972).
2. A. L. Hammond, *ibid.*, p. 1079.

G. A. Lincoln recently retired as director of the Office of Emergency Preparedness, Executive Office of the President, Washington, D.C., and is currently professor of economics and international studies, University of Denver.

From *Science*, Vol. 180, No. 4082, pp. 155–161, April 13, 1973. Reprinted by permission of the author and *Science*. Copyright 1973 by the American Association for the Advancement of Science.

general introduction of available energy when and where it was needed. In previous ages man used energy largely when and where it was found: he sailed when the wind blew; he forged his metals by the forests where firewood was plentiful. With the advent of combustion engines, however, man was freed to travel without (or even against) the wind, and at speeds which animals could not match. He could transmit large amounts of controlled mechanical power throughout a mill by use of shafts and pulleys. And finally, the understanding of electricity completed the revolution by permitting not only mechanical power but also information made available far from the originating source.

First wood and then coal was used to satisfy the increasing demands for manageable sources of thermal energy. Both served the purpose admirably; however, both wood and coal presented certain problems. Then came the discovery of oil and gas and how to use them with greatly increased versatility and flexibility in conversion of fuel to thermal energy. The internal combustion engine arrived and flourished, automatically fired boilers became the norm, and our modern mechanized society was at hand.

Increases in the convenience and economy of harnessed energy have led to additional applications, which, in turn, have increased the demand and, coming full circle, fostered further technological advances in the convenience and economy of harnessing energy. The use of energy in the United States today is not only growing but accelerating. With only 5 per cent of the world's population, this nation already consumes about one-third of the energy produced in the world. And the current annual appetite for about 70×10^{15} British thermal units (BTU) is projected to double within 20 years to 140 quadrillion BTU (that is, from 1.7×10^{19} to 3.5×10^{19} calories).* Unfortu-

nately, the finding and production of domestic energy supplies is not keeping pace with this rapidly growing demand. Thus, careful attention must now be addressed to the adequacy of our remaining domestic energy resources. More than 10 per cent of our present requirements are met by importing foreign oil (about 4.6×10^6 barrels per day). Even the more conservative projections indicate that the level of oil imports, and also of some gas in liquefied form, must increase a great deal within the next decade alone to compensate for the projected depletion of available domestic fuels. This situation suggests serious problems both for the national security and for the balance of payments.

Furthermore, recent practices and trends in methods of fuel extraction, energy conversion, and energy utilization have caused pollution problems affecting the nation's health and natural environment. Yet, because of extensive interdependence among these important national concerns—problems related to energy and problems related to the environment—measures to alleviate one can easily aggravate another. For example, estimates derived from a recent study indicate that the removal of lead from gasoline for pollution control as presently planned will cause an increase of about 1 million barrels per day in our gasoline needs by 1975, thus worsening the nation's supply situation.** Some of the state implementa-

$\times 10^6$ BTU = 1.5×10^9 cal; 1000 cubic feet of gas = 1.0×10^6 BTU = 2.5×10^8 cal; 1 ton of bituminous coal = 2.5×10^7 BTU = 6.3×10^9 cal; 1 kilowatt-hour of electricity = 3.4×10^3 BTU = 9.6×10^5 calories.
**"An economic analysis of proposed schedules for removal of lead additives from gasoline," a study prepared by Bonnor and Moore Associates, Inc., Houston, Texas, for the Environmental Protection Agency in June 1971. They estimate that lead removal would cause a 12 percent increase in gasoline consumption. Since refinery output is approximately 50 percent gasoline, and 1975 crude runs are estimated by the National Petroleum Council [Committee on U.S. Energy Outlook, *U.S. Energy Outlook* (National Petroleum Council, Washington, D.C., 1971), Vol. 1, p. 27] at 18.4 million barrels per day, the resulting increase in gasoline would be approximately 0.5 × 0.12 × 18.4 = 1.1 million barrels per day in 1975.

*Energy industries generally employ the British thermal unit (BTU) as a common denominator among the various specialized fuel and energy units of measure. Approximate conversion factors between the major measures are: 1 barrel of oil = 5.8

tion plans to meet the requirements of the 1970 Clean Air Act[3] provide other examples of such conflicting demands. Several of these plans project a demand for quantities of fuels of low sulfur content (gas, oil, coal, nuclear power), which will simply not be available on the time schedule envisaged. In addition, the shifts in equipment and fuel types and the processing costs to manufacture such quality fuel can impose a severe economic penalty for the consumer. Shifts in types of fuel and their timing can have severe impacts on our industrial and economic structure, our foreign economic arrangements, including our balance of payments, our relations with other countries, and the security of our own country. Energy security is now a critical component of our national security and over-all foreign policy. Clearly, no one of these problems can be addressed apart from, or to the neglect of, the others; any reasonable solution must take all into account.

In order to lessen our potential dependence on foreign supply, we must increase our domestic energy supplies. Increased field exploration and extraction of oil and gas, including shale oil, are necessary. Measures should include expanded coal mining and the gasification and liquefaction of coal, more rapid introduction of electrical power generated from nuclear fuels, and greater emphasis on the development and exploitation of unconventional energy sources such as solar radiation and geothermal power. All of these will be undertaken as the demand for energy rises in relation to the supply, and fuel prices inevitably follow suit. Most of these measures take years, even decades, however, and even though successful, may leave a continuing energy gap. Certainly that energy gap is going to exist for a long while.

Although the increasing of energy supplies is essential, it is also important to reduce consumption or at least to ease its growth rate. This approach is intuitively appealing from the standpoint of assuaging those problems of environmental pollution

which are related to energy consumption. Nevertheless, some of the potential approaches to reduced energy demand call for technology which, if available at all, is not yet advanced to an economically viable level. Equally or more difficult, some of the approaches may depend on fundamental changes in national attitudes toward living style, and even if the process of mass application of social incentives were well understood—which it is not—its ethical implications would require careful attention.

Energy conservation needs to be viewed both from the standpoint of the consumer and from the standpoint of broad national policy. The viewpoints are not necessarily conflicting, but at times may be. The useful but simplistic approach of achieving the same economic and social objectives with less energy needs to be combined with an approach of using the types of fuels that best further our national objectives, while emphasizing the conservation of those fuels creating policy problems. For example, the consumption of oil is now beginning to pose policy problems. So also is the consumption of gas, since our domestic shortfall in production is made up by imported oil and gas. Hence, that conservation holding back on consumption of oil and gas is most broadly useful. The most desirable way is through absolute reduction in consumption of energy from gas and oil. But just the substitution of more domestically available fuels, such as coal and nuclear fuel, is a plus in solving our energy problems and a logical component of an energy conservation endeavor.

Unfortunately, the concept of energy conservation through substitution of domestically more abundant fuels for the less abundant does run directly into the continuing friction, and sometimes direct confrontation, between environmental programs and energy utilization programs. Coal, for example, is abundant but often does not conform to environmental objectives. Coal now poses a challenge to science and technology of the same importance as that posed by oil and gas in the early period of their utilization for producing thermal energy.

3. Public Law 91-604.

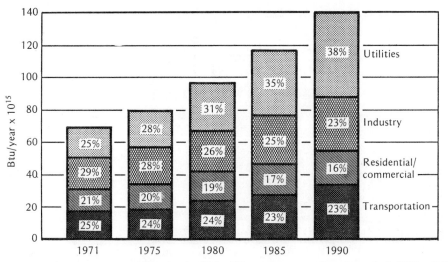

Figure 5-2. Energy consumption in the United States by consuming sectors;[4] 1 BTU = 0.259 calories; 10^{15} BTU ≅ 172 × 10^6 barrels of oil, 970 × 10^9 cubic feet of natural gas, 41.7 × 10^6 tons of coal.

The issue of energy conservation is an important and complex one. To provide the appropriate setting for its discussion, the general patterns of the United States energy supply and demand will be outlined, and then the four major categories of energy consumption—transportation, residential/commercial, industry, and electric utilities—will be examined somewhat more closely to reveal present trends and suggest possibilities for improved conservation. Finally, the complex problems of environmental pollution and economic investment will be introduced briefly.

PATTERNS OF UNITED STATES ENERGY SUPPLY AND DEMAND

The Bureau of Mines, Department of the Interior, has made careful projections of energy consumption by consuming sector (Figure 5-2) and by source (Figure 5-3) for the period 1971 to 1990.[4] The major projected change between now and 1990 in the consuming sector is a tripling in the energy used in generating electric power in order to meet increases in projected demand. Electrical generation is expected to increase by 72 per

cent from 1971 to 1980 and by 78 per cent from 1980 to 1990. Transportation is expected to hold its current share of the market, with projected increases of 35 per cent from 1971 to 1980 and 41 per cent from 1980 to 1990. For the entire period 1971 to 1990, industrial use of fossil fuel is expected to increase by 53 per cent and residential/commercial use by 41 per cent. The major projected change in the sources of energy between now and 1990 is that nuclear power will significantly increase its proportionate contribution, but the consumption of fossil fuels will also increase a great deal. Projections to 1990 indicate that the sources of

4. Bureau of Mines, *U.S. Energy Through the Year 2000* (Department of the Interior, Washington, D.C., December 1972).
*For the purposes of this article, electric energy is accounted for solely in the electric utility sector and is not distributed to the other sectors. In addition to extrapolating current trends in energy consumption, the Bureau of Mines projections assume continued improvements in the efficiency of fossil fuel plants, improved insulation in new home construction based on raised Federal Housing Authorities standards, and a continued increase in the proportion of steel produced by the more energy-efficient basic oxygen furnace process.

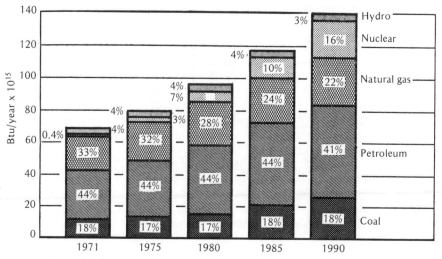

Figure 5-3. Energy consumption in the United States by source.[4] See Figure 5-2 caption.

United States energy in that year will be distributed as follows: coal, 18 per cent; petroleum, 41 per cent; natural gas, 22 per cent; nuclear power, 16 per cent; hydropower, 3 per cent.

In terms of dollar expenditures for energy, the patterns and trends differ considerably between intermediate demand and final demand. The final demand consists of purchases for end uses such as automobile fuel, residential heating, and exports, whereas the intermediate demand consists of industrial and commercial purchases to produce products and services for end consumers. Intermediate demand expenditures for energy are related very closely to the gross national product; the ratio of the two has varied less than a quarter of a percent over more than a decade. Although not correlated as closely with the gross national product, the final demand for energy has grown substantially during the same period. About 85 per cent of the energy final demand during the last five years (Figure 5-4) has consisted of personal consumption expenditures, which represent domestic consumption of fuel for such uses as private cars, home heating and air conditioning, and electric appliances.[5]

TRANSPORTATION

In 1970, transportation consumed 16.4 × 10^{15} BTU—one-quarter of the total energy used in this country, a share which is expected to continue. Petroleum accounted for 96 per cent of the fuel consumed for transportation in 1970. This amounted to 2,830,000,000 barrels of crude oil, or a rate of about 7,750,000 barrels per day (roughly equivalent to daily dissipation of a dozen 100,000-ton tanker loads). Automobiles are the leading consumer, using 55 per cent of the transportation energy in 1970 (14 per cent of total national energy consumption), with trucks second at 21 per cent, and aircraft third at 7.5 per cent. The remaining 16 per cent is made up of rail, bus, waterway, pipeline, and other categories.[6]

5. A. A. Schulman, *DITT Data Estimates and Input-Output Review* (OEP Report IST-103, Office of Emergency Preparedness, Executive Office of the President, Washington, D.C., 1972). Values are derived by the use of the demand impact transformation tables (DITT).

6. E. Hirst, *Energy Consumption for Transportation in the U.S.* (Report ORNL-NSF-EP-15, Oak Ridge National Laboratory, Oak Ridge, Tenn., 1972).

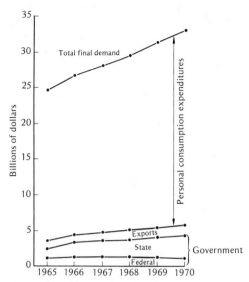

Figure 5-4. Final demand expenditures for energy consumption according to the category of the consumer.[5]

Major trends among transportation modes include railroads and waterways giving way to pipelines and trucks for intercity freight movement; buses and railroads giving way to aircraft and automobiles for intercity passenger traffic; and mass transit, especially buses and trains, giving way to private automobiles for urban passenger traffic.

As Table 5-8[7] shows, enormous differences exist in the energy efficiencies of these transportation modes.[8] For passenger travel, airplanes are less efficient users of energy than automobiles, which are in turn less efficient than buses and railroads. For freight movement, airplanes are less energy efficient than trucks and considerably less efficient than pipelines, waterways, and railroads.

7. R. A. Rice, "System energy as a factor in considering future transportation," *Amer. Soc. Mech. Eng. Pap. 70-WA/ENER-8* (1970).
8. R. A. Rice, *Technol. Rev. 74*, 31 (January 1972).

Table 5-8. Transportation Propulsion Efficiency.[7] Conversions are: 1 foot ≅ 0.305 m; 1 mile ≅ 1.609 km; 1 horsepower (hp) ≅ 746 watts; 1 knot ≅ 1.85 km/hr.

Passenger		*Freight*	
Transport Type	*Passenger Miles Per Gallon*	*Transport Type*	*Cargo Ton Miles Per Gallon*
Large jet plane (Boeing 747)	22	One-half of a Boeing 707 (160 tons, 30,000 horsepower)	8.3
Small jet plane (Boeing 704)	21	One-fourth of a Boeing 747 (360 tons, 60,000 horsepower)	11.4
Automobile (sedan)	32	Sixty 250-hp, 40-ton trucks	50.0
Cross-country train[a]	80	Fast 3,000-ton, 40-car freight train	97.0
Commuter train[b]	100	Three 5,000-ton, 100-car freight trains	250.0
Large bus (40 foot)	125	Inland barge tow, 60,000 gross tons	220.0
Small bus (35 foot)	126	Large pipeline, 100 miles, two pumps	500.0
Suburban train (two-deck)[c]	200	100,000-ton supertanker, 15 knots	930.0

[a]One 150-ton locomotive and four 70-seat coaches plus diner lounge and baggage coach.
[b]Ten 65-ton cars and two 150-ton 2000-hp diesel locomotives.
[c]A ten-car gallery-car commuter train, 160 seats per car.

Energy efficiency within transportation modes also varies substantially and, unfortunately, has tended to decrease with time. The quest for ever-increasing speed and convenience has been succeeding at the cost of increased energy consumption. The low average occupancy of commuter cars combines with short distances traveled and traffic congestion to lower drastically the energy efficiency of the automobile. In 1971, for example, 55 per cent of automobile energy consumption went for urban trips of 10 miles (16 kilometers) or less, and 56 per cent of all commuting was by automobiles containing only one occupant.[9] Emission controls also contribute significantly to lower energy efficiency; those currently being installed and projected will result in an additional gasoline consumption by 1980 of the order of 2 million barrels per day.

A large number of factors, including government policy, social and environmental concerns, the low cost of energy, uncontrolled urban growth, and the demand for increased mobility and transportation service (speed, comfort, reliability, and convenience) have contributed to a continuing shift toward the use of less energy-efficient modes and to a continuing decline in the energy efficiency of all transportation modes.

Furthermore, this discouraging trend shows every indication of persisting. Projections of growth in aircraft and automobile use show that these two modes alone will account for 22×10^{15} BTU in 1985, more than 73 per cent of the total transportation energy consumption for that year.

Several actions could be taken over the short and midterm periods (within three years and within ten years) to increase energy efficiency, improve the balance between transportation modes, and decrease total demand for transportation. Incentives for using smaller automobiles, subsidized mass transit, and improved traffic flow through traffic metering systems and priority bus lanes, would encourage greater use of more energy efficient transportation modes. Improved communications facilities, the development of urban clusters, and the construction of attractive walkways and bicycle paths will all help to reduce total transportation demand for energy.

Estimates indicate that the implementation of short and midterm conservation measures could result in savings of 15 to 25 per cent of the projected transportation energy demand by the early 1980's.[10] These estimates are predicated on the assumption that curtailment of passenger and freight movement is largely unacceptable, so that emphasis is placed not on restrictions but on approaches designed to improve energy efficiency relative to present standards. Even though many of the measures have been tried to some extent, experience is insufficient to allow a full analysis of the effects of all plausible actions. Nevertheless, these short and midterm estimates assume no significant advances in technology.

Over the long term, however, technology and urban design stand out as the areas of greatest promise in reducing transportation energy demands. For example, the development of practical hybrid energy storage systems, such as a system combining a gas turbine with electricity, could significantly increase the operating efficiency of urban automobiles.[11] Similarly, the quest for solutions to urban social, economic, and environmental problems is yielding concepts that have important implications for transportation. For example, the development of urban clusters can reduce drastically the need for transportation. In general, trans-

9. Automobile Manufacturers Association, *Automobile Facts and Figures* (Automobile Manufacturers Association, Detroit, Mich., 1971), p. 53.

10. Office of Emergency Preparedness, *The Potential for Energy Conservation: A Staff Study* (Government Printing Office, Washington, D.C., 1972).
11. W. E. Fraize and R. K. Lay, *A Survey of Propulsion Systems for Low Emission Urban Vehicles* (Report No. M70-45, MITRE Corporation, McLean, Va., September 1970). Hybrid energy storage systems allow constant maximum efficiency load to be placed on the engine, for example, by combining a heat engine, storage batteries, and electric drive.

portation energy efficiency could be greatly increased by providing incentives to separate people from automobiles, for example, rapid transit trunk lines between clusters, moving walk ways, and bicycle paths. Increased understanding and appropriate coordination in transportation and urban planning can yield enormous dividends not only for energy conservation but also for the environment, and for the health and mobility of the American people.

Many of these conservation measures have costs—economic, political, and social. Government, industry, and consumers, however, must come to grips with some of the difficult choices that will have to be made over the next several decades. For example, should the federal government institute financial disincentives to encourage the use of small automobiles, as through higher use taxes on large automobiles or taxes on engine size or automobile weight? The advantages of promoting the use of smaller automobiles are numerous—savings approaching 3 million barrels per day in 1985, reduced import costs, and less pollution. On the other hand, by forcing consumers to choices they may not desire, such measures may be highly unpopular. Moreover, they could produce serious, adverse economic consequences for the automobile industry and related industries.

Another hard question concerns the extent to which the federal government should encourage or subsidize mass transit. With a much greater efficiency on a passenger-mile basis, effective mass transit systems would not only materially reduce fuel consumption but also significantly ameliorate our balance of payments and environmental and traffic problems. On the other hand, to make such systems effective would probably require restrictions on consumers in the form of bans on automobile use in the inner city or high parking taxes. Moreover, adequate efforts would require an enormous capital outlay, and mass transit would have to compete with other important social needs for these funds.

Transportation serves a number of na-tional and social goals that must be balanced against the objective of energy conservation. Nevertheless, energy efficiency must be given proper emphasis in the design, development, and utilization of our transportation systems. Any truly practical program will require a blend of actions carefully balanced and timed to avoid disruption of needed traffic flows, upheavals in life styles, damage to industries dependent on transportation, and aggravation of problems concerning the international balance of payments.

RESIDENTIAL AND COMMERCIAL SECTORS

Private residences and commercial establishments account for about one-fifth of the total United States energy consumption. Space heating and cooling, water heating, refrigeration, and cooking represent somewhat more than 75 per cent of the commercial energy use and more than 85 per cent of the residential use.[12] By far the largest portion of this is due to space heating and cooling.

By 1980 the annual energy requirement for household space heating and cooling is expected to reach about 11×10^{16} BTU, or about 63 per cent of the total projected residential and commercial energy consumption.

Some reduction in residential energy consumption could be achieved through a nation-wide educational program encouraging good energy conservation practices in the home. For example, the setting of all residential thermostats 2 degrees higher during summer and 2 degrees lower during winter could produce in 1980 energy savings of about 1.3×10^{15} BTU.

The National Bureau of Standards estimates that improvements in insulation and construction can reduce the energy consumed in heating and air conditioning by 40

12. "Patterns of energy consumption in the United States," a study prepared by Stanford Research Institute, Menlo Park, Calif., for the Office of Science and Technology in 1972.

to 50 per cent from present norms. Improved insulation technology can be readily adopted in new homes. Unfortunately, the high cost of introducing such improvements into existing homes by present methods makes widespread introduction there less likely (except for storm windows). The discovery and development of inexpensive insulating techniques for reducing the energy loss from existing houses is therefore an urgent need. Furnaces of higher efficiency, with provision for easy periodic cleaning by the home owner, can and should be developed also.

An important step was taken in 1971 with the revision to the minimum property standard of the Federal Housing Authority, which significantly tightened insulation requirements for single-family houses. Unfortunately, many apartment houses and single-family homes built under conventional loans do not meet desirable minimum standards. Further tightening of insulation requirements for single- and multiple-family homes, offices, and other buildings would provide an additional and important contribution in reducing unnecessary energy consumption.

Higher fuel costs may make economically feasible the wider use of common district heating, total energy systems, and heat pumps. More urban buildings could be heated and cooled by using the steam rejected by a central power-generating station. Municipal waste could be burned in power-plant boilers along with coal, thus reducing fossil fuel requirements for power generation by an estimated 8 per cent[13] (and also helping in solid waste disposal). One consulting engineer has estimated that a saving of 15 to 20 per cent can be achieved in the amount of energy normally used in office and commercial buildings by such energy conservation measures as using better insulation to recover the heat in winter and the cooling effect in summer from the exhaust

air of the building, recovering heat from lighting, reducing heating (or cooling) and lighting levels in corridors and certain other spaces, and reducing the amount of outdoor air drawn into the building.[14]

All of this suggests that by 1990 the nation's over-all space heating and cooling requirements could be reduced by perhaps 30 per cent from the projected demand levels in that year. Even within 10 years a more modest 20 per cent reduction (about 2×10^{15} BTU) should be possible, most of it through improved insulation. Additional energy savings in water heating, refrigerating, cooking, and lighting systems and in air conditioning equipment are also possible. Improved technology, however, is necessary in most cases to make the introduction and operation of the improvements economically feasible.

INDUSTRY

Industrial energy consumption constituted about 29 per cent of the total domestic energy consumption in 1971.* The primary metal industries, chemicals and allied products, and petroleum refining and related industries together accounted for more than half of that. Natural gas was the most rapidly growing and largest source of industrial energy used (46.5 per cent), followed by coal (26.0 per cent), petroleum (16.8 per cent), and electricity (10.6 per cent).

There will undoubtedly be a change in the relative amounts of the various energy sources used by industry in the future. Many gas pipelines are now having to curtail shipments or existing contracts as a result of the developing shortage of natural gas. Gas prices will increase, especially as the pipelines seek to turn to such expensive sources as liquefied natural gas and synthetic natural gas. Natural gas usage certainly cannot continue to increase its share of the industrial market, and most likely will be cut back.

13. Federal Power Commission, *The 1970 National Power Survey* (Government Printing Office, Washington, D.C., 1971).

14. F. S. Dubin, professional engineer, Dubin-Mindell-Bloome Associates, New York, personal communication.
*Electrical power is excluded.

Today, however, plans for new petrochemical complexes call for the use of heavy oil feedstocks, primarily naphtha, and a new regulation of the mandatory oil import program will assist petrochemical producers to acquire naphtha produced from imported oil. This should enable these producers to remain competitive with foreign producers, which have traditionally used these feedstocks.

The major industrial sectors have all achieved a general decline in energy used per unit output over the last decade.[12] Nevertheless, more rapid improvement could almost certainly be effected by focusing attention on energy conservation.

Given a sufficient incentive, industry as a whole could probably cut energy demand by 5 to 10 per cent of the projected demand by 1980, primarily by replacing old equipment, demanding more energy-conscious design, and increasing maintenance on boilers, heat exchangers and so forth. Underpriced energy does not help in encouraging efficient energy use and results in inadequate exploration of avenues for improvement. Any deliberate economic incentive designed to cut energy demand would be made more effective by an accompanying "energy awareness" information program directed at trade associations, professional societies, equipment advertisers, and engineering design companies.

The general introduction of planning for the total life cycle in the use of resources, to include energy-conserving recycling or reusing, could also make noticeable contributions to energy conservation as well as conservation of other resources. Secondary recovery of materials is often less energy consuming than primary extraction and production; for many nonferrous metals, recycling energy requirements are only 20 per cent or less of primary processing requirements. This approach is also a prime target for research and development.

Elimination of wasteful practices is of course important. Energy is a sufficiently large expense for the major industrial energy consumers that they already try to use it as economically as possible. Thus, wasteful practices are likely to be found primarily in marginal operators or in industries for which energy is a relatively minor expense.

ELECTRIC UTILITIES

Electric utilities represent the most rapidly growing consumer of energy among the four major groups. Their 1971 use of about 17×10^{15} BTU is projected to expand by a factor of almost 4 within 20 years, representing an increase from 25 to 38 per cent of the growing national energy consumption. Thus, any improvements that can be achieved in the energy efficiency of electric power generation are of major importance.

Utilities already concentrate heavily on efficient use of fuel because it constitutes a major operating expense. Over the past quarter of a century, industry efficiency has improved by more than 30 per cent; that is, the national average for energy input (in British thermal units) per kilowatt-hour output has declined from nearly 16,000 to about 11,000.[15] Nevertheless, efficiency has undergone a slight decline recently, a trend that the National Coal Association attributes partly to the heavy use of inefficient, old plants and peaking generators and partly to the use of more environmentally acceptable fuels with lower energy contents.[16] An additional factor is the increasing installation of energy-consuming pollution control equipment.

It is sometimes argued that electrical power generation as a whole is a wasteful use of energy because of the large energy conversion losses involved (the over-all conversion efficiency is about one-third). This alleged wastefulness must properly be measured against the alternative of direct fuel use. Except in some large industrial applications

15. National Economic Research Associates, *Energy Consumption and Gross National Product in the U.S.* (National Economic Research Associates, Washington, D.C., 1971).
16. National Coal Association, *Steam-Electric Plant Factors* (National Coal Association, Washington, D.C., 1971).

such as aluminum production, electrical power is distributed from large central generating plants to many small customers. Thus, the comparison is between one large energy consumer and many small ones. Aside from efficiencies due to the economics of size—which are substantial—the central generating plant generally offers much better equipment maintenance, a factor which bears directly on efficiency (a half-millimeter of soot in an oil burner can reduce its efficiency by 50 per cent). Power distribution by electric transmission grids is generally more efficient than transportation of coal and oil by trucks,[13] the usual method for small consumers. Furthermore, large electric power plants permit centralized pollution control, which is both more efficient and more easily monitored. On balance, it is not at all clear that electric power generation and distribution is as wasteful from an overall point of view as some of its detractors would claim.

Regardless of its relative efficiency, electrical power is increasingly in demand because of its cleanliness, convenience, and ease of control. An important question, then, is how to increase the energy efficiency of the centralized generation of electrical power by the utilities. Certainly, increased rates of replacement for obsolete equipment and reduced delays in bringing more efficient new plants on stream would help. Accelerated introduction of nuclear plants would ameliorate substantially the demand on fossil fuels, whether from domestic or foreign sources. Smoothing the daily demand cycle in order to reduce heavy peak loads would significantly lessen the use of inefficient peaking generators.

The mechanisms for implementing these various measures are primarily economic, and perhaps also regulatory, in nature. In electric power generation, however, there are also long-term opportunities for significant improvements from new technology, some of it requiring extensive research and development. Historically, the electric utility industry has lacked an organized industry-wide research program. Equipment manufacturers have performed most of the research and development that has been carried out, and this has been limited. In recognition of this problem, the Electric Research Council, representing all segments of the electric utility industry, recently established a research corporation to be supported financially on a shared basis. This corporation is intended to address critical electric power problems at both applied and fundamental levels.

The more promising approaches to long-range improvement in the efficiency of generating electrical power include advanced power cycles, magnetohydrodynamics, various types of nuclear reactors, geothermal sources, cryogenic transmission, improved "waste heat" utilization, and total energy systems. The advanced power cycle, involving the combination of coal gasification with a gas turbine-steam turbine power plant, is currently under study by the Environmental Protection Agency. It promises improved flexibility in plant size and location, relatively low cost, and potentially high efficiency. Nevertheless, full benefits depend on the development of an efficient coal gasification process. Magnetohydrodynamic techniques offer greater efficiency as well as low maintenance and substantially reduced cooling requirements. The practical realization of any of several breeder reactor principles under study would reduce not only fossil fuel demands but also nuclear fuel demands—a factor of equal importance in the long run. Geothermal power already enjoys some limited use, but large-scale exploitation (which would make available a huge new energy source) awaits extensive investigation of several alternative approaches. Cryogenic transmission systems suggest exceedingly high distribution efficiencies but present difficult problems in practical realization beyond the laboratory scale. Various uses suggested for the waste heat dissipated from electrical power plants (two-thirds of the energy value of the fuel) include district heating for nearby residential or commercial installations, hothouse support for increased agricultural production, and prevention of ship channel freezing. All require considerable additional investigation.

A particularly appealing idea is the total energy system—an integrated package for electrical generation, air conditioning, water heating, steam generation, and any other energy functions required by a residential complex or shopping center. The greatest deterrent appears to be the problem of balancing demand among the various functions, but the potential is sufficiently promising that the National Bureau of Standards is conducting a carefully controlled experiment with a pilot total energy system in an apartment-shopping complex.[17]

SOME THORNY ISSUES—POLLUTION, INVESTMENT

Most energy sources can have significant environmental impacts, for example, strip mining and the disposal of heated water from power generating plants. Conversely, pollution controls have significant impacts on energy consumption. They can and do result in the additional use of energy, and may contribute to shortages of desired fuels. For example, motor-vehicle exhaust systems have been proposed to reduce automobile pollution by over 90 per cent; when implemented, they will introduce fuel penalties of 5 per cent or more (under 1970 performance) for *each* technique needed to control different emissions; a fully equipped car will probably experience at least a 15 per cent fuel penalty.[18]

To meet environmental quality standards industry invested some $9.3 billion in pollution control in 1970, and this investment rate is expected to double by 1975. The percentages of total annual capital expenditures invested in pollution control ranged from a high of 10 per cent for the iron and steel industries down to 0 per cent for the

communications industry, with 2.6 per cent for transportation and 3.8 per cent for electric utilities.[19] It is not clear, however, that all these investments produce energy-efficient pollution control systems. Firms in which energy costs are significant with respect to profits have probably installed efficient pollution controls. For other firms, federal standards may need to be considered.

Relating traditional profit incentives to efficient pollution control will be difficult until the principle of pollution control itself has become common practice. The capital investment and the institutional policies (such as environmental protection and licensing policies) take substantial time to implement on a regional or a national basis.

In some cases energy conservation is the fortuitous result of investment in other programs. For example, urban mass transit systems promoted to shorten commuter time and reduce highway costs also produce several times less pollutants per passenger mile than the automobile and reduce energy consumption considerably. Effective urban mass transit, for whatever goals, however, remains experimental in scope for lack of the capital necessary to modernize and integrate such systems as a substitute for the ubiquitous private automobile.

At any rate, environmental studies and programs should now have the included task of considering energy costs (and benefits). An explicit treatment should provide decision-makers with the tradeoffs which are the needed bases for thinking and actions by all of us.

We must begin to ask whether a slight relaxation in environmental standards, many of which may have been arbitrarily set with little thought to their full ramifications, could permit significant energy savings. For example, we should carefully examine the question of whether the nation's environment would be better served by obtaining emission reduction through policies designed

17. Department of Commerce, *The Application of Total Energy Systems to Housing Developments* (Department of Commerce, Washington, D.C., March 1972).
18. Environmental Protection Agency, *The Economics of Clean Air* (Environmental Protection Agency, Washington, D.C., February 1972).

19. Council on Environmental Quality, *Second Annual Report* (Government Printing Office, Washington, D.C., August 1971).

to reduce automobile use and to increase mass transit use rather than by maintaining current strict emission standards which increase engine inefficiency.[1] We also have to balance the extremely high costs which consumers are paying to obtain the last small increment of environmental protection against the potential energy and dollar savings from more energy efficient pollution control systems.

In summary, the close interdependence of these issues makes it essential that programs be coordinated in meeting where possible the objectives of both pollution control and energy conservation. The needed objectives include making pollution control systems more energy efficient and encouraging greater energy conservation by choice of those options (applications) which limit both energy consumption and pollution.

Improvements to energy conservation could also have significant implications for the national economy. Capital investment in the United States energy industry has been projected at $566 billion (in 1971 dollars) for the period 1971 to 1985. This enormous capital investment requirement raises several potentially serious problems with respect to all areas of our economy. The amount of money available for capital investment is rather inflexible—it depends strongly on the rate of savings, which tends to be very stable. An increase in capital needs for one sector, that is, the energy industry, could significantly affect the money market through increased interest rates. Hence, if conservation can reduce energy consumption, the sizable capital requirement in this area would also be lessened. One estimate indicates that the savings in energy consumption, as a result of conservation measures, could lead to potential savings of $97 billion, or 17 per cent of the projected capital investment of $566 billion during the fifteen-year period from 1971 to 1985.[10]

SUMMARY

We no longer can afford to ignore the serious potential consequences of our lavish use of energy. Continuation of the present rate of increase, particularly with the trend to imported fuels, will lead in short order to a level of dependency on imports that is disturbing for both the national security and the balance of payments.

The inevitable rise in the price of energy will presumably lead to some increases in the domestic energy supply. But our reserves, particularly in the preferred forms of petroleum, gas, and even low-sulfur coal are finite. Thus, the energy problem must also be attacked from the standpoint of energy conservation. The forthcoming rise in fuel prices will, of course, make more attractive some forms of conservation, which are at present economically marginal. Nevertheless, consumers, industry, and government will have to make difficult choices in the years ahead—between greater convenience and lower energy bills, between the high capital costs of energy conservation measures and the long-term dollar savings from increased energy efficiency, and between environmental protection and the availability of needed energy supplies.

Existing capabilities and technology, on which short- and mid-term improvements must be based, appear to offer substantial possibilities for reducing United States energy consumption within the next decade.[10] Long-term solutions to the energy problem, however, will depend to a considerable extent on the continuing appearance of new technological capabilities for increased efficiency of energy utilization and increased integration of energy applications. The capacity for continuing technological advances is, of course, dependent in turn on a strong relevant scientific base.

A word of caution is necessary. Recent experience has shown that technological advances alone will not solve the problem. The problem spans not only the traditional physical and engineering sciences but also those sciences which deal with human attitudes and actions, that is, the social sciences, and includes a more fundamental understanding of underlying economic principles. The challenge to all sectors of American science should be clear.

ACKNOWLEDGMENTS

I thank Robert H. Kupperman of the Office of Emergency Preparedness, chairman of the interagency working group on energy conservation. This article draws heavily from the findings of that group's report. I am also grateful to Felix Ginsburg, Frederick McGoldrick, and Richard Wilcox for their review and thoughtful observations and to Philip Essley and Robert Shepherd for their critical reading of the article.

The Impact of Technical Advice on the Choice for Nuclear Power

JOHN S. STEINHART   1974

In 1954, the first nuclear-powered electric generating plant was placed in service near Moscow. That same year the United States launched the first nuclear submarine, *Nautilus*. At the time these developments appeared promising. If there was some nagging worry that, as a military ship, the *Nautilus* represented another step in an arms race, it also might be a step toward the widespread economical nuclear-powered transportation that many hoped for. The accomplishment of the Soviet Union seemed to show that our own hopes for cheap nuclear electric power were possible.

This kind of tangible achievement must have been welcome to politicians as well as the public, because the 1954 budget assigned more than 1.2 billion dollars to the Atomic Energy Commission, and the Atomic Energy Commission had been spending about 1 in every 7 research dollars expended from the public purse. Nor were these expenditures new. By then more than $6 billion of public funds had been invested in nuclear reactor research and development.[1] If the long-promised peaceful uses of nuclear energy did not come to fruition, we would have been left with the most terrible weapons of war ever devised as the principal result of the vast expenditure of public funds and technical talent. It would not have seemed like much of a bargain in those days of the politics of "brinkmanship."

By 1957, when the first United States nuclear power station was put into service in Shippingport, Pennsylvania, the commitments of public funds and the efforts of many of our finest scientists had continued for more than a decade. It is a fair question to ask: What forces carried this commitment forward for so long, at such public expense, without more in the way of practical results? Clearly, our society made a decision to pursue nuclear energy as a power source. It is my purpose to examine the circumstances and machinery under which this decision was made.

1. U.S. Bureau of the Census, Historical Statistics of the United States, Colonial Times to 1957, Washington, D.C., 1960.

Original article by Dr. John S. Steinhart. Dr. Steinhart is professor of geology and geophysics and environmental studies at the Univerity of Wisconsin-Madison. He has engaged in extensive science advisory work, having served in the Office of Science and Technology, Executive Office of the President, 1968–1970. He is associate director of the Marine Studies Center at the University of Wisconsin-Madison and coauthor of a recent book, *Energy: Source, Use, and Role in Human Affairs.*

Two disclaimers seem necessary at the outset. First, the desirability or hazards of a nuclear powered future have been discussed elsewhere in this volume in considerable detail. I do not propose to enter this argument, nor to pass final judgment on the merits of the case. It is sufficient to note that there is a real dispute among competent, sober men. Second, since the following discussion often must examine the performance of a scientific advisory apparatus made up of real people, I must state that hindsight may reveal things not obvious at the time, and that these advisors (many of whom are colleagues, acquaintances and friends) are, as far as I know, honest and able men of good intentions.

The question is this: How have we embarked on a nuclear-powered future, what mechanisms have continued the commitment in the face of the dispute about the decision, and how may future policy for energy be affected by his history? In a larger sense, these questions could and should be asked for any technical decision that may be made in a democratic society. If we are unable to anticipate the consequences of our technical decisions with the institutional machinery at hand, we stand continually vulnerable in a society increasingly shaped by technological change.

THE EARLY DAYS OF TRAGEDY AND EUPHORIA

On August 7, 1945 the atom bomb exploded in the air over Hiroshima. The grisly results may have been exaggerated by the press that day, but subsequent radiation deaths and genetic damage produced a toll that is still being counted. It is not that the results were quantitatively unprecedented, for one massive fire bomb raid on Tokyo produced as much or more death and tragedy, but, thought most of us, "a single bomb from a single plane. . . ." In a different way the bomb burst upon the American people and most of the elected officials in Washington. Awe at the technical achievement, horror at the human suffering, optimism about the

imminent end of the trauma of World War II, and just plain bewilderment underlay the talk at the corner tavern and in congressional cloakrooms. German nuclear physicists, interned in England, at first refused to believe it. Among them, Otto Hahn (who later received a Nobel Prize for the discovery of fission) considered suicide.[2] The American pilot of the plane that dropped the bomb in Hiroshima went slowly and publicly mad.

In the years just following World War II, the awe won out among these conflicting emotions, at least in the government of the United States. Scientists fresh from the wartime technical successes were summoned before the Congress and sought out as evening speakers or special attractions at Washington meetings. David Lilienthal, who became the first chairman of the Atomic Energy Commission, described the situation this way:

The scientist's mastery of nature, as so dramatically evidenced by the Atom, was expected to lead to the conquest of hunger, of poverty, and of the greatest of human ills: war. As "Master of the Atom," the scientist had transformed the world. His views on all subjects were sought by newspapermen, by Congressional committees, by organizations of all kinds; he was asked in effect to transfer his scientific mastery to the analysis of the very different questions of human affairs: peace, world government, population control, military strategy, and so forth. And his authority in these nonscientific areas was, at least at first, not strongly questioned.[3]

Almost all the scientists were strongly affected by their wartime experiences. Much has been written about the wartime scientific projects of the Office of Scientific Research and Development, and especially about the building of the bomb.[4] The paths

2. Jungk, Robert, *Brighter Than a Thousand Suns*, Harcourt Brace Jovanovich, Inc., New York, 1958, p. 220.
3. Lilienthal, D. E., *Change, Hope and the Bomb*, Princeton Univ. Press, Princeton, N. J., 1963, p. 64.
4. Among the numerous accounts of the building of the bomb, three especially illuminating and well-written ones are: Jungk, Robert, *op. cit.*; Laurence, W. L., *Men and Atoms*, Simon &

of a number of extraordinarily intelligent physicists had been crossing and recrossing from early days at Gottingen and elsewhere. They and their teachers restructured physics almost completely. These dizzy heights of discovery and achievement were lived out in the presence of (but separate from) the gathering storm of World War II. Then, increasingly, their private lives were uprooted as European colleagues fled to Britain and the United States. Colleagues in the Soviet Union became silent and were sometimes imprisoned.

As World War II began, these scientists shared with almost everyone else, an overwhelming sense of the morality of the Allied cause and the grimness of the Axis threat. When they were mobilized into the projects of the Office of Scientific Research and Development, many found a cause for their concern and conviction—as well as one for their talents.

In the next few years these scientists worked in relative isolation from the society around them. The Office of Scientific Research and Development laboratories—the Chicago Metallurgical Laboratory, the Massachusetts Institute of Technology's Rad Lab, Los Alamos, and several others—were a heady and purposeful atmosphere, especially for the young scientists that became the post-war advisors to the government. It was like living and working amid a continual meeting of the world's finest physical scientists. No university, before or since, could match the concentration of talent. Budgets, too, were almost unlimited if one could convince the group that an idea was worth pursuing. Many participants later spoke of these times with a fondness reserved for the great adventures of life.

Schuster, New York, 1962, 319 p.; Davis, N. P., *Lawrence and Oppenheimer*, Simon & Schuster, New York, 1968, 384 p.

An account of all the major technical achievements of the Office of Scientific Research and Development during World War II is: Baxter, J. P., 3rd, *Scientists Against Time*, Little, Brown, 1946, MIT Press, Boston, 1968, 473 p. Personal memoirs of these events have been written by many of the participants (see Davis *op. cit.*, for an extensive bibliography).

It is not that there were no doubts or second thoughts. As the bomb neared completion and testing (and as it became clear that the Germans had not succeeded in developing a fission bomb), protests were heard. In June of 1945, the Franck petition was forwarded to Secretary of War Henry Stimson opposing the military use of the bomb. Among its signers was Glenn Seaborg—later Chairman of the Atomic Energy Commission—then just turned 33. When the successful bomb test was detonated at Alamagordo, the heady feeling of accomplishment was joined by forebodings of what the risks for civilization might be. Oppenheimer distilled the feeling for many: "There floated through my mind a line from the Bhagavad-Gitā in which Krishna is trying to persuade the Prince that he should do his duty: 'I am become death, the shatterer of worlds.' I think we all had this feeling more or less."[4]

Against this background a new estate in public affairs, the science advisors, were catapulted into government circles at the end of the war.

THE NEW-OLD PRIESTHOOD

Not all the wartime scientists and engineers involved themselves in public affairs at the end of World War II. The leaders of the successful proximity fuse project returned to their research in physics—some even abjuring the use of government funds. Some left physics altogether and made important contributions to the field that later became molecular biology. Yet another group of the wartime scientists involved themselves in the public debates about nuclear arms control without entering the growing government science advisory apparatus. Leo Szilard, for example, coauthor of the famous Einstein letter that had begun the bomb project (and author of the Franck petition opposing its use in 1945), devoted a substantial portion of his remaining years to the search for controls to nuclear weapons. Before he died in 1963, Szilard founded the Council for a

Livable World, which continues the search for an end to nuclear arms.

The group that became the government science advisors were thus largely self-selected. Those who moved to other fields, whether in search of new challenges or seeking escape from the horror of their creation of the bomb, removed themselves. Those who chose the public forum for opposition to governmental nuclear weapons policies were not invited into the inner circles. Known pacifists, like Einstein, were rarely consulted by the new scientific advisors about anything.

But what has all this about bombs and their builders to do with energy policy? It is one of my principal theses that, to the extent scientific advice played any role in creating the momentum that carries us into a nuclear power based future, the scientists that provided this advice (and still do) were a very special and biased subset of the scientific community. They had these things in common: (a) they were drawn from a narrow set of specialties in the physical sciences (nuclear physics or chemistry and related fields); (b) they shared a set of intensely moving experiences during World War II; (c) they were not among those who objected to weapons work or whose values dictated that they leave the field; (d) they were born between 1890 and 1920 (suggesting a common generational bias); (e) they had enough liking or tolerance for the quasi-political world of science advising to be extensively involved in the heady atmosphere of deliberating on crucial public issues at the highest levels of government. These men, and their self-selected successors have run the science advisory apparatus ever since.

None of this is news to the participants themselves. Harvey Brooks, a long time member of the inner circles, writes:

The advisory role tends to become self-perpetuating, and constitutes a kind of sub-profession within the scientific professions. Certain administrative skills and some degree of political sophistication are factors almost as vital as scientific competence and reputation in the selection of members for top committees. Experience in one of the major wartime laboratories, especially the MIT Rad Lab and the laboratories of the Manhattan (bomb) Project, or an apprenticeship on one of more of the military 'summer studies,' still appears to be a useful qualification for scientific advising. There is as yet little sign of a change in generations that would affect this pattern. Even the relatively few younger scientists that have filtered into the higher level advisory committees are often students of one of the wartime giants like Rabi, Teller, Oppenheimer, or Fermi.[5]

What is this "political sophistication" that is so valuable? It cannot be a sophistication born of the experience of elective politics, for none of these men is known to have stood for elective political office. With few exceptions, none is known to have participated in the scholarly study of political science. Brooks' statement suggests that "political sophistication" is, in part, a shared world view with those already there. This view includes a center or slightly left-of-center political posture. Neither the radical or extreme liberal scientists nor the politically conservative scientists are invited into the self-perpetuating apparatus of science advising (although such scientists are sometimes summoned to testify before congressional committees by elected officials of similar political views).

One article of faith among those who hold centrist or liberal political views—whether scientists or not—has been that reasoned discussion will produce a consensus plan that will solve both technical and social problems. Similarly, this group has shown a consistent preference for, and faith in, technological remedies for problems rather than politically negotiated ones. Eugene Wigner, Nobel Laureate in Physics, asked, "why scientists so consistently overestimate the realizibility of what appears to them the rational solution? It is, in my opinion," he continued, "because they are not sufficiently

5. Brooks, Harvey, "The Scientific Advisor," in R. Gilpin and C. Wright, eds., *Scientists and National Policy-Making*, Columbia Univ. Press, New York, 1964, pp. 73–96.

aware of the phenomenon of the conflict of desires."[6] As long as the self-perpetuating scientific advisors share a common political position on the political spectrum, as well as a common background, officials receiving advice from this group can be deluded into believing that there is genuine consensus among the technical community.

In a larger sense, many students of politics conclude that for a decision-maker adequate representation of conflicting viewpoints is more useful in choosing among alternatives than is the "platitudinous consensus"[5] that is often the result of the present system. Yet the faith in reason to resolve all political and value differences is so strong that Brooks justifies exclusion of some dissenters on the grounds that:

People with very strong viewpoints which are impervious to rational argument or compromise merely tend to lead to a hung jury which does not help the decision-maker. A majority vote is much less useful than a well-reasoned consensus in providing scientific advice.[5]

Yet Walter Heller, writing from the experience as Chairman of the Council of Economics Advisors, concludes that value judgments are "inescapable" and "obligatory."[7] At the very least, it is hard to see how the elimination of contradictory views from the scientific advisory apparatus improves the quality of government decisions.

WHO IS IN AND WHO IS NOT

It has already been mentioned that the World War II leaders of the scientific advisory apparatus represent a narrow collection of specialties. By looking at the composition of the President's Scientific Advisory Committee, the extent of this concentration may be easily seen. Between its origin in 1951 and 1966,[8][*] there were 65

members of the President's Scientific Advisory Committee. Of these 32 are physicists, 8 are from closely related fields of physical chemistry, nuclear chemistry, biophysics, or nuclear engineering, and 5 others are veterans of the World War II Manhattan projects (though they identify themselves as practitioners of other disciplines). The next-largest grouping are physicians and biologists, together including 10 members. The same analysis performed on the General Advisory Committee of the Atomic Energy Commission, not surprisingly, reveals even more concentration.

Since the special interest here is in energy and, particularly, the choice of nuclear energy, it is important to note who has not been in these advisory councils. There has never been a geologist or a resource expert on the President's Scientific Advisory Committee.[**] When the National Aeronautics and Space Administration and the National Science Foundation assembled a panel of solar energy experts, it was clear that none had ever been on the President's Scientific Advisory Committee. Other fields in the sciences and social sciences have been similarly not represented (economics, ecology, and the agricultural sciences come immediately to mind), but they are of less immediate interest to the subject of this essay.

It is quite true that when the President's Scientific Advisory Committee undertook studies, experts relevant to the question under study were sought out, but the make-up of this group usually determined the questions that led to such studies. In any

6. Wigner, E. P., as quoted in Lilienthal (3) *op cit.*
7. Heller, W. W., "Economic Policy Advisors," in T. E. Cronin and S. D. Greenberg, eds., *The Presidential Advisory System*, Harper and Row, New York, 1969, pp. 29–39.
8. U. S. Government, *The Office of Science and*

Technology, A report to the Committee on Government Operations, House of Representatives, U. S. Government Printing Office, 1967.
[*]The President's Science Advisory Committee obtained that title in 1957, when the office of President's Science Advisor was created. A somewhat less influential committee, the Science Advisory Committee, was set up in 1951 in the Executive Office of the President. This analysis considers members of both committees.
[**]There have been two members of the President's Scientific Advisory Committee from the related field of geophysics, but neither of them has ever claimed to be geologists, nor are they regarded as geologists or resource experts by the geological profession.

case, it is not surprising that the scientific advisory apparatus favored nuclear energy as a power source for society. To put it bluntly, if you ask an economist for a solution to the problems of society, he will propose an economic solution in most cases—and if you ask a nuclear physicist about energy sources, do not be surprised if nuclear power is suggested.

It can be argued that much of this is beside the point. After all, nuclear power does work, in the sense that the heat generated by a reactor does supply electric power (although up to 1972 we had still supplied more energy to the nuclear power program than we had obtained from it). What is more, the arguments displayed earlier in this book show that the risks of nuclear power are under active discussion. The question of energy sources for society, however, is never a qualitative question. Many sources of energy—sun and wind, for example—have been demonstrated to work centuries ago, but are little used today. The question of advantage or risk associated with the choice of an energy source can only be dealt with if one asks: In comparison to what? Such comparative discussions must involve economic, health, and environmental considerations as well as technical ones. Usually a good many aesthetic and value considerations are important as well. With the advantage of hindsight it is clear that these discussions did not take place. Meanwhile, public money has been heavily committed to nuclear-power development. Would things have been different if a broader mix of backgrounds had characterized the scientific advisory apparatus? That question cannot be answered directly, but a retracing of the steps in the decision may help readers make their own estimates.

THE DECISION FOR NUCLEAR POWER

It would be all too easy to credit or blame scientists solely for the decision to pursue nuclear power. History is much more complex than that. The overwhelming impact of the atomic bomb on the public has made many forget that the romance of "splitting the atom" and the possible results of doing so had, even in 1945, a long history in the popular press. Magazines like *Popular Mechanics* and *Popular Science* had frequent articles on both peaceful and weapons uses of nuclear power all through the 1930's. Ernest O. Laurence's experiments with the early cyclotron led to newspaper headlines like "DEATH RAY" and "POSSIBLE CURE FOR CANCER" in 1934, and it was written in the *New York Times* that "transmutation [of the elements] and the release of atomic energy are no longer mere romantic possibilities."[9]

After World War II, the public notice was much intensified. David Lilienthal wrote that:

... it is well to recall the temper of those early days of the Atom. No predictions seemed too fantastic, whether of the doom of civilization through nuclear holocaust or of a world beneficently transformed through the peaceful use of this great new source of energy. Men were convinced that they were living in a world in which only the Atom counted, and man was almost incidental.[3]

Writing of the 1955 International Conference on the Peaceful Uses of Atomic Energy, W. L. Laurence (easily the most knowledgeable journalist in the field at that time) listened to "one scientific report after another, presented by world authorities in the field" and concluded that "it thus became clear ... that man was on the eve of the greatest industrial, social and economic revolution in the millions of years of his evolution on earth. From a civilization limited and controlled by scarcity, he is about to emerge into the green pastures of a civilization built upon plenty."[10] Laurence quoted the confident predictions of scientists that the power of the hydrogen bombs would be tamed "possibly no more than twenty years from now." The twenty years have now elapsed, and fusion power is expected in twenty years—or never.

9. Davis, N. P., *op. cit.*, p. 63.
10. Laurence, W. L., *op. cit.*, p. 243.

The bureaucracy of nuclear power had somehow acquired similar ideas. The leaders of Congress's powerful Joint Committee on Atomic Energy made rosy predictions, and Lewis Strauss, chairman of the Atomic Energy Commission, said in 1954 that "it is not too much to expect that our children will enjoy in their homes electrical power too cheap to meter, will know of great periodic regional famines in the world only of matters of history. . . ." Now we struggle to produce nuclear electric power at prices even competitive with other methods, and stand on the brink of world famine.

The past is littered with many foolish predictions about the future—on many subjects. The point here is that the unfulfilled predictions were based on what scientists said, or did not deny, and they are the very same scientists who have occupied central positions in the scientific advisory apparatus ever since.

It is often said that science has become the secular religion of our time. Despite this it comes as no surprise to most that scientists on political and ethical matters are right or wrong with about the same frequency as everyone else, but Brooks says that "scientists are often in a position to exercise their political and ethical judgments as citizens in a more realistic and balanced manner than other citizens."[5] Less often understood is the fact that scientists are, occasionally, completely wrong on technical and scientific matters. In a celebrated case, Lord Rutherford, pioneer leader in studies of atomic physics, asserted near the end of his life in 1936 that nuclear fission was a long way off and probably impossible. The following year Hahn, Strassman, and Meitner achieved fission. To add to the irony, fission had already been obtained in 1934 by Fermi and his co-workers, but he failed to identify the results of his own experiment. Fermi was, by common agreement, among the greatest physicists of the twentieth century, yet did not recognize his achievement, even after a German chemist, Ida Noddack, pointed out the result in a paper of the time.[11]

In a more recent, and less well-known,

case in 1962, a British astronomer objected to a planned high-altitude nuclear explosion (named Starfish) on the grounds that it would disrupt the Van Allen radiation belts. In the resulting dispute, the Atomic Energy Commission and the Department of Defense assembled a panel of distinguished scientists who concluded that, "these planned U.S. explosions in the upper atmosphere will not greatly disturb conditions for the magnetic orbits of the particles of the Van Allen belt. Perturbations produced on the inner belt will be minor if detectable at all." James Van Allen, discoverer of the radiation belts, agreed with the panel. In the face of these reassurances, President Kennedy told a news conference that his advisors had deliberated carefully and that "Van Allen says it is not going to affect the belt and it is his."[12] Nevertheless, when the bomb went off, on July 9, 1962, the resulting trapped electrons seriously damaged five United States satellites and Britain's first satellite, Ariel. Clearly, committees as well as individuals in the scientific community can be very wrong.

None of the foregoing should be interpreted as a veiled implication the scientists are always wrong—quite the reverse. Because most good scientists are right most of the time about matters within their field, they have little experience, or expectation, of having events prove them wrong. The best clue an outsider to a scientific dispute can use is that disagreement among qualified and serious men is likely to signal that the answers are uncertain. Unfortunately, simple observations such as the foregoing have never penetrated to some of our political leaders. Richard Nixon, upon awarding the National Medals of Science, May 21, 1971, said that he had read the citations accompanying the awards and went on: "I have read them, and I want you to know that I do not understand them, but I want you to know, too, that because I do not understand

11. Segre, E., *Enrico Fermi: Physicist*, Univ. of Chicago Press, Chicago, 1970, 273 p.
12. Cox, D. W., *America's New Policy Makers*, Chilton Books, Philadelphia, 1964, 298 p.

them, I realize how enormously important their contributions are to this nation."[13]

If this attitude prevails, and the government credits the inside science advisors more heavily than others—rather than struggle to understand the facts of the case—we are always open to the admittedly rare colossal blunder.

THE MOMENTUM FOR NUCLEAR POWER

By the late 1950's or the early 1960's, expansion of nuclear electric power plants was well underway. The most visible and powerful advocates of nuclear power were no longer the research and development scientists. Political figures on Congress's Joint Committee on Atomic Energy welcomed supporters of expanded nuclear power and treated those expressing doubts about the nuclear power program as hostile witnesses in an adversary court proceeding. By 1971, in authorizing hearings for the budget of the Atomic Energy Commission, Congressman Holifield suggested that the only work in radiation safety worth considering was that of the Atomic Energy Commission, and went on to say that "there is nothing else for them [the environmentalists] to lean on unless they want to lean on the fairy tales that are sold to *Esquire, Look,* and a few other of our popular magazines."[14] The statement is startling partly because it ignores the long and detailed dispute in technical books and journals, and partly because none of these critics of nuclear policy was called to testify before the committee that year.

In elementary mechanics momentum is defined as mass times velocity. These remarks suit that definition well. The mass in this case is composed of three parts: (a) a well established government bureaucracy includ-

ing the Joint Committee on Atomic Energy and the Atomic Energy Commission with a total budget of more than $2 billion; (b) the electric utility industry with vastly larger sums already spent or committed to nuclear electric-generating plants; and (c) the oil industry, the entry of which into all aspects of nuclear fuel supply and processing and to reactor manufacture has been documented elsewhere.[15] The political lobbyists of the oil industry have been the most powerful in Washington for decades. The analogue of velocity in nuclear energy development arises from the times required to provide solid, unchallenged results about medical risk, safety of fuel transport, and storage as compared to the rate of expansion of nuclear generating plants. Thus, we are building new nuclear power plants even though the Atomic Energy Commission has announced that there is not yet a satisfactory way to store nuclear wastes.

The decision for nuclear power is, for most of these vested interests, a closed issue. Money, jobs, prestige, and political power provide a momentum that overshadows mere scientific uncertainty. It is no surprise, then, that many outsiders, who feel the commitment to nuclear power is a mistake, seek to apply political pressure to gain their ends. In the absence of some abrupt catastrophe, the reversal of the nuclear power commitment on the basis of some suggested technical risk seems very unlikely.

CONCLUSION: WHO DECIDED WHAT?

Scholarly analyses of scientists effect on public decisions emphasize the enormous differences in impact, depending upon what is being decided. In a carefully reasoned study, Schooler[16] concludes that scientists have had greatest impact in governmental "entrepreneurial" programs for which atomic energy and the space program are

13. Reported in Science and Government Report, Vol. 1, No. 9, June 1971, Washington, D.C., p. 3.
14. U.S. Government, *AEC Authorizing Legislation, Fiscal Year 1972,* hearings before the Joint Committee on Atomic Energy, U.S. Congress, Part 1, Feb. 3, 4, and March 2, 1971. U.S. Government Printing Office, 1971.

15. Steinhart, C. E. and J. S., *Energy: Sources, Use and Role in Human Affairs,* Duxbury Press, North Scituate, Mass., 1974, 362 p.
16. Schooler, D., Jr., *Science, Scientists and Public Policy,* Free Press, New York, 1971, 338 p.

typical. These programs are seen initially as largely technological, and provide growing benefits to special groups (who in turn become advocates for the program) and do not, at first, challenge any vested interests. Haberer[17] emphasizes the ambivalence of scientists concerning responsibility for applications of their scientific results and the "prudential acquiescence" of the leaders of science to political pressure.[17]

In the unique setting of the weapons successes of post World War II we found a public and a government in awe of the physical scientists. The prospect of harnessing nuclear energy seemed attractive indeed and even a political conservative like Enrico Fermi wrote in 1945, without thinking it needed justification: "A few remarks about the peaceful possibilities of atomic energy. There is little doubt that the applications both to industry and to sciences other than physics will develop rapidly."[11] At that time, few asked the question, that became relevant later: "What other options than nuclear power have we, and how do they compare with nuclear power?"

As the 1950's began and wore on, there were more nuclear weapons developments, but the peaceful applications failed one by one. After $1.1 billion, the nuclear airplane was given up, partly at the recommendation of the science advisors. The nuclear cargo ship *Savannah* was launched with great fanfare about a new age in commercial shipping, only to be put quietly into mothballs a few years later. It was not competitive commercially. Schemes to increase natural gas flow, extract oil from shale and tar sands, and create underground storage reservoirs from underground nuclear explosions have had modest-scale tests and even some mixed success, but met increasing resistance from local residents. Plans for creating new harbors and canals from nuclear bomb excavation are seldom even spoken of anymore. Medical and research uses of radioactive isotopes

continue, but, by the 1960's, had taken a place among other tools and treatments. The glittering promise of a "cure for cancer" did not materialize.

We have seen a generation of brilliant scientists thrust into a position of prominence that they did not seek. Those who tolerated the role, or liked it, have spent a large portion of their lives ever since advising the government.[18] * Always living with their part in creating the bomb, the advisor scientists must have noted the peaceful uses slipping away or falling far short of the rosy hopes of the 1940's. Finally, only nuclear generation of electric power is left. We should not be surprised if the old guard of advisors cling to the hopes for this peaceful use, even in the face of doubts.

Alvin Weinberg, bomb project physicist and presently White House energy advisor, is forthright about the choice. Admitting some disadvantages of nuclear power to be weighed against the advantages, he calls the choice a "Faustian bargain" for society and urges for nuclear power. But, according to the legend of Faust, such a bargain must be made by each person with Mephistopheles. When the technical issues raised in this book are understood we must each make a choice. The momentum that now drives the expansion of nuclear power has removed the decision from the hands of the scientists, however; if and when the disputes are settled among technical people, or even if the disputes are not settled, the choice of society's energy source will be fought out in the political and economic arenas. If the cost of nuclear electric power continues to increase as it has in recent years, still more public subsidy will be required. The breeder-reactor

17. Haberer, J., *Politics and the Community of Science,* Van Nostrand Reinhold, New York, 1969, 337 p.

18. Atomic Energy Commission, *In the Matter of J. Robert Oppenheimer,* Transcript of hearing before the Personnel Security Board, April 12 through May 6, 1954, U.S. Government Printing Office, 1954, p. 451.
*I. E. Rabi noted in the Oppenheimer hearings that he spent 120 days a year on advisory committees. Allowing some nonworking time and a little time for homework for these advisory services, 120 days must represent more than half his working time. Such commitments of time are not unusual among the inside advisors.

and fusion programs will require public money. The machinery for making these choices and commitments has not always been responsive.

As for the technical advisory machinery of government, the President's Scientific Advisory Committee, President's Science Advisor, Office of Science and Technology mechanism has been dismantled by the Nixon administration. Many scientists, not among the insiders, had already become disillusioned with that machinery. Paul Ehrlich, for example, says that "these Washington committees also seem to wind up having an impact that's zero . . . or bad."[19] Yet the same group and their hand picked successors are still in control of the advisory apparatus. With a tenure longer than all but four of the Supreme Court Justices in the history of the United States, they recommend, in a recent National Academy of Sciences report,[20] a

reestablishment of the old system with minor cosmetic changes. Eugene Skolnikoff, a long-time student of the scientific advisory apparatus, concludes sadly that "the report . . . reflects the attitudes and arguments of the 50's and 60's, and rather obstinately refuses to recognize some of the important lessons of OST and PSAC."[21] The scientific advisory apparatus, like most human institutions, seems to find self-criticism and reform difficult.

The history recalled in this essay suggests that, though scientists must bear a fundamental share of the praise or blame for the choice of nuclear power, there is no nefarious conspiracy to be found. Many scientists still work—as this volume shows—to clarify the advantages and risks in our choice of energy sources. From a world-wide point of view, it is not too late to end the race toward a nuclear-powered future, but, with the commitment and momentum now driving the choice, it will become more difficult each year to change.

19. Interview with Paul Ehrlich, *Mother Earth News*, No. 28, July 1974, p. 12.
20. National Academy of Sciences, *Science and Technology in Presidential Policy-making: A Proposal*, Report of the ad hoc Committee on Science and Technology, National Academy of Sciences, Washington, D.C., June, 1974.

21. Skolnikoff, E., *Science and the President: A New Debate with Old Solutions, Public Science*, Vol. 5, No. 6, 1974, pp. 1–9.

Role of the Government in Resolving Energy Problems

STANLEY YORK   1974

One question is fundamental to any discussion of energy: "Will there be enough fuel, in environmentally acceptable forms, to meet future needs of the world's population?"

No one can flatly answer "yes" to that question, and that fact neatly summarizes the nature of our energy problem. We simply do not know if there will be enough energy to go around. The material in this book of readings states that we have the *potential* to find adequate supplies, but what we do not know is how, and whether, that potential will be translated into reality.

Private enterprise and government both will play major roles in the quest for adequate energy resources. The purpose here is to identify what functions government should perform, and, implicitly, what role will be played by the private sector. Before doing that, however, a review of the important points made elsewhere in the book is in order to put the suggested role of government in perspective.

M. King Hubbert's article on world energy resources puts into meaningful terms the finite nature of our existing fossil-fuel resources. His widely respected estimates indicate that roughly one-half of the world's petroleum and natural gas will be exhausted by the end of this century—just twenty-five years from now. From that point on, the gap between supply and demand will grow dramatically unless alternate sources are available. The implications are clear, and quite serious—we will soon start running out of the fuels on which we rely most.

As Dr. Hubbert and others point out, there are alternatives to oil and gas, including a more abundant fossil fuel (coal), or truly unlimited energy resources, such as certain forms of nuclear reactions, hydrogen, and the sun. But none of these alternates will appear magically on the scene—there simply is no guarantee that the great potential of alternate fuels will be realized by the time oil and gas production starts to irreversibly decline.

Thus, an obvious role for government is to take those actions that expedite development of environmentally acceptable energy resources to replace fossil fuels, particularly oil and gas.

But what about today, and the foresee-

Original article by Mr. Stanley York. Mr. York has served as director of the Wisconsin Office of Emergency Energy Assistance. A minister and former state legislator in Wisconsin, he was appointed director of the state energy office in November, 1973, and has been instrumental in formulating energy policy for the state.

able future, when relatively vast quantities of fossil fuels still exist? Do these relatively ample supplies mean that for the rest of this decade and beyond we will face few problems?

Hardly—in the more immediate future we face critical problems in bringing fossil fuels out of the ground and putting them to use in environmentally acceptable forms. We further face the task of reconciling the conflicting opinions of respected experts on the question of nuclear fission.

It is highly unlikely that these nearer-term problems will be resolved so as to assure adequate energy supplies. To be sure, no imminent collapse of the economy can be realistically foreseen, but it does appear certain that continued imbalances will exist between available supplies and legitimate needs. Even during periods when supply and need are in rough balance, there will be recurring supply *dislocations,* situations in which acute shortages exist in particular geographical areas even though the over-all availability appears adequate.

Before considering the specific steps that government might take to supplement and guide the actions of private enterprise in resolving these problems, let us look in more detail at some immediate dilemmas.

Take coal, for example. By almost any standard, the world, and the United States, in particular, has vast supplies. A myriad of problems prevent coal from becoming an easy solution, however—a partial list includes labor problems, safety rules, environmental considerations, the cost of money, availability of mining equipment, and transportation problems.

As for natural gas, the supply will continue to be short in the foreseeable future, in large part owing to artificially imposed government price controls. When those controls are relaxed, as appears inevitable, many users will switch to other fuels, notably petroleum, which already is in short supply.

Our rubber-tired and air-borne economy will continue to put pressure on petroleum, causing an ever-increasing dependence on foreign producers. An increasing world demand for petroleum, with the producers getting a better price in other markets, and the possible political implications in the availability from foreign producers, will make foreign supplies questionable at best. Inadequate domestic drilling and refinery, pipeline, and storage capacity will continue to restrict the flow of petroleum. Furthermore, the use of natural gas and petroleum for other uses, such as the manufacture of fertilizers, pharmaceuticals, and plastics, will cut into the availability of these products for transportation, heating, and industrial use.

As for electricity, the likelihood of adequate supplies is greatly diminished by unresolved safety and environmental issues, the financial problems of most utility companies, and the long lead time needed to build generating plants.

It is thus clear that there are no easy ways to significantly increase the supply of the fuels and sources of power upon which we now rely heavily. The supply most likely will lag behind traditional per capita demand levels, creating the constant threat of shortages and dislocations.

Because adequate supplies of energy are vital to the general welfare of our society, this likelihood of shortages mandates a role for all levels of government—federal, state, and local. The precise functions to be performed at each level must relate in some fashion to a national energy policy, in an attempt to meet the following broad goals:

(1) *Development of a true conservation ethic:* There is substantial waste and inefficiency in our current methods of energy consumption. Reducing this wasteful demand can go a long way toward alleviating near-term supply shortages and dislocations. In addition, conservation can extend, over the longer term, the lifetime of our scarce fossil fuel reserves and provide more time for the development of alternate fuels.

(2) *Perfection of technologically feasible, environmentally sound, and economically usable alternate sources of energy:* Government must take a major role in this area, primarily because the vast amounts of capi-

tal required, and the uncertainties that exist as to which is the "best" alternate, will continue to minimize private leadership in this field.

(3) *Development of existing energy resources to meet* legitimate *demands and needs:* This is not the same as urging rapid exploitation of known reserves; a government policy is needed that provides for a modest growth in demand to reflect legitimate needs.

(4) *Creation of operational strategies to be used in alleviating serious national or regional shortages/dislocations.*

There is a desperate need for a coordinated national energy policy. It will take time, and it will probably be done on a piecemeal basis, but that in no way diminishes the need.

The policy must encompass such questions as federal expenditures for research in development of alternate fuel sources; the distribution of fuel among competing states and regions; the recognition that some taxation policies encourage the use of more fuel, whereas others discourage consumption; and the role of economic policy, for example, the present policy that works toward cheap fuel as a result of the strict regulation of the price of natural gas at the well head.

A national energy policy must also consider the questions of safety and protection of the environment and ensure that the cure is not worse than the disease. This need for a national policy is made clear by the existing governmental response to energy matters. At the federal level, in fact, some of our problems are caused, rather than alleviated, by government action.

The Federal Power Commission regulates the flow of natural gas and electricity. The Atomic Energy Commission has major responsibilities relating to the nuclear generation of electricity. The Federal Energy Administration is responsible for petroleum products. A host of other agencies are in varying degrees responsible as they relate to either the production or the consumption of our energy resources. Many times, the right

hand does not know what the left hand is doing, or more important, the right hand sees as its objectives and methods those things that run directly counter to what the left hand is trying to accomplish. The net result is confusion, at best, which allows the providers of energy resources to play one agency off against the other to accomplish their own ends.

If the federal situation is bad, the state situation is worse. Every state has its own array of regulations to accomplish what that state felt was worthwhile, and often little attention is paid to the implications for other states.

In addition, the interests of energy-producing states may run directly counter to the interests of consuming states. States that produce petroleum, but must import electricity, face problems and pressures different from those states having vast quantities of low-sulfur coal lying just below the surface but minimal water resources for the production of electricity at the mine mouth.

To speak of the meaningful management of energy resources for the next twenty-five years is to talk of the urgent necessity for government at all levels to see each of our energy resources as an integral part of a total network and to see the absolute necessity for government to approach that network in a coordinated fashion.

Because a national policy in this country frequently represents the application of successful state actions, the need for more leadership in energy at the state level is a pressing requirement. Among other things, this means an immediate need exists for a review of legislation and regulations relating to gathering information about and control of energy resources. In many states, we have developed departments of human resources and departments of natural resources. Perhaps the time has come to develop a department of energy resources as a way to approach these problems.

The most important function of such a department would not be its most visible nor have its largest staff. The crying need today is for a "Division of Thinking and Planning," which would identify the issues in energy resource management, the policy alternatives available to government in dealing with those issues, and the powers and potentials of government presently available and those needed but not available. The second major function of the department would be to monitor the flow of energy resources. The agency would be required to develop a monitoring system for all parts of the energy network and would be responsible to see that the public and the decision makers had access to that information in a useful form. It is impossible to deal with a shortage of any kind without understanding the industry producing the product, the nature of the product, and the normal flow of that product through the market place. Not only is the information important, but, in order to be usable, it must be based on standard information-gathering techniques and standard definitions. In the summer of 1974, for example, state energy offices were getting one set of figures from large oil companies, another set from the Federal Energy Administration, and yet a third from their own state tax departments. The need for the development of standardized information is critical. A carefully conceived reporting system—coordinated at the federal level—recognizing the marketing and distribution system of the various energy industries is essential to the ability of government to act in a shortage situation. Without such a system, government will find itself unable to act rationally or affectively in the face of crisis.

Another critical responsibility of this agency would relate to public education regarding the production and use of energy. At the present, the general public and many in government know too little about the energy industries; the bits and pieces we have learned in recent months more often serve to confuse, anger, and puzzle, rather than clarify.

A small sampling of the questions for which the public demands and needs answers include:

(1) What is the oil depletion allowance and what is the real effect of such tax policies on the decisions of oil companies in relation to reinvestment?

(2) What is the relationship between American oil companies and international oil companies?

(3) What is the relationship between oil companies and oil-producing nations?

(4) Can and should this nation become independent in its capacity to produce energy resources?

(5) What is the relationship between producers of natural gas and producers of petroleum? Don't both products come from the same well?

(6) What are our reserves of various energy resources and how much of those reserves can we reasonably expect to extract? What is a known reserve? What is a proven reserve?

The list goes on endlessly.

Clear answers to these kinds of questions must be provided to the general public for a very pragmatic reason: the actions of government generally occur in response to public opinion, whether informed or not. Citizens will either demand a "Band-Aid" approach, which ignores the real problem or makes it worse, or they will call for reasoned policies, which will meet true short- and long-range energy needs.

In addition to the broad functions outlined above, government energy policy must come to grips with the many "people problems" that have developed in this era of energy shortages and higher prices. The following reflects some of the major problems and unresolved issues that fit in this category.

Perhaps the most obvious and pressing problem is the effect of rapidly rising prices on people with low, fixed incomes. Government will have to face the question of whether to deal with this problem by regulating prices, with its damaging corollaries, or by letting prices rise and then supplementing the incomes of those who cannot afford to heat their homes or drive to work. Welfare grants for heating and transportation are inadequate today for the prices now being charged for energy.

A great many social services in this country are provided today by volunteers, such as those furnishing transportation for disadvantaged groups. Unless the government is willing to provide for the additional fuel cost, such a program as "meals on wheels" could fall apart completely. Medical services for many dependent children could not be provided without volunteer drivers.

Apart from cost, there is the basic question of supply. In a shortage situation, a fuel oil distributor will be reluctant to put 10, 15, or even 50 gallons in the tank of a person with a $300 unpaid bill, even when the payment for the immediate fill is guaranteed. Many of the rural poor use No. 1 heating oil for outside supply tanks. Who will guarantee the availability of that fuel as refiners and distributors phase out its production?

In the area of regulated monopolistic utilities, there is the question of "disconnect" policies. Will government prohibit the disconnection of electricity from the home of the elderly man living alone who has not paid his bill? Will government require that the utility notify a social service agency before that disconnection takes place? What power will the social service agency have to deal with the situation when they are notified? If government prohibits such a disconnection, will the cost be borne by the utility, that is, by all of the other customers, or will tax funds be used to pay the bill? If disconnections are prohibited, what incentive will there be for anyone to pay his utility bill?

Another issue directly facing the nation is the question of improving utility rate structures. Many present rate structures encourage the additional use of electricity and natural gas by decreasing the price per unit as the quantity of use increases. This has encouraged the use of energy-intensive equipment and has contributed to our shortage situation.

If these rate structures are inverted or otherwise changed, what will be the impact on the small user in terms of his utility bill? More important, what will be the impact in terms of the price he must then pay for all other goods and services which will reflect the higher utility rate? Should government tax increased use? This might help the utility bill of the individual consumer but would do nothing about the increased cost of goods and services.

Such "people problems" are significant and, at this point, have not seriously been faced by government. There are several things government might do that could make a difference. One would be to establish a set of standards against which a householder could judge his use of fuel. It could itemize the relative energy cost of: (a) frost-free refrigerators as opposed to the do-it-yourself kind; (b) air conditioners, with or without the use of an attic fan or supplemental dehumidifying equipment; (c) insulation standards, and so forth. In states in which there is a state building code, they could be written into the code. In those states without a code, the state could promulgate a voluntary standard for the use of the householder.

States can develop a directory of resources for those people, who, squeezed in the energy cost crunch, turn to government for help and make it clear as to who has the authority to do what and also specify those situations in which no agency of government is able to act. This not only will provide a useful directory for persons in trouble and agencies trying to help, but also will become a guide for legislators studying areas where there are no current remedies.

States can publish current information on energy costs so that those on tight budgets, and particularly those on fixed incomes, can plan ahead. Various kinds of budget prepayment plans are available from many companies in the energy industry. These can be publicized, and those most needing them can be encouraged to participate.

One major problem plaguing the states is the question of what to do with those persons temporarily out of work because of the energy crisis, especially in those situations where they will probably be called back to their olds jobs in 3 to 6 months after a dislocation has passed. If public employment is a possibility, on a short-term basis, unskilled or semiskilled people could be trained quickly and easily to make energy assessments of private homes on request from a householder. A check list can be devised in two parts, one showing an owner what he can do, and the other showing a renter what he can do. The "energy conservation aide" (that is, an unemployed worker) then can be trained to take the check list into a home, review the home against the check list, make suggestions for things the householder can do to improve his situation, and actually demonstrate for the householder, a series of do-it-yourself actions that can be taken, such as changing the filters on a furnace. The aide could be equipped with several brochures published by the state that would indicate such things as the benefits of regular furnace maintenance. If the householder is an owner, he can be shown permanent improvements and repairs. If he is a renter, he can be shown temporary improvements and repairs and be supplied with a list of things he might want to discuss with his landlord. Very often those least able to afford the high cost of fuel drive automobiles and live in homes that are very wasteful of energy resources. They frequently also know the least about the

things they can do to make the situation better. Such an energy conservation aide program would have the dual effect of saving money for the householder and conserving much needed fuel for the public at large.

It will take imaginative action and the careful expenditure of funds on the part of the states to affect major conservation results on a voluntary basis. Such activity will help address our "people problems" as well as our fuel problems.

This recitation of problems and issues with which government must deal is by no means all inclusive. In fact, the issue still is so new that unforeseen problems develop literally on a daily basis.

In terms of opportunities, then, the energy situation creates substantial areas in which government leadership and initiative will be required. The record of past governmental response to various crisis situations provides no firm assurance that this leadership and initiative will be forthcoming. The response of government in large part will reflect the degree to which the general public understands the energy problem and demands constructive action.

The Recycle Society of Tomorrow

GLENN T. SEABORG   1974

These days it is difficult enough to forecast what the world will be like in 1974, let alone predict what kind of a world it may be in 1994. Although I enjoy future forecasting sessions, I cannot help but feel that many of us are projecting the kind of futures we would like to see, knowing that the world could be that way, hoping it will, rather than trying to base our forecasts on that combination of progress and the obstruction to it due to human foibles and follies that inevitably contribute to future conditions.

In the past we physical scientists have been especially prone to the blue skies approach to the future, tending to see the possibilities of gaining and applying new knowledge, of the "technological fix," and of the value of human cooperation. Those in the social and political sciences and in the business world are apt to be a bit more realistic because they deal more directly with the perversity and irrationality of human nature as well as its admirable features.

In recent years we have also seen the rise of forecasting by systems analysts, using elaborate computer models and warning us of total collapse based on the projections of current trends. Their studies offer some serious warnings of what could be. But I am inclined to agree with Rene Dubos' statement that "trend is not destiny."

My own thoughts about where we might be, and might be going, in the 1990's are based on what I consider to be a number of imperatives. That is, I think there are things that will have to happen, conditions that will have to prevail, given the physical limitations we face, but also given man's creativity and will to survive. In other words, sooner or later we will stabilize our population; we will minimize our environmental impact and efficiently manage our use of natural resources; and we will achieve a relatively peaceful world with a more equitable distribution of goods, services and opportunities throughout the world.

The difficulty is not in predicting that we

Dr. Glenn T. Seaborg is currently University Professor of Chemistry, Lawrence Berkeley Laboratory, University of California. He previously served as chairman of the U.S. Atomic Energy Commission and chancellor of the University of California. He won the Nobel Prize for Chemistry in 1951 as codiscoverer of plutonium and other elements.

From *The Futurist*, Vol. VIII, No. 3, pp. 108–115. Reprinted by permission of the author and *The Futurist*. Copyright ©, 1974 World Future Society.

will arrive at these points, or even how we will arrive, but when, and how much disruption, deprivation, and destruction will take place in the interim. This, in turn, will depend to a large extent on how quickly we grasp and apply certain principles of constructive human behavior, how we balance self-interest with mutual interest, to what degree and how soon we greatly improve cooperation between people and nations. It will also depend—somewhat fortuitously—on the kind of leadership that rises around the world. That is a most important catalyzing agent over which we seem to have little control.

Bearing all this in mind, I will try to give some approximation of where we might be in 1994—assuming that most things turn out right, that cool heads, kind hearts, and common sense prevail and guide all our other human assets. I will not speculate on what might happen if "the ghost in the machine," as Arthur Koestler refers to man's self-destructive flaws, takes over.

First, let me cover some general conditions that I think will have arrived by 1994, or will be well along in their formation.

TOWARD A "STEADY-STATE" WORLD

Broadly speaking, the 1990's will be a period characterized mainly by the need to stimulate maximum creativity in a tightly controlled social and physical environment. The reason for this is that by the mid-1990's we should be well on our way to making the transition from an "open-ended" world to a "steady-state" one. The United States will be in the forefront of this movement. Others will be following with various degrees of enthusiasm and reluctance, much depending on the sacrifice and cooperation of the advanced nations.

Some of the major characteristics of this transition will be:

(1) *Movement toward a highly disciplined society with behavior self-modified and modified by social conditions.* On the surface, and by the standards of many young people today, it will be a "straight society," but a happier, well adjusted one with a much healthier kind of freedom, as I will explain later.

(2) *Organization of a "recycle society" using all resources with maximum efficiency and effectiveness and a minimum of environmental impact.*

(3) *A mixed energy economy, depending on a combination of several energy sources and technologies and highly conservation conscious.* During this time we will still be searching for the best ways to phase ourselves out of the fossil fuel age.

(4) *Greater progress toward a successful international community spearheaded by the economics of multinational industry, new international trade arrangements that improve the distribution of resources, and a high degree of scientific and technical cooperation.*

Let me elaborate a bit on these characteristics, beginning with a few words about social attitude and behavior, as I believe these will be among the biggest determinants of where we are and where we will be going.

A HIGHLY DISCIPLINED SOCIETY IN THE 1990'S?

By the 1990's I suspect we will be a society almost 180 degrees different from what we are today, or some think we will be in the future. I see us in 1994 as a highly disciplined society with behavior self-modified by social and physical conditions already being generated today. The permissiveness, violence, self-indulgence and material extravagance which seem to be some of the earmarks of today will not be characteristic of our 1990 society. In fact, we will have gone through a total reaction to these.

We, therefore, will have a society that on the whole exercises a quiet, non-neurotic self-control, displays a highly cooperative public spirit, has an almost religious attitude toward environmental quality and resource conservation, exercises great care and ingenuity in managing its personal belongings and shows an extraordinary degree of reliability in its work. Furthermore, I see such a society as being mentally and physically healthier and enjoying a greater degree of freedom, even though it will be living in a more crowded, complex environment.

All this will not come about by making everyone subscribe and live up to the Boy Scout oath. I think it will come about as an outgrowth of a number of painful shocks—shocks of recognition, not future shocks—we will undergo over the coming years, one of which we are already getting in our current energy situation. The energy crisis is just the forerunner of a number of situations that we will be facing that will change our attitudes, behavior, and institutions—although I do not in any way minimize its importance or its far-reaching effect on all aspects of our lives. We will face a number of critical materials shortages and some failures in our technological systems that will force us into fairly radical changes in the way we are conducting our lives and managing our society.

My reasons for projecting the "straight society" I have mentioned for the 1990's spring from the series of reactions that the forthcoming shocks will elicit. The reactions will come in sequence but will also widely overlap. The first period will be one mainly emphasizing conservation and cooperation. Of course, there will be some degree of negativism about and noncompliance with the required changes. And there will be those who, with the usual amount of hindsight, blame others for not being able to anticipate current problems.

But by and large, most people will respond positively as they have in the past in the time of crisis. In fact, after the extended period of comparative affluence and self-indulgence most people have enjoyed in this country, we may witness something of a quiet pride and spartan-like spirit in facing some shortages and exercising both the stoicism and ingenuity to face and overcome them. What is important, though, is that the emphasis will shift from stoicism to ingenuity as we come up with new ideas and technologies, to overcome our problems. By

the mid-1990's we should be a good way along in this shift, the results of which I will discuss in a moment. But the results of the changes and transitions we face will have left their effect on our society, for we will have realized that we will never again live in a society where so much is taken for granted—where so many apparently "knew the price of everything and the value of nothing." The environmental movement, the energy crisis, and the problems yet to come will have changed all that well before 1990.

Oddly enough, the kind of general outlook that will prevail in 1994 will be a synthesis of ideas coming out of today's low technology communes and high technology industries. We will not see complexity for its own sake. But neither will we be able to maintain the desired quality of life for the number of people present by depending on something akin to handicraft and cottage industry. High technology, much better planned and managed, and important scientific advances will still be the basis for progress.

But that progress will be guided by many of the new values being expressed by young people today. We will be more of a functional and less of a possessive society, more apt to enjoy a less cluttered life, more inclined to share material things and take pleasure in doing so. This will bring us a different kind of freedom, one more closely related to Hegel's definition when he said, "Freedom is the recognition of necessity," but one also allowing more people to "do their own thing" within the framework of a cooperative society.

Let me turn now to some of the physical changes that will be taking place that will accompany these social changes as we move toward and through the 1990's.

HOW THE "RECYCLE SOCIETY" WILL WORK

As I mentioned before, we will be creating a "recycle society." By this I do not mean simply one in which beer cans and Coke bottles are all returned to the supermarket, but one in which virtually all materials used are reused indefinitely and virgin resources become primarily the "make-up" materials to account for the amounts lost in use and production and needed to supplement new production to take care of any new growth that would improve the quality of life.

In such a society the present materials situation is literally reversed; all waste and scrap—what are now called "secondary materials"—become our major resources, and our natural, untapped resources become our backup supplies. This must eventually be the industrial philosophy of a stabilized society and the one toward which we must work.

To many who have not thought about it, this idea may sound simple, or a bit confusing. To many who have given it considerable thought, it can be mind boggling and sound physically and economically impossible, given our current state of industry. And to some it may even appear morally objectionable, given the state of development in many parts of the world. To clarify the concept, let me explain some of the things it will and will not involve.

First, it involves a shift in industry to the design and production of consumer goods that are essentially nonobsolescent. This means that products will be built to be more durable; easily repairable with standardized, replaceable parts; accessible; and able to be repaired with very basic tools. (Along these lines, I understand that at a recent international auto show in Frankfurt, Porsche displayed a car designed to have a twenty-year life, or 200,000 miles—but then quickly assured the industry it had no intentions of putting the car into production!)

In the recycle society, all products and parts will be labeled in such a way that their use, origin and material content can be readily identified; and all will have a regulated trade-in value. Many items of furniture, housewares, appliances, and tools, in addition to their low-maintenance qualities, will be multifunctional, modular, and designed for easy assembly and breakdown to be readily moved and set up in a different location when necessary. Their design and construc-

tion will also allow for their reassembly and redesign into essentially new products when their owners have different uses for them or seek a change.

When a consumer (it would be more correct to call him a "user") wishes to replace an item or trade up for something better or different, he can return the old item for the standard trade-in price. All stores will have to accept these trade-ins. They thus will become collection centers as well as selling outlets in the recycle society.

Manufacturers in turn will receive and recondition the used products, use their parts as replacement parts in "new" products, or scrap them for recycled material. Since literally everything will be coded and tagged for material content, much of the high cost of technological materials separation will be eliminated. Materials that were mixtures and alloys would be color coded, magnetically or isotopically tagged to facilitate optical or electromagnetic separation.

Recycling and reprocessing will also apply to software—clothing, bedding, carpeting, and all other textile materials, organic and synthetic, and, of course, to paper products.

The industrial processing aspects of the recycle society may be the easiest to achieve, as there are already one or two major companies that claim the ability to recycle totally the waste products of selected plants without any economic penalty. By 1994 we should also see extensive recycle of organic material from agriculture and forest industries. Animal waste will find many uses as fertilizer, fuel, and feed. Protein will be grown on petroleum waste and extracted from otherwise inedible plants and agricultural products.

PEOPLE WILL BE BETTER OFF IN THE 1990'S

To make the transition to an economic recycle society, much will be required in the way of new legislation, regulation, tax incentives, and other measures that will make the use of secondary materials more economic than that of virgin resources. The opposite situation prevails today. In addition to the setting up of new methods of marketing and management required to operate a recycle society there will be the necessity for a long-term consumer education program. A whole new public outlook will have to be acquired.

An entire society reusing and recycling almost all its possessions, especially after an extended era of conspicuous consumption and waste, will take a great deal of pride in a life style that is extremely creative and varied and based on a new degree of human ingenuity and innovation. The "recycle society" of the 1990's will be better off than the affluent society of the 1970's.

Several types of assets will accrue from the movement toward the kind of society I have been describing. One lies in the fact that it will be far less energy-intensive. For example, recycled steel requires 75 per cent less energy than steel made from iron ore; 70 per cent less energy is used in recycling paper than in using virgin pulp; 12 times as much energy is needed to produce primary aluminum as to recover aluminum scrap. A society set up to reuse most of its resources systematically and habitually could effect enormous energy savings.

Perhaps even greater would be the reduction in the environmental impact of a recycle society. This would be true for several reasons. An over-all one is that such a society would have developed by the 1990's an environmental and conservationist ethic, due to the scarcity of resources as well as the new value placed on land, water, and air. We can see this ethic already in the making today. There is some fear that because of our energy crisis we will severely compromise this ethic, or even abandon it. I do not think this will be the case. Rather I believe we will be making substantial sacrifices in the years ahead to change our life style in order to match our economic and environmental needs.

By the 1990's our industrial and power systems will be much more efficient users of energy; hence, they will not be rejecting as great a percentage of waste heat to the environment. In fact, most systems will probably be planned and designed to make maximum use of waste heat, using it for space heating or possibly agriculture or aquaculture. Waste water and sewage water will also be recycled in industrial and perhaps even municipal water systems. What water is returned to the environment will be as clean, if not cleaner, than it was when it entered the man-made system.

It would be foolish to believe that by 1994, even with a recycle society as a reality, we will not still be drawing on a substantial amount of new resources. Even with population growth leveling off and economic growth cooling off we can expect substantial growing demands for new materials resources well into the next century. What this means is that by the 1990's we will have to develop a new level of ingenuity in materials substitution and in what Buckminster Fuller calls "ephemeralization"—the process of doing more with less. Fuller uses as an example of this, the Telstar satellite, which, while weighing only one-tenth of a ton, outperforms 75,000 tons of transatlantic cable.

Communications, of course, have offered the best examples of this "more with less" phenomenon, as in electronics we have seen the size of basic devices reduced by a factor of ten roughly every five years. Today, a single chip of silicon a tenth of an inch square may hold microscopic units that perform the functions of as many as 1,000 separate electronic components. Furthermore, silicon is one of the most abundant substances in the earth's crust.

COMMUNICATIONS MAY SUBSTITUTE FOR TRANSPORTATION

By the 1990's substitution not only in materials but in functions—the way we conduct our business and personal affairs—may vastly alter our lives, effecting many savings in energy and time, and therefore affecting how we otherwise spend our energy and time. For example, communications as a substitution for transportation can effect such savings to a great extent.

Shopping by telephone and having good

home delivery is a very old example of this, one which is largely out of style today. But if one could survey local supermarkets and department stores via videophone and a computer to do some quick comparative shopping and then have the selections delivered, one would have more time, money, and energy (personal and automotive) left for other things. Another aspect of this type of shopping involves a considerable saving in space. A simple warehouse with a small fleet of trucks could service thousands of customers, eliminating the paving over of large parking areas and the operation of an elaborate market.

A society that exercises this option of using communication in place of transportation in many of its activities—whether in shopping, business, or educational activities—can conserve many resources. But it must be one that has learned to substitute other activities for the social and entertainment value that we have come to find in our more random way of life.

In conducting this kind of society, questions that loom larger than those of the technological possibilities are as follows. Assuming a 1994 liberated housewife (if that is not a contradiction in terms) is able to do most of her shopping by video computer and her other chores so efficiently, how will she spend her extra free time? When it is possible to hold national and international conferences via home holography, will we miss the luncheons, banquets, and corridor talk? And what will happen to the Willy Lomans of the world when they can sell their lines long distance in living color, through similar electronic techniques, without covering the territory in person? These are only a sample of the kinds of questions that can be raised when the matter of substituting communication for travel becomes a viable option.

It is possible to speculate that in 1994 we may find a situation in which our working world will be served mainly by communication and public transportation, and the savings from this will allow us to use private transportation in a limited way for recreation and vacations.

WILL URBAN SPRAWL DESTROY THE COUNTRYSIDE?

Much of what I have said to this point will be influenced by, and influence, how we use our space here on Earth—how we manage our land, develop our urban areas, place our industry, locate our power system. Do we build out, up, or down? Do we draw people back into the cities, continue to disperse them around cities, or cluster them in new areas, in new cities around new industries? And how much of a planned, concerted effort do we make to do any of these? We are, in fact, just beginning to take a serious look at land management and the control of our populated areas in this country. In the past we have seen our population explode and implode with both good and bad effects, but certainly without much conscious control on our part.

It is difficult to speculate on how far we will have gone by 1994 in having effected any widespread control or change over today's patterns of growth. Twenty years is not a long time to institute and carry out major changes in land use and population distribution. And yet, unless we do, some of the major effects of the current style of growth could (according to Environmental Protection Agency estimates) lead to some 20 million additional acres being covered by urban sprawl (an area equivalent to New Hampshire, Vermont, Massachusetts, and Rhode Island); more than 3 million acres paved over for highways and airports; and about 5 million acres of agricultural land lost to public facilities, second-home development, and waste control projects. In addition, the approximately 1,000 power stations that may be built by the 1990's, together with their cooling facilities, fuel storage and safety exclusion areas, and right-of-way land for power lines, could require another two to three million acres.

MEASURES TO COUNTER THE DESTRUCTION OF THE COUNTRYSIDE

Much of this is inevitably going to take place

before 1994. But I see some of the following as countermeasures and countertrends that may be initated, or well under way, by the 1990's:

(1) *A large shift toward clustered, attached, and high rise housing surrounded by community-owned open lands.* This planned housing would rise as "new cities" and as neighborhood communities within larger urban areas. It would help eliminate today's suburban sprawl and preserve more open land for recreation, agriculture, or natural reserves.

(2) *Increased use of underground space made possible by advances in excavation technology:* underground in shopping centers, warehouses, recreation and entertainment complexes, rapid mass-transit lines, and power and communications cables.

(3) *Offshore power plants with extrahigh voltage, and superconducting transmission cables carrying electricity greater distances inland.* Cables would be underground and might occupy the same rights-of-way as rapid rail transit systems and cable communication systems. In the 1990's such offshore powerplants would be nuclear electric. In the next century they might include nuclear hydrogen-producing plants and solar-powered electric and hydrogen-producing plants.

(4) *Integrated industrial complexes planned to concentrate energy sources, materials, and manufacturing in single locations.* This would reduce long shipments of fuel and material resources, make more efficient use of waste heat and materials, confine and control environmental impact.

IMPROVEMENT IN ENERGY SITUATION BY 1994?

Concerning the energy situation, which is uppermost on people's minds today and will certainly have a major bearing on our future, I believe we will see a turning point in our difficulties by 1994. But the intervening years will necessitate bearing some difficulties and hardships because we have not

given ourselves the necessary lead time to make an orderly transition to new energy technologies and resources. There is no doubt that we have been shortsighted and complacent about supplies and have overestimated our ability to develop and shake down new technologies and to get them on line economically.

In the 1990's we will still have a very mixed energy economy. By then, oil and gas will be giving way to coal—but grudgingly, as it will take some time to develop and build economic coal gasification and liquefaction systems. Oil shale may be a significant factor by then. And we may even have found a way to retort the oil from the shale via underground heating and explosives (chemical) to avoid stripping and excavation. We will have a growing amount of electricity in some parts of the country supplied by geothermal energy. Solar energy equipment for home heating and cooling will be prevalent on new single-family dwellings and incentives will be introduced to encourage home owners to retrofit their houses with such equipment if possible. Solar energy may supply a small amount of home use electricity, but the large-scale production of solar electricity may still be a few years off. However, by the 1990s we should see some prototype "solar farms" in the Southwestern United States testing out the economic, large-scale conversion of solar energy.

I am confident that by the 1990's we will be well over the difficulties and resistance facing nuclear power today and that more than a third of our electric power will be generated by nuclear plants. The liquid-metal fast-breeder reactor will have been tested out to everyone's satisfaction by then and coming on line commercially. Other systems, such as the high-temperature gas-cooled reactor, will be adding to our national electric capacity. We may have achieved laboratory success in controlled fusion by 1994 and be building prototype fusion reactors.

We simply must pay the price of pursuing all possibilities in the energy field, and at the same time pursue the energy conservation ethic I mentioned before. Any energy bonuses that would come our way through new breakthroughs would not give us energy to squander but would allow a well-planned, equitable increase in the quality of life on a worldwide basis.

A DESIRABLE FUTURE FOR AN INTERDEPENDENT WORLD

This brings me to my concluding thoughts on 1994, which center on global cooperation. In the midst of an energy crisis aggravated by the withholding of oil as a political weapon, this does not seem to be a popular topic. It is quite natural to want to act from a position of strength. With some sacrifices we can.

And yet, in our immediate reaction to strive for energy self-sufficency, we should not overreact, not delude ourselves into believing that self-sufficiency in energy and other matters is the total solution to national security and well being. This would lead to a dangerous neo-isolationism at a time when we must move in the other direction—toward greater international cooperation, no matter how difficult and painstaking the process seems at some times. The harsh facts are that we live in a highly interdependent world—one in which there will continue to be some hard bargaining but in which cooperation is growing increasingly important.

By 1994, I see the scales tipping more and more in favor of cooperation over competition. My travels over the world in the past dozen years to more than sixty countries—and most recently to the People's Republic of China—have led me to believe that all nations of the world need each other, that all have something to offer, and all could benefit by a greater exchange of human and material resources, of knowledge and goods.

Over the next twenty years we will have to make enormous strides—together—in controlling population, increasing food production, managing our environment, investigating and controlling the resources of the seas, conducting global research, developing methods to reduce the human impact of natural disasters, and generally uplifting the economic conditions of a large number of the world's peoples. There are no alternatives to these measures—except a tremendous increase in human misery that will ultimately affect all the world's peoples.

Forecasting the world of 1994 involves much more than projecting the trends at hand or even reciting the possibilities ahead. That future will be determined in large part by our considering and choosing values, examining and deciding among alternatives, exercising great will and perseverance, and searching for the leadership that will assemble and catalyze the proper resources to construct a chosen future. To the extent that we can do this we will either be drifting toward the world of 1994 or building it. Most likely, we will be doing a little of each, but I hope we will not be trusting to luck that which we could achieve by a new and concerted human effort.

Finally, 1994 is only twenty years off. We had better get moving!